VOLUME THREE HUNDRED AND SEVEN

INTERNATIONAL REVIEW OF CELL AND MOLECULAR BIOLOGY

New Models of the Cell Nucleus: Crowding, Entropic Forces, Phase Separation, and Fractals

International Review of Cell and Molecular Biology

VOLUME THREE HUNDRED AND SEVEN

INTERNATIONAL REVIEW OF CELL AND MOLECULAR BIOLOGY

New Models of the Cell Nucleus: Crowding, Entropic Forces, Phase Separation, and Fractals

Edited by

RONALD HANCOCK

Laval University Cancer Research Centre
CRCHUQ-Oncology, Québec, Canada, and
Biosystems Group, Biotechnology Centre
Silesian University of Technology
Gliwice, Poland

KWANG W. JEON

Department of Biochemistry, University of Tennessee
Knoxville, Tennessee

ELSEVIER

AMSTERDAM • BOSTON • HEIDELBERG • LONDON
NEW YORK • OXFORD • PARIS • SAN DIEGO
SAN FRANCISCO • SINGAPORE • SYDNEY • TOKYO
Academic Press is an imprint of Elsevier

Academic Press is an imprint of Elsevier
525 B Street, Suite 1800, San Diego, CA 92101-4495, USA
225 Wyman Street, Waltham, MA 02451, USA
The Boulevard, Langford Lane, Kidlington, Oxford, OX5 1GB, UK
32 Jamestown Road, London, NW1 7BY, UK
Radarweg 29, PO Box 211, 1000 AE Amsterdam, The Netherlands

First edition 2014

British Library Cataloguing in Publication Data
A catalogue record for this book is available from the British Library

Library of Congress Cataloging-in-Publication Data
A catalog record for this book is available from the Library of Congress

ISBN: 978-0-12-800046-5
ISSN: 1937-6448

For information on all Academic Press publications
visit our website at store.elsevier.com

CONTENTS

CONTRIBUTORS

William M. Aumiller Jr.
Department of Chemistry, Pennsylvania State University, University Park, Pennsylvania, USA

Aurélien Bancaud
CNRS GDR 3536, University Pierre et Marie Curie-Paris 6, Paris; CNRS, LAAS, and Université de Toulouse, LAAS, Toulouse, France

Victor M. Bolanos-Garcia
Faculty of Health and Life Sciences, Department of Biological and Medical Sciences, Oxford Brookes University, Oxford, United Kingdom

Pascal Carrivain
CNRS GDR 3536, and LPTMC UMR 7600, CNRS, Université Pierre et Marie Curie-Paris 6, Paris, France

Bradley W. Davis
Department of Chemistry, Pennsylvania State University, University Park, Pennsylvania, USA

Alan R. Denton
Department of Physics, North Dakota State University, Fargo, North Dakota , USA

Ronald Hancock
Laval University Cancer Research Centre, CRCHUQ-Oncology, Québec, Canada, and Biosystems Group, Biotechnology Centre, Silesian University of Technology, Gliwice, Poland

Dieter W. Heermann
Institute for Theoretical Physics; Interdisciplinary Center for Scientific Computing (IWR), Heidelberg, Germany; The Jackson Laboratory, Bar Harbor, Maine, USA, and Shanghai Center for Bioinformation Technology (SCBIT), Shanghai, PR China

Sébastien Huet
CNRS, UMR 6290, Institut Génétique et Développement de Rennes; Université Rennes 1, Université Européenne de Bretagne, Structure fédérative de recherche, Biosit, Faculté de Médecine, Rennes, and CNRS GDR 3536, University Pierre et Marie Curie-Paris 6, Paris, France

Hansjoerg Jerabek
Institute for Theoretical Physics, and Interdisciplinary Center for Scientific Computing (IWR), Heidelberg, Germany

Christine D. Keating
Department of Chemistry, Pennsylvania State University, University Park, Pennsylvania, USA

Jun Soo Kim
Department of Chemistry and Nano Science, Global Top5 Research Program, Ewha Womans University, Seoul, Republic of Korea

Christophe Lavelle
CNRS GDR 3536, University Pierre et Marie Curie-Paris 6; National Museum of Natural History; CNRS UMR 7196, and INSERM U565, Paris, France

Andrew Mugler
FOM Institute AMOLF, Amsterdam, The Netherlands

Thoru Pederson
Program in Cell and Developmental Dynamics, Department of Biochemistry and Molecular Pharmacology, University of Massachusetts Medical School, Worcester, Massachusetts, USA

Hubert Ranchon
CNRS GDR 3536, University Pierre et Marie Curie-Paris 6, Paris; CNRS, LAAS, and Université de Toulouse, LAAS, Toulouse, France

Angelo Rosa
Scuola Internazionale Superiore di Studi Avanzati, Trieste, Italy

Takahiro Sakaue
Department of Physics, Graduate School of Sciences, Kyushu University, Fukuoka, Japan

Naoki Sugimoto
Frontier Institute for Biomolecular Engineering Research (FIBER) and Faculty of Frontiers of Innovative Research in Science and Technology (FIRST), Konan University, Kobe, Japan

Igal Szleifer
Department of Biomedical Engineering; Department of Chemistry, and Chemistry of Life Processes Institute, Northwestern University, Evanston, Illinois, USA

Pieter Rein ten Wolde
FOM Institute AMOLF, Amsterdam, The Netherlands

Jean-Marc Victor
CNRS GDR 3536, and LPTMC UMR 7600, CNRS, Université Pierre et Marie Curie-Paris 6, Paris, France

Matthias Weiss
Department of Experimental Physics I, University of Bayreuth, Bayreuth, Germany

Miho Yanagisawa
Department of Physics, Graduate School of Sciences, Kyushu University, Fukuoka, Japan

Kenichi Yoshikawa
Faculty of Life and Medical Sciences, Doshisha University, Kyotanabe, Kyoto, Japan

Christophe Zimmer
Institut Pasteur, Unité Imagerie et Modélisation; CNRS URA 2582, Paris, France

PREFACE

Interest in the cell nucleus has been heightened in recent years, and the principles which determine structure and functions have become considerably clearer during the nearly 30 years since the publication of a volume in this series with a similar theme (Berezney, R., Jeon, K.W. (Eds.), 1995. Nuclear matrix—structural and functional organization. Int. Rev. Cytol. 162). It is noteworthy that this progress has been due largely to ideas from the fields of biophysics, physical chemistry and polymer, colloid, and soft-matter science, and also to the development of powerful simulation methods for modeling the conformations and interactions of macromolecules.

In presenting this volume, our hope is to encourage cell and molecular biologists to become more familiar with and understand these new concepts and methods, and the crucial contributions which they are making to our perception of the nucleus. They are likely to be essential factors in all the systems in the nucleus and in their activities.

It is a great pleasure to thank the authors for their willing and friendly cooperation and for finding the time to prepare their contributions in spite of their busy schedules. We also thank workers at Elsevier Academic Press for having produced an excellent volume.

Ronald Hancock and Kwang W. Jeon
July 2013

CHAPTER ONE

The Nuclear Physique

Thoru Pederson[1]
Program in Cell and Developmental Dynamics, Department of Biochemistry and Molecular Pharmacology, University of Massachusetts Medical School, Worcester, Massachusetts, USA
[1]Corresponding author: e-mail address: thoru.pederson@umassmed.edu

Contents

Abstract

This volume brings together a number of perspectives on how certain physical phenomena contribute to the functional design and operation of the nucleus. This collection could not be more timely, resonating with an increasing awareness of the opportunities that lie at the interface of cell biology and the physical sciences. For example, this was a major theme in the 2012 and 2013 annual meetings of the American Society for Cell Biology, and one that the Society aims to emphasize even further going forward. In addition, the emerging canonical relevance of the physical sciences to cell biology has in recent summers made a most conspicuous appearance in the curriculum (lectures and intense labs) of the famed Physiology Course at the Marine Biological Laboratory in Woods Hole. So, much credit is due to Ronald Hancock and Kwang Jeon, the coeditors of this volume, and all the authors for creating a work that is so *au courant.*

It all started with the Big Bang

Robertson (2007)

1. INTRODUCTION: A BRIEF HISTORY OF BIOPHYSICS

The English word "physics" deerives, as the plural, from the Greek noun "physic" (Φίζικ) and has historically been applied to both medicine (the practice of "physik") and, more durably, natural phenomena (excluding

International Review of Cell and Molecular Biology, Volume 307
ISSN 1937-6448
http://dx.doi.org/10.1016/B978-0-12-800046-5.00001-1

life). The former term is no longer used in medicine except on occasion when one sees that a professor at some venerable medical school holds a title such as "*The* (insert an august name here, such as Vesalius or Paracelsus) *Professor of Physik*." No one can quite be sure when the prefix "bio" was first attached to physics. The combined word does not appear in any of Aristotle's brilliant essays on either physics or natural history, for example, though his treatise "Movement of Animals" (translated by Farquharson, 1984) makes for fascinating reading, particularly bearing in mind that he wrote more than two millenia before the time of Newton.

In the mid-eighteenth century, the Italian physician and physicist Luigi Galvani (1737–1798) discovered that the leg muscle of a recently sacrificed frog could conduct static electricity from a piece of metal, resulting in a twitch. This allowed him to make the leap that electricity is related to muscle activity in the living animal. The subsequent discovery by the German physician and physiologist Emil du Bois-Reymound (1818–1896) of what came to be known as the nerve action potential paved the way for what could be called the "pre-biophysics era." In due course, as the field of biochemistry evolved, a parallel line of thinking arose that "mechanical" and indeed even "engineering" principles were as much at play in living systems as chemistry—a doctrine promoted particularly by the German–American biologist Jacques Loeb (1859–1924).

By the 1920s, the arrival of biophysics was at hand, at least in a form that took up that name. It was both an approach (particular equipment) and a discipline (a vision of mechanism) and was driven, at least to a considerable degree by physiologists continuing to bump into physical underpinnings of biological processes. But there was a second, portentous domain of biophysics in the wind.

Back in 1912, the German physicist, Max von Laue (1879–1960), had discovered the diffraction of X-rays by crystalline materials, and subsequently he and others realized that this diffraction from the lattice would allow the crystal's structure to be deduced by calculating the pattern back through reciprocal space. In due course, it occurred to several people that such an approach might be applied to more complex molecules than crystalline minerals. The undisputed leader of this movement was the Irish physicist John Desmond Bernal (1901–1971). He was a formidable genius who, even as an undergraduate at Cambridge, presented one of his professors with a penetrating mathematical analysis of the 230 crystal space groups, an achievement even more striking given that he had not done well on the mathematics tripos, leading him to turn to the natural sciences curriculum

(Brown, 2005). Later, having become a leading crystallographer, arguably *the* leading one, Bernal encouraged his former student Dorothy Crowfoot Hodgkin (1910–1994) to take up the larger task (as both the size of the unit cell and magnitude of the achievement, if reached) of tackling biological molecules. By the age of 35, Hodgkin had got the structure of penicillin before it had been determined by chemical means. This was a monumental achievement, and she followed it in due course by solving the structures of vitamin B_{12} and insulin.

In parallel with these triumphs, there arrived the era of X-ray diffraction analysis of both crystalline proteins and (wet or dry) biological fibers as in the case of collagen, wool, and DNA, and these endeavors, in particular in the X-ray crystallography field, increasingly took on the name "biophysics" in many venues. Thus, for example, the laboratory at King's College, London, where Maurice Wilkins and, later, Rosalind Franklin undertook X-ray diffraction of DNA had long been named the Biophysics Unit of the Medical Research Council. The term biophysics had become, by the end of the war, part of the scientific lexicon. The crystallographic axis of biophysics soon reached its first "post-Hodgkin" pinnacle in the DNA work of Wilkins and Franklin at King's, and of Watson and Crick at the University of Cambridge Department of Physics' Cavendish Laboratory, and, subsequently, in the structural solutions of myoglobin and hemoglobin by John Kendrew and Max Perutz, respectively, at the Medical Research Council's Laboratory of Molecular Biology in Cambridge. Despite the name of this latter institute, its founding was a direct result of the *biophysics* era that Bernal and Hodgkin had pioneered, and which England so proudly led for more than three decades.

In the 1950s, another dimension of biophysics coming to fruition was the study of cell structure. Phase contrast microscopy, a major advance in light microscopy, had been discovered in 1930 by the Dutch physicist Frits Zernike (1888–1966), for which he received the 1953 Nobel Prize in Physics, unshared. Meanwhile, also in the early 1930s, the German physicist Ernst Ruska (1906–1988) had developed the principles and prototypes of the electron microscope, for which he received the 1988 Nobel Prize in Physics, shared with Gerd Binning and Heinrich Rohrer for their invention of the scanning tunneling microscope. By the time phase-contrast and electron microscopy arrived, many cytologists, biochemists, and even physcicts (*vide infra*) were trying to define "protoplasm." Rivals seemed ready to bet their first-born children on "gel" versus "sol" models of protoplasm. In his very first work with live material, Francis Crick measured the recoil of intracellular iron particles when subjected to a magnetic field, as a probe of the rheological and, thus, cross-sectional density of cytoplasm (Crick, 1950; Crick and Hughes, 1950). I recently reread the first paper (the second is beyond my expertise in the physics required) and would recommend it to all cell biologists as a model of experimental elegance for its time, as well as a marvelous example of fine writing.

Except for the obvious risk of confusing (or annoying) readers, this chapter might have been titled "Nuclear Physics" and this point reminds us, of course, of another major development in biology that sprang from pure physics, namely, the codiscovery of radioactivity by Marie Curie and Henri Becquerel. The subsequent understanding of this phenomenon (i.e., "nuclear physics," literally) led to the use of unstable isotopes by the Hungarian chemist George de Hevesy (1885–1966; Nobel Prize in Chemistry, 1943) and Rudolf Schőnheimer (1898–1941) as tracers to pursue the biosynthesis and flux of molecules in living systems. Later, the stable isotope nitrogen-15 was the key to the Meselson–Stahl experiment demonstrating the semiconservative replication of DNA, dubbed "the most beautiful experiment in biology" (Holmes, 2001). The entry of radioisotopes to biological research was pure biophysics, although they became such standard tools that their use was not labeled as biophysics *per se*. So too was the allied field of radiation biology.

Coming to the modern era (1975–present), biophysics might be defined by what is published in the *Biophysical Journal* (reminding me of a professor who told the class that lipids are defined as substances that are soluble in a lipid solvent). Huge advances have taken place in our ability to measure things such as the diffusion-based transport of molecules in cells—thanks mainly to the advent of fluorescent dyes and proteins. More recently, in just the last decade, advances have cracked the Abbé limit (~200 nm) for spatial resolution in diffraction-limited optical microscopy. And meanwhile, the introduction of "optogenetics," pioneered by Gero Miesenbőck and refined by Karl Deisseroth, has brought new meaning to "biophysics." This method exploits the biophysical properties of opsins as sensors, combined with reporting fluorescent proteins, to allow cell physiology to be observed in real time. This past May, I attended a talk in which the speaker, Adam Cohen of Harvard University, showed videos of an action potential running down a neuron and culminating in synaptic transmission. Never did I dream that I would see such a thing in live cells, as opposed to on an oscilloscope screen. Driving home I pondered: *Sic transit gloria biophysica.*

2. THE BIOPHYSICAL NUCLEUS

Let us now turn to specifically the nucleus, the subject of this volume. For the purposes of this introductory chapter, three key questions can be posed: What is the structure of the nucleus at the mesoscopic scale? To what extent do certain phenomena in the nucleus obey purely physical laws? What are the emerging methods that can tackle the physics of the nucleus going forward?

Before we move on, I want to mention a guideline that the biophysicist Daniel Branton taught me. At the 1982 meeting of the American Society for Cell Biology, I was in charge of the program and invited him to chair one of the symposia. In his introductory remarks, he said he had instructed the speakers to emphasize any places in their work where the formation of covalent bonds *had not been* at play. I thought this was a most enabling device at the time (and his speakers indeed obeyed). It remains a valuable construct for us all, as we delve into any part of biology.

2.1. Nuclear structure at mesoscopic scale

By the time cytochemical studies had revealed the very large proportion of the nuclear volume occupied by DNA, at least in most cells, it became logical to assume that the physical properties of the nucleus derive mostly from a very high concentration of these extraordinarily long polymers (The discovery of uninemy was just around the corner but, in any case, no one doubted that the DNA molecules were *very* long, whether chromosome-size or not.). Later, Thomas and Christoph Cremer and colleagues insightfully deduced, from chromosome breaks induced in interphase cells by UV irradiation, that each chromosome occupied a distinct portion of the nuclear volume, rather than being intertwined (Cremer et al., 1982). This and subsequent work by them and others brought the concept of chromosomes territories (Pederson, 2003, 2004) and also, more subtly, hinted that while the chromosomes were not entangled, the amount of interchromosomal space in the nucleus was relatively small. Nonetheless, it was natural for the curious to ponder whether the interchromosome space itself possessed some kind of structure.

Meanwhile, in the 1960s, Ilya Zbarsky in Moscow was subjecting isolated nuclei to escalating concentrations of NaCl and noted that, after the histone and much other nuclear protein had been extracted, a residual structure remained. In a refinement of his work, Berezney and Coffey used DNase and salt extraction to obtain a residual fraction that contained only traces of DNA. They called this fraction a "nuclear matrix" and launched a major field that delved into the possibility that many nuclear processes occur on such a scaffold structure. Problematically, however, this notion was so attractive to many that there was much less effort directed at asking the most important question of all, namely, whether a nuclear matrix exists *in vivo*. I have expressed my doubts about the nuclear matrix as an arborized scaffold extending throughout the nucleus *in vivo* (Pederson, 1998, 2000a, 2010a) and will say nothing more here. What is not ruled out, however, is the possibility of a relatively short-range ordered structure in the nucleus, likely to be the case for nuclear actin (Pederson, 2008; McDonald et al., 2006; Belin et al., 2013), but this is a fundamentally different concept and bears no epistemological relationship to the earlier notion of a nuclear matrix. What *is* germane, both here and for this volume, is to ask what is known as to the physical structure of the nucleus *in live cells* as interrogated by biophysical methods.

The viscosity of the nuclear interior has been measured by the mobility of green fluorescent protein (GFP) using the powerful method of fluorescence correlation spectroscopy (FCS), with the conclusion that it is approximately five times greater than water (Wachsmuth et al., 2000). Of course, this and other such measurements (Wedemeier et al., 2007) have presumed that GFP or other fluorescent probes used are "inert" and thus have insufficient affinity for chromatin or other nuclear bodies to be significantly influenced in their mobility. In all these studies, the observed mobility, plotted as the mean square displacement versus time, clearly constituted the mode of diffusion termed "anomalous," signifying frequent collisions with relatively immobile structures and/or temporary corralling in *cul de sacs*.

It is to be borne in mind that in FCS, the interrogation volumes are on the order of picoliters and mobility is measured by the rapid flux of fluorescent molecules in and out of the sampling volume. It is thus only the viscosity of the medium at these very small sites that is at play. Other methods, notably particle tracking, offer opportunities to observe longer trajectories. For example, by microinjecting ~100 nm diameter fluorescent nanospheres into the nuclei of live mouse fibroblasts, Tseng et al. (2004) measured a mean shear viscosity of the nuclear interior much higher than that observed with FCS, and they were also in a position to estimate the elasticity of the nuclear interior. These two types of methods, FCS and particle tracking, operate at the microscopic and mesoscopic scales, respectively, and there are plausible interpretations for the differences observed. Even larger probes, namely, platinum–nickel nanorods ~200 nm × 1.5 μm, gunned into live cell nuclei and then subjected to magnetic fields, have been used to induce micro-scale displacements of the nuclear interior (Celedon et al., 2011). At this scale one measures a culmintaed nuclear structure, one in which these investigators implicated nuclear lamins. Finally, there have been courageous studies in which the deformability of nuclei in live cells has been measured by micropipette-applied negative pressure (Rowat et al., 2005, 2006). The major force being observed there is the resistance of the nuclear envelope, but it is likely that the structure of the nuclear interior is at play, in both the deformation and recoil. At presen, there is considerable interest in the degree to which the nuclear lamina interacts with the underlying nucleoplasm (e.g., Funkhouser et al., 2013).

FCS has taught us much about the nucleus, but an equally valuable method has been fluorescence recovery after photobleaching (FRAP). FRAP measures the kinetics and extent of entry of fluorescent molecules into a zone in which its fluorescent siblings were previously bleached. The first application of FRAP to the nucleus was in a study of the dynamics of DNA repair machinery (Houtsmuller et al., 1999) and immediately thereafter, a boomlet of contemporaneous studies on the mobility of numerous other nuclear proteins appeared, notably ones in chromatin and nucleoli (Pederson, 2000b, 2001). These studies ushered in a new view of the nucleus, one in which many of its key functional proteins spend much time roaming around instead of being locked down at functional sites. These findings immediately erased assumptions (ones based on no evidence in hindsight) about the presumed stasis of chromatin and the nucleolus and constituted a major advance in the understanding of the nucleus as a highly dynamic organelle with respect to proteins that reside, at steady-state, at sites of high concentration (Pederson, 2010a).

All these studies revealing the high degree to which diffusion governs nuclear dynamics came as a surprise. But a major clue had already been revealed. Several years before the explosion of FRAP-based studies of the mobility of nuclear proteins, Marshall et al. (1997) had shown that the interphase chromosomes themselves are dynamically moving by diffusion, although of course within confinement volumes. I read this paper with great enthusiasm and satisfaction. Why would they not be moving? Like all particles in the universe since the Big Bang, they have their inherited kinetic energy.

2.2. Physical laws governing nuclear function

The discovery of a DNA polymerase by Arthur Kornberg (though later found not to be the enzyme that replicates DNA) and that of an RNA polymerase independently by Audrey Stevens, Samuel Weiss, and Jerard Hurwitz served to advance the concept of the gene as an enzyme substrate. But the logisitics of how the double helix accommodated these maneuvers, and how the nascent product strands were handled in envisioned spatial gymnastics gave birth to a strong role of biophysics in the replication and transcription fields. As one example, following the monumental formulation by Jacob and Monod of the repressor in gene regulation, the question arose as to how this protein would find its appropriate binding site, amidst so much competing DNA (Lin and Riggs, 1975). It took many years for biophysicists such as Peter von Hippel to figure out that repressors walk along DNA by thermal motion until they arrive at high affinity sites.

Biophysical approaches have also been of major importance in the recent era of gene transcription. Faced with the hypothesis that mRNA transport out of the nucleus might be a process that required metabolic energy, my laboratory employed FCS, FRAP, and tracking methods to establish that mRNA moves within the nucleus by diffusion, with no dependence on ATP (Politz and Pederson, 2000; Politz et al., 1998, 1999). We later showed that diffusion was also the process by which ribosomes are transported within the nucleus (Politz et al., 2003). Our studies were subsequently confirmed and extended by several groups (Mor et al., 2010). As an aside, when presenting this work of ours, arriving at the data demonstrating no dependence on ATP, I always used the phrase "no dependence on *metabolic* energy" (emphasis added). The mRNAs and ribosomes we were tracking were, of course, being moved by energy, the kinetic energy they possessed. All such energy in the universe, distributed among many entities, comes from a single, ancient source—hence the song lyric given at the beginning of this chapter.

Another aspect of nuclear function in which biophysical thinking has come into play is the question of how large-scale rearrangements of gene loci observed in certain situations of cellular responses to external signals may involve directed movements, that is, ones propelled by an active motor of some kind. As attractive an idea as this is, the initial claim (Hu et al., 2008) has failed to be replicated (Belmont, 2010; Kocanova et al., 2010) and so we must await further work.

Kinetic modeling has had a long history in research on nuclear processes and while such data decomposition is not necessarily "biophysics" (since metabolic energy and covalent bond formation are involved and hence, by both criteria, it is also chemistry), there are reasons to believe that seeking a detailed kinetic framework for any nuclear process would be instructive.

Two initial studies as regards transcription of the genes for ribosomal RNA came to diametrically opposed conclusions (Dundr et al., 2002; Schneider and Nomura, 2004). Similar kinetic approaches for RNA polymerase II transcription of mRNA genes are just underway (e.g., Larson et al., 2011). Other biophysical models of transcription have been proposed (Canals-Hamann et al., 2013) and await experimental test.

Another field that is just underway is the genesis of nuclear bodies. The most familiar of these, the nucleolus, arises from the genome itself—the repeated genes for ribosomal RNA (Pederson, 2010b). But whether other nuclear bodies are dependent on the genome for their formation is unknown. Cajal bodies have been claimed to arise spontaneously (Kaiser et al., 2008) and this provocative finding needs to be confirmed and expanded. Another recent study has revealed that the extra-chromosomal nucleoli in the nucleus of *Xenopus* oocytes possess a liquid-like structure (Brangwynne et al., 2011), a finding made possible only by a rigorous biophysical analysis. This study set a standard for how observations of nuclear body dynamics, combined with physical principles, can provide new insights (Marko, 2012).

Macromolecular crowding, borne in classical physical chemistry and brought into biological conceptualization most notably by Allen Minton at the U.S. National Institutes of Health (Ellis and Minton, 2003) is another biophysical concept that deserves more attention in the nucleus field than it has received. Ronald Hancock has been among the intrepid few who have advanced this idea and he summarizes his incisive experiments in the chapter that follows. I have been intrigued by this notion as one of the factors governing nuclear organization ever since I was a postdoc, when my mentor and I found that exposure of cells to mild hypertonicity induced dramatic but entirely reversible effects on the interphase nucleus, including condensation of the chromosomes, dissolution of nucleoli, and disassembly of the nuclear envelope, all resulting in an appearance virtually indistinguishable from normal prophase (Robbins et al., 1970). We reasoned that the immediacy of these effects intimated a physical basis and that the degree of water egress under this osmotic condition would not change the concentrations of intranuclear regulatory molecules very much. Whether these observations were a manifestation of macromolecular crowding was left in the dustbin as we moved on to other work. But the concept has had an enduring appeal for me and important advances have recently been made, both experimental and theoretical (Bancaud et al., 2009; Hancock, 2004, 2007; Iborra, 2007; Richter et al., 2008; Tjong, 2012).

2.3. What lies ahead and the methods that may be in play

Single-molecule detection in the nucleus of live cells reigns currently as a major endeavor. Initial efforts (Grűnwald et al., 2008; Kues et al., 2001) have set the stage for advances and innovation, now underway (Grűnwald et al., 2011). Issues of fluor brightness, photobleaching and blinking, the need for acquisition rates high enough to capture intranuclear molecules moving as fast as 10 μm/s, and the desirability of imaging two or three colors simultaneously at ~20-ms time resolution, all loom large. Fluorescence cross-correlation microscopy will likely play a major role going forward, given its power to interrogate at the mesoscopic scale. The tremendous advances in structured illumination microscopy that we have seen in the last decade, bringing the spatial resolution of light microscopy down from the ~200 nm Abbé limit by a hitherto unimagined order of magnitude, are playing a major role as these instruments, in the right hands, are turned to the nucleus. Sheet illumination, in which extraneous fluorescence in the z-axis is eliminated (Pederson, 2012), has already been demonstrated to have great potential for observing nuclear events (Siebrasse et al., 2012), and this method is likely to play a major role in the fluorescence microscopy of the nucleus going forward. Finally, we should keep our eyes on nuclear actin (Belin et al., 2013), particularly if it can be brought into a very high spatial resolution, as now seems at hand.

Beyond these plausible perspectives, any other speculations would be sheer sophistry. What can be predicted, with some confidence, is that beyond emerging studies of genome arrangement as a possible fractal structure (Dekker et al., 2013) and the application of systems biology to gene expression as pioneered by Eric Davidson and colleagues (Tu et al., 2013), integrating the perspectives of both physics and biophysics in the next era of research on the nucleus will be productive, perhaps to an unanticipated degree. That said, it is important to bear in mind that the reductionist hypothesis, which states that all phenomena in the universe can in principle be understood from the fundamental laws of physics, does not mean that there is a converse "constructionist" hypothesis, positing that we can necessarily build on our understanding of higher order scales from what we know from beneath (see Anderson, 1972 for a particularly lucid exposition of this point). The approach that seems most likely to succeed in the case of the nucleus, and cell biology in general, is to keep drilling down, from the molecular, cell, and systems biology perspectives, but at the same time *to have physical laws in mind* while we are on this research journey. None of us expect that we can anytime soon understand the operation of the nucleus as quantum electrodynamics, vaulting over a large gap and probably ill-prepared for surprises along the way, as there are likely to be. Instead, the most modest but, I think, enabling goal should be for us who are cell biologists to do one good thing as we go: remember physics. This volume is a wonderful step on that good vector and the authors offer a rich tapestry of ideas and approaches that will help to define this emerging field.

In closing, it has now been a decade since a white paper appeared on how physics and biophysics might advance our understanding of the nucleus (O'Brien et al., 2003), and it is striking how many of the perspectives and proposed lines of investigation set forth in that paper have come to fruition. Let us hope that this volume will set a new bar for where biophysics can, once again, contribute to advancing our understanding of the nuclear physique.

ACKNOWLEDGMENTS

This chapter is dedicated to Thomas Cremer, a pioneer in the nucleus field, whose work has been at the forefront both conceptually and empirically.

Work cited from the author's laboratory was supported by grants from the U.S. National Institute of General Medical Sciences (R01-60551), the U.S. National Science Foundation (MCB-044581 and MCB-1051398), and the Human Frontier Research Program Organization.

REFERENCES

Anderson, P.W., 1972. More is different. Science 177, 393–396.

Bancaud, A., et al., 2009. Molecular crowding affects diffusion and binding of nuclear proteins in heterochromatin and reveals the fractal organization of chromatin. EMBO J. 28, 3785–3798.

Belin, B.J., et al., 2013. Visualization of actin filaments and monomers in somatic cell nuclei. Mol. Biol. Cell 24, 982–994.

Belmont, A.S., 2010. Estrogen fueled, nuclear kiss. Did it move for you? Nucleus 1, 440–443.

Brangwynne, C.P., et al., 2011. Active liquid-like behavior of nucleoli determines their size and shape in Xenopus laevis oocytes. Proc. Natl. Acad. Sci. U. S. A. 108, 4334–4349.

Brown, A., 2005. J.D. Bernal. The Sage of Science. Oxford University Press, Oxford, U.K.

Canals-Hamann, A.Z., et al., 2013. A biophysical model for transcription factories. BMC Biophys. 6, 2–6.

Celedon, A., Hale, C.M., Wirtz, D., 2011. Magnetic manipulation of nanorods in the nucleus of living cells. Biophys. J. 101, 1880–1886.

Cremer, T., et al., 1982. Analysis of chromosome positions in the interphase nucleus of Chinese hamster cells by laser-UV-microirradiation experiments. Hum. Genet. 62, 201–209.

Crick, F.H.C., 1950. The physical properties of cytoplasm. A study by means of the magnetic particle method. Part II. Theoretical treatment. Exp. Cell Res. 1, 505–533.
Crick, F.H.C., Hughes, A.F.W., 1950. The physical properties of cytoplasm. A study by means of the magnetic particle method. Part I. Experimental. Exp. Cell Res. 1, 37–60.
Dekker, J., Marti-Remon, M.A., Mirny, L.A., 2013. Exploring the three-dimensional organization of genomes: interpreting chromatin interaction data. Nat. Rev. Genet. 14, 390–403.
Dundr, M., et al., 2002. A kinetic framework for a mammalian RNA polymerase in vivo. Science 298, 1623–1626.
Ellis, R.J., Minton, A.P., 2003. Join the crowd. Nature 425, 27–28.
Farquharson, A.S.L., 1984. Movement of animals. In: Barnes, J. (Ed.), The Complete Works of Aristotle. Bollingen Series LXXI, vol. 1. Princeton University Press, Princeton, NJ, pp. 1087–1096 (Revised Oxford translation).
Funkhouser, C.M., et al., 2013. Mechanical model of blebbing in nuclear lamin meshworks. Proc. Natl. Acad. Sci. U. S. A. 110, 3248–3253.
Grűnwald, D., et al., 2008. Probing intranuclear environments at the single-molecule level. Biophys. J. 94, 2847–2858.
Grűnwald, D., Singer, R.H., Rout, M.P., 2011. Nuclear export dynamics of RNA-protein complexes. Nature 475, 333–341.
Hancock, R., 2004. A role for macromolecular crowding effects in the assembly and function of compartments in the nucleus. J. Struct. Biol. 146, 281–290.
Hancock, R., 2007. Packing of the polynucleosomne chain in interphase chromosomes. Evidence for a contribution of crowding and entropic forces. Semin. Cell Dev. Biol. 18, 668–675.
Holmes, F.L., 2001. Meselson, Stahl, and the Replication of DNA. A History of "The Most Beautiful Experiment in Biology" Yale University Press, New Haven, CT.
Houtsmuller, A., et al., 1999. Action of DNA repair endonuclease ERCC/XPF in living cells. Science 284, 958–961.
Hu, Q., et al., 2008. Enhancing nuclear receptor-induced transcription requires nuclear motor and LSD1-dependent gene networking in interchromatin granules. Proc. Natl. Acad. Sci. U. S. A. 105, 19199–19204.
Iborra, F.J., 2007. Can visco-elastic phase separation, macromolecular crowding and colloidal physics explain nuclear organization? Theor. Biol. Med. Model. 4, 15.
Kaiser, T.E., Intine, R.V., Dundr, M., 2008. De novo formation of a subnuclear body. Science 322, 1713–1717.
Kocanova, S., et al., 2010. Activation of estrogen-responsive genes does not require their nuclear co-localization. PLoS Genet. 6, 1000922.
Kues, T., et al., 2001. Visualization and tracking of single protein molecules in the cell nucleus. Biophys. J. 80, 2954–2967.
Larson, D.R., et al., 2011. Real-time observation of transcription initiation and elongation on an endogenous yeat gene. Science 332, 475–478.
Lin, S., Riggs, A.D., 1975. The general affinity of lac repressor for E. coli DNA: implications for gene regulation in prokaryotes and eucaryotes. Cell 4, 107–111.
Marko, J.F., 2012. The liquid drop nature of nucleoli. Nucleus 3, 115–117.
Marshall, W.F., et al., 1997. Interphase chromosomes undergo constrained diffusional motion in living cells. Curr. Biol. 7, 930–939.
McDonald, D., et al., 2006. Nucleoplasmic beta-actin exists in a dynamic equilibrium between low-mobility polymeric species and rapidly diffusing populations. J. Cell Biol. 172, 541–552.
Mor, A., Ben-Yishay, R., Shav-Tal, Y., 2010. On the right track: following the nucleo-cytoplasmic path of an mRNA. Nucleus 1, 492–498.
O'Brien, T.P., et al., 2003. Genome function and nuclear architecture: from gene expression to nanoscience. Genome Res. 13, 1029–1041.

Pederson, T., 1998. Thinking about a nuclear matrix. J. Mol. Biol. 277, 147–159.
Pederson, T., 2000a. Half a century of "the nuclear matrix" Mol. Biol. Cell 11, 799–805.
Pederson, T., 2000b. Diffusional transport in the nucleus: a message in the medium. Nat. Cell Biol. 2, E73–E74.
Pederson, T., 2001. Protein mobility within the nucleus—what are the right moves? Cell 104, 635–638.
Pederson, T., 2003. Gene territories and cancer. Nat. Genet. 34, 4–5.
Pederson, T., 2004. The spatial organization of the genome in mammalian cells. Curr. Opin. Genet. Dev. 14, 203–209.
Pederson, T., 2008. As functional nuclear actin comes into view, is it globular, filamentous, or both? J. Cell Biol. 180, 1061–1064.
Pederson, T., 2010a. The nucleus introduced. In: Misteli, T., Spector, D.L. (Eds.), The Nucleus. Cold Spring Harbor Laboratory Press, Cold Spring Harbor, NY, pp. 1–16.
Pederson, T., 2010b. The nucleolus. In: Misteli, T., Spector, D.L. (Eds.), The Nucleus. Cold Spring Harbor Laboratory Press, Cold Spring Harbor, NY, pp. 209–223.
Pederson, T., 2012. Nuclear export, enlightened. Proc. Natl. Acad. Sci. U. S. A. 109, 9228–9229.
Politz, J.C., Pederson, T., 2000. Movement of mRNA from transcription site to nuclear pores. J. Struct. Biol. 129, 252–257.
Politz, J.C., et al., 1998. Intranuclear diffusion and hybridization state of oligonucleotides measured by fluorescence correlation spectroscopy. Proc. Natl. Acad. Sci. U. S. A. 95, 6043–6048.
Politz, J.C., et al., 1999. Movement of poly(A) RNA throughout the interchromatin space in living cells. Curr. Biol. 9, 285–291.
Politz, J.C., Tuft, R.A., Pederson, T., 2003. Diffusion-based transport of nascent ribosomes in the nucleus. Mol. Biol. Cell 14, 4805–4812.
Richter, K., Neesling, M., Lichter, P., 2008. Macromolecular crowding and its potential impact on nuclear function. Biochim. Biophys. Acta 1783, 2100–2107.
Robbins, E., Pederson, T., Klein, P., 1970. Comparison of mitotic phenomena and effects induced by hypertonic solutions in HeLa cells. J. Cell Biol. 44, 400–416.
Robertson, E., 2007. Music and lyrics. Theme for "The Big Bang Theory" television show.
Rowat, A.C., et al., 2005. Characterization of the elastic properties of the nuclear envelope. J. R. Soc. Interface. 2, 63–69.
Rowat, A.C., Lammerling, J., Ipsen, J.H., 2006. Mechanical properties of the cell nucleus and the effect of emerin deficiency. Biophys. J. 91, 4649–4664.
Schneider, D.A., Nomura, M., 2004. RNA polymerase I remains intact without subunit exchange through multiple rounds of transcription in Saccharomyces cerevisiae. Proc. Natl. Acad. Sci. U. S. A. 101, 15112–15117.
Siebrasse, J.P., Kaminski, T., Kubitscheck, U., 2012. Nuclear export of single native mRNA molecules observed by light sheet fluorescence microscopy. Proc. Natl. Acad. Sci. U. S. A. 109, 9426–9431.
Tjong, H., 2012. Physical tethering and volume exclusion determine higher-order genome organization in budding yeast. Genome Res. 22, 1295–1305.
Tseng, Y., et al., 2004. Micro-organization and visco-elasticity of the interphase nucleus revealed by particle tracking. J. Cell Sci. 117, 2159–2167.
Tu, S., Pederson, T., Weng, Z., 2013. Networking development by Boolean logic. Nucleus 4, 89–91.
Wachsmuth, M., Waldeck, W., Langowski, J., 2000. Anomalous diffusion of fluoresecent probes inside living cell nuclei investigated by spatially resolved fluorescence correlation spectroscopy. J. Mol. Biol. 298, 677–689.
Wedemeier, A., et al., 2007. Modeling diffusional transport in the interphase cell nucleus. J. Chem. Phys. 127, 045102.

CHAPTER TWO

The Crowded Nucleus

Ronald Hancock[*,†,1]
[*]Laval University Cancer Research Centre, CRCHUQ-Oncology, Québec, Canada
[†]Biosystems Group, Biotechnology Centre, Silesian University of Technology, Gliwice, Poland
[1]Corresponding author: e-mail address: ronald.hancock@crhdq.ulaval.ca

Contents

Abstract

The principles that determine the organization of the nucleus have become clearer in recent years, largely because of new insights into polymer, colloid, and soft-matter science. Macromolecules, together with the giant linear polymers that form the chromosomes, are confined at high concentrations within the nuclear envelope and their interactions are influenced strongly by short-range depletion or entropic forces which are negligible in dilute systems, in addition to the more familiar van der Waals, electrostatic, steric, hydrogen bonding, and hydrophobic forces. The studies described in this volume are consistent with the model that this complex and concentrated mixture of macromolecules is maintained in a delicate equilibrium by quite simple although unsuspected physicochemical principles. The sensitivity of this equilibrium to perturbation may underlie the controversies about the existence of a nuclear matrix or scaffold. In this volume, we underline the importance for cell biologists of being familiar with current work in colloid, polymer, soft matter, and nanoscience. This chapter presents a brief background to the aspects of the nucleus that are considered in detail in subsequent chapters.

1. INTRODUCTION

Nearly 20 years ago it was pointed out that the concentration of macromolecules in the cytoplasm is so high that it would inevitably become compartmentalized, in a paper that opened with the sentences: "Mixtures

International Review of Cell and Molecular Biology, Volume 307
ISSN 1937-6448
http://dx.doi.org/10.1016/B978-0-12-800046-5.00002-3

of aqueous solutions of structurally distinct macromolecules, above certain concentrations, undergo phase separation. The phases formed are generally solutions which differ in concentration and physical properties. Cytoplasm contains high concentrations of proteins and nucleic acids. Can these mixtures exist *without* undergoing phase separation?" (Walter and Brooks, 1995). Surprisingly, it was not realized until 10 years later that similar effects are expected to occur in the nucleus.

The molecular environment in the nucleus differs greatly from the dilute and homogeneous solutions employed in laboratory experiments. Macromolecules, together with the giant linear polymers that form the chromosomes, are confined at high concentrations within the nuclear envelope. This milieu can be considered as a complex colloidal system in which "molecules or polymolecular particles with a dimension between 1 nm and 1 μm are dispersed in a continuous phase of a different composition" (IUPAC, 2001), and nuclei were regarded from this viewpoint in some earlier studies (Anderson and Wilbur, 1952; Harding, 1949). Colloid, polymer, soft matter, and nanoscientists have developed powerful simulation methods that allow the responses of macromolecules in these conditions to be modelled and predicted (Bishop et al., 2009; Yodh et al., 2001) (Chapters 3, 4, 9, 10, 12, and 13).

2. MACROMOLECULAR CROWDING IN THE NUCLEUS

The first measurements using interference microscopy, reported nearly 50 years ago, showed a dry mass concentration in nuclei in the range of 100–200 mg/ml (Lampert et al., 1966; Viola and Puccinelli, 1965), and subsequent studies using more sophisticated methods have confirmed values in this range. A compilation of published values is presented in Table 2.1.

3. ENTROPIC (DEPLETION) FORCES IN THE NUCLEUS

In crowded cellular milieux where macromolecules are in close proximity, their interactions are influenced strongly by short-range forces termed depletion or entropic forces which are usually negligible in dilute systems, in addition to the more familiar van der Waals, electrostatic, hydrogen bonding, and hydrophobic forces (Leckband and Israelachvili, 2001). Entropic forces favor contact between larger particles in a mixture of different-sized particles, because this allows the excluded volumes surrounding them to overlap and increases the accessible volume and the entropy for smaller

Table 2.1 Concentration of macromolecules in the nucleus and its compartments

Nuclei				**Nuclear compartments**			
	Concentration (mg/ml)	**Method**	**References**		**Concentration (mg/ml)**	**Method**	**References**
Global				*Nucleoli*			
HeLa	96	A	Lampert et al. (1966)	HeLa	200	A	Lampert et al. (1966)
HeLa	170–400	D	Nolin et al. (2013)	HeLa	370	C	Pliss et al. (2010)
Human liver	165	A	Ranek (1976)	Mesothelial, newt	220	A	Blumenstein and Amenta (1981)
Human glial	165	A	Viola and Puccinelli (1965)	*Xenopus* oocytes (dense fibrillar)	215[a]	A	Handwerger et al. (2005)
Drosophila (polytene)	65	A	Hartmann-Goldstein and Goldstein (1979)				
Nucleoplasm *Xenopus* oocytes	106[a]	A	Handwerger et al. (2005)	*Transcription sites*, HeLa	140	C	Pliss et al. (2010)
Chromatin				*Speckles*, HeLa	165	C	Pliss et al. (2010)
HeLa, centromere	150	C	Pliss et al. (2010)				
HeLa, metaphase	190	C	Pliss et al. (2010)				
Euglena, heterochromatin	400	B	Bohrmann et al. (1993)				

[a]Value expressed as protein.
Methods: A, dry mass by interference microscopy; B, quantitative STEM; C, confocal Raman microspectrometry; D, quantitative cryoSTEM.

particles (Asakura and Oosawa, 1954; Chapters 3 and 4). The strength of the resulting attractive forces is of the order of piconewtons (pN) and their range is of the order of nanometers (Bishop et al., 2009; Kleshchanok et al., 2008; Knoben et al., 2006). In addition to favoring macromolecular interactions (Yodh et al., 2001), entropic forces can confer high specificity on interactions of a "lock and key" type (Kinoshita and Oguni, 2002). The effects of crowding on polymorphic nucleic acid structures have important implications for the control of transcription (Chapter 8). Current models suggest that entropic forces are crucial for the dense packing of chromatin to form chromosomes (Grosberg, 2012; Chapters 9, 10, and 13).

A further consequence of crowding arises because the effective volume of a macromolecule exceeds its intrinsic volume; for example, a spherical molecule excludes others of a similar size from a volume eight times larger than its intrinsic volume. Thus even when macromolecules appear to occupy only a modest part of a volume there may be no space available for others of the same size, and their effective concentrations and thermodynamic activities are enhanced significantly (Schnell and Hancock, 2008; Zimmerman and Trach, 1991).

4. COMPARTMENTALIZATION IN THE NUCLEUS

Nucleoli, Cajal (coiled) and PML bodies, speckles, paraspeckles, gems, and replication and transcription factories appear to be ubiquitous compartments in all nuclei, while numerous other types are observed in particular conditions or cell types (Table 2.2). Compartments, generally termed "nuclear bodies," share a number of characteristic properties including the absence of a bounding membrane, movement within the nucleus (Platani et al., 2000), dynamic exchange of their components with the surrounding nucleoplasm (Handwerger et al., 2005; Weidtkamp-Peters et al., 2008), the ability to divide and to fuse (Brangwynne et al., 2011; Platani et al., 2000), and localization in the interchromatin region. Localization of a macromolecule within a compartment may not be essential for it to carry out its function; for example, the specific RNAs that are normally found in Cajal bodies function even when Cajal bodies cannot be formed (Deryusheva and Gall, 2009).

Exogenous macromolecules may also form discrete bodies if they accumulate in the nucleus, for example, oligoribonucleotides (Lorenz et al., 1998; Shoeman et al., 1997), overexpressed proteins (Kovalsky et al., 2001; Ueno et al., 1988), and proteins whose export from the nucleus is

Table 2.2 Classes of compartments in nuclei

Ubiquitous	In particular cells	References	In particular cells	References
Nucleoli	Polycomb bodies	Pirrotta and Li (2012)	Rad51 bodies	Haaf et al. (1995)
Cajal bodies	Histone locus bodies	Liu et al. (2006)	Tax bodies	Bex and Gaynor (1998)
PML bodies	Virus replication factories	Erickson et al. (2012) and Shishido-Hara et al. (2012)	Sam68 bodies	Chen et al. (1999)
Speckles	Marinesco bodies	Kumada et al. (2002)	Inclusions in neuro-degenerative diseases	Woulfe (2007)
Paraspeckles	SUMO-1 bodies	Navascués et al. (2007)	Clastosomes	Lafarga et al. (2002)
Gems	53BP1 bodies	Lukas et al. (2011)	Cleavage bodies	Schul et al. (1996)
Replication factories	NAC1 bodies	Wu et al. (2011)	SUMO bodies	Navascués et al. (2007)
Transcription factories	SP2 bodies	Moorefield et al., 2006	OPT domains	Pombo et al., 1998

inhibited (Fabbro et al., 2002; Lain et al., 1999; Wada et al., 1998), which suggests that the environment within the nucleus favors the self-association of macromolecules.

Components of some compartments self-associate *in vitro* (Hebert and Matera, 2000; Kovalsky et al., 2001; Morse et al., 2007; Passon et al., 2012), and their assembly to form compartments *in vivo* may be driven by amplification of these affinities caused by crowding (Hancock, 2004; Richter et al., 2007; Chapters 3, 4, and 6) with possible structuring by specific RNAs (Caudron-Herger and Rippe, 2012). Concentrated protein solutions may contain clusters of molecules whose internal concentration reaches up to 700 mg/ml (Johnston et al., 2012).

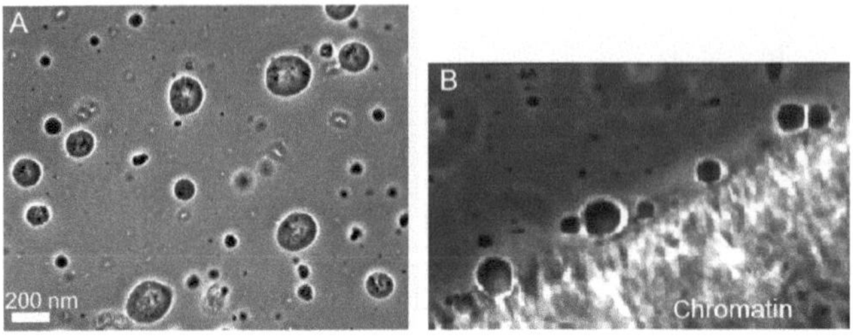

Figure 2.1 Particles formed in a crowded nuclear extract and their localization at the surface of chromatin. Nuclei of K562 cells were broken by gentle sonication and this extract concentrated to ~20 mg protein/ml was incubated with dextran (10.5 kDa, final concentration 15% w/v). Quasi-spherical particles were formed (A, phase contrast image), which collected on the surface of chromatin (B, merged phase contrast and fluorescence image, chromatin labeled with DAPI) (R. Hancock, unpublished data).

Certain nuclear bodies, particularly Cajal and PML bodies, are observed to contact the surface of chromosomes frequently (Platani et al., 2000; Wang et al., 2004), while transcription and replication factories are permanently localized on the periphery of chromosomes (Cremer et al., 2004; Cseresnyes et al., 2009; Jaunin et al., 2000). The histone locus body in the nuclei of *Drosophila* cells is localized on the chromosome locus which contains the histone genes (Liu et al., 2006). Localization of particles at a surface is favored by entropic effects (Dinsmore et al., 1998), and particles that are formed in a crowded extract of nuclei are similarly attracted to the surface of chromatin (Fig. 2.1).

5. PHASE SEPARATION IN THE NUCLEUS

In a mixture of particles or of particles and polymers that have a large difference in properties such as shape and flexibility, depletion and other forces may cause one component to separate and form a discrete phase (Buitenhuis et al., 1995; Tanaka and Araki, 2006; Tuinier et al., 2003). It has been proposed that the separation between the chromosomes and the interchromatin space in the nucleus is a result of this effect (Iborra, 2007). A chromatin-rich phase separates when chromatin is incubated *in vitro* with an inert crowding macromolecule (Hancock, 2008; Fig. 2.2).

6. CONCLUDING REMARKS

The studies presented in this volume are consistent with the model that the complex and concentrated mixture of macromolecules in the

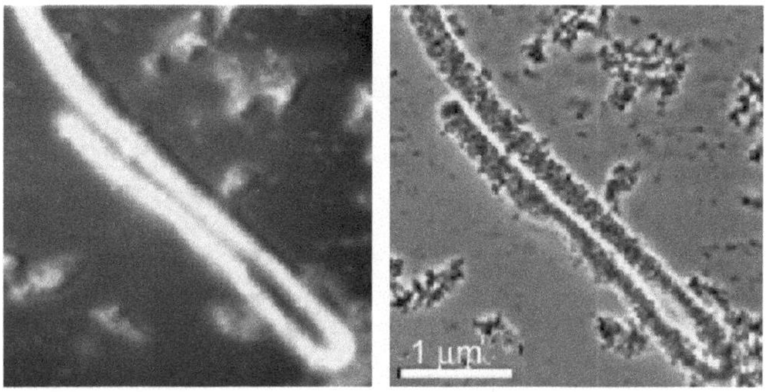

Figure 2.2 Phase separation of polynucleosomes in a crowded solution. Polynucleosomes (~20 nucleosomes) electroeluted from nuclei digested with *Hae*III and concentrated to ~100 μg DNA/ml were layered onto a solution of dextran (10.5 kDa, 25% w/v). Left, DNA labeled with YOYO-1; right, corresponding phase contrast image (Hancock, 2008). *Reproduced from Hancock, 2008 with the permission of Elsevier.*

nucleus is maintained in a delicate equilibrium by physicochemical principles which are basically quite simple. The sensitivity of this equilibrium may explain the controversies about the existence of a matrix or scaffold in the nucleus (Pederson, 1998); the effects of electrostatic forces on model systems (Buzzaccaro et al., 2010; Denton and Schmidt, 2005) suggest that the equilibrium would be perturbed when nuclei are exposed to the high concentration of salt used to prepare a matrix (Tan et al., 2000), while the chaotrope lithium 3,5-diiodosalicylate used to isolate a scaffold (Mirkovitch et al., 1984) causes a number of proteins to aggregate (Choules et al., 1973).

Experimentalists are faced with the challenge of reproducing these crowded conditions *in vivo* when isolating and studying nuclei or nuclear compartments. Nuclei are commonly isolated in buffers that contain Na^+, K^+, and/or Mg^{2+} at mM concentrations, but the concentrations of *free* ions in the cytoplasm and in the nucleus *in vivo* are unknown (Hancock and Hadj-Sahraoui, 2009). Nuclei can be isolated in a medium without cations but containing an inert crowding macromolecule (PEG, dextran, or Ficoll) to mimic the crowding exerted by cytoplasmic macromolecules *in vivo*, and their ultrastructure, compartmentalization, and ability to elongate nascent RNA are maintained (Hancock and Hadj-Sahraoui, 2009). The structure of metaphase chromosomes is also conserved in these conditions (Hancock, 2012; Fig. 2.3). Thus, Kornberg's seventh commandment "Correct for extract dilution with molecular crowding" (Kornberg, 2000) is crucial when isolating and studying nuclei and their components.

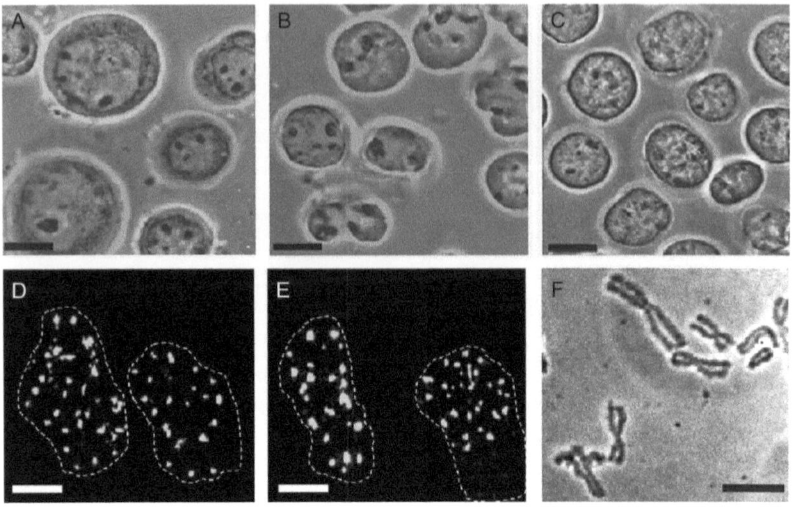

Figure 2.3 Nuclei and metaphase chromosomes isolated in solution containing a crowding macromolecule in 100-μM K-Hepes buffer with no cations. A, nuclei in growing K562 cells; B, C, nuclei isolated from K562 cells by homogenization in 70-kDa Ficoll (50% w/v) or 70-kDa dextran (35% w/v), respectively (phase-contrast); D, E, conservation of coiled bodies (GFP-coilin) and PML bodies (GFP-PML isoform IV) in nuclei isolated in 70-kDa Ficoll (50% w/v) from HeLa or U2OS cells, respectively (maximum intensity projections from confocal series); F, metaphase chromosomes isolated by homogenization of mitotic CHO cells in 8-kDa PEG (12% w/v) (phase-contrast). All scale bars 5 μm. *A–E reproduced from Hancock and Hadj-Sahraoui (2009) and F reproduced from Hancock (2012).* (For the color version of this figure, the reader is referred to the online version of this chapter.)

Other chapters in this volume describe how new insights and methods are contributing to a better understanding of the structure and properties of the nucleus and its components. Our objective is to familiarize cell biologists with these concepts, and one message that we hope to convey is the value of being aware of new work and ideas in the fields of colloid, polymer, soft matter, and nanoscience.

REFERENCES

Anderson, N.G., Wilbur, K.M., 1952. Studies on isolated cell components IV. The effects of various solutions on the isolated rat liver nucleus. J. Gen. Physiol. 35, 781–796.

Asakura, S., Oosawa, F., 1954. On interaction between two bodies immersed in a solution of macromolecules. J. Chem. Phys. 22, 1255–1256.

Bex, F., Gaynor, R.B., 1998. Regulation of gene expression by HTLV-I Tax protein. Methods 16, 83–94.

Bishop, K.J.M., Wilmer, C.E., Soh, S., Grzybowski, B.A., 2009. Nanoscale forces and their uses in self-assembly. Small 5, 1600–1630.

Blumenstein, R., Amenta, P.S., 1981. An interferometric analysis of nucleoli. Anat. Rec. 201, 13–21.
Bohrmann, B., Haider, M., Kellenberger, E., 1993. Concentration evaluation of chromatin in unstained resin-embedded sections by means of lowdose ratio-contrast imaging in STEM. Ultramicroscopy 49, 235–251.
Brangwynne, C.P., Mitchison, T.J., Hyman, A.A., 2011. Active liquid-like behavior of nucleoli determines their size and shape in Xenopus laevis oocytes. Proc. Natl. Acad. Sci. U. S. A. 108, 4334–4339.
Buitenhuis, J., Donselaar, L.N., Buining, P.A., Stroobants, A., Lekkerkerker, H.N.K., 1995. Phase separation of mixtures of colloidal boehmite rods and flexible polymer. J. Colloid Interface Sci. 175, 46–56.
Buzzaccaro, S., Piazza, R., Colombo, J., Parola, A., 2010. Enhancement of depletion forces by electrostatic depletant repulsion. J. Chem. Phys. 132, 124902.
Caudron-Herger, M., Rippe, K., 2012. Nuclear architecture by RNA. Curr. Opin. Genet. Dev. 22, 179–187.
Chen, T., Boisvert, F.M., Bazett-Jones, D.P., Richard, S., 1999. A role for the GSG domain in localizing Sam68 to novel nuclear structures in cancer cell lines. Mol. Biol. Cell 10, 3015–3033.
Choules, G.L., Sandberg, R.G., Steggall, M., Eyring, E.M., 1973. Analysis of noncovalent bonding in Mycoplasma membranes. Kinetics of solubilization in sodium dodecyl sulfate and lithium diiodosalicylate solutions. Biochemistry 12, 4544–4550.
Cremer, T., Küpper, K., Dietzel, S., Fakan, S., 2004. Higher order chromatin architecture in the cell nucleus: on the way from structure to function. Biol. Cell. 96, 555–567.
Cseresnyes, Z., Schwarz, U., Green, C.M., 2009. Analysis of replication factories in human cells by super-resolution light microscopy. BMC Cell Biol. 10, 88.
Denton, A.R., Schmidt, M., 2005. Mixtures of charged colloid and neutral polymer: influence of electrostatic interactions on demixing and interfacial tension. J. Chem. Phys. 122, 244911.
Deryusheva, S., Gall, J.G., 2009. Small Cajal body-specific RNAs of Drosophila function in the absence of Cajal bodies. Mol. Biol. Cell 20, 5250–5259.
Dinsmore, A.D., Wong, D.T., Nelson, P., Yodh, A.G., 1998. Hard spheres in vesicles: curvature-induced forces and particle-induced curvature. Phys. Rev. Lett. 80, 409–412.
Erickson, K.D., Bouchet-Marquis, C., Heiser, K., Szomolanyi-Tsuda, E., Mishra, R., Lamothe, B., Hoenger, A., Garcea, R.L., 2012. Virion assembly factories in the nucleus of polyomavirus-infected cells. PLoS Pathog. 8, e1002630.
Fabbro, M., Rodriguez, J.A., Baer, R., Henderson, B.R., 2002. BARD1 induces BRCA1 intranuclear foci formation by increasing RING-dependent BRCA1 nuclear import and inhibiting BRCA1 nuclear export. J. Biol. Chem. 277, 21315–21324.
Grosberg, A.Yu., 2012. How two meters of DNA fit into a cell nucleus: polymer models with topological constraints and experimental data. Polym. Sci. Ser. C 54, 1–10.
Haaf, T., Golub, E.I., Reddy, G., Radding, C.M., Ward, D.C., 1995. Nuclear foci of mammalian Rad51 recombination protein in somatic cells after DNA damage and its localization in synaptonemal complexes. Proc. Natl. Acad. Sci. U. S. A. 92, 2298–2302.
Hancock, R., 2004. A role for macromolecular crowding effects in the assembly and function of compartments in the nucleus. J. Struct. Biol. 146, 281–290.
Hancock, R., 2008. Self-association of polynucleosome chains by macromolecular crowding. Eur. Biophys. J. 37, 1059–1064.
Hancock, R., 2012. Structure of metaphase chromosomes: a role for effects of macromolecular crowding. PLoS ONE 7, e36045.
Hancock, R., Hadj-Sahraoui, Y., 2009. Isolation of cell nuclei using inert macromolecules to mimic the crowded cytoplasm. PLoS ONE 4, e7560.

Handwerger, K.E., Cordero, J.A., Gall, J.G., 2005. Cajal bodies, nucleoli, and speckles in the *Xenopus* oocyte nucleus have a low-density, sponge-like structure. Mol. Biol. Cell 16, 202–211.

Harding, C.V., 1949. Colloidal properties of nucleus; effect of temperature on nuclear viscosity in the starfish egg. Proc. Soc. Exp. Biol. Med. 70, 705–708.

Hartmann-Goldstein, I.J., Goldstein, D.J., 1979. Effect of temperature on dry mass of polytene nuclei in *Drosophila*. J. Cell Sci. 38, 405–416.

Hebert, M.D., Matera, A.G., 2000. Self-association of coilin reveals a common theme in nuclear body localization. Mol. Biol. Cell 11, 4159–4171.

Iborra, F.J., 2007. Can visco-elastic phase separation, macromolecular crowding and colloidal physics explain nuclear organisation? Theor. Biol. Med. Model. 4, 15.

International Union of Pure and Chemistry, 2001. Colloid and Surface Chemistry Including Catalysis. http://old.iupac.org/reports/2001/colloid_2001/manual_of_s_and_t/node3.html.

Jaunin, F., Visser, A.E., Cmarko, D., Aten, J.A., Fakan, S., 2000. Fine structural in situ analysis of nascent DNA movement following DNA replication. Exp. Cell Res. 260, 313–323.

Johnston, K.P., Maynard, J.A., Truskett, T.M., Borwankar, A.U., Miller, M.A., Wilson, B.K., Dinin, A.K., Khan, T.A., Kaczorowski, K.J., 2012. Concentrated dispersions of equilibrium protein nanoclusters that reversibly dissociate into active monomers. ACS Nano 6, 1357–1369.

Kinoshita, M., Oguni, T., 2002. Depletion effects on the lock and key steric interactions between macromolecules. Chem. Phys. Lett. 351, 79–84.

Kleshchanok, D., Tuinier, R., Lang, P.R., 2008. Direct measurements of polymer-induced forces. J. Phys. Condens. Matter 20, 073101.

Knoben, W., Besseling, N.A., Cohen Stuart, M.A., 2006. Long-range depletion forces induced by associating small molecules. Phys. Rev. Lett. 97, 068301.

Kornberg, A., 2000. Ten commandments: lessons from the enzymology of DNA replication. J. Bacteriol. 182, 3613–3618.

Kovalsky, O., Lung, F.T., Roller, P.P., Fornace, A.J., 2001. Oligomerization of human Gadd45a protein. J. Biol. Chem. 276, 39330–39339.

Kumada, S., Uchihara, T., Hayashi, M., Nakamura, A., Kikuchi, E., Mizutani, T., Oda, M., 2002. Promyelocytic leukemia protein is redistributed during the formation of intranuclear inclusions independent of polyglutamine expansion: an immunohistochemical study on Marinesco bodies. J. Neuropathol. Exp. Neurol. 61, 984–991.

Lafarga, M., Berciano, M.T., Pena, E., Mayo, I., Castaño, J.G., Bohmann, D., Rodrigues, J.P., Tavanez, J.P., Carmo-Fonseca, M., 2002. Clastosome: a subtype of nuclear body enriched in 19S and 20S proteasomes, ubiquitin, and protein substrates of proteasome. Mol. Biol. Cell 13, 2771–2782.

Laín, S., Midgley, C., Sparks, A., Lane, E.B., Lane, D.P., 1999. An inhibitor of nuclear export activates the p53 response and induces the localization of HDM2 and p53 to U1A-positive nuclear bodies associated with the PODs. Exp. Cell Res. 248, 457–472.

Lampert, F., Hofmann, D., Sandritter, W., 1966. Interference microscopic dry weight determination of nucleus, cytoplasm and nucleolus of HeLa and amnion cells in tissue culture. Histochemie 6, 370–371.

Leckband, D., Israelachvili, J., 2001. Intermolecular forces in biology. Q. Rev. Biophys. 34, 105–267.

Liu, J.L., Murphy, C., Buszcza, K.M., Clatterbuck, S., Goodman, R., Gall, J.G., 2006. The Drosophila melanogaster Cajal body. J. Cell Biol. 172, 875–884.

Lorenz, P., Brenda, F., Baker, B.F., Bennett, C.F., Spector, D.L., 1998. Phosphorothioate antisense oligonucleotides induce the formation of nuclear bodies. Mol. Biol. Cell 9, 1007–1023.

Lukas, C., Savic, V., Bekker-Jensen, S., Doil, C., Neumann, B., Pedersen, R.S., Grøfte, M., Chan, K.L., Hickson, I.D., Bartek, J., Lukas, J., 2011. 53BP1 nuclear bodies form around DNA lesions generated by mitotic transmission of chromosomes under replication stress. Nat. Cell Biol. 13, 243–253.

Mirkovitch, J., Mirault, M.-E., Laemmli, U.K., 1984. Organization of the higher-order chromatin loop: specific DNA attachment sites on nuclear scaffold. Cell 39, 223–232.

Moorefield, K.S., Yin, H., Nichols, T.D., Cathcart, C., Simmons, S.O., Horowitz, J.M., 2006. Sp2 localizes to subnuclear foci associated with the nuclear matrix. Mol. Biol. Cell 17, 1711–1722.

Morse, R., Shaw, D.J., Todd, A.G., Young, P.J., 2007. Targeting of SMN to Cajal bodies is mediated by self-association. Hum. Mol. Genet. 16, 2349–2358.

Navascués, J., Bengoechea, R., Tapia, O.P., Vaqué, J.P., Lafarga, M., Berciano, M.T., 2007. Characterization of a new SUMO-1 nuclear body (SNB) enriched in pCREB, CBP, c-Jun in neuron-like UR61 cells. Chromosoma 116, 441–451.

Nolin, F., Michel, J., Wortham, L., Tchelidze, P., Balossier, G., Banchet, V., Bobichon, H., Lalun, N., Terryn, C., Ploton, D., 2013. Changes to cellular water and element content induced by nucleolar stress: investigation by a cryo-correlative nano-imaging approach. Cell. Mol. Life Sci. 70, 2383–2394.

Passon, D.M., Lee, M., Rackham, O., Stanley, W.A., Sadowska, A., Filipovska, A., Fox, A.H., Bond, C.S., 2012. Structure of the heterodimer of human NONO and paraspeckle protein component 1 and analysis of its role in subnuclear body formation. Proc. Natl. Acad. Sci. U. S. A. 109, 4846–4850.

Pederson, T., 1998. Thinking about a nuclear matrix. J. Mol. Biol. 277, 147–159.

Pirrotta, V., Li, H.B., 2012. A view of nuclear Polycomb bodies. Curr. Opin. Genet. Dev. 22, 101–109.

Platani, M., Goldberg, I., Swedlow, J.R., Lamond, A.I., 2000. In vivo analysis of Cajal body movement, separation, and joining in live human cells. J. Cell Biol. 151, 1561–1574.

Pliss, A., Kuzmin, A.N., Kachynski, A.V., Prasad, P.N., 2010. Nonlinear optical imaging and raman microspectrometry of the cell nucleus throughout the cell cycle. Biophys. J. 99, 3483–3491.

Pombo, A., Cuello, P., Schul, W., Yoon, J.-B., Roeder, R.G., Cook, P.R., Murphy, S., 1998. Regional and temporal specialization in the nucleus: a transcriptionally-active nuclear domain rich in PTF, Oct1 and PIKA antigens associates with specific chromosomes early in the cell cycle. EMBO J. 17, 1768–1778.

Ranek, L., 1976. The nuclear dry weight of liver cells from patients with virus hepatitis and from controls. Acta Cytol. 20, 58–61.

Richter, K., Nessling, M., Lichter, P., 2007. Experimental evidence for the influence of molecular crowding on nuclear architecture. J. Cell Sci. 120, 1673–1680.

Richter, K., Nessling, M., Lichter, P., 2008. Macromolecular crowding and its potential impact on nuclear function. Biochim. Biophys. Acta 1783, 2100–2107.

Schnell, S., Hancock, R., 2008. The intranuclear environment. Methods Mol. Biol. 463, 3–19.

Schul, W., Groenhout, B., Koberna, K., Takagaki, Y., Jenny, A., Manders, E.M.M., Raska, I., van Driel, R., de Jong, L., 1996. The RNA 3′ cleavage factors CstF 64 kDa and CPSF 100 kDa are concentrated in nuclear domains closely associated with coiled bodies and newly synthesized RNA. EMBO J. 15, 2883–2892.

Shishido-Hara, Y., Ichinose, S., Uchihara, T., 2012. JC virus intranuclear inclusions associated with PML-NBs. Analysis by electron microscopy and structured illumination microscopy. Am. J. Pathol. 180, 1095–1106.

Shoeman, R.L., Hartig, R., Huang, Y., Grüb, S., Traub, P., 1997. Fluorescence microscopic comparison of the binding of phosphodiester and phosphorothioate (antisense) oligodeoxyribonucleotides to subcellular structures, including intermediate filaments,

the endoplasmic reticulum, and the nuclear interior. Antisense Nucleic Acid Drug Dev. 7, 291–308.
Tan, J.H., Wooley, J.C., LeStourgeon, W.M., 2000. Nuclear matrix-like filaments and fibrogranular complexes form through the rearrangement of specific nuclear ribonucleoproteins. Mol. Biol. Cell 11, 1547–1554.
Tanaka, H., Araki, T., 2006. Viscoelastic phase separation in soft matter: numerical-simulation study on its physical mechanism. Chem. Eng. Sci. 61, 2108–2141.
Tuinier, R., Rieger, J., de Kruif, C.G., 2003. Depletion-induced phase separation in colloid-polymer mixtures. Adv. Colloid Interface Sci. 103, 1–31.
Ueno, K., Katoh, K., Kondoh, H., 1988. Subnuclear localization and antitransforming activity of N-myc:beta-galactosidase fusion proteins. Mol. Cell. Biol. 8, 4529–4532.
Viola, M.P., Puccinelli, E., 1965. Determination of the dry mass of nervous and glial cell nuclei by interference microscopy. Experientia 21, 320–321.
Wada, A., Fukuda, M., Mishima, M., Nishida, E., 1998. Nuclear export of actin: a novel mechanism regulating the subcellular localization of a major cytoskeletal protein. EMBO J. 17, 1635–1641.
Walter, H., Brooks, D.E., 1995. Phase separation in cytoplasm, due to macromolecular crowding, is the basis for microcompartmentation. FEBS Lett. 361, 135–139.
Wang, J., Shiels, C., Sasieni, P., Wu, P.J., Islam, S.A., Freemont, P.S., Sheer, D., 2004. Promyelocytic leukemia nuclear bodies associate with transcriptionally active genomic regions. J. Cell Biol. 164, 515–526.
Weidtkamp-Peters, S., Lenser, T., Negorev, D., Gerstner, N., Hofmann, T.G., Schwanitz, G., Hoischen, C., Maul, G., Dittrich, P., Hemmerich, P., 2008. Dynamics of component exchange at PML nuclear bodies. J. Cell Sci. 121, 2731–2743.
Woulfe, J.M., 2007. Abnormalities of the nucleus and nuclear inclusions in neurodegenerative disease: a work in progress. Neuropathol. Appl. Neurobiol. 33, 2–42.
Wu, P.-H., Hung, S.-H., Ren, T., Shih, Ie.M., Tseng, Y., 2011. Cell cycle-dependent alteration in NAC1 nuclear body dynamics and morphology. Phys. Biol. 8, 015005.
Yodh, A.G., Lin, K., Crocker, J.C., Dinsmore, A.D., Kaplan, P.D., 2001. Entropically driven self-assembly and interaction in suspension. Phil. Trans. R. Soc. Lond. A 359, 921–937.
Zimmerman, S.B., Trach, S.O., 1991. Estimation of macromolecule concentrations and excluded volume effects for the cytoplasm of Escherichia coli. J. Mol. Biol. 222, 599–620.

CHAPTER THREE

Crowding in Polymer–Nanoparticle Mixtures

Alan R. Denton[1]
Department of Physics, North Dakota State University, Fargo, North Dakota , USA
[1]Corresponding author: e-mail address: alan.denton@ndsu.edu

Contents

Abstract

The cell nucleus is a highly crowded environment, filled with a multicomponent, polydisperse mixture of biopolymers and nuclear bodies dispersed in a viscous solvent. With volume fractions approaching 20%, excluded-volume interactions play a key role in determining the structure, dynamics, and function of macromolecules *in vivo*. Under such constraints, the ensembles of macromolecular conformations can differ substantially from those prevailing in dilute solutions. Crowding thus can affect protein and RNA folding, conformational stability, and reaction kinetics, as well as phase stability of macromolecular mixtures. From the perspective of soft matter physics, this chapter reviews

International Review of Cell and Molecular Biology, Volume 307
ISSN 1937-6448
http://dx.doi.org/10.1016/B978-0-12-800046-5.00003-5

recent studies on crowding in polymer–nanoparticle mixtures, seeking to demonstrate the utility of simple physical models for addressing challenging issues in cell biology. The focus is on applications of free-volume theory and Monte Carlo simulation, based on geometrical models of polymers as fluctuating spheres or ellipsoids. Ideal polymer coils respond to hard-sphere crowding agents by compactifying, reducing their radius of gyration, and becoming more spherical. At sufficiently high concentrations, polymers and crowders phase-separate. The goal of this review is to identify universal principles governing macromolecular crowding and to establish a general framework for future explorations of more realistic models that may include nonsteric (e.g., electrostatic) interactions.

1. INTRODUCTION

1.1. Polymers and nanoparticles in cell biology

Biological cells are complex, highly crowded environments, hosting a veritable zoo of molecules of varying size, shape, interactivity, and function (Phillips et al., 2009; van der Maarel, 2008). The nuclei of eukaryotic cells, aside from harboring the genome, are replete with myriad biopolymers (e.g., proteins, RNA) and nuclear bodies that jostle and share a tightly restricted space, typically less than 1000 μm^3 in volume. Confined within the nuclear envelope, macromolecules mutually interact and crowd one another as they diffuse through the viscous nucleoplasm. The significance of crowding for cell biology, especially for macromolecular structure within the cytoplasm, was recognized over three decades ago (Minton, 1981, 1992, 1997). In recent years, the implications of crowding for the organization and function of the cell and, in particular, of the cell nucleus have gained increasing attention (Elcock, 2010; Ellis, 2001; Hancock, 2012; Richter et al., 2007, 2008).

At concentrations of the order of 100 g/l, macromolecules in the nucleus can occupy in excess of 20% of the available volume, collectively excluding volume to one another. Under such constraints, macromolecular conformations, diffusion rates, and reaction kinetics may differ substantially from those prevailing in dilute solutions (Minton, 2001). Biopolymers in cell nuclei respond to their environment by adjusting their size and shape, adopting states ranging from extended (unfolded) random walks to more compact (folded, native, globular) assemblies resembling hard nanoparticles. The influence of crowding on folding and stability of proteins has been intensively explored recently by theoretical modeling (Minton, 2000a,b, 2005), simulations (Cheung, 2013; Cheung et al., 2005; Hoppe and

Yuan, 2011), and experiments (Benton et al., 2012; Chen et al., 2012; Koutsioubas et al., 2012; Schlesinger et al., 2011; Wang et al., 2012). At a sufficiently high concentration, excluded-volume interactions can drive changes of conformation and demixing, which may account for the observed subdivision of cell nuclei into compartments that differ in composition, despite the absence of intranuclear membranes (Hancock, 2004a,b). Moreover, inert crowding agents, by simply occupying volume, can shift statistical distributions of microstates and promote protein-ligand binding, as can be deduced from simple lattice models (Phillips et al., 2009).

Beyond the realm of cell biology, mixtures of polymers and nanoparticles are of widespread interest in several fields. In pharmaceutical science, interactions between nanoscale drug molecules and polyelectrolyte capsules or microgels have important applications in drug delivery (Kim et al., 2005). In food science, aqueous mixtures of proteins (polymers) and starch (nanoparticles) are of interest as the basis of numerous foodstuffs, for example, gelatin-dextrose solutions (Butler and Heppenstall-Butler, 2003; Tolstoguzov, 1991). Throughout materials science and nanotechnology, nanoparticles are intensely studied for their potential to modify and enhance mechanical, thermodynamic, optical, and electronic properties of polymeric materials (Balazs et al., 2006). Of biomedical relevance is crowding-induced aggregation of proteins, as occurs in the condensation of crystallin proteins within the lens of the eye, leading to cataracts (Stradner et al., 2007). Crowding by nanoparticles can also promote crystallization of proteins, whose primary structure can then be sequenced by X-ray scattering.

1.2. Confinement and excluded volume in the cell nucleus

In the crowded environment of the cell nucleus, the conformations of macromolecules and their interactions with one another are subject to two distinct but related constraints. First, macromolecules are confined to a finite volume within the nuclear membrane. While not rigid, the membrane nonetheless restricts the configurational freedom of neighboring molecules. A relevant model of worm-like polymer chains confined within a flexible enclosure (fluid vesicle) was recently investigated via Monte Carlo (MC) simulations by Fošnarič et al. (2013), who found that both the polymer and the vesicle adjust their shape (asphericity) in response to confinement. Second, each macromolecule can access only the free (available) volume unoccupied by its neighbors. In the cell nucleus, confinement and excluded volume combine to constrain macromolecules and limit their configurational and conformational degrees of freedom.

The concepts of excluded volume and free volume are usually associated with steric repulsions between two or more bodies, some of which are hard (impenetrable). The historical origins of excluded volume in physics and chemistry date back to at least the 1870s and the classic van der Waals equation of state, which corrects the ideal-gas law by incorporating molecular volume. While polymers in extended conformations are compressible, they may, in a crowded environment, adopt compact, relatively incompressible states. For example, proteins in their native (folded) state may be approximated as impenetrable globules. Moreover, the concept of excluded volume generalizes to soft (penetrable) bodies by accounting for the entropic cost of penetration.

The volumes that hard bodies exclude to one another depend on the geometries and configurations of the bodies. In the simplest case of a fluid of hard spheres, monodisperse in size, the only relevant parameter is the particle radius, each sphere excluding to its neighbors a volume eight times its own volume. In a polydisperse mixture, each sphere excludes to another the volume of a sphere whose radius is the sum of the radii of the two spheres. In the case of aspherical particles, the excluded volume depends on both the shape parameters and orientations of the bodies involved. At concentrations sufficiently high for the excluded volumes surrounding individual particles to intersect, the total excluded volume can have a complex shape.

In fluid phases, excluded volume usually entails a loss of free volume available to particles and, hence, a decrease in entropy or, equivalently, a gain in free energy. At a sufficient concentration, however, the competition for free volume can induce a phase transition from a disordered to an ordered phase. For example, a hard-sphere fluid can increase its total entropy upon crystallization, the loss in translational entropy incurred by localizing particles about the sites of a periodic lattice being outweighed by a gain in local configurational entropy. Similarly, a disordered fluid of hard rods can undergo a phase transition to an orientationally ordered nematic liquid crystal, the loss in orientational entropy being balanced by a gain in configurational entropy (Frenkel, 1991). In both examples, the ordered phase is stabilized by a net gain in entropy.

1.3. Central issues and open questions

Recent studies and reviews of macromolecular crowding highlight several key issues and unresolved questions: How do crowding agents, such as globular proteins and nanoscale particles, act to modify conformations of macromolecules in the cell nucleus and other confined spaces? How do intramolecular interactions, for example, repulsion between segments within a polymer in a good solvent, affect conformations of macromolecules in crowded environments? How do nonsteric interactions (e.g., long-ranged electrostatic interactions) between crowding agents and between crowders and macromolecules, modify crowding mechanisms? More generally, what is the interplay between excluded-volume and other nonspecific interactions? For example, by varying ionic strength or other solution conditions, can conformational changes of macromolecules be steered? How do crowding agents influence protein folding and reaction kinetics, for example, protein-ligand binding rates? Is crowding alone sufficient to drive demixing of macromolecules in cell nuclei and stabilize compartments? This list summarizes merely a few of the open questions that inspire research in the interdisciplinary field of macromolecular crowding.

1.4. Goals and outline

This chapter seeks to establish a minimal framework for addressing at least some of the previously mentioned questions. To this end, our aim is first to review relevant structural properties of polymer–nanoparticle mixtures, focusing on polymer conformations, and second, to identify a few general principles that may apply to the crowded environment within cell nuclei. In the spirit of the proverbial "spherical cow" approximation, we broach the subject within the context of tractable models, idealizing crowding agents as hard spheres (Minton, 2005) and polymers as soft, effective spheres or ellipsoids with variable size, shape, and penetrability (Eurich and Maass, 2001; Murat and Kremer, 1998; Schmidt and Fuchs, 2002).

While such generic models are certainly caricatures of reality, they nonetheless capture some essential features of real macromolecules. From statistical mechanical theories and simulations (Lu and Denton, 2011), predictions follow for the response of random-walk coils to crowding by nanoparticles. It is hoped that insights gained from this basic approach may help interpret and guide simulations of more realistic models, *in vitro* experiments involving synthetic crowders (e.g., Ficoll 70, dextran, PEG), and, ultimately, *in vivo* observations of cell nuclei.

In the next section, we introduce physical and mathematical models—from random-walk macromolecules to hard-nanoparticle crowders—progressing (or regressing) from microscopic to coarse-grained models. In

Section 3, we outline several theoretical and computational methods that have proved useful in the modeling of polymers in crowded solutions. These methods, which are firmly rooted in statistical mechanics, range from theories of effective interactions to free-volume theory and MC simulation. In Section 4, we review results, both established and preliminary, for crowding and demixing in polymer–nanoparticle mixtures, emphasizing their relevance to cell biology. Several general guiding principles emerge: (1) Crowding can significantly modify macromolecular conformations (e.g., compressing polymer coils and stabilizing folded states of proteins) in a manner that is sensitive to the size and volume fraction of crowders, the solvent conditions, and the interplay between entropy and energy. (2) For a given volume fraction of crowders, the smaller the crowder relative to the radius of gyration of the polymer, the greater the crowding effect. (3) In a crowded environment, the conformations of polymer coils shift not only toward smaller radii of gyration, but also toward more spherical average shapes. (4) Crowding can induce demixing of macromolecules, assuming that at least one of the species involved is relatively soft and penetrable (e.g., unfolded). Finally, in Section 5, we summarize and conclude with an outlook for future work.

2. MODELS OF MACROMOLECULES: POLYMERS AND NANOPARTICLES

2.1. Interactions between macromolecules: Excluded volume and electrostatics

Macromolecules in solution interact via a variety of forces that vary in strength and range (Israelachvili, 1992). Most ubiquitous are van der Waals attractions, which arise from interactions between fluctuating atomic dipoles. Acting between any two bodies, these forces are strong at short range but weak at long range, falling off rapidly with separation. Steric forces originate in Pauli exclusion and Born repulsion between overlapping electron clouds. These repulsive forces vary rapidly at short range and are directly responsible for excluded-volume repulsion, for example, between pairs of nanoparticles, between polymer segments in a good solvent, and between polymers and nanoparticles.

In mixtures of macromolecules, excluded-volume interactions can give rise to depletion-induced attractive forces. In a mixture of polymers and nanoparticles—for example, unfolded and folded (native) proteins—exclusion or depletion of polymers (depletants) from the space between

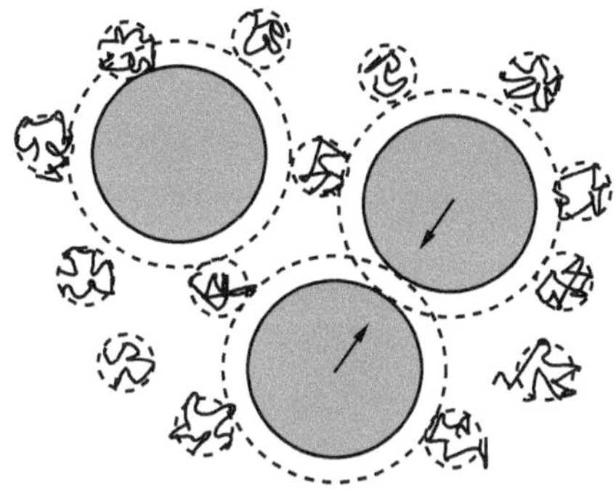

Figure 3.1 Hard nanoparticles (crowders) exclude volume to polymer coils, preventing the center of a coil from coming closer than the coil radius to the crowder surface. When excluded-volume shells surrounding crowders (dashed circles) overlap, polymer coils are depleted from the space between the crowders. Depletion modifies polymer conformations (size and shape) and induces effective attractions between crowders. (For color version of this figure, the reader is referred to the online version of this chapter.)

a pair of nanoparticles can generate an imbalance in osmotic pressure that draws the nanoparticles together (see Fig. 3.1). Equivalently, configurations in which nanoparticles are close enough to exclude polymer coils create more free volume for the polymers and so are entropically favored.

Electrostatic forces between charged particles in solution can vary widely in strength and range. The strength typically varies in proportion to the charges and inversely with the dielectric constant of the solvent. Most importantly, while bare Coulomb interactions are long ranged, they are typically strongly screened by small ions under physiological conditions. Since cells can regulate their salt content and pH via ion channels, they can tune the range of electrostatic screening between proteins and other charged macromolecules. As shown later, screening also affects the effective range of depletion-induced attractive interactions.

In addition to these ubiquitous intermolecular forces, various specific interactions, from strong covalent bonding to relatively weak hydrogen bonding, affect the large-scale structure of macromolecules (e.g., folding of proteins). We focus here on nonspecific interactions, primarily steric and electrostatic interactions, since these are the most relevant to macromolecular crowding. To facilitate computational modeling of such complex systems, however, it is necessary to adopt *effective* intermolecular interactions derived by coarse-graining, that is, averaging over a subset of the vast number of degrees of freedom (Section 2.3).

2.2. Conformations of macromolecules: Size and shape of a random coil

The cell nucleus is home to a multitude of macromolecules, including proteins, RNA, and DNA (packaged as chromatin fibers), with a broad polydispersity in size and shape. Depending on environmental conditions (concentration, ionic strength, pH, etc.), these molecules may adopt conformations characteristic of random-walk coils or of compact nanoparticles. In the case that species with random coil structures (e.g., unfolded proteins) mingle with species having more compact structures (e.g., folded proteins), crowding may induce changes of conformation and even drive demixing at sufficiently high concentrations.

In this review, we adopt a conceptual framework in which molecules of one species ("polymers") are modeled as random coils, while molecules of another species ("nanoparticles") are modeled as hard, possibly charged, spheres. For simplicity, we focus on binary mixtures, although this model is easily generalized to multi-component mixtures. In this picture (Fig. 3.2), the nanoparticle crowding agents are distinguished simply by their fixed radius R_c and, if charged, their valence Z. The polymers, on the other hand, are ramified, penetrable coils that fluctuate in size and shape. Indeed, MC simulations of bead-spring models of polymers have shown that polymer coils can be penetrated by smaller nanoparticles (Doxastakis et al., 2004, 2005). To understand how polymers are affected by crowding, we must establish a baseline by characterizing the conformations of free polymer coils.

We first ask what the size of a polymer coil is. One way to quantify the size of a linear polymer is by the root-mean-square (rms) end-to-end

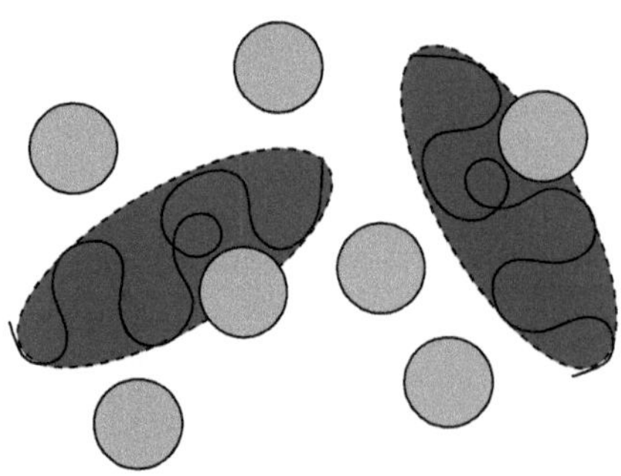

Figure 3.2 Model mixture of macromolecules (polymers and crowders) in solution. Polymers are modeled as penetrable coils (spheres or ellipsoids) and crowders as impenetrable spheres, possibly charged. (For color version of this figure, the reader is referred to the online version of this chapter.)

displacement, which is typically much shorter than the contour length of a random coil. A more practical measure of size, independent of polymer architecture and so applicable also to branched and ring polymers, is the radius of gyration. For a polymer of N segments, the radius of gyration is defined as

$$R_p \equiv \left(\frac{1}{N}\sum_{i=1}^{N}|\vec{r}_i - \vec{r}_{cm}|^2\right)^{1/2} = \left(\frac{1}{2N^2}\sum_{i,j=1}^{N}|\vec{r}_i - \vec{r}_j|^2\right)^{1/2} \tag{3.1}$$

that is, the root-mean-square deviation of the segment positions $\vec{r}_i(i=1,\ldots,N)$ from the center-of-mass position $\vec{r}_{cm}$. While a single random walk (i.e., static conformation) has a fixed size (R_p), polymers in solution fluctuate in conformation, their radii of gyration following a broad probability distribution $P_r(R_p)$. The subscript r here distinguishes the size distribution of a *reservoir* solution of pure polymer from the modified distribution $P(R_p)$ (introduced later) of polymers in the presence of crowders (nanoparticles). The quantity that is accessible to experiments (e.g., via light or neutron scattering) is the rms radius of gyration R_g^r, defined as

$$R_g^r \equiv \left(\frac{1}{N}\sum_{i=1}^{N}\left\langle|\vec{r}_i - \vec{r}_{cm}|^2\right\rangle\right)^{1/2} = \left(\int_0^\infty dR_p R_p^2 P_r(R_p)\right)^{1/2} \tag{3.2}$$

where the angular brackets $\langle\ \rangle$ denote a statistical average over conformations.

To further simplify the model, we approximate the polymers as ideal (self-intersecting) random walks, implying weak effective intramolecular (segment-segment) interactions. The only correlations among segments then arise from the topological constraint of connectedness. This neglect of excluded-volume interactions between segments is justified for polymers in a theta solvent (de Gennes, 1979). For polymers in a good solvent, segment excluded volume can be incorporated by adopting the size distribution for self-avoiding chains (Hadizadeh et al., 2011; Lhuilier, 1988; Schäfer, 1999; Sciutto, 1996). Assuming ideal polymers, the radius of gyration scales with the square-root of the segment number $\left(R_g^r \sim \sqrt{N}\right)$.

Approximations for the probability distribution of the radius of gyration of an ideal, freely-jointed chain were proposed by Fixman (1962) and Flory and Fisk (1966). The *exact* distribution was subsequently derived (Fujita and Norisuye, 1970; Yamakawa, 1970) and is given by

$$
\begin{aligned}
P_r(R_p) = &\frac{1}{\sqrt{2}\pi R_g^r t^3} \sum_{k=0}^{\infty} \frac{(2k+1)!}{(2^k k!)^2} (4k+3)^{7/2} \exp(-t_k) \\
&\times \left[\left(1 - \frac{5}{8t_k}\right) K_{1/4}(t_k) + \left(1 - \frac{3}{8t_k}\right) K_{3/4}(t_k) \right]
\end{aligned}
\tag{3.3}
$$

where $t = (R_p/R_g^r)^2$, $t_k = (4k+3)^2/(8t)$, and K_n are modified Bessel functions of the second kind. In practice, it proves convenient to approximate $P_r(R_p)$ by the accurate, analytical ansatz of Eurich and Maass (2001):

$$
P_r(u) = \frac{1}{2uK_0} \exp\left(-\frac{u}{a} - d^2 \frac{a}{u}\right) \tag{3.4}
$$

where $u \equiv R_p^2/N$, $K_0 = 0.015923$, $a = 0.0802$, and $d = 1.842$. As Fig. 3.3 illustrates, the function $P_r(R_p)$ is a broad distribution and differs from the end-to-end displacement distribution, which is Gaussian. Thus, even polymers that (hypothetically) are monodisperse in molecular weight are highly polydisperse in radius of gyration.

We next ask what the shape of a polymer coil is. The answer depends, of course, on the level of detail desired. The conformation of a polymer can be defined by specifying the coordinates of all of its monomers or of a smaller number of segments, each comprising multiple monomers. This definition corresponds conceptually to the secondary structure of a protein, involving alpha-helices and beta-sheets stabilized by hydrogen bonds between peptide groups. A much coarser, but nevertheless useful, measure of shape can be quantified by the gyration tensor

$$
\mathbf{T} = \frac{1}{N} \sum_{i=1}^{N} \left\langle \vec{r_i}\, \vec{r_i} \right\rangle = \begin{pmatrix} \langle x^2 \rangle & \langle xy \rangle & \langle xz \rangle \\ \langle xy \rangle & \langle y^2 \rangle & \langle yz \rangle \\ \langle xz \rangle & \langle yz \rangle & \langle z^2 \rangle \end{pmatrix} \tag{3.5}
$$

where the segment coordinates (x, y, z) are relative to the center of mass, and the angular brackets again denote an average over conformations. If the average is performed relative to a fixed (laboratory) frame of reference, the resulting average tensor is symmetric with identical eigenvalues, thus describing a sphere. As expected, the rotational average washes out any anisotropy, since an arbitrarily shaped object, when randomly tumbled, sweeps out a sphere. On the other hand, if the conformational average is performed in a frame of reference that rotates with the principal axes of the polymer, the resulting average tensor is asymmetric, thus describing an anisotropic object (Kuhn, 1934). The latter averaging procedure

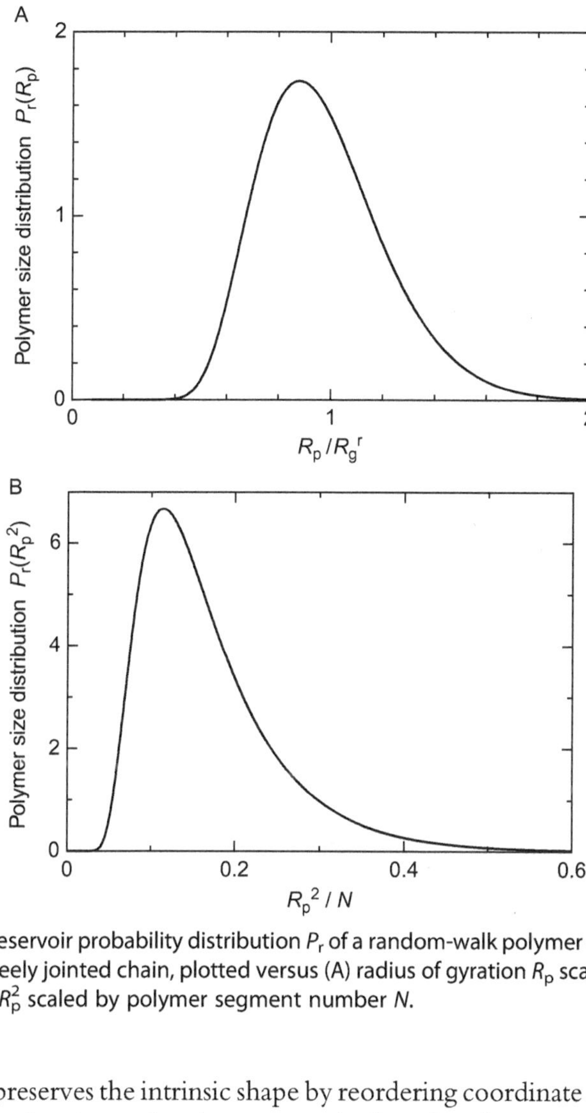

Figure 3.3 Reservoir probability distribution P_r of a random-walk polymer coil, modeled as an ideal, freely jointed chain, plotted versus (A) radius of gyration R_p scaled by its rms value R_g^r; (B) R_p^2 scaled by polymer segment number N.

effectively preserves the intrinsic shape by reordering coordinate axes so as to always order the eigenvalues by magnitude ($\lambda_1 > \lambda_2 > \lambda_3$). The eigenvalues, which are proportional to the squares of the principal radii, then determine the ellipsoid that best fits the shape of the polymer coil (Rudnick and Gaspari, 1987).

The best-fit ellipsoid corresponds roughly to the tertiary structure of a protein, as determined by nonspecific hydrophobic interactions. For reference, the moment of inertia tensor $\mathbf{I}$, familiar from mechanics of rigid bodies, is related to the gyration tensor and the unit tensor I via $\mathbf{I} = (R_g^r)^2 I - \mathbf{T}$.

As it turns out, the average shape of a random walk, in the principal-axis frame, is a significantly elongated (prolate) ellipsoid, whose principal radii are approximately in the ratio 3.4 : 1.6 : 1 (Šolc, 1971, 1973).

As with the size distribution, the probability distribution for the shape of an ideal, freely-jointed chain in a reservoir solution of pure (uncrowded) polymer $P_r(\lambda_1, \lambda_2, \lambda_3)$ is well approximated by the analytical ansatz of Eurich and Maass (2001), who extended earlier work by Murat and Kremer (1998):

$$P_r(\vec{\lambda}) = \prod_i P_{ir}(\lambda_i) \tag{3.6}$$

with

$$P_{ir}(\lambda_i) = \frac{(a_i d_i)^{n_i - 1} \lambda_i^{-n_i}}{2K_i} \exp\left(-\frac{\lambda_i}{a_i} - d_i^2 \frac{a_i}{\lambda_i}\right) \tag{3.7}$$

where $K_1 = 0.094551$, $K_2 = 0.0144146$, $K_3 = 0.0052767$, $a_1 = 0.08065$, $a_2 = 0.01813$, $a_3 = 0.006031$, $d_1 = 1.096$, $d_2 = 1.998$, $d_3 = 2.684$, $n_1 = 1/2$, $n_2 = 5/2$, and $n_3 = 4$. Here, the eigenvalues are implicitly scaled by the segment number N, since the polymers are assumed to be ideal. As Fig. 3.4 shows, each of the three eigenvalues is broadly distributed, implying significant shape fluctuations. In reality, the eigenvalues of a random walk are not

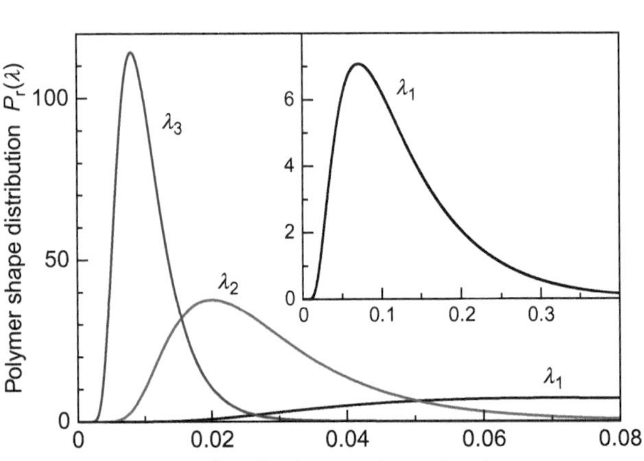

Figure 3.4 Probability distributions $P_r(\lambda_i)$ for the eigenvalues λ_1 (black), λ_2 (red), and λ_3 (blue) of the gyration tensor of a polymer coil, modeled as an ideal, freely-jointed chain, in a reservoir solution of pure polymer (no crowders) (Eurich and Maass, 2001). Compare with Fig. 3.8. (For interpretation of the references to color in this figure legend, the reader is referred to the online version of this chapter.)

independent, since an extension in one direction affects the probability of an orthogonal extension. The exact eigenvalue probability distribution, then, should not strictly factorize. For random walks sufficiently long to model real polymers, however, conformations that significantly violate the factorization ansatz (Eq. 3.6) are relatively rare and, in practice, can be neglected. To quantify a polymer's deviation from a perfect sphere, it is useful to define the *asphericity* (Rudnick and Gaspari, 1987):

$$A = \frac{\sum_{i>j=1}^{3} \left\langle (\lambda_i - \lambda_j)^2 \right\rangle}{2\left\langle \left(\sum_{i=1}^{3} \lambda_i \right)^2 \right\rangle} \tag{3.8}$$

While the results presented later in Section 4 pertain to ideal (freely-jointed chain) polymers, the same general approach outlined here can be applied to polymers with excluded-volume intramolecular interactions (Schäfer, 1999) by substituting the appropriate size and shape distributions. For such applications, it is worth noting that accurate approximations for the radius of gyration distribution of self-avoiding random walks exist (Lhuilier, 1988; Sciutto, 1996). For applications to proteins, Goldenberg (2003) has calculated size distributions of several unfolded proteins from simulations, while Dima and Thirumalai (2004) have analyzed the shapes of proteins modeled in Langevin dynamics simulations, finding both folded and unfolded conformations to be significantly asymmetric.

2.3. Coarse-grained models of macromolecular solutions

Modeling macromolecular solutions in full microscopic detail is a daunting challenge. Considering that the number of realizations of a random walk of N steps increases exponentially with N, a single polymer chain can adopt an enormous number of possible conformations. The ongoing mystery, of course, is how a protein efficiently finds the folding pathway to a particular conformation that favors a certain function. Moreover, the solvent (water) molecules and ions are too vast in number to model explicitly in large-scale simulations. Consider that a volume of only 1 μm^3 (10^{-15} l) contains of order 10^{13} water molecules and 10^8 ions at physiological salt concentrations (0.1 M).

To make progress in modeling such complex systems, we resort to coarse-grained models derived by averaging probability distributions or partition functions over microscopic degrees of freedom. In modeling polymer–nanoparticle mixtures, we reduce rather drastically the complexity of the polymer coils by regarding them, in the simplest case, as effective spheres that are mutually interpenetrable and can fluctuate in size. To explore crowding-induced variations in shape, we also consider a model of effective ellipsoidal coils whose principal radii can fluctuate.

At first glance, approximating a polymer coil by a sphere, or even by an ellipsoid, may appear overly crude. It is important to note, however, that our main interest here is in the interaction between polymers and hard nanoparticles, representing crowders. From the shape distributions in Fig. 3.4, we see that significantly distorted shapes rapidly become less probable with growing deviation from the typical (most probable) shape. In the presence of a hard obstacle, a random coil may distort in shape, but only at the expense of conformational entropy. Indeed, simulations of bead-spring models of polymers (Doxastakis et al., 2004, 2005; Murat and Kremer, 1998) confirm that the shapes of polymer coils are well described by ellipsoids. Moreover, as shown in Section 4, crowding by nanoparticles tends to compress ideal polymers, narrowing and sharpening the polymer shape distributions and reducing the asphericity of the average shape. The effective spherical/ellipsoidal polymer model thus isolates and highlights essential geometrical features of random coils and may capture, at least qualitatively, the response of macromolecular conformations to a crowded environment.

To elaborate on the coarse-graining procedure, we may start from a microscopic model of a polymer, governed by a Hamiltonian (total energy) function $H(\vec{r}_1, \ldots, \vec{r}_N)$, which depends on the coordinates $\vec{r}_i$ of all N segments. The segment coordinates are not independent, of course, but are strongly constrained by the topology (connectedness) of the polymer. At absolute temperature T, a coil adopts a particular conformation with probability

$$P_{\mathrm{r}}(\vec{r}_1, \cdots, \vec{r}_N) = \frac{\mathrm{e}^{-\beta H(\vec{r}_1, \ldots, \vec{r}_N)}}{\mathcal{Z}_{\mathrm{p}}} \tag{3.9}$$

where $\beta \equiv 1/k_B T$,

$$\mathcal{Z}_{\mathrm{p}} = \frac{1}{N! \Lambda^{3N}} \int \mathrm{d}\vec{r}_1 \cdots \int \mathrm{d}\vec{r}_N \mathrm{e}^{-\beta H(\vec{r}_1, \ldots, \vec{r}_N)} = \left\langle \mathrm{e}^{-\beta H(\vec{r}_1, \ldots, \vec{r}_N)} \right\rangle \tag{3.10}$$

is the partition function of a single polymer, Λ is the de Broglie thermal wavelength, and the angular brackets denote a trace over polymer segment degrees of freedom:

$$\langle\,\rangle \equiv \frac{1}{N!\Lambda^{3N}}\int d\vec{r}_1 \cdots \int d\vec{r}_N \tag{3.11}$$

The single-polymer partition function may also be formally expressed as

$$\begin{aligned}\mathcal{Z}_p &= \int_0^\infty dR_p^2 \left\langle \delta\left(\frac{1}{N}\sum_{i=1}^{N}|\vec{r}_i - \vec{r}_{cm}|^2 - R_p^2\right) e^{-\beta H(\vec{r}_1,\ldots\vec{r}_N)} \right\rangle \\ &= \int_0^\infty dR_p^2 e^{-\beta H_{\text{eff}}(R_p)}\end{aligned} \tag{3.12}$$

where the Dirac δ-function imposes a size constraint on each conformation and

$$H_{\text{eff}}(R_p) \equiv -k_B T \ln \left\langle \delta\left(\frac{1}{N}\sum_{i=1}^{N}|\vec{r}_i - \vec{r}_{cm}|^2 - R_p^2\right) e^{-\beta H(\vec{r}_1,\ldots,\vec{r}_N)} \right\rangle \tag{3.13}$$

is the *effective Hamiltonian* of a polymer coil with radius of gyration R_p. A conformation with a given radius of gyration R_p then occurs with probability,

$$P_r(R_p) = \frac{e^{-\beta H_{\text{eff}}(R_p)}}{\mathcal{Z}_p} \tag{3.14}$$

In a similar manner, by integrating over all but a limited number of degrees of freedom $\vec{\lambda} \equiv (\lambda_1, \lambda_2, \ldots)$, the single-polymer partition function may be expressed in the form

$$\mathcal{Z}_p = \int d\vec{\lambda} \left\langle e^{-\beta H(\vec{r}_1,\ldots,\vec{r}_N)} \right\rangle = \int d\vec{\lambda}\; e^{-\beta H_{\text{eff}}(\vec{\lambda})} \tag{3.15}$$

In the case that $\vec{\lambda}$ represents the three eigenvalues of the gyration tensor, characterizing the shape of a polymer coil (Section 2.2), the eigenvalue (i.e., shape) probability distribution may be expressed as

$$P_r(\vec{\lambda}) = \frac{e^{-\beta H_{\text{eff}}(\vec{\lambda})}}{\mathcal{Z}_p} \tag{3.16}$$

where $H_{\rm eff}(\vec{\lambda})$ now represents the effective Hamiltonian of a polymer coil whose shape is described by a given set of eigenvalues. The point of these manipulations is to illustrate how a complex system with many degrees of freedom may be mapped onto a conceptually simpler system with fewer degrees of freedom, but governed by an effective Hamiltonian that involves effective interparticle interactions (Denton, 2007).

The same coarse-graining procedure may be applied to any mixture of species. For example, a mixture of crowders (nanoparticles) and polymers, with Hamiltonian $H = H_c + H_p + H_{cp}$, where H_c and H_p are the Hamiltonians of the crowders (c) and polymers (p) alone and H_{cp} is the crowder–polymer interaction energy, has a partition function that may be expressed as

$$\mathcal{Z} = \left\langle \langle \exp(-\beta H) \rangle_p \right\rangle_c = \langle \exp(-\beta H_{\rm eff}) \rangle_c \tag{3.17}$$

where $\langle\ \rangle_p$ and $\langle\ \rangle_c$ denote traces over polymer and crowder degrees of freedom, $H_{\rm eff} = H_c + F_p$ is the effective Hamiltonian of the mixture, and

$$F_p = -k_B T \ln \left\langle e^{-\beta\left(H_p + H_{cp}\right)} \right\rangle_p \tag{3.18}$$

is the free energy of the polymers in the presence of the crowders. The latter may be approximated as

$$F_p = F_0 + \Pi_p V_{\rm exc}[\rho_c(\vec{r})] \tag{3.19}$$

where F_0 and Π_p are the free energy and osmotic pressure of a pure polymer solution (absent crowders) and $V_{\rm exc}[\rho_c(\vec{r})]$ is the total volume excluded by the crowders to the polymers, which is a functional of the (nonuniform) crowder number density $\rho_c(\vec{r})$. By relating $V_{\rm exc}$ to overlaps of excluded-volume shells surrounding the crowders, the effective Hamiltonian can be expressed as a sum of depletion-induced effective pair (and many-body) interactions between crowders (Denton, 2007). These effective interactions can then be input into simulations of a one-component model. Alternatively, Eq. (3.19) provides a foundation for formulating a free-volume theory, as described later in Section 3.1. In this context, the $\Pi_p V_{\rm exc}$ term may be physically interpreted as the work performed against the polymer osmotic pressure to inflate cavities of volume $V_{\rm exc}$ sufficiently to accommodate the crowders.

For a mixture of *charged* crowders and microions (counterions and salt ions) dispersed in a solvent, the Hamiltonian can be expressed as $H = H_c + H_\mu + H_{c\mu}$, where H_c and H_μ are the crowder (c) and microion (μ) Hamiltonians and $H_{c\mu}$ is the crowder–microion interaction energy. The coarse-graining procedure now renders the partition function in the form

$$\mathcal{Z} = \left\langle \langle \exp(-\beta H) \rangle_\mu \right\rangle_c = \langle \exp(-\beta H_{\text{eff}}) \rangle_c \tag{3.20}$$

where $H_{\text{eff}} = H_c + F_\mu$ is the effective Hamiltonian of the crowder–microion mixture and

$$F_\mu = -k_B T \ln \left\langle e^{-\beta(H_\mu + H_{c\mu})} \right\rangle_\mu \tag{3.21}$$

is the free energy of the microions in the presence of the macroions. As with the polymer free energy, the microion free energy depends on the configuration of the crowders. In a linear-response approximation for the microion density (Denton, 2000, 2007), the effective Hamiltonian includes a one-body volume energy E_0 and a pair energy, which can be expressed as a sum over pairs of crowders of an effective electrostatic pair potential $v_{\text{eff}}(r)$,

$$H_{\text{eff}} = E_0 + \sum_{i<j} v_{\text{eff}}\left(\left|\vec{r}_i - \vec{r}_j\right|\right) \tag{3.22}$$

where $\vec{r}_i$ denotes the position of the center of particle i. The volume energy, which depends on the average density of the crowders, can influence the thermodynamic properties of de-ionized colloidal suspensions. At the relatively high salt concentrations prevailing within cells, however, the volume energy is essentially constant and can be ignored. The effective pair potential can be modeled within a mean-field (random phase) approximation for the microion correlations, which is valid especially for monovalent microions and proves equivalent to linearized Poisson-Boltzmann (Debye-Hückel) theory (Denton, 2010). The result is a repulsive screened-Coulomb (Yukawa) potential:

$$v_{\text{eff}}(r) = \frac{Z^2 e^2}{\varepsilon} \left(\frac{e^{\kappa R_c}}{1 + \kappa R_c} \right)^2 \frac{e^{-\kappa r}}{r}, \quad r \geq 2R_c \tag{3.23}$$

where r is the center-to-center distance, Z is the effective charge of the crowders, ε is the dielectric constant of the solvent (water), and κ is the

Debye screening constant (inverse screening length). The screening constant is given by $\kappa = \sqrt{4\pi\lambda_B\rho_\mu}$, where $\lambda_B = e^2/\varepsilon k_B T$ is the Bjerrum length and ρ_μ is the total number density of microions in the free volume, that is, the volume not occupied by the crowders (Denton, 2000).

2.4. Models of polymer–nanoparticle mixtures

We consider a mixture of free (nonadsorbing) polymers and macromolecular crowders dispersed in a solvent (e.g., the nucleoplasm) and confined (e.g., in the cell nucleus) within a volume V. The mixture is assumed to be in osmotic equilibrium with a reservoir of polymer (e.g., the cytoplasm) that can exchange polymers with the system to maintain constant polymer chemical potential. At temperature T, the thermodynamic state of the mixture is characterized by the crowder number density ρ_c and the polymer number density in the reservoir ρ_p^r. Chemical equilibrium between the system and reservoir determines the polymer density in the system ρ_p.

A widely studied model of polymer-particle mixtures is that of Asakura and Oosawa (1954), elaborated by Vrij (1976), which treats the particles as hard (impenetrable) spheres and the polymers as effective spheres, which are mutually ideal (noninteracting). For fixed particle (crowder) radius R_c and (instantaneous) polymer radius of gyration R_p, the crowder–crowder (cc) and crowder–polymer (cp) interactions are described by hard-sphere pair potentials:

$$v_{cc}(r) = \begin{cases} \infty, & r < 2R_c \\ 0, & r \geq 2R_c \end{cases} \tag{3.24}$$

$$v_{cp}(r) = \begin{cases} \infty, & r < R_c + R_p \\ 0, & r \geq R_c + R_p \end{cases} \tag{3.25}$$

For ideal polymers, the polymer–polymer interaction vanishes ($v_{pp}(r) = 0$) for all r, which is valid for theta solvents (de Gennes, 1979), in which the polymer second virial coefficient is negligible. In the original AOV model, the system is thus specified by a single parameter—the ratio $q \equiv R_p/R_c$ of the polymer radius to the crowder radius.

In modeling the response of compressible polymers to crowding, Denton and Schmidt (2002) modified the AOV model to allow the polymer radius of gyration to fluctuate. The system is then specified by the reservoir polymer size distribution $P_r(R_p)$ (Eq. 3.9) and the ratio of the rms polymer radius of gyration in the reservoir to the crowder radius, $q_r \equiv R_g^r/R_c$, a parameter that can be determined experimentally. The ratio q_r is to be distinguished from the ratio $q = R_p/R_c$ in the presence of crowders, which may vary with crowder concentration.

The assumption of hard-sphere polymer–particle interactions is reasonable in the "colloid" limit ($q_r < 1$), in which crowders are too large to penetrate polymer coils. A random-walk coil may, of course, extend around a larger particle, but such a conformation costs significant entropy and, therefore, is quite improbable. In the "protein" or "nanoparticle" limit ($q_r > 1$), however, small crowders may penetrate the polymer coils. In this limit, Schmidt and Fuchs (2002) considered a variation of the AOV model in which the polymer–particle hard-core interaction is replaced by an average penetration energy:

$$v_{cp}(r) = \begin{cases} \varepsilon, & r < R_c + R_p \\ 0, & r \geq R_c + R_p \end{cases} \tag{3.26}$$

In this model, the system is specified by two parameters: the size ratio q and the penetration energy ε. In the colloid limit, impenetrable polymers can be modeled by taking $\varepsilon \to \infty$. In the protein or nanoparticle limit, polymer field theory predicts (Eisenriegler et al., 1996; Hanke et al., 1999) that hard-sphere crowders can penetrate ideal polymer coils with an average cost in free energy:

$$\beta\varepsilon = \frac{3}{q} \tag{3.27}$$

The loss in conformational entropy suffered by a polymer coil upon engulfing a crowder is thus modeled by a free energy penalty, which increases as the polymer coil contracts. (For comparison, self-avoiding polymer coils follow $\beta\varepsilon \simeq 4.1/q^{1.3}$.)

Recently, Lu and Denton (2011) combined the Denton-Schmidt and Schmidt-Fuchs models to study polymer–nanoparticle mixtures with polymers modeled as effective spheres that are *both* compressible *and* penetrable. In this generalized model, the system is specified by the polymer size distribution, the reservoir size ratio q_r, and the penetration energy ε. In the next section, we review methods of free-volume theory and MC simulation, which have been developed and applied to the above-mentioned models to explore the response of polymer conformations to crowding and the demixing of polymer–crowder mixtures.

3. THEORETICAL AND COMPUTATIONAL METHODS

Among a variety of theoretical and computational methods developed for modeling mixtures of polymers and particles, two of the most powerful are density-functional theory (Forsman and Woodward, 2009; Woodward and Forsman, 2008) and integral-equation theory (Chatterjee and Schweizer, 1998a,b; Fuchs and Schweizer, 2000, 2002; Ramakrishnan et al., 2002), both firmly rooted in classical liquid-state theory (Hansen and McDonald, 1990). Based on connected-bead models of polymer chains, these methods are designed to accurately predict the structure (e.g., correlation functions) of polymers in dense polymer–particle mixtures. We focus here instead on a relatively simple free-volume theory of mixtures. Although based on an even coarser-grained model of polymers, this theory is straightforward to implement, requires negligible computational resources, and yields immediate conceptual insights into crowding-induced changes in polymer conformation.

3.1. Free-volume theory of crowding in polymer–nanoparticle mixtures

Over the last two decades, a free-volume theory proposed by Lekkerkerker et al. (1992) has been widely applied to the AOV model and related models of colloid–polymer mixtures. The theory is based on an exact separation of the total Helmholtz free energy,

$$F(\rho_c\rho_p) = F_c(\rho_c) + F_p(\rho_p\rho_c) \tag{3.28}$$

into colloid (F_c) and polymer (F_p) free energies. The colloid free energy density, $f = F_c/V$, can be decomposed as

$$f_c(\phi_c) = f_{id}(\phi_c) + f_{ex,c}(\phi_c) \tag{3.29}$$

where $\phi_c = (4\pi/3)\rho_c R_c^3$ is the colloid (crowder) volume fraction,

$$f_{id}(\phi_c) = k_B T \rho_c(\ln \phi_c - 1) \tag{3.30}$$

is the exact free energy density of an ideal gas of colloids, and $f_{ex,c}$ is the excess free energy density of the crowders, which is very accurately approximated by the Carnahan–Starling expression for a hard-sphere fluid (Hansen and McDonald, 1990):

$$f_{ex,c}(\phi_c) \simeq f_{hs}(\phi_c) = k_B T \rho_c \frac{\phi_c(4-3\phi_c)}{(1-\phi_c)^2} \tag{3.31}$$

The polymer free energy is approximated by the free energy of ideal polymers confined to the *free* volume, $V_f = V - V_{exc}$, that is, the volume that is not excluded by the colloid hard cores. Thus, the polymer free energy density, $f_p = F_p/V$, can be expressed as

$$f_p(\phi_c, \rho_p) = k_B T \rho_p \left[\ln\left(\frac{V\phi_p}{V_f(\phi_c)} \right) - 1 \right] = f_{id}(\phi_p) + f_{ex,p}(\phi_c, \rho_p) \tag{3.32}$$

where $\phi_p = (4\pi/3)\rho_p R_p^3$ (with fixed R_p) is the effective polymer volume fraction,

$$f_{id}(\phi_p) = k_B T \rho_p (\ln \phi_p - 1) \tag{3.33}$$

is the exact polymer ideal-gas free energy density, and (from Eqs. 3.19 and 3.32)

$$f_{ex,p}(\phi_c, \rho_p) = \Pi_p[1 - \alpha(\phi_c)] = -k_B T \rho_p \ln \alpha(\phi_c) \tag{3.34}$$

is the excess free energy density of the polymers, which depends on the colloid density via the polymer free-volume fraction $\alpha(\phi_c) \equiv V_f(\phi_c)/V$. The assumption that the polymer free-volume fraction depends only on the average colloid density, and not also on the polymer density (even though, in reality, the polymers can affect colloid configurations) is a form of mean-field approximation that amounts to entirely neglecting colloid–polymer correlations. Equality of chemical potentials of ideal polymers in the system and reservoir implies a direct relation between the corresponding densities of polymers: $\rho_p = \rho_p^r \alpha(\phi_c)$. Assembling the pieces, the total free energy density $f = F/V$ can then be expressed (in $k_B T$ units, and to within an arbitrary constant) as

$$f(\phi_c, \rho_p^r) = \rho_c(\ln \phi_c - 1) + f_{hs}(\phi_c) + \rho_p^r \alpha(\phi_c)(\ln \phi_p^r - 1) \tag{3.35}$$

where $\phi_p^r \equiv (4\pi/3)\rho_p^r R_p^3$ is the effective polymer volume fraction in the reservoir.

The free-volume theory of Lekkerkerker et al. has been generalized to models of compressible polymers with a *fluctuating* radius of gyration (Denton and Schmidt, 2002). The polymer free energy must now be augmented by the conformational free energy $f_{\mathrm{conf}}(R_p)$, associated with internal degrees of freedom of a polymer coil, and averaged over radius of gyration:

$$f_p = \rho_p \int_0^\infty \mathrm{d}R_p P(R_p; \phi_c) \left[\ln \left(\frac{P(R_p; \phi_c)\phi_p}{\alpha(R_p; \phi_c)} \right) - 1 + f_{\mathrm{conf}}(R_p) \right] \tag{3.36}$$

where $\alpha(R_p; \phi_c)$ is the free-volume fraction, and $P(R_p; \phi_c)$ the probability distribution of polymer coils of radius of gyration R_p amidst crowders of volume fraction ϕ_c. Assuming that a polymer has the same conformational entropy in the system as in the reservoir,

$$f_{\mathrm{conf}}(R_p) = -\ln P_r(R_p) \tag{3.37}$$

Equating chemical potentials of ideal polymers *of a given size* (radius of gyration R_p) in the system and reservoir now implies

$$\rho_p(R_p; \phi_c) = \rho_p^r(R_p)\alpha(R_p; \phi_c) \tag{3.38}$$

where $\rho_p(R_p; \phi_c) = \rho_p P(R_p; \phi_c)$ and $\rho_p^r(R_p) = \rho_p^r P_r(R_p)$. Combining Eqs. (3.36)–(3.38), the total free energy density may be expressed as

$$f(\phi_c, \rho_p^r) = \rho_c(\ln \phi_c - 1) + f_{\mathrm{hs}}(\phi_c) + \rho_p^r \alpha_{\mathrm{eff}}(\phi_c)(\ln \phi_p^r - 1) \tag{3.39}$$

where now $\phi_p^r \equiv (4\pi/3)\rho_p^r (R_g^r)^3$ is the effective polymer volume fraction in the reservoir, R_g^r is the rms radius of gyration of polymer in the reservoir (Eq. 3.2), and

$$\alpha_{\mathrm{eff}}(\phi_c) \equiv \int_0^\infty \mathrm{d}R_p P_r(R_p)\alpha(R_p; \phi_c) \tag{3.40}$$

is an *effective* polymer free-volume fraction, expressed as an average of free-volume fraction over polymer size. The corresponding size distribution of crowded polymers is

$$P(R_p; \phi_c) = P_r(R_p)\frac{\alpha(R_p; \phi_c)}{\alpha_{\mathrm{eff}}(\phi_c)} \tag{3.41}$$

Equation (3.39) can also be applied to the ellipsoidal polymer model, with the effective free-volume fraction then generalized to

$$\alpha_{\mathrm{eff}}(\phi_c) \equiv \int \mathrm{d}\vec{\lambda}\, P_r(\vec{\lambda})\alpha(\vec{\lambda}; \phi_c) \tag{3.42}$$

where $\alpha(\vec{\lambda};\phi_c)$ denotes the free-volume fraction of polymers with shape eigenvalues $\vec{\lambda} \equiv (\lambda_1, \lambda_2, \lambda_3)$ amidst crowders of volume fraction ϕ_c. The crowded polymers have a three-dimensional shape distribution that varies with ϕ_c according to

$$\frac{P(\vec{\lambda};\phi_c) = P_r(\vec{\lambda})\alpha(\vec{\lambda};\phi_c)}{\alpha_{\text{eff}}(\phi_c)} \tag{3.43}$$

Averaging over two of the eigenvalues yields a one-dimensional probability distribution for just one principal axis of the corresponding ellipsoid. For example,

$$P_1(\lambda_1;\phi_c) = \frac{1}{\alpha_{\text{eff}}(\phi_c)} \int d\lambda_2 \int d\lambda_3 P_r(\vec{\lambda})\alpha(\vec{\lambda};\phi_c) \tag{3.44}$$

Note that in the dilute crowder limit ($\phi_c \to 0$), the free-volume fractions α, $\alpha_{\text{eff}} \to 1$ and the shape distribution reduces to that of the reservoir: $P(\vec{\lambda}) \to P_r(\vec{\lambda})$.

The polymer free-volume fraction is accurately modeled by the geometry-based approximation of Oversteegen and Roth (2005). This approach generalizes scaled-particle theory (Lebowitz et al., 1964) to arbitrary shapes by using fundamental-measures density-functional theory (Rosenfeld, 1989; Rosenfeld et al., 1997; Schmidt et al., 2000) to separate the thermodynamic properties of the crowders from the geometric properties of the depletants (polymers). The result is

$$\alpha(\vec{\lambda};\phi_c) = (1-\phi_c)\exp\left[-\beta\left(pv_p + \gamma a_p + \kappa c_p\right)\right] \tag{3.45}$$

where p, γ, and κ are the bulk pressure, surface tension at a planar hard wall, and bending rigidity of the crowders, respectively, while v_p, a_p, and c_p are, respectively, the volume, surface area, and integrated mean curvature of a polymer. For a spherical polymer, $v_p = (4\pi/3)R_p^3$, $a_p = 4\pi R_p^2$, and $c_p = R_p$. A general ellipsoid polymer, with principal radii R_1, R_2, R_3, has volume $v_p = (4\pi/3)R_1R_2R_3$, while a_p and c_p can be numerically evaluated. Finally, the thermodynamic properties of hard-sphere crowders are accurately approximated by the Carnahan–Starling expressions (Oversteegen and Roth, 2005):

$$
\begin{aligned}
\beta p &= \frac{3\phi_c}{4\pi R_c^3}\frac{1+\phi_c+\phi_c^2-\phi_c^3}{(1-\phi_c)^3} \\
\beta\gamma &= \frac{3}{4\pi R_c^2}\left[\frac{\phi_c(2-\phi_c)}{(1-\phi_c)^2}+\ln(1-\phi_c)\right] \\
\beta\kappa &= \frac{3\phi_c}{R_c(1-\phi_c)}
\end{aligned}
\tag{3.46}
$$

Recently, Lu and Denton (2011) generalized the above free-volume theory to the nanoparticle limit, modeling the polymers as effective spheres that are both compressible *and* penetrable by the smaller (nanoparticle) crowders. In implementing the theory, the only practical change is replacement of the crowder volume fraction ϕ_c on the right sides of Eqs. (3.45) and (3.46) by an *effective* volume fraction $\tilde{\phi}_c \equiv [1-\exp(-\beta\varepsilon)]\phi_c$, where ε is the penetration energy from Eqs. (3.26) and (3.27) (Schmidt and Fuchs, 2002).

In earlier work, Denton and Schmidt (2005) explored another variation of the AOV model in which the nanoparticle crowders are *charged* and interact via effective electrostatic interactions. Assuming that the screened-Coulomb effective pair potential between crowders (Eq. 3.23) is steeply repulsive ($\kappa R_c > 1$), which is generally true at physiological salt concentrations, the free energy of the crowders may be reasonably approximated by that of a fluid of effective hard spheres of radius R_c' (Andersen et al., 1971; Hansen and McDonald, 1990; Weeks et al., 1971).

As illustrated in Fig. 3.5, the mixture of polymers and charged crowders can then be mapped onto an equivalent AOV model by introducing an effective (fictitious) polymer radius R_p', such that $R_c'+R_p'=R_c+R_p$. In this way, the polymer radius is rescaled to recover the correct excluded-volume interaction between crowders and polymers. The system is thus specified by the rescaled size ratio $q' \equiv R_p'/R_c'$ (ratio of effective polymer and crowder radii). As the effective crowders fill a larger (rescaled) volume fraction $\phi_c'=\phi_c(R_c'/R_c)^3$, so the effective polymers fill a correspondingly smaller volume fraction $\phi_p'=\phi_p(R_p'/R_p)^3$.

For the effective crowder size, Denton and Schmidt (2005) adopted a simple thermal criterion, defining the effective hard-sphere diameter $\sigma_c'=2R_c'$ as the distance at which the effective pair interaction equals the typical thermal energy, that is, $v_{\rm eff}(\sigma_c')=k_BT$. A more accurate criterion invokes a variational perturbation theory, with a hard-sphere reference system (Hansen and McDonald, 1990):

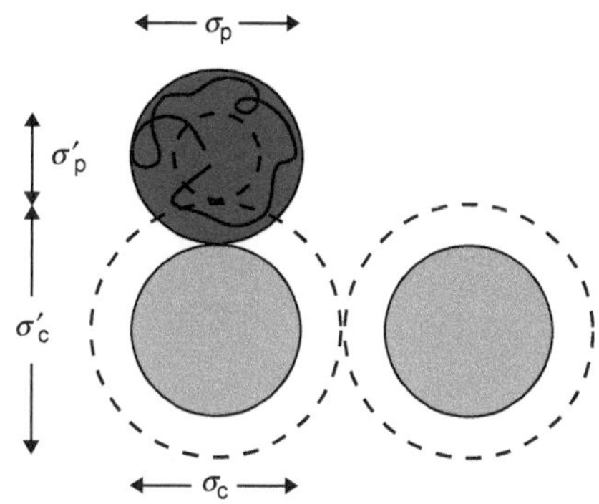

Figure 3.5 Model of charged crowders (blue circles) and uncharged polymer (red circle). Because of electrostatic repulsion between crowders, the effective crowder diameter σ'_c is larger than the bare diameter σ_c. Correspondingly, the effective polymer diameter σ'_p is smaller than the bare diameter σ_p, such that $\sigma'_c+\sigma'_p=\sigma_c+\sigma_p$. (For interpretation of the references to color in this figure legend, the reader is referred to the online version of this chapter.)

$$f_{\mathrm{ex,c}}(\phi_c)=\min_{\sigma'_c}\left\{f_{\mathrm{hs}}(\phi'_c)+2\pi\rho_c^2\int_{\sigma'_c}^{\infty}\mathrm{d}r\,r^2 g_{\mathrm{hs}}(r;\phi'_c)v_{\mathrm{eff}}(r)\right\} \tag{3.47}$$

where $g_{\mathrm{hs}}(r;\phi'_c)$ is the radial distribution function of the hard-sphere fluid and σ'_c is varied to minimize the excess free energy of the crowders. At physiological salt concentrations, the volume energy in Eq. (3.22) is essentially independent of density and, thus, adds to the total free energy a trivial constant that can be ignored. The free-volume theory is modified by simply substituting the rescaled particle sizes and volume fractions in the expressions for the polymer free-volume fraction (Eqs. 3.45 and 3.46) and the crowder and polymer excess free energies:

$$f(\phi_c,\rho_p^r)=\rho_c(\ln\phi_c-1)+f_{\mathrm{ex,c}}(\phi_c)+\rho_p^r\alpha_{\mathrm{eff}}(\phi'_c)(\ln\phi_p^r-1) \tag{3.48}$$

Equations (3.35), (3.39), and (3.48) provide a basis for applying free-volume theory to the phase behavior of polymer–nanoparticle mixtures within the original AOV model or one of its several variations. The phase diagram can be straightforwardly calculated from a coexistence analysis (Maxwell common-tangent construction) on the free energy. Correspondingly, Eqs. (3.41) and (3.43) allow calculation of crowding-induced modifications to polymer size and shape distributions (see Section 4).

3.2. Computer simulations of polymer–nanoparticle mixtures

Over the last three decades, computer simulation has been an essential complement to experiment and theory in elucidating the properties of polymer–particle mixtures. The two main methods deployed have been MC and molecular dynamics (MD), including variations, such as Langevin and Brownian dynamics. MD methods have been used to simulate relatively simple models of spherical particles (Cho and Kim, 2012) and are especially valuable for simulating folding of biopolymers (Cheung, 2013, Cheung et al., 2005; Denesyuk and Thirumalai, 2011, 2013; Kudlay et al., 2012). MC methods have proved especially appropriate for modeling mixtures of nanoparticles (or colloids) and nonadsorbing polymers, in which excluded-volume interactions and entropy dominate and kinetics can be slowed by proliferation of metastable states (Binder, 1995; Binder and Heermann, 2010; Frenkel and Smit, 2001). Given their many degrees of freedom, polymer coils are typically represented in simulations by coarse-grained models, ranging in complexity from the effective-sphere (AOV) model (Dijkstra et al., 1999; Moncho-Jordá et al., 2003; Vink and Horbach, 2004; Vink et al., 2005) to the soft-ellipsoid model (Eurich and Maass, 2001; Murat and Kremer, 1998) to lattice models (Bolhuis et al., 2003; Dickman and Yethiraj, 1994; Hoppe and Yuan, 2011; Meijer and Frenkel, 1994) to bead-spring models (Bolhuis et al., 2001a, 2001b, 2002; Cheung et al., 2005; Denesyuk and Thirumalai, 2011, 2013; Doxastakis et al., 2004, 2005; Fritsch and Langowski, 2011; Hoppe and Yuan, 2011; Kudlay et al., 2012; Louis et al., 2000, 2002; Murat and Kremer, 1998) to all-atom force field models (Linhananta et al., 2011, 2012).

3.3. Monte Carlo simulations of crowding in polymer–nanoparticle mixtures

To test the free-volume theory described in Section 3.1, Lu and Denton (2011) recently developed a simple MC algorithm for simulating coarse-grained models of compressible, penetrable macromolecules mixed with hard crowding agents. Applying the method to binary mixtures of ideal polymer coils and hard-sphere nanoparticles, they explored the response of coil size to crowding and the stability of polymer–nanoparticle mixtures. To facilitate demixing into polymer-rich and polymer-poor phases, Lu and Denton also implemented a variation of the Gibbs ensemble MC (GEMC) method (Panagiotopoulos, 1987, 1989, 1992, 1995, 2000), which obviates any need for interfaces by allowing each coexisting phase to occupy its own simulation box. The GEMC method provides a computationally efficient means of computing fluid phase coexistence curves (binodals), although it does not accurately resolve critical points.

The semigrand ensemble, which corresponds closely with experimental conditions, fixes the temperature, total nanoparticle number, and total volume, as well as the polymer chemical potential via exchange of polymers with a reservoir. The conventional GEMC algorithm involves four trial moves: (1) displacements of particles within each box to ensure thermal equilibrium; (2) exchanges of volume between boxes to equalize pressures; (3) transfers of particles between boxes to equalize chemical potentials of each species; and (4) transfers of polymers between each box and the reservoir. The acceptance probabilities for these trial moves are standard (Frenkel and Smit, 2001; Panagiotopoulos, 1989, 1995). To this list, Lu and Denton (2011) added a new trial move that varies the polymer radius of gyration, allowing the polymer size distribution to adjust to a crowded environment. In the Metropolis algorithm, the probability of a polymer coil having radius of gyration R_p is $P(R_p) \propto P_r(R_p)\exp(-\beta U)$, where $U = \varepsilon N_{cp}$ is the potential energy of N_{cp} polymer-penetrating crowders. (In the colloid limit, $U = \infty$ and overlaps are forbidden.) A trial change in radius of gyration of a given coil, from an old value R_p^o to a new value R_p^n, is then accepted with probability

$$\mathcal{P}_{\text{size}}(R_p^o \to R_p^n) = \min\left\{ \frac{P_r\left(R_p^n\right)}{P_r\left(R_p^o\right)} \exp(-\beta\Delta U), 1 \right\} \tag{3.49}$$

where ΔU is the associated change in potential energy. Thus, any move that decreases the potential energy is automatically accepted, while a move that increases the potential energy is accepted only with a probability equal to the Boltzmann factor for the change in potential energy multiplied by the ratio of the new to the old size probabilities. Via trial changes in radius of gyration, polymers expand and contract to achieve their equilibrium size distribution in the presence of crowders. In a similar manner, the response of polymer *shape* to a crowded environment could be explored within the ellipsoidal polymer model by implementing trial moves that change the eigenvalues of the gyration tensor (i.e., principal radii of an ellipsoidal coil):

$$\mathcal{P}_{\text{shape}}\left(\vec{\lambda}^o \to \vec{\lambda}^n\right) = \min\left\{ \frac{P_r\left(\vec{\lambda}^n\right)}{P_r\left(\vec{\lambda}^o\right)} \exp(-\beta\Delta U), 1 \right\} \tag{3.50}$$

where $\vec{\lambda}^{o}$ and $\vec{\lambda}^{n}$ represent the old and new sets of gyration tensor (shape) eigenvalues.

While the results presented in the next section are limited to systems with hard spherical crowders (nanoparticles), the same theoretical and simulation approaches outlined earlier can be generalized to nonspherical crowders. For reference, Kudlay et al. (2012) recently performed Langevin dynamics simulations of bead-spring polymers to explore the dependence of structural transitions (e.g., coil-helix-helical hairpin) on the shape of crowding particles (spherical vs. rodlike).

4. RESPONSE OF POLYMER CONFORMATIONS TO NANOPARTICLE CROWDING

To explore the influence of crowding on polymer size distributions and phase stability, Denton and Schmidt (2002) and Lu and Denton (2011) applied the free-volume theory and MC methods described in Section 3 to mixtures of hard-sphere nanoparticles and spherical, compressible polymers. The polymer–particle interactions were modeled by Eq. (3.25) in the colloid limit ($q_r \leq 1$) and by Eqs. (3.26) and (3.27) in the protein limit ($q_r > 1$). Recently, Denton et al. (2013) investigated the response of polymer *shape* fluctuations to crowding within the ellipsoidal, compressible polymer model. The impact of electrostatics on the phase stability of (incompressible) polymers mixed with *charged* nanoparticles has been examined by Denton and Schmidt (2005), using for the particle–particle interactions an effective screened-Coulomb potential (Eq. 3.23). Results of these and related studies are reviewed later.

4.1. Ideal polymers: Theoretical predictions and simulation data

Denton and Schmidt (2002) first implemented the free-volume theory for mixtures of spherical, compressible polymers and hard-sphere (nanoparticle) crowders in the colloid limit, with reservoir size ratio $q_r = 1$. For the same system, Lu and Denton (2011) performed standard *NVT* (canonical ensemble) simulations, with polymers and crowders confined to a single cubic box subject to periodic boundary conditions. While Denton and Schmidt modeled the reservoir polymer size distribution using the exact Fujita-Norisuye formula (Eq. 3.3), Lu and Denton applied the near-exact Eurich-Maass formula (Eq. 3.4). In the protein limit ($q_r = 3$), Lu and Denton performed both free-volume theory calculations and GEMC simulations in the semigrand ensemble. In the simulations, the polymers and crowders were confined to *two* boxes and the fixed parameters were the total crowder number (in both boxes), total volume, and the number density of polymers in the reservoir.

The simulations were initialized by randomly distributing crowders and spherical polymers, initially monodisperse in radius, over the sites of an fcc lattice. The initial total number of polymers was chosen to yield a desired effective polymer volume fraction, defined as $\phi_p = (4\pi/3)\rho_p(R_g^r)^3$, a quantity that can be experimentally controlled. From these initial states, the simulations executed the various trial moves outlined in Section 3.3, tolerances being adjusted to yield reasonable acceptance ratios for each move. During the simulations, the polymer sizes fluctuated as a result of trial changes in radius of gyration (Eq. 3.49). For comparison with free-volume theory, a histogram of polymer size $P(R_p)$ was accumulated.

Figure 3.6 compares results from MC simulations with predictions of free-volume theory for the probability distribution of polymer radius of gyration in a mixture of hard-sphere crowders (nanoparticles) and spherical,

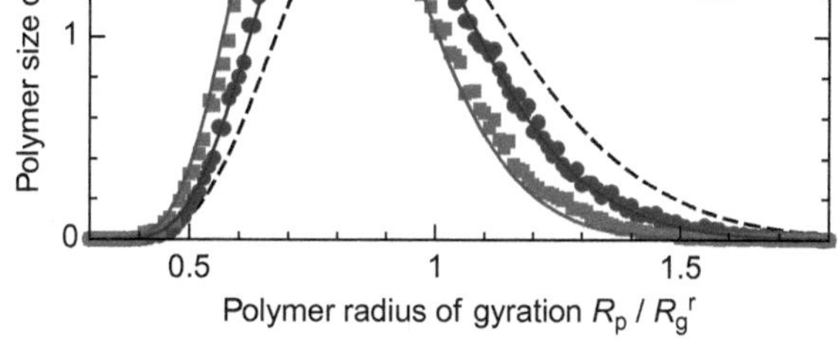

Figure 3.6 Probability distribution of polymer radius of gyration R_p in a mixture of nanoparticle crowders and spherical, impenetrable, compressible polymer at polymer volume fraction $\phi_p = 0.1$. The rms radius of gyration of polymer in the reservoir is equal to the nanoparticle radius ($q_r = 1$). Simulation data (symbols) and free-volume theory predictions (curves) are shown for increasing crowder volume fraction $\phi_c = 0$ (reservoir), 0.1, and 0.2 (right to left). *Adapted from Lu and Denton (2011).* (For color version of this figure, the reader is referred to the online version of this chapter.)

compressible (but impenetrable) polymers. The polymers here have a reservoir rms radius of gyration equal to the nanoparticle radius ($q_r = 1$). For crowder volume fractions ϕ_c ranging up to 20%—typical of the nuclear environment—both theory and simulation indicate a significant compression of polymer, the distribution shifting to smaller polymer size with increasing nanoparticle crowding (Lu and Denton, 2011). For $\phi_c = 0.1$ and 0.2, the fractional reductions in polymer rms radius of gyration are, respectively, 0.915 and 0.826.

Figure 3.7 shows a similar comparison of MC simulation results and free-volume theory predictions in the protein limit, the polymer reservoir rms radius of gyration now being three times the nanoparticle radius ($q_r = 3$). In this case, with the nanoparticles small enough to penetrate the polymer coils, the response of the polymers to crowding is not obvious. On the one hand, by contracting, the polymers could possibly avoid penetration by nanoparticles; on the other hand, by swelling, the polymers could reduce the overall penetration energy. As Fig. 3.7 illustrates, both simulation and free-volume theory

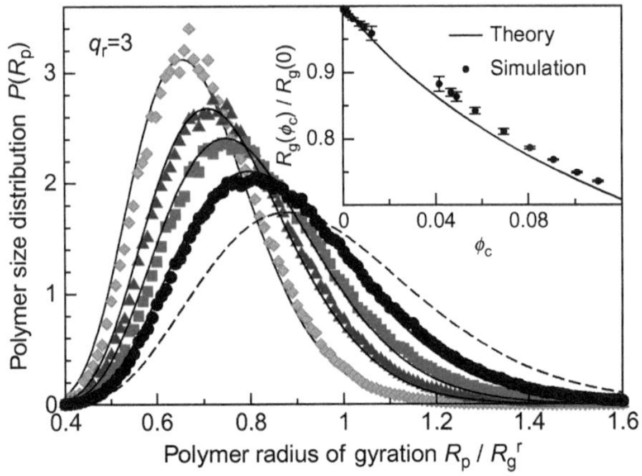

Figure 3.7 Probability distribution of polymer radius of gyration in a mixture of nanoparticle crowders and spherical, penetrable, compressible polymers in osmotic equilibrium with a polymer reservoir. The rms radius of gyration of polymer in the reservoir is three times the nanoparticle radius ($q_r = 3$). Polymer size distributions are shown for polymer chemical potentials $-\beta\mu_p = 0.9$, 1.3, 1.5, and 1.7 (left to right), corresponding to mean crowder volume fractions $\phi_c = 0.120$, 0.0809, 0.0572, and 0.0312, respectively, along the liquid and vapor binodals. The symbols are simulation data, the solid curves are free-volume theory predictions, and the dashed curve is the reservoir size distribution ($\phi_c = 0$). Inset: Ratio of rms polymer radius of gyration in the system $R_g \equiv \langle R_p^2 \rangle^{1/2}$ (Eq. 3.2) to that in the reservoir $R_g(0) = R_g^r$. *Adapted from Lu and Denton (2011).* (For color version of this figure, the reader is referred to the online version of this chapter.)

indicate that crowding compresses, rather than expands, the polymer coils, the rms radius of gyration decreasing monotonically with increasing concentration of crowders. Moreover, as revealed by the inset to Fig. 3.7, the fractional reduction in polymer rms radius of gyration is substantially greater in the protein limit, the compression ratio being roughly 0.7 for a crowder volume fraction as low as $\phi_c = 0.12$. These results qualitatively align with predictions of scaling theory (Odijk, 2000) and neutron scattering experiments on polymer chains trapped in arrays of obstacles (Briber et al., 1995; Horkay et al., 1995). In passing, we note that the close agreement between theory and simulation, evident from these comparisons, provides a consistency check that validates both the approximations underlying the free-volume theory and the algorithms driving the MC simulations.

The reduction in polymer rms radius of gyration in response to crowding, predicted by theory and confirmed by simulation, leaves open the question of how the shape of a polymer coil is affected by crowders. Denton et al. have addressed this question by applying the generalized free-volume theory described in Section 3.1 to a model mixture of hard-sphere nanoparticles and *ellipsoidal*, compressible polymer coils and calculating the probability distributions of the gyration tensor eigenvalues. As shown in Figs. 3.8 and 3.9, for the case of equally sized polymers and nanoparticles ($q_r = 1$), the theory

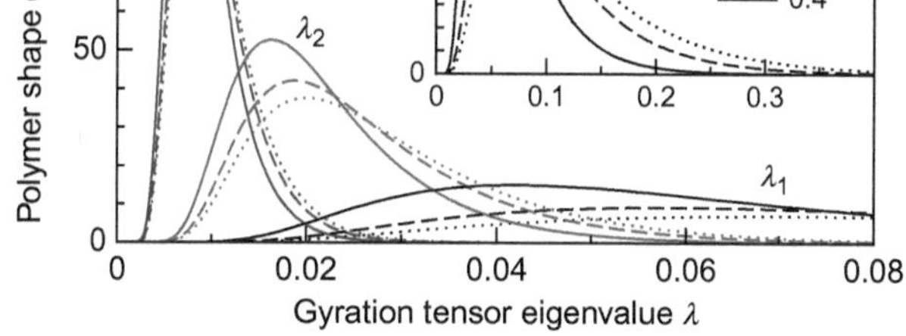

Figure 3.8 Probability distributions $P_i(\lambda_i, \phi_c)$, as predicted by the free-volume theory with ellipsoidal polymer (Eq. 3.44), for the eigenvalues λ_1 (black), λ_2 (red), and λ_3 (blue) of the gyration tensor of a polymer coil, modeled as an ideal, freely-jointed chain, in the presence of crowders (nanoparticles) of volume fraction $\phi_c = 0$ (dotted), 0.2 (dashed), and 0.4 (solid). The rms radius of gyration of polymer in the reservoir is equal to the nanoparticle radius ($q_r = 1$). *Adapted from Denton et al. (2013).* (For interpretation of the references to color in this figure legend, the reader is referred to the online version of this chapter.)

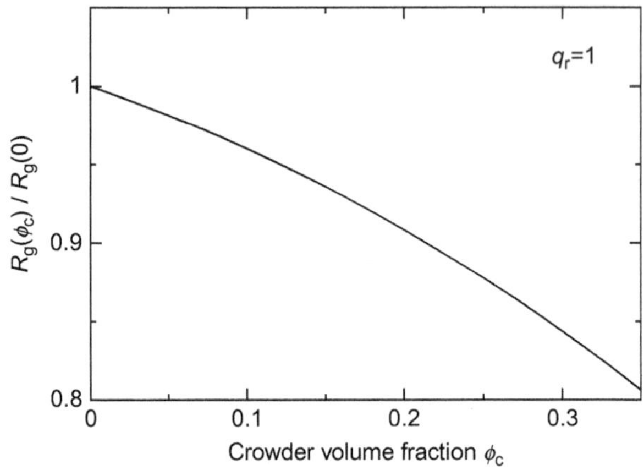

Figure 3.9 Normalized radius of gyration of ellipsoidal polymers versus volume fraction ϕ_c of hard-sphere crowders, as predicted by free-volume theory. The rms radius of gyration in the polymer reservoir equals the crowder radius ($q_r \equiv R_g^r/R_c = 1$). *Adapted from Denton et al. (2013).*

predicts that, with increasing volume fraction of crowders, the distributions of all three eigenvalues shift to lower values and, correspondingly, the radius of gyration decreases. Interestingly, the greatest fractional shift occurs in the two largest eigenvalues, implying that the best-fit ellipsoids tend to become less elongated. This implication can be quantitatively assessed by calculating the asphericity A, defined by Eq. (3.8). As Fig. 3.10 confirms, the asphericity *decreases*—the ellipsoids become more spherical—with increasing crowder volume fraction. Thus, not only do polymer coils contract as their environment grows more crowded, but they also become significantly more compact.

4.2. Comparisons with related studies

While Denton et al. confined their studies to ideal, random-walk polymer coils, Minton (2000b, 2005) recently explored the impact of crowding on native-denatured (folded-unfolded) transitions of real polypeptides. Based on a previously proposed effective two-state model of proteins (Minton, 2000b), in which properties of the denatured state were defined as averages over conformations—in somewhat the same spirit as free-volume theory—Minton (2005) calculated the excluded-volume interactions between unfolded proteins and cosolutes, which were modeled as hard spheres or rods. Using radius of gyration probability distributions of four actual proteins, computed by Goldenberg (2003) via MC simulations, which take into account steric interactions between adjacent amino acid residues, Minton calculated chemical potentials and radii of gyration of the unfolded proteins

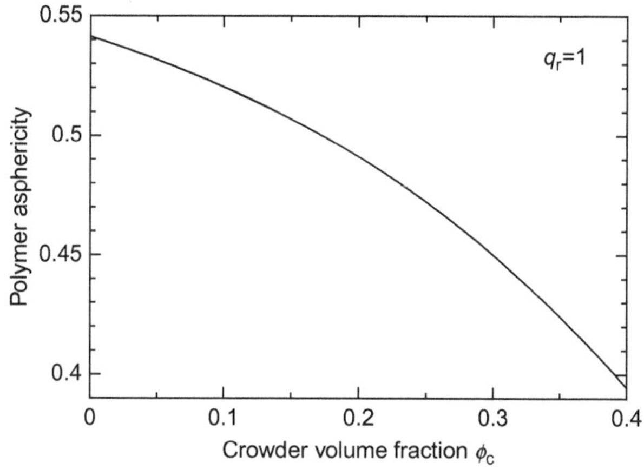

Figure 3.10 Asphericity of ellipsoidal polymers versus nanoparticle crowder volume fraction (calculated from Eq. 3.8), illustrating that crowding compactifies polymer coils.

as functions of crowder volume fraction. From his studies, Minton concluded that long-range intramolecular steric interactions significantly increase the radii of gyration of unfolded polypeptides in crowded fluids.

In related work, Cheung et al. (2005) analyzed the influence of crowding on folding kinetics and stabilization of folded states of globular proteins by conducting Brownian dynamics simulations of coarse-grained models of proteins. To focus on the role of excluded-volume interactions between crowders and polypeptide chains, the crowders were modeled as hard spheres of radius equal to the radius of gyration of the folded polypeptide. Their main conclusions are that crowding enhances the stability of the native (folded) state but that the structure of the folded state is significantly modified only for relatively high crowder volume fractions ($\phi_c > 0.25$).

Denesyuk and Thirumalai (2011, 2013) recently examined the role of crowding in the folding of RNA, focusing on human telomerase enzyme. RNA folding is distinguished from protein folding by the active participation of counterions in tertiary folding of RNA. In agreement with Cheung et al. (2005), Denesyuk and Thirumalai confirmed that crowding stabilizes the folded state. They also found that crowding is most effective when crowders are smaller than the RNA radius of gyration and that large macromolecules serve mainly to increase the effective volume fraction of relatively small crowders. Denesyuk and Thirumalai further investigated the influence of ionic strength on RNA folding, concluding that stabilization of folded RNA states by crowding is enhanced with decreasing ionic strength in the range 0.1–1 M. They attribute this effect to the expanded conformations of RNA at lower ionic strengths. The latter conclusion is particularly intriguing given that cells can tune ionic strength and, thus, screening of electrostatic interactions, by controlling transport of ions across membrane channels. A possibly related phenomenon was reported recently by Richter et al. (2007), who observed, using fluorescence and electron microscopy, compaction of chromatin in response to added salt.

Recent experiments have applied light and neutron scattering techniques to probe conformations of polymer coils in crowded solutions. Fluorescence spectroscopy has been used to image single molecules in solution (Mazouchi et al., 2013). Neutrons provide the unique ability to distinguish between isotopically labeled species via the strong difference in coherent neutron scattering length between hydrogen and deuterium. This isotope contrast permits extraction of the wave-vector dependent contribution to the scattering intensity caused by single marked (deuterated) polymer chains. While the studies cited later involve synthetic polymers and crowders, their general conclusions may be relevant to macromolecular crowding *in vivo*.

Kramer et al. (2005a,b,c) performed both light scattering and small-angle neutron scattering (SANS) experiments on solutions of polystyrene (PS) chains of various molecular weights, and substantially smaller crowding agents—PEMA microgel particles, silsesquioxane molecules, and hydroxy-functionalized silica particles (OHSil)—in good solvents. Well within the protein limit, they measured a substantial shrinkage of the PS chains with increasing volume fraction of crowders. Although the crowders in these experiments are not hard spheres, the observed trends are qualitatively consistent with those found by Lu and Denton (2011).

Le Coeur et al. (2009, 2010) used SANS to measure the influence of a crowding agent (Ficoll 70) on the radius of gyration of deuterated PEG polymer in water. Ficoll 70 is a ramified sucrose polysaccharide—a near-spherical macromolecule—that is highly soluble in water. PEG is a water-soluble polymer that adopts a Gaussian coil conformation in solution. (In other contexts, PEG itself serves as a synthetic crowder for proteins.) Focusing initially on aqueous solutions of polymers and crowding agents of comparable radii of gyration (50–57 Å), Le Coeur et al. measured the PEG radius of gyration for varying concentrations of Ficoll 70. By extrapolating to zero polymer concentration, they were able to extract the radius of gyration of a single PEG coil. Most importantly, with increasing mass fraction of the crowding agent, Le Coeur et al. observed a significant compression of the PEG coils, as illustrated in Fig. 3.11. Moreover, the fractional compression increased with the molecular weight of the PEG, that is, the size ratio q_r (Le Coeur et al.,

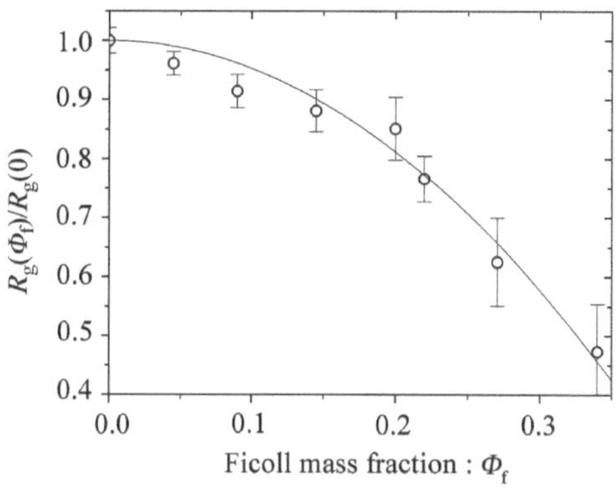

Figure 3.11 Experimental (SANS) measurements of normalized radius of gyration of polymers (deuterated-PEG) versus mass fraction of crowders (Ficoll 70 polysaccharide). The rms radius of gyration in the absence of crowders ($\Phi_f = 0$) is approximately equal to the crowder radius ($q_r \equiv R_g^r/R_c \simeq 1$). *Reprinted Fig. 5 with permission from Le Coeur et al. (2009). Copyright 2009 by the American Physical Society.*

2010), consistent with the findings of Denesyuk and Thirumalai (2011, 2013) for RNA.

A comparison of Figs. 3.9 and 3.11 shows that free-volume theory for the ellipsoidal-polymer model qualitatively predicts the observed decrease in polymer radius of gyration with increasing crowder volume fraction. The measured dependence on polymer molecular weight is also captured by the theory, as seen from an inspection of Figs. 3.6 and 3.7. In a quantitative comparison, however, the theory significantly underestimates the extent of compression. While this discrepancy will require further study to resolve, an obvious point of difference between the model and the experimental samples is that Ficoll 70, a flexible branched molecule, is rather far from the ideal of a hard sphere and, therefore, may partially penetrate the polymer coils. Moreover, since water is a good solvent for PEG, the polymers may be more accurately modeled as self-avoiding coils. Also conceivable is that enthalpy here plays a role, rivaling entropy, in determining the polymer conformations.

In closing this section, we note that, while much effort has been devoted recently to understanding the influence of crowding on conformations of polymers in solution, other workers in parallel have explored conformations of polymers in nanocomposite materials, such as polymer melts and blends filled with nanoparticles, in which the chains are entangled, or even cross-linked, with one another. Because polymer nanocomposites present crowding environments and mechanisms that differ somewhat from protein solutions, we mention only a few representative studies and make no attempt to systematically survey the extensive literature on this broad class of materials. The experimental technique of choice is again SANS. Nakatani et al. (2001) reported SANS and MC simulation results for chain dimensions of isotopic blends of PDMS polymers and polysilicate filler particles. Mackay et al. (2006), and Tuteja et al. (2008) used SANS to explore swelling of PS chains induced by dispersed PS nanoparticles. Sen et al. (2007) applied SANS and PRISM integral-equation theory to study conformations of PS chains mixed with silica nanoparticles. Nusser et al. (2010) used SANS to explore shrinking of PEP chains induced by silica nanoparticles. Despite some conflicting conclusions, likely the result of differing experimental conditions, these studies converge on the result that polymer conformations are significantly perturbed only by nanoparticles smaller than the polymer radius of gyration.

4.3. Demixing of polymer–nanoparticle mixtures

Beyond modifying the conformations of polymer coils, crowding agents can also affect the thermodynamic phase stability of polymer solutions. It is well known that mixtures of colloids and nonadsorbing polymers in the colloid limit, when sufficiently concentrated in both components, separate into stable colloid-rich (polymer-poor) and colloid-poor (polymer-rich) phases, if the polymer radius of gyration exceeds about 25% of the colloid radius (Poon, 2002; Pusey, 1991; Tuinier et al., 2003). As discussed in Section 2.1, Asakura and Oosawa (1954) and Vrij (1976) explained this demixing behavior via polymer depletion, which induces effective attractions between colloids (Fig. 3.1).

In the protein or nanoparticle limit, a mixture of free polymer coils and smaller nanoparticles is even less stable toward demixing, assuming the particles are larger than the correlation length (blob radius) ξ of the polymer chain, that is, $\xi < R_c < R_g^r$ (de Gennes, 1979; Odijk, 2000). In this limit, MC simulations of the AOV model (Moncho-Jordá et al., 2003), a lattice model (Bolhuis et al., 2003), and the AOV model with penetrable, compressible polymers (Lu and Denton, 2011) all yield demixing phase diagrams in qualitative agreement with experiments (Hennequin et al., 2005; Zhang and van Duijneveldt, 2006) and predictions (Schmidt and Fuchs, 2002; Sear, 2001).

As an example of a typical demixing phase diagram in the colloid limit, Fig. 3.12 shows, for a size ratio $q_r = 1$, results calculated from free-volume

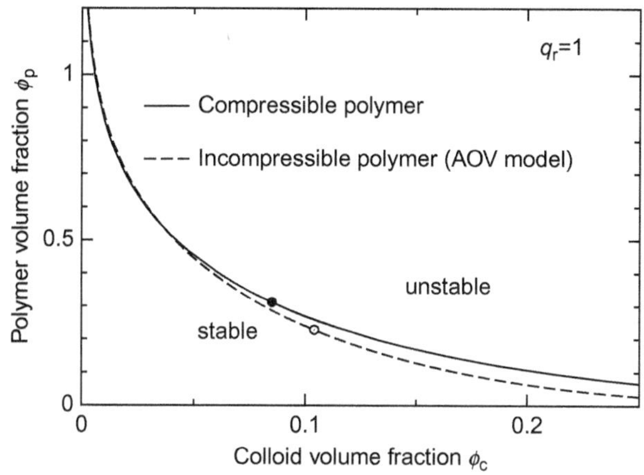

Figure 3.12 Phase diagram of a mixture of polymers and hard-sphere nanoparticles (crowders) in osmotic equilibrium with a reservoir of polymers whose rms radius of gyration equals the nanoparticle radius ($q_r = 1$). Curves are predictions of free-volume theory for crowder-rich and -poor binodals. The polymers are modeled as impenetrable spheres that are either compressible (solid curve) or incompressible (dashed curve). Circles denote predicted critical points. *Adapted from Denton and Schmidt (2002).*

theory for the AOV model and the variation of that model with compressible (but impenetrable) polymers (Denton and Schmidt, 2002). Similarly, Fig. 3.13 shows a phase diagram in the nanoparticle limit, now for $q_r = 3$, calculated from free-volume theory and MC simulation (using the methods described in Section 3) for the AOV model and its variation with penetrable, compressible polymers (Lu and Denton, 2011). A comparison of the results for the various models confirms that polymer–nanoparticle mixtures tend to be less stable in the nanoparticle limit than in the colloid limit and demonstrates that polymer compressibility and penetrability have stabilizing effects.

Finally, as an illustrative example of the influence of electrostatic interactions, Fig. 3.14 shows the demixing phase diagram, now in the colloid limit, for an AOV-like model of incompressible, spherical polymers mixed with *charged* crowders that interact via a screened-Coulomb pair potential (Eq. 3.23). As explained in Section 3.1, long-range repulsion between crowders weakens polymer depletion, thus diminishing depletion-induced attraction between crowders (Denton and Schmidt, 2005). As seen in Fig. 3.14, a thermodynamic consequence is enhanced stability of the mixture with increasing strength of repulsion. Whether the demixing observed in these simple models of polymer–nanoparticle mixtures relates at all to the separation of macromolecules into compartments observed in cell nuclei will require further study to resolve.

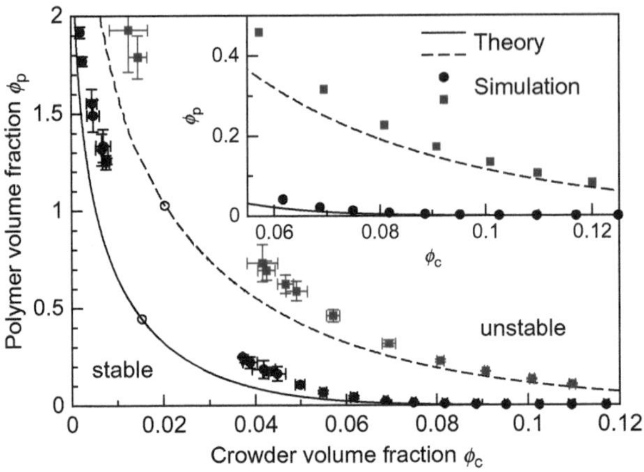

Figure 3.13 Phase diagram of a mixture of polymers and hard-sphere nanoparticles (crowders) in osmotic equilibrium with a reservoir of polymers whose rms radius of gyration is three times the nanoparticle radius ($q_r = 3$). Symbols represent simulation data and curves predictions of free-volume theory for crowder-rich and -poor binodals (open circles are predicted critical points). Results are shown for two polymer models: impenetrable, incompressible polymer (black circles, solid curve) and penetrable, compressible polymer (blue squares, dashed curve). Inset: expanded view of concentrated crowder phase. *Adapted from Lu and Denton (2011).* (For interpretation of the references to color in this figure legend, the reader is referred to the online version of this chapter.)

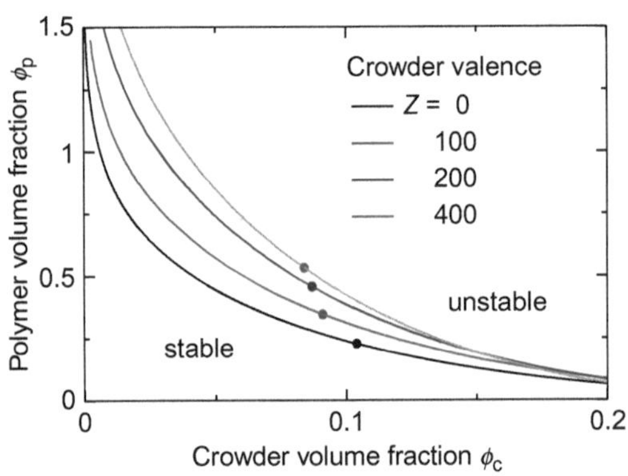

Figure 3.14 Phase diagram of a mixture of neutral polymers, modeled as incompressible spheres, and charged crowders, modeled as charged hard spheres, equal in size ($q_r = 1$), interacting via the effective screened-Coulomb pair potential of Eq. (3.23). With increasing effective valence of the crowders, $Z = 0, 100, 200, 400$ (bottom to top curves), the stability of the mixture is enhanced. *Adapted from Denton and Schmidt (2005).* (For color version of this figure, the reader is referred to the online version of this chapter.)

5. CONCLUDING REMARKS

Since the potential importance of macromolecular crowding in cell biology was first recognized (Minton, 1981), interest in crowding phenomena has permeated the biochemistry, biophysics, and molecular biology communities. From the perspective of soft matter physics, this chapter partially surveys recent work on crowding, keying on polymer–nanoparticle mixtures as generic models of crowded solutions typical of the interiors of cells and cell nuclei. With the aim of identifying universal principles of crowding, we have adopted coarse-grained conceptions of polymers as random coils and nanoparticles as hard spheres.

Assuming only excluded-volume (steric repulsive) interactions and abstracting polymer coils as penetrable spheres or ellipsoids, we have outlined theoretical and computational strategies for modeling polymer–nanoparticle mixtures, focusing on practical methods of free-volume theory and MC simulation. Within this simplified framework, a polymer's conformation is characterized entirely by the gyration tensor, whose eigenvalues are the squares of the principal radii of the best-fit ellipsoid. The eigenvalues fluctuate according to well-defined probability distributions, which naturally adjust to the free (available) volume that is not excluded by the nanoparticles.

Results for size and shape distributions, and demixing phase diagrams of ideal, freely-jointed polymer chains as functions of volume fraction of nanoparticle crowders, predicted by theory and simulations (Denton and Schmidt, 2002; Denton et al., 2013; Lu and Denton, 2011), were reviewed and set in a broader biophysical context. Based on idealized models and computationally inexpensive methods, these studies isolate essential structural features of macromolecules in crowded environments. Several general principles, possibly applicable in cell biology, emerge: Hard-sphere crowders both compress *and* compactify ideal polymer chains in solution. At sufficiently high concentrations, hard-sphere crowders induce demixing of ideal polymers into polymer-rich (crowder-poor) and polymer-poor (crowder-rich) phases. For a given volume fraction of crowders, the smaller the crowding particle relative to the polymer radius of gyration, the greater the crowding effect—in both compression ratio and demixing instability. Electrostatic repulsion between crowding particles weakens the depletion-induced attraction between polymers and promotes phase stability (inhibits demixing).

These conclusions inevitably raise a number of questions for future exploration. How do self-avoiding polymer coils (in good solvents) respond to crowded environments? In cases where steric and nonsteric interactions compete (as in protein and RNA folding), how is the interplay between entropy and enthalpy modified by crowding agents? To what extent are polymer conformations sensitive to the shapes of crowding agents and to nonsteric interactions between them? How do electrostatic interactions, tunable by varying ionic strength and pH of the solution, affect conformations and folding of charged biopolymers (e.g., proteins, RNA, chromatin fibers) in the presence of charged crowders? Such challenging questions will continue to draw researchers to the frontiers of the rapidly growing (yet still uncrowded) field of macromolecular crowding.

ACKNOWLEDGMENTS

It is a pleasure to thank Ben Lu, Sylvio May, Emmanual Mbamala, and Matthias Schmidt for their inspiring discussions and fruitful collaborations, and Stéphane Longeville for kindly providing Fig. 3.11. This work was supported by the National Science Foundation (Grant No. DMR-1106331) and by the Donors of the American Chemical Society Petroleum Research Fund (Grant No. PRF 44365-AC7).

REFERENCES

Andersen, H.C., Weeks, J.D., Chandler, D., 1971. Relationship between the hard-sphere fluid and fluids with realistic repulsive forces. Phys. Rev. A 4, 1597–1607.

Asakura, S., Oosawa, F., 1954. Surface tension of high-polymer solutions. J. Chem. Phys. 22, 1255.

Balazs, A.C., Emrick, T., Russell, T.P., 2006. Nanoparticle polymer composites: where two small worlds meet. Science 314, 1107–1110.

Benton, L.A., Smith, A.E., Young, G.B., Pielak, G.J., 2012. Unexpected effects of macromolecular crowding on protein stability. Biochemistry 51, 9773–9775.

Binder, K., 1995. Monte Carlo and Molecular Dynamics Simulations in Polymer Science. Oxford, New York.

Binder, K., Heermann, D.W., 2010. Monte Carlo Simulation in Statistical Physics: An Introduction, 5th ed. Springer, Berlin.

Bolhuis, P.G., Louis, A.A., Hansen, J.P., Meijer, E.J., 2001a. Accurate effective pair potentials for polymer solutions. J. Chem. Phys. 114, 4296–4311.

Bolhuis, P.G., Louis, A.A., Hansen, J.P., 2001b. Many-body interactions and correlations in coarse-grained description of polymer solutions. Phys. Rev. E. 64, 021801-1-12.

Bolhuis, P.G., Louis, A.A., Hansen, J.-P., 2002. Influence of polymer-excluded volume on the phase-behavior of colloid-polymer mixtures. Phys. Rev. Lett. 89, 128302-1-4.

Bolhuis, P.G., Meijer, E.J., Louis, A.A., 2003. Colloid-polymer mixtures in the protein limit. Phys. Rev. Lett. 90, 68304-1-4.

Briber, R.M., Liu, X., Bauer, B.J., 1995. The collapse of free polymer chains in a network. Science 268, 395–397.

Butler, M.F., Heppenstall-Butler, M., 2003. Phase separation in gelatin/dextran and gelatin/maltodextrin mixtures. Food Hydrocoll. 17, 815–830.

Chatterjee, A.P., Schweizer, K.S., 1998a. Microscopic theory of polymer-mediated interactions between spherical particles. J. Chem. Phys. 109, 10464–10476.
Chatterjee, A.P., Schweizer, K.S., 1998b. Correlation effects in dilute particle-polymer mixtures. J. Chem. Phys. 109, 10477–10488.
Chen, E., Christiansen, A., Wang, Q., Cheung, M.S., Kliger, D.S., Wittung-Stafshede, P., 2012. Effects of macromolecular crowding on burst phase kinetics of cytochrome c folding. Biochemistry 51, 9836–9845.
Cheung, M.S., 2013. Where soft matter meets living matter—protein structure, stability, and folding in the cell. Current Opin. Struct. Biol. 23, 1–6.
Cheung, M., Klimov, D., Thirumalai, D., 2005. Molecular crowding enhances native state stability and refolding rates of globular proteins. Proc. Natl. Acad. Sci. 102, 4753–4758.
Cho, E.J., Kim, J.S., 2012. Crowding effects on the formation and maintenance of nuclear bodies: insights from molecular-dynamics simulations of simple spherical model particles. Biophys. J. 103, 424–433.
de Gennes, P.G., 1979. Scaling Concepts in Polymer Physics. Cornell, Ithaca.
Denesyuk, N.A., Thirumalai, D., 2011. Crowding promotes the switch from hair-pin to pseudoknot conformation in human telomerase RNA. J. Am. Chem. Soc. 133, 11858–11861.
Denesyuk, N.A., Thirumalai, D., 2013. Entropic stabilization of the folded states of RNA due to macromolecular crowding. Biophys. Rev. 5, 225–232.
Denton, A.R., 2000. Effective interactions and volume energies in charged colloids: linear response theory. Phys. Rev. E. 62, 3855–3864.
Denton, A.R., 2007. Effective interactions in soft materials. In: Zvelindovsky, A.V. (Ed.), Nanostructured Soft Matter: Experiment, Theory, Simulation and Perspectives. Springer, Dordrecht, pp. 395–433.
Denton, A.R., Schmidt, M., 2002. Colloid-induced polymer compression. J. Phys. Condens. Matter 14, 12051–12062.
Denton, A.R., Schmidt, M., 2005. Mixtures of charged colloid and neutral polymer: influence of electrostatic interactions on demixing and interfacial tension. J. Chem. Phys. 122, 244911-1-7.
Denton, A.R., 2010. Poisson-Boltzmann theory of charged colloids: limits of the cell model for salty suspensions. J. Phys. Condens. Matter 22, 364108-1-12.
Denton, A.R., Mbamala, E., May, S., 2013. Unpublished.
Dickman, R., Yethiraj, A., 1994. Polymer-induced forces between colloidal particles. A Monte Carlo simulation. J. Chem. Phys. 100, 4683–4689.
Dijkstra, M., Brader, J.M., Evans, R., 1999. Phase behavior and structure of model colloid-polymer mixtures. J. Phys. Condens. Matter 11, 10079–10106.
Dima, R.I., Thirumalai, D., 2004. Asymmetry in the shapes of folded and denatured states of proteins. J. Phys. C: Solid State Phys. B 108, 6564–6570.
Doxastakis, M., Chen, Y.-L., Guzmán, O., de Pablo, J.J., 2004. Polymer-particle mixtures: depletion and packing effects. J. Chem. Phys. 120, 9335–9342.
Doxastakis, M., Chen, Y.-L., de Pablo, J.J., 2005. Potential of mean force between two nanometer-scale particles in a polymer solution. J. Chem. Phys. 123, 34901-1-8.
Eisenriegler, E., Hanke, A., Dietrich, S., 1996. Polymers interacting with spherical and rodlike particles. Phys. Rev. E. 54, 1134–1152.
Elcock, A.H., 2010. Models of macromolecular crowding effects and the need for quantitative comparisons with experiment. Curr. Opin. Struct. Biol. 20, 1–11.
Ellis, R.J., 2001. Macromolecular crowding: an important but neglected aspect of the intracellular environment. Curr. Opin. Struct. Biol. 11, 114–119.
Eurich, F., Maass, P., 2001. Soft ellipsoid model for Gaussian polymer chains. J. Chem. Phys. 11, 7655–7668.

Fixman, M., 1962. Radius of gyration of polymer chains. J. Chem. Phys. 36, 306–310.
Flory, P.J., Fisk, S., 1966. Effect of volume exclusion on the dimensions of polymer chains. J. Chem. Phys. 44, 2243–2248.
Forsman, J., Woodward, C.E., 2009. Colloidal interactions in thermal and athermal polymer solutions: the Derjaguin approximation, and exact results for mono- and polydisperse ideal chains. J. Chem. Phys. 131, 044903-1-10.
Fošnarič, M., Iglič, A., Kroll, D.M., May, S., 2013. Monte Carlo simulations of a polymer confined within a fluid vesicle. Soft Matter 9, 3976–3984.
Frenkel, D., 1991. Statistical mechanics of liquid crystals. In: Hansen, J.-P., Levesque, D., Zinn-Justin, J. (Eds.), Liquids, Freezing and Glass Transition, Les Houches Session 51. North-Holland, Amsterdam, pp. 763–931.
Frenkel, D., Smit, B., 2001. Understanding Molecular Simulation, 2nd ed. Academic, London.
Fritsch, C.C., Langowski, J., 2011. Chromosome dynamics, molecular crowding, and diffusion in the interphase cell nucleus: a Monte Carlo lattice simulation study. Chromosome Res. 19, 63–81.
Fuchs, M., Schweizer, K.S., 2000. Structure and thermodynamics of colloid-polymer mixtures: a macromolecular approach. Europhys. Lett. 51, 621–626.
Fuchs, M., Schweizer, K.S., 2002. Structure of colloid-polymer suspensions. J. Phys. Condens. Matter 14, R239–R270.
Fujita, H., Norisuye, T., 1970. Some topics concerning the radius of gyration of linear polymer molecules in solution. J. Chem. Phys. 52, 1115–1120.
Goldenberg, D.P., 2003. Computational simulation of the statistical properties of unfolded proteins. J. Mol. Biol. 326, 1615–1633.
Hadizadeh, S., Linhananta, A., Plotkin, S.S., 2011. Improved measures for the shape of a disordered polymer to test a mean-field theory of collapse. Macromolecules 44, 6182–6197.
Hancock, R., 2004a. A role for macromolecular crowding effects in the assembly and function of compartments in the nucleus. J. Struct. Biol. 146, 281–290.
Hancock, R., 2004b. Internal organisation of the nucleus: assembly of compartments by macromolecular crowding and the nuclear matrix model. Biol. Cell 96, 595–601.
Hancock, R., 2012. The crowded environment of the genome. In: Rippe, K. (Ed.), Genome Organization and Function in the Cell Nucleus. Wiley-VCH, Weinheim, pp. 169–184.
Hanke, A., Eisenriegler, E., Dietrich, S., 1999. Polymer depletion effects near mesoscopic particles. Phys. Rev. E 59, 6853–6878.
Hansen, J.-P., McDonald, I.R., 1990. Theory of Simple Liquids. Academic, New York.
Hennequin, Y., Evens, M., Quezada Angulo, C.M., van Duijneveldt, J.S., 2005. Miscibility of small colloidal spheres with large polymers in good solvent. J. Chem. Phys. 123, 054906-1-8.
Hoppe, T., Yuan, J.-M., 2011. Protein folding with implicit crowders: a study of conformational states using the Wang-Landau method. J. Phys. Chem. B 115, 2006–2013.
Horkay, F., Stanley, H.B., Geissler, E., King, S.M., 1995. Free chains trapped in a swollen gel under different solvent conditions. Macromolecules 25, 678–681.
Israelachvili, J., 1992. Intermolecular and Surface Forces. Academic, London.
Kim, B.-S., Fan, T.-H., Lebedeva, O.V., Vinogradova, O.I., 2005. Superswollen ultrasoft polyelectrolyte microcapsules. Macromolecules 38, 8066–8070.
Koutsioubas, A., Lairez, D., Combet, S., Fadda, G.C., Longeville, S., Zalczer, G., 2012. Crowding effect on helix-coil transition: beyond entropic stabilization. J. Chem. Phys. 136, 215101-1-7.
Kramer, T., Schweins, R., Huber, K., 2005a. Small-angle neutron scattering of dilute polystyrene chains at the protein limit of a colloid-polymer mixture. J. Chem. Phys. 123, 014903-1-7.

Kramer, T., Schweins, R., Huber, K., 2005b. Silsesquioxane molecules and polystyrene chains as a model system for colloid-polymer mixtures in the protein-limit. Macromolecules 38, 151–159.

Kramer, T., Schweins, R., Huber, K., 2005c. Coil dimensions of polystyrene chains in colloid-polymer mixtures at the protein limit: a SANS study. Macromolecules 38, 9783–9793.

Kudlay, A., Cheung, M.S., Thirumalai, D., 2012. Influence of the shape of crowding particles on the structural transitions in a polymer. J. Phys. C: Solid State Phys. B 116, 8513–8522.

Kuhn, W., 1934. Gestalt fadenförmiger Moleküle in Lösungen (the shape of fibrous molecules in solutions). Kolloid-Zeitschrift 68, 2–15.

Le Coeur, C., Demé, B., Longeville, S., 2009. Compression of random coils due to macromolecular crowding. Phys. Rev. E 79, 031910-1-4.

Le Coeur, C., Teixeira, J., Busch, P., Longeville, S., 2010. Compression of random coils due to macromolecular crowding: scaling effects. Phys. Rev. E 81, 061914-1-4.

Lebowitz, J.L., Helfand, E., Praestgaard, E., 1964. Scaled particle theory of fluid mixtures. J. Chem. Phys. 43, 774–779.

Lekkerkerker, H.N.W., Poon, W.C.K., Pusey, P.N., Stroobants, A., Warren, P.B., 1992. Phase behaviour of colloid+polymer mixtures. Europhys. Lett. 20, 559–564.

Lhuilier, D., 1988. A simple model for polymeric fractals in a good solvent and an improved version of the Flory approximation. J. Phys. 49, 705–710.

Linhananta, A., Hadizadeh, S., Plotkin, S.S., 2011. An effective solvent theory connecting the underlying mechanisms of osmolytes and denaturants for protein stability. Biophys. J. 100, 459–468 (18 pages of supplementary material).

Linhananta, A., Amadei, G., Miao, T., 2012. Computer simulation study of folding thermodynamics and kinetics of proteins in osmolytes and denaturants. J. Phys. Conf. Ser. 341, 012009-1-12.

Louis, A.A., Bolhuis, P.G., Hansen, J.P., Meijer, E.J., 2000. Can polymer coils be modeled as "soft colloids"? Phys. Rev. Lett. 85, 2522–2525.

Louis, A.A., Bolhuis, P.G., Meijer, E.J., Hansen, J.-P., 2002. Polymer induced depletion potentials in polymer-colloid mixtures. J. Chem. Phys. 117, 1893–1907.

Lu, B., Denton, A.R., 2011. Crowding of polymer coils and demixing in nanoparticle-polymer mixtures. J. Phys. Condens. Matter 23, 285102-1-9.

Mackay, M.E., Tuteja, A., Duxbury, P.M., Hawker, C.J., Van Horn, B., Guan, Z., Chen, G., Krishnan, R.S., 2006. General strategies for nanoparticle dispersion. Science 311, 1740–1743.

Mazouchi, A., Bahram, A., Gradinaru, C.C., 2013. Sub-diffusion decays in fluorescence correlation spectroscopy: dye photophysics or protein dynamics? J. Phys. Chem. B 117, 11100–11111.

Meijer, E.J., Frenkel, D., 1994. Colloids dispersed in polymer solutions. A computer simulation study. J. Chem. Phys. 100, 6873–6887.

Minton, A.P., 1981. Excluded volume as a determinant of macromolecular structure and reactivity. Biopolymers 20, 2093–2120.

Minton, A.P., 1997. Influence of excluded volume upon macromolecular structure and associations in 'crowded' media. Curr. Opin. Biotechnol. 8, 65–69.

Minton, A.P., 2000a. Implications of macromolecular crowding for protein assembly. Curr. Opin. Struct. Biol. 10, 34–39.

Minton, A.P., 2000b. Effect of a concentrated "inert" macromolecular cosolute on the stability of a globular protein with respect to denaturation by heat and by chaotropes: a statistical thermodynamic model. Biophys. J. 78, 101–109.

Minton, A.P., 2001. The influence of macromolecular crowding and macromolecular confinement on biochemical reactions in physiological media. J. Biol. Chem. 276, 10577–10580.

Minton, A.P., 2005. Models for excluded volume interaction between an unfolded protein and rigid macromolecular cosolutes: macromolecular crowding and protein stability revisited. Biophys. J. 88, 971–985.

Minton, A.P., Colclasure, G.C., Parker, J.C., 1992. Model for the role of macromolecular crowding in regulation of cellular volume. Proc. Natl. Acad. Sci. 89, 10504–10506.

Moncho-Jordá, A., Louis, A.A., Bolhuis, P.G., Roth, R., 2003. The Asakura-Oosawa model in the protein limit: the role of many-body interactions. J. Phys. Condens. Matter 15, S3429–S3442.

Murat, M., Kremer, K., 1998. From many monomers to many polymers: soft ellipsoid model for polymer melts and mixtures. J. Chem. Phys. 108, 4340–4348.

Nakatani, A.I., Chen, W., Schmidt, R.G., Gordon, G.V., Han, C.C., 2001. Chain dimensions in polysilicate-filled poly(dimethyl siloxane). Polymer 42, 3713–3722.

Nusser, K., Neueder, S., Schneider, G.J., Meyer, M., Pyckhout-Hintzen, W., Willner, L., Radulescu, A., Richter, D., 2010. Conformations of silicapoly(ethylenepropylene) nanocomposites. Macromolecules 43, 9837–9847.

Odijk, T., 2000. Remarks on the depletion interaction between nanoparticles and flexible polymers. Phys. A 278, 347–355.

Oversteegen, S.M., Roth, R., 2005. General methods for free-volume theory. J. Chem. Phys. 122, 214502-1-12.

Panagiotopoulos, A.Z., 1987. Direct determination of phase coexistence properties of fluids by Monte Carlo simulation in a new ensemble. Mol. Phys. 61, 813–826.

Panagiotopoulos, A.Z., 1989. Exact calculations of fluid-phase equilibria by Monte Carlo simulation in a new statistical ensemble. Int. J. Thermophys. 10, 447–457.

Panagiotopoulos, A.Z., 1992. Molecular simulation of phase equilibria: simple, ionic and polymeric fluids. Fluid Phase Equilib. 76, 97–112.

Panagiotopoulos, A.Z., 1995. Gibbs ensemble techniques. In: Baus, M., Rull, L.R., Ryckaert, J.P. (Eds.), Observation, Prediction, and Simulation of Phase Transitions in Complex Fluids, NATO ASI Series C **460**. Kluwer, Dordrecht, pp. 463–501.

Panagiotopoulos, A.Z., 2000. Monte Carlo methods for phase equilibria of fluids. J. Phys. Condens. Matter 12, R25–R52.

Phillips, R., Kondev, J., Theriot, J., 2009. Physical Biology of the Cell. Garland Science, New York.

Poon, W.C.K., 2002. The physics of a model colloid-polymer mixture. J. Phys. Condens. Matter 14, R859–R880.

Pusey, P.N., 1991. Colloidal suspensions. In: Hansen, J.-P., Levesque, D., Zinn-Justin, J. (Eds.), Liquids, Freezing and Glass Transition, Les Houches Session 51. North-Holland, Amsterdam, pp. 763–931.

Ramakrishnan, S., Fuchs, M., Schweizer, K.S., Zukoski, C.F., 2002. Entropy driven phase transitions in colloid-polymer suspensions: tests of depletion theories. J. Chem. Phys. 116, 2201–2212.

Richter, K., Nessling, M., Lichter, P., 2007. Experimental evidence for the influence of molecular crowding on nuclear architecture. J. Cell Sci. 120, 1673–1680.

Richter, K., Nessling, M., Lichter, P., 2008. Macromolecular crowding and its potential impact on nuclear function. Biochim. Biophys. Acta 1783, 2100–2107.

Rosenfeld, Y., 1989. Free-energy model for the inhomogeneous hard-sphere fluid mixture and density-functional theory of freezing. Phys. Rev. Lett. 63, 980–983.

Rosenfeld, Y., Schmidt, M., Löwen, H., Tarazona, P., 1997. Fundamental-measure free-energy density functional for hard spheres: dimensional crossover and freezing. Phys. Rev. E 55, 4245–4263.

Rudnick, J., Gaspari, G., 1987. The shapes of random walks. Science 237, 384–389.

Schäfer, L., 1999. Excluded volume effects in polymer solutions as explained by the renormalization group. Springer, Berlin.

Schlesinger, A.P., Wang, Y., Tadeo, X., Millet, O., Pielak, G.J., 2011. Macromolecular crowding fails to fold a globular protein in cells. J. Am. Chem. Soc. 133, 8082–8085.
Schmidt, M., Fuchs, M., 2002. Penetrability in model colloid-polymer mixtures. J. Chem. Phys. 117, 6308–6312.
Schmidt, M., Löwen, H., Brader, J.M., Evans, R., 2000. Density functional for a model colloid-polymer mixture. Phys. Rev. Lett. 85, 1934–1937.
Sciutto, S.J., 1996. The shape of self-avoiding walks. J. Phys. A: Math. Gen. 29, 5455–5473.
Sear, R.P., 2001. Phase separation in mixtures of colloids and long ideal polymer coils. Phys. Rev. Lett. 86, 4696–4699.
Sen, S., Xie, Y., Kumar, S.K., Yang, H., Bansal, A., Ho, D.L., Hall, L., Hooper, J.B., Schweizer, K.S., 2007. Chain conformations and bound-layer correlations in polymer nanocomposites. Phys. Rev. Lett. 98, 128302-1-4.
Šolc, K., 1971. Shape of a random-flight chain. J. Chem. Phys. 55, 335–344.
Šolc, K., 1973. Statistical mechanics of random-flight chains. IV, size and shape parameters of cyclic, star-like, and comb-like chains. Macromolecules 6, 378–385.
Stradner, A., Foffi, G., Dorsaz, N., Thurston, G., Schurtenberger, P., 2007. New insight into cataract formation: enhanced stability through mutual attraction. Phys. Rev. Lett. 99, 198103-1-4.
Tolstoguzov, V.B., 1991. Functional properties of food proteins and role of protein-polysaccharide interaction. Food Hydrocoll. 4, 429–468.
Tuinier, R., Rieger, J., de Kruif, C.G., 2003. Depletion-induced phase separation in colloid-polymer mixtures. Adv. Coll. Int. Sci. 103, 1–31.
Tuteja, A., Duxbury, P.M., Mackay, M.E., 2008. Polymer chain swelling induced by dispersed nanoparticles. Phys. Rev. Lett. 100, 077801-1-4.
van der Maarel, J.R.C., 2008. Introduction to Biopolymer Physics. World Scientific, Singapore.
Vink, R.L.C., Horbach, J., 2004. Grand canonical Monte Carlo simulation of a model colloid-polymer mixture: coexistence line, critical behavior, and interfacial tension. J. Chem. Phys. 121, 3253-1-6.
Vink, R.L.C., Horbach, J., Binder, K., 2005. Critical phenomena in colloid-polymer mixtures: interfacial tension, order parameter, susceptibility, and coexistence diameter. Phys. Rev. E 71, 011401-1-10.
Vrij, A., 1976. Polymers at interfaces and the interactions in colloidal dispersions. Pure Appl. Chem. 48, 471–483.
Wang, Y., Sarkar, M., Smith, A.E., Krois, A.S., Pielak, G.J., 2012. Macromolecular crowding and protein stability. J. Am. Chem. Soc. 134, 16614–16618.
Weeks, J.D., Chandler, D., Andersen, H.C., 1971. Role of repulsive forces in determining the equilibrium structure of simple liquids. J. Chem. Phys. 54, 5237–5247.
Woodward, C.E., Forsman, J., 2008. Density-functional theory for polymer fluids with molecular weight polydispersity. Phys. Rev. Lett. 100, 98301-1-4.
Yamakawa, H., 1970. Modern Theory of Polymer Solutions. Harper & Row, New York.
Zhang, Z., van Duijneveldt, J.S., 2006. Experimental phase diagram of a model colloid-polymer mixture in the protein limit. Langmuir 22, 63–66.

CHAPTER FOUR

Crowding-Induced Formation and Structural Alteration of Nuclear Compartments: Insights from Computer Simulations

Jun Soo Kim[*,1], **Igal Szleifer**[†,‡,§]
[*]Department of Chemistry and Nano Science, Global Top5 Research Program, Ewha Womans University, Seoul, Republic of Korea
[†]Department of Biomedical Engineering, Northwestern University, Evanston, Illinois, USA
[‡]Department of Chemistry, Northwestern University, Evanston, Illinois, USA
[§]Chemistry of Life Processes Institute, Northwestern University, Evanston, Illinois, USA
[1]Corresponding author: e-mail address: jkim@ewha.ac.kr

Contents

Abstract

Our understanding of the structural and dynamical characteristics of nuclear structures such as chromosomes and nuclear bodies (NBs) has increased significantly in recent days owing to advances in biophysical and biochemical techniques. These techniques

International Review of Cell and Molecular Biology, Volume 307
ISSN 1937-6448
http://dx.doi.org/10.1016/B978-0-12-800046-5.00004-7

include the use of computer simulations, which have provided further physical insights complementary to findings from experiments. In this chapter, we review recent computer simulation studies on the structural alteration of chromosome subcompartments and the formation and maintenance of NBs in the highly crowded cell nucleus. It is found that because of macromolecular crowding, the degree of chromosome compaction changes significantly and the formation of NBs is facilitated. We further discuss the physical consequences of these phenomena, which may be of critical importance in understanding genome processes.

1. INTRODUCTION

Cell nuclei are highly compartmentalized and contain numerous structures such as chromosomes and nuclear bodies (NBs). Chromosomes are the most prominent structures, and most genetic processes occur in chromosomes including transcription, DNA replication and repair, and repression of gene expression, which are modulated by interactions with gene-regulating proteins. NBs such as nucleoli, Cajal bodies, PML bodies, and speckles are distinct organelles that play functional roles in transcription and RNA processing (Brangwynne, 2011; Dundr, 2012; Mao et al., 2011b; Matera et al., 2009).

Nuclei are crowded with DNA, RNA, and proteins, and in this chapter, we first examine the effect of crowding on the structure of chromosome subcompartments (Kim et al., 2011b) and then review its effects on the formation and maintenance of NBs (Cho and Kim, 2012b). In addition, the physical consequences and biological implications of these structural changes are discussed. The compaction and decompaction of chromosome subcompartments change the chromosome surface area that is accessible to gene-regulating proteins, which may change the processes mediated by binding of these proteins to specific DNA sequences. The condensation of NBs influences the transport of proteins within and out of them to a different extent and enhances the macromolecular associations that occur within them. We conclude with a discussion of the potential use of computational approaches to further enhance our understanding of the role of crowding in determining the structure and function of nuclear compartments.

2. STRUCTURAL PROPERTIES OF NUCLEAR COMPARTMENTS

2.1. Chromosome subcompartments

Despite the advances in our knowledge of one-dimensional genome structure, that is, gene sequences, the three-dimensional structure of chromosomes and their dynamical alterations during specific cellular processes remain poorly understood (Cremer et al., 2006; Naumova and Dekker, 2010). To further our understanding of genome biology, the physical structure of chromosomes and factors that regulate these structures at different stages in normal and pathological cells should be thoroughly studied. The DNA in each human cell is about 2 m long and is condensed into a nucleus with a dimension of about 10–20 μm. This dramatic compaction is driven by association with architectural proteins such as histone complexes (forming nucleosomes), SatB1 proteins (mediating chromatin loops), and SMC proteins (involved in chromosome condensation and maintenance of mitotic chromosomes) (Cai et al., 2006; Hirano, 2006). In contrast to the highly compacted chromosomes in metaphase, those in interphase are more dispersed, but each occupies a rather limited volume called a chromosome territory (Cremer and Cremer, 2006; Cremer et al., 2006; Naumova and Dekker, 2010). The structure of chromosomes in metaphase has been relatively well studied, but much less is known about their structure in the interphase nucleus. The most valuable information has been revealed using experimental techniques such as fluorescent *in situ* hybridization (FISH) (Cremer and Cremer, 2001; Croft et al., 1999; Sachs et al., 1995) and chromosome conformation capture (3C) and its variants (Dekker et al., 2002; Lieberman-Aiden et al., 2009).

It has been suggested that within each chromosome territory, there exist largely decondensed domains termed euchromatin, which represent gene-rich regions, whereas condensed heterochromatin is correlated with gene-poor stretches of the genome. The density of euchromatin is typically 0.100 g/ml, and that of heterochromatin is about 0.2–0.4 g/ml (Bancaud et al., 2009). Therefore, the open, less dense euchromatin is more accessible to gene-regulating proteins than the dense heterochromatin, and this accessibility has been correlated with gene activity and silencing in these domains (Beato and Eisfeld, 1997; Verschure et al., 2003). Although this correlation is still controversial, since large macromolecules can also penetrate deep into heterochromatin (Handwerger et al., 2005; Verschure et al., 2003), the chance of penetration drops significantly in the case of more dense heterochromatin and hence it must still be a critical factor in determining the structure–function relation of chromatin domains.

The primary structure of chromatin is a nucleosome array called a beads-on-a-string filament, which is about 10 nm in thickness and in which DNA wraps around histone complexes. This fiber can be further condensed by forming a highly compact fiber about 30 nm in thickness observed when chromatin fragments are isolated or released from nuclei (Tremethick, 2007). Although the existence of this 30 nm fiber *in vivo* has been challenged recently (Maeshima and Eltsov, 2008), compaction of the nucleosome array, irrespective of whether it forms a regular 30 nm fiber or an irregularly condensed aggregate, should result in the formation of a more condensed higher order chromatin domain. Higher levels of chromatin structure are believed to be determined by various arrangements of loops formed by an underlying fiber (see Chapter 10). Domains of about 1 Mbp in size have been observed that are physically separate and are maintained over several cell cycles (Albiez et al., 2006; Cremer et al., 2006; Naumova and Dekker, 2010), and these are referred to here as chromosome subcompartments.

Recently, computer simulations of chromosome models have been employed to provide more detailed information on the physical properties of such structures. Early models assumed the formation of long-range loops or multiple short-ranged loops (Münkel et al., 1999; Sachs et al., 1995). Recently, a model incorporating loops in a random fashion proved to be successful in predicting the three-dimensional structure of the chromosome in terms of the mean square distance, consistent with those from FISH experiments. (Mateos-Langerak et al., 2009). This random-loop (RL) polymer model of the chromosome was employed to explain the structural change in chromosome subcompartments induced by crowding (Kim et al., 2011b). More recently, a fractal-like structure of chromosome domains was suggested based on Hi-C data (Lieberman-Aiden et al., 2009). Recent simulations employing a decondensing polymer model in a confined volume, which mimics the decondensation of a mitotic chromosome to an interphase chromosome, successfully reproduced both FISH and Hi-C data and explained the physical origin of chromosome territories as a nonequilibrium and kinetic behavior in a topologically confined volume (Rosa and Everaers, 2008; Rosa et al., 2010).

2.2. Nuclear bodies

NBs such as nucleoli, Cajal bodies, PML bodies, and speckles (Brangwynne, 2011; Dundr, 2012; Mao et al., 2011b; Matera et al., 2009) are morphologically distinct and have a roughly spherical shape, with a size between several hundred nanometers and several micrometers (see Chapters 2 and 5). They are not surrounded by physical boundaries such as membranes, and thus dynamic exchange of constituent macromolecules (i.e., body-specific proteins and RNAs) occurs between NBs and the nucleoplasmic space.

Functional roles of NBs in transcription and RNA processing have been suggested based on the role of their constituent proteins in these processes and the association of NBs with specific gene loci. However, the physics of their formation and maintenance are not well understood, with fundamental questions regarding their formation remaining unclear. For instance, it is not known whether their assembly occurs in a specific order or in a random fashion, and only recently have some studies provided evidence supporting the idea that NBs are formed by a random assembly of their constituent proteins and RNAs (Kaiser et al., 2008; Mao et al., 2011a; Shevtsov and Dundr, 2011). It has been reported that NBs are preferentially formed on the surface of chromatin, for reasons that remain unclear (Matera et al., 2009).

The density of NBs has been determined for *Xenopus* oocyte nuclei (Handwerger et al., 2005), and compared to the density of nucleoplasm of 0.106 g/ml, those of Cajal bodies, speckles, and dense fibrillar regions of nucleoli were 0.136, 0.162, and 0.215 g/ml, respectively. Large macromolecules were also shown to penetrate the nucleoli, which shows that they have a sponge-like structure. Recent experiments suggest that NBs are liquid-like assemblies with rapid exchange of their components with the nucleoplasmic space (Brangwynne, 2011). Similarly, cytoplasmic P granules in *Caenorhabditis elegans* show liquid-like properties (Brangwynne et al., 2009).

NBs are believed to be regions where macromolecules involved in the function of NBs congregate (Brangwynne, 2011; Dundr, 2012; Mao et al., 2011b; Matera et al., 2009). For example, Cajal bodies play a role in the assembly and modification of a variety of different small RNPs; each RNP consists of a small U-rich RNA complexed with a number of specific RNP proteins, and the assembly of RNPs is significantly enhanced in Cajal bodies although this also occurs in their absence (Klingauf et al., 2006). It has been suggested that the rates of macromolecular association processes can be regulated in several other types of NB (Klingauf et al., 2006).

3. CROWDED NATURE OF CELL NUCLEUS

3.1. Macromolecular crowding in nucleus

The cell nucleus is crowded with DNA, RNA, and proteins. Chromatin concentrations range from ~0.1 g/ml in euchromatin to ~0.2–0.4 g/ml in heterochromatin (see Chapter 2) (Bancaud et al., 2009). Protein concentrations measured in the nuclei of Xenopus oocytes range from 0.106 g/ml in the nucleoplasm to 0.136 g/ml in Cajal bodies, 0.162 g/ml in speckles, and 0.215 g/ml in the dense fibrillar region of nucleoli (Handwerger et al., 2005) (see Table 2.1 of Chapter 2). Assuming a partial specific volume of 0.73 ml/g for proteins (Brown et al., 2011), the total macromolecular concentrations reach up to more than 15% and up to 45% of the total nuclear volume.

Nuclear macromolecules implicated in common pathways often concentrate together in distinct nuclear compartments. For example, numerous factors essential for rDNA processing and ribosome biogenesis are concentrated in nucleoli, while others for snRNP assembly and modification are concentrated in Cajal bodies. Thus, nuclear proteins can be categorized into different groups according to their localization in specific compartments (Bickmore and Sutherland, 2002; Sutherland et al., 2001). Many other proteins are also found at the nuclear periphery and diffusely in the nucleoplasm. In studies that examine the effect of crowding on association reactions, macromolecules that do not directly participate in the association reaction are usually considered as inert crowding particles exerting nonspecific, volume-exclusive forces. For example, proteins categorized as diffuse nucleoplasmic proteins as well as those associated with other NBs may be considered as inert crowding particles in determining the structure of chromosome subcompartments.

Nuclear compartments are formed in an intrinsically crowded environment, as described above. The kinetics and equilibria of genome processes in the crowded nucleus are therefore expected to be different from those in dilute buffer solutions and, thus, a great deal of experimental and computational effort has been directed toward understanding the role of crowding in these processes (Ellis, 2001; Marenduzzo et al., 2006; Minton, 2001; Zhou et al., 2008; Zimmerman, 1993; Kim and Yethiraj, 2009, 2010, 2011). Furthermore, the crowding condition in the nucleus can be altered significantly by abrupt changes in either the cell volume or protein concentration under normal and pathological conditions (Lang et al., 1998). The corresponding changes in the crowding condition may markedly influence the functions of proteins and thus inhibit or promote biochemical processes, as described below. Therefore, a proper description of crowding effects can be very important for understanding the regulation of structure and function of nuclear compartments under changing environmental conditions.

The effects of macromolecular crowding on nuclear compartments have been studied mostly by incubating cells or isolated nuclei in hypertonic or hypotonic media that induce volume changes (Albiez et al., 2006; Hancock,

2004; Richter et al., 2007). Morphological changes in nuclear compartments were observed by microscopy and were attributed to changes in the effective concentration of macromolecules, that is, changes in the crowding conditions, and it was concluded that chromosomes are compacted and the formation of NBs is facilitated by crowding. When the nuclear volume increases in hypotonic media, disassembly of nucleoli and PML bodies is observed and these NBs are recovered by incubating in normal nuclear buffers (Hancock, 2004). When the cell volume decreases in hypertonic media, hypercondensation of chromatin occurs and normal chromatin compaction is recovered by incubation in normal buffers (Richter et al., 2007). Interestingly, the hypercondensation of chromatin and reassembly of NBs also occur when inert crowding molecules, such as polyethylene glycol and dextran, are introduced, which provides critical evidence for the effect of macromolecular crowding on chromatin structure and formation of NBs (Hancock, 2004; Richter et al., 2007). The effect of crowding could be very important in the regulation of biological functions in these compartments. In the experiments mentioned earlier, nucleolar transcript elongation decreased by ~85% when the nuclear volume increased (Hancock, 2004) and cell proliferation ceased when the cell volume was significantly decreased, which induced hypercondensation of chromatin under the large osmotic pressures in hypertonic media (Richter et al., 2007). Although these studies provide useful information on regulation of structures by crowding, these observations are limited by the resolution of detection, and detailed structural changes are still not clearly understood.

To better understand the effects of crowding, we recently performed computer simulation studies on the crowding-induced structural regulation of nuclear compartments (Cho and Kim, 2012b; Kim et al., 2011b), and in this chapter we discuss the detailed findings obtained from these simulations. Computer simulations provide a very useful tool to study the effect of macromolecular crowding, since they enable separation of the role of each of the different components under optimal conditions; simulations are carried out under optimal conditions with the proper choice of only the elements of interest. It should be stressed, however, that this advantage is also a disadvantage of the methodology, since there is a large distance between the model systems simulated and the complex biological environments. However, as will be discussed in detail throughout this chapter, one can gain insightful information that is not accessible experimentally. Therefore, computer simulations provide one more tool, together with experimental observations, for a comprehensive understanding of the role of macromolecular crowding in cell nuclei. It has been suggested that macromolecular crowding can also function as a signal for cell volume control, and since changes in cell volume have been associated with a variety of biological functions including regulation of metabolism, hormone release, cell proliferation, and cell death, a change in the crowding condition may also be involved in the regulation of these processes (Lang et al., 1998). The crowding condition can be significantly varied by a sudden change in cell volume or in the concentration of a variety of proteins. An acute change in cell volume may occur because of changes in osmolarity under normal and pathological conditions. For example, excessive alterations in the extracellular osmolarity occur in kidney medulla cells during the transition between antidiuresis and diuresis, where cells shrink and the effective concentration of proteins increases (Lang et al., 1998). In addition, the cell volume increases exponentially during most of the cell cycle (Edgar and Kim, 2009). Unless the rate of protein synthesis keeps pace with that of cell growth, the effective protein concentration may be subjected to a significant change. The concentration of a variety of proteins changes during the cell cycle by regulation of their synthesis and degradation (Rustici et al., 2004). Posttranslational modification of certain NB marker proteins induces the assembly or disassembly of NBs, dramatically changing the total amount of proteins in the nucleoplasm; for example, hyperphosphorylation of coilin, a marker protein of Cajal bodies, results in their disassembly (Carmo-Fonseca et al., 1993), and the dispersed macromolecular components could exert a crowding force on other nuclear compartments. On the whole, the crowding conditions vary in normal and pathological cells and it is, therefore, of critical importance to understand crowding-induced structural alterations in varying nuclear environments.

3.2. General understanding of effects of macromolecular crowding on bimolecular associations

Since the cell nucleus is crowded with macromolecules, it has been of interest to see how this crowding affects the structure of chromosomes and NBs. Many studies have focused on the effects of macromolecular crowding on association equilibria between biomolecules (Ellis, 2001; Minton, 2001; Zhou et al., 2008; Zimmerman, 1993). In crowded media, the self-association of spectrin, actin, and FtsZ are enhanced, the affinity of DNA-binding proteins for DNA is increased, and the formation of a decamer of bovine pancreatic trypsin inhibitor is enhanced (Zhou et al., 2008). Such enhancement of biomolecular associations can be explained either in terms of an increase in the equilibrium constant caused by crowding, or in terms of an enhanced effective attraction between biomolecules termed the depletion attraction (Asakura and Oosawa, 1958).

A thermodynamic framework was developed by Minton in terms of the nonideality factor describing the change in equilibrium constant induced by macromolecular crowding (Minton, 1998; Zhou et al., 2008). Consider a prototypical bimolecular association of A+B=AB. The standard free energy change in a dilute buffer solution and in a crowded medium are ΔG^* and ΔG, respectively, and the corresponding equilibrium constants are K^* and K, respectively, as depicted in Fig. 4.1A. Standard free energy changes and equilibrium constants are related as $\Delta G^* = -RT \ln K^*$ and $\Delta G = -RT \ln K$, where R is the gas constant and T is the absolute temperature. The change in equilibrium constants because of crowding can be written as $K = K^* \exp(-\Delta\Delta G/RT)$, where $\Delta\Delta G = \Delta G - \Delta G^*$, the difference in standard free energy change of the association reaction in crowded medium and in dilute buffer solution. The ratio K/K^* for the association in simple models has been calculated theoretically and computationally and was shown to increase with the degree of crowding (Kim and Yethiraj, 2011). The increase as a function of the volume fraction of crowding particles ϕ_c is shown in Fig. 4.1. In the association of two spherical reactants to form a single two-tangent-sphere product, the equilibrium constant increases by a factor of 1.8 for a change of ϕ_c between 0.0 and 0.2 and by a factor of 2.5 between 0.0 and 0.3. Thus, it was concluded that the equilibrium is shifted to induce more association in crowded media. It is important to emphasize that the role of crowders is completely characterized by excluded volume interactions in this model.

The induced association can also be explained in terms of the depletion attraction, which has been extensively investigated in colloid–polymer mixtures. This phenomenon is explained in terms of an entropy increase in crowding particles when two reactants associate (Asakura and Oosawa, 1958; Kim and Szleifer, 2010; Marenduzzo et al., 2006). The effective depletion interactions are shown in Fig. 4.2 for a variety of crowder densities. More explicitly, consider two associating reactants with a diameter of d_r in a sea of crowding particles whose diameter is σ, for which the minimum distance between a reactant and any crowding particle is $r_{min} = (d_r + \sigma)/2$. When the two reactants are separated, each occupies a spherical volume

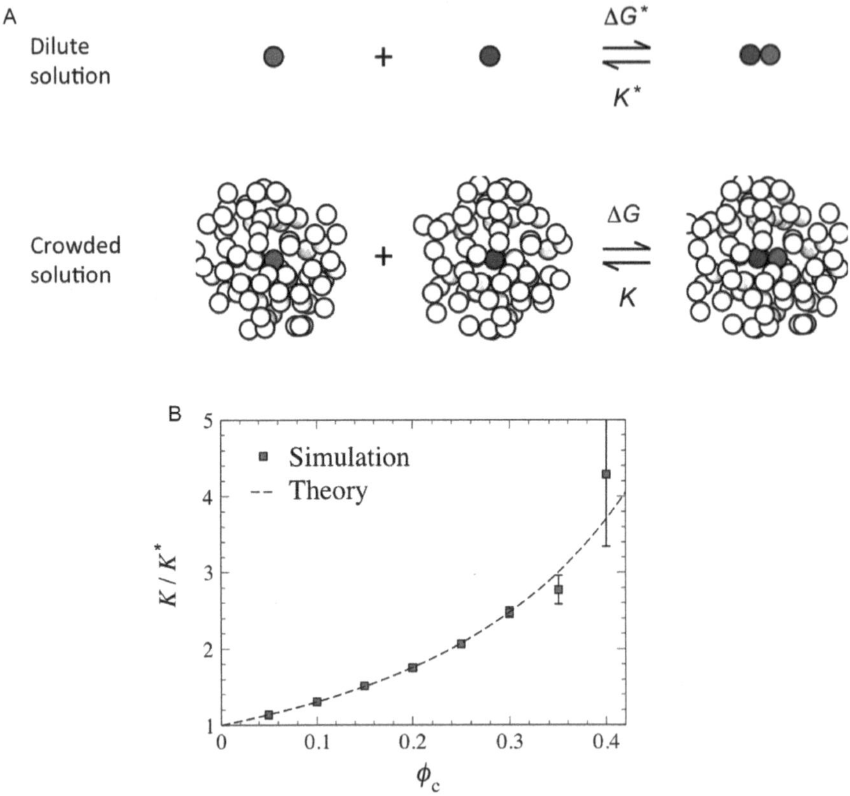

Figure 4.1 Comparison of equilibrium constants in dilute and crowded solutions. The reaction schemes in the dilute and crowded solutions are presented in panel (A). The equilibrium constants K^* and K represent those in dilute and crowded solutions, respectively. In panel (B), the ratio of K/K^* calculated using Monte Carlo simulations and theoretical expressions are presented. The ratio increases, implying that the equilibrium constant increases as the solution becomes more crowded. *Data in (B) were adapted from Kim and Yethiraj (2011) and replotted.* (For color version of this figure, the reader is referred to the online version of this chapter.)

inaccessible to crowding particles defined by $V_{\mathrm{excl},1} = (4/3)\pi r_{\mathrm{min}}^3$. However, when the two reactants are in contact with each other, the inaccessible volumes around each overlap, and thus the total volume inaccessible to crowding particles, $V_{\mathrm{excl},2}$, becomes less than the sum of each, resulting in $V_{\mathrm{excl},2} < 2V_{\mathrm{excl},1}$. This provides more space accessible to crowding particles, amounting to $V_{\mathrm{gain}} = 2V_{\mathrm{excl},1} - V_{\mathrm{excl},2}$, and increases the configurational freedom of the crowding particles and thus their entropy. The entropy increase favors the formation of a product rather than separate

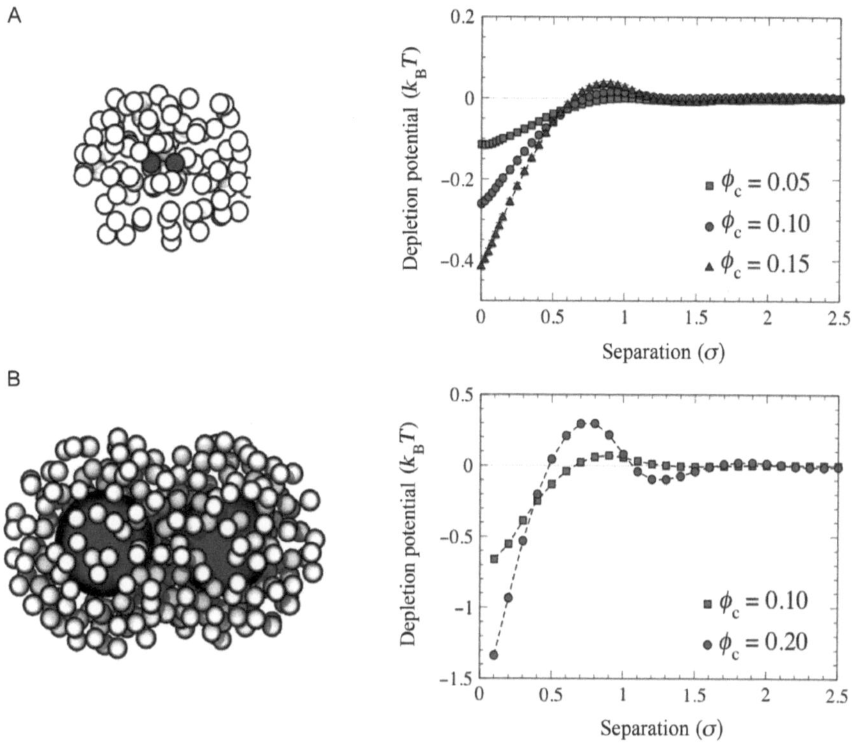

Figure 4.2 The change in the interactions between two central particles (colored blue) in the presence of a crowding agent. Panels (A) and (B) are distinguished by the size of the two central particles relative to the size of crowders. The effective potentials are drawn to different scales on the *y*-axis. (For interpretation of the references to color in this figure legend, the reader is referred to the online version of this chapter.)

reactants and induces an effective attraction between two reactants as exemplified in Fig. 4.2. This entropic effect becomes more significant for larger reactants because the increase in the volume, V_{gain}, for crowding particles when the reactants are in contact is larger. The effective attraction at a crowding volume fraction of $\phi_c = 0.10$ is compared in Fig. 4.2A and B. For reactants whose size is the same as the crowding particles ($d_r = \sigma$), the effective attraction amounts to $-0.26\ k_B T$ when they are in contact. However, when the size of the reactants is five times larger than that of the crowding particles ($d_r = 5\sigma$), it amounts to $-0.67\ k_B T$. Obviously, the crowding-induced attraction becomes more significant as the degree of crowding increases. For $d_r = \sigma$, the effective attraction increases from $-0.12\ k_B T$ at $\phi_c = 0.05$ to $-0.26\ k_B T$ at $\phi_c = 0.10$ and to $-0.42\ k_B T$ at

$\phi_c=0.15$ (Fig. 4.2A). Interestingly, at high crowding volume fractions, effective repulsion are observed at a reactant separation of $\sim\sigma$ as well as effective attraction at the contact of reactants, as shown in Fig. 4.2B. This repulsion can be attributed to the presence of crowding particles between two reactants; the reactants must squeeze out these crowding particles for them to get close, which explains the effective repulsion. One of the important results presented in Fig. 4.2 is that depletion interactions show nonmonotonic behavior with separation, a result that arises from the finite density of crowders (purely attractive depletion interactions occur only in the very dilute limit). As discussed later, this nonmonotonic nature of the depletion interactions plays an important role in models of chromosome compaction.

The increase in the equilibrium constant and the induced effective attraction are both nonspecific effects induced by macromolecular crowding, irrespective of specific interactions between the associating macromolecules. When the crowder density is increased, macromolecular associations are expected to increase because of the increase in the equilibrium constant or equivalently because of the amplified effective attraction. Based on this understanding, we can expect that the volume exclusion effect caused by macromolecular crowding induces compaction of chromosome subcompartments (Hancock, 2007, 2012) and a stronger association between the components of NBs. A more detailed discussion of the effects of crowding, which are more complex than indicated by the simple statement above, is presented throughout the rest of this chapter.

Most theoretical and computational studies on crowding effects have focused on volume exclusion effects caused by the presence of crowding particles. It is obvious, however, that the macromolecules in the nucleus not only exert volume-exclusive, repulsive interactions but also have attractive interactions such as van der Waals interactions, electrostatic interactions, hydrogen bonding interactions, and hydrophobic interactions. In our recent work, we also investigated the effect of crowding on the stability of NBs in the presence of attractive crowders (Cho and Kim, 2012a). When crowding particles have an intrinsic attraction to the reactants, the crowding effect becomes less significant and even unfavorable for molecular associations in the case of very strong attractions. However, experiments demonstrate that nuclear compartments are not disassembled by crowding; therefore, we exclude the possibility of very strong attractions between crowding particles and components of nuclear compartments. Even in the presence of crowding particles having a mild attraction to nuclear compartment components mixed with volume-exclusive, nonattractive crowders, the repulsive effects of the nonattractive crowders were found to be significant with regard to determining the structure of nuclear compartments. Therefore, in this chapter we focus on the excluded volume effect exerted by inert, repulsive crowding particles.

4. STRUCTURAL ALTERATIONS OF CHROMOSOME SUBCOMPARTMENTS BY MACROMOLECULAR CROWDING

4.1. Computational model for chromosome subcompartments

We recently examined structural alterations of chromosome subcompartments under different crowding conditions, using computer simulations of a model chromosome subcompartment in the presence of volume-excluding crowders (Kim et al., 2011b). A subcompartment was modeled as a RL polymer in which a single linear polymer, mimicking the size of 1 Mbp, was looped between randomly chosen pairs of polymer segments (Fig. 4.3A). This model was originally proposed by Mateos-Langerak et al. (2009), and successfully reproduced the distance distribution between chromosome segments when compared with experimental FISH data.

Recently, a fractal-like structure of chromosomes in the size range of 500 kbp to 5 Mbp was suggested based on contact probabilities in Hi-C experiments (Lieberman-Aiden et al., 2009). In other simulation studies, a linear polymer that is initially compacted as a mitotic chromosome and then decondenses with time was used to mimic decondensation in interphase (Rosa and Everaers, 2008; Rosa et al., 2010). This model explains the kinetically maintained chromosome territories and predicts a fractal-like structure of the chromosomes, and is consistent with both FISH and Hi-C data despite its simplicity. Although the decondensing linear model seems more consistent with current experimental data, it is a nonequilibrium, kinetic model controlled by topological constraints and, therefore, the ensemble-averaged effect of crowding could not be captured readily; therefore, we utilized an RL model. It should be noted that the interphase chromosome is, in fact, in a nonequilibrium state and its structure is constantly influenced by the changing environment. However, we confirmed that our conclusions on crowding-induced structural changes are independent of the specific chromosome model, since the same qualitative conclusion was obtained from simulations of a simple linear polymer without loops

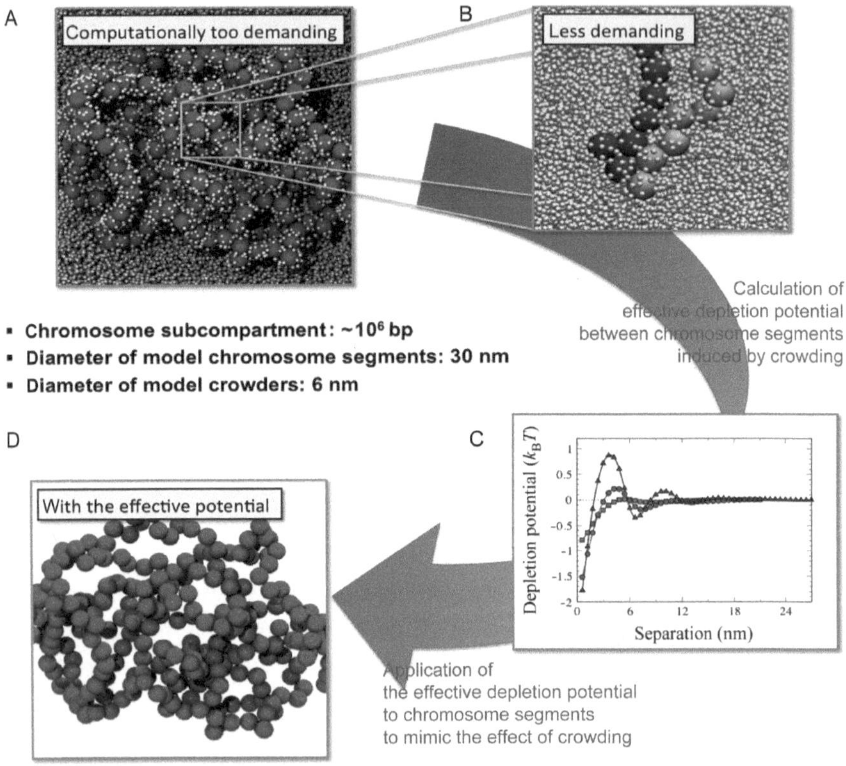

Figure 4.3 The computational strategy used to investigate the effect of crowding on the structural alteration of an RL chromosome model. (A) Straightforward simulation of the RL polymer in the explicit presence of crowding particles is computationally too demanding and impractical. (B, C) Effective depletion potential between polymer segments induced by crowding was obtained from simulations of short polymers in the presence of model crowders. (D) The effective potential was applied to each RL segment to investigate the effect of crowding without the explicit presence of crowders. (For color version of this figure, the reader is referred to the online version of this chapter.)

(Kim et al., 2011b). Therefore, the conclusions presented in this section are not attributable to the specific model chosen, but rather to the change of effective potential induced by the presence of crowding particles. It should also be noted that these models are coarse-grained and that we assumed chromosomes have highly packed 30-nm chromatin fibers (Tremethick, 2007). These fibers have not been observed *in vivo*, and their presence is controversial (Maeshima and Eltsov, 2008). However, compaction of a nucleosome array into a higher order structure *in vivo* is indispensable irrespective of whether it forms a regular 30-nm fiber or an irregular aggregate. Further investigation is necessary to better understand the crowding-induced structural changes at a finer level, that is, in a nucleosome array.

The computational strategy adopted in our work is illustrated in Fig. 4.3. In order to investigate chromosome structure under different degrees of crowding, one can imagine computer simulations of an RL polymer in the presence of various concentrations of crowders (Fig. 4.3A). However, because of the large number of model crowders that should be included, such simulations are time-consuming and practically impossible, especially when the ensemble-averaged structural properties are to be determined. Since each RL polymer had randomly chosen pairs of looping segments, the structure of each may be very different depending on the looping segments. To obtain statistically reliable average structures, a large number of simulation sets should be performed that take into account different combinations of randomly looped polymer segments. Therefore, although simulations of an RL polymer in the explicit presence of model crowders seem straightforward, they are not quite adequate for studying the effects of crowding on its structure.

Instead, we incorporated the effect of crowders implicitly as shown in Fig. 4.3. First, simulations of short polymers were performed in the presence of explicit model crowders (Fig. 4.3B), from which the effective interactions between polymer segments were calculated at different degrees of crowding (Fig. 4.3C). The effective interaction in each crowding condition was then applied to the RL polymer segments without the explicit presence of crowders in the simulation system (Fig. 4.3D), and the changes in the RL polymer structure were investigated.

We first performed simulations of two short polymers in the presence of model crowders as shown in Fig. 4.3B, where the diameters of the polymer segments and crowders were set at 30 and 6 nm, respectively. As explained earlier, the polymer diameter was chosen assuming that the chromatin forms a compact 30-nm fiber. The average molecular weight of the proteins distributed in the nucleoplasm is 67.7 kDa (Bickmore and Sutherland, 2002), and while the polydispersity of the size distribution may play an important role, we focus only on the volume exclusion effect induced by the same size of crowders, whose diameter was set at 6 nm with a spherical shape and a partial specific volume of 0.73 ml/g, the typical value for proteins (Brown et al., 2011). Simulations were performed at three crowder volume fractions of 0.10, 0.20, and 0.30. The effective (depletion) potential because of crowding was determined as a function of segment–segment separation. The effective interactions and their details, for example, strength of attractions and repulsions, were shown to depend on the crowder volume fraction (Fig. 4.3C). These potentials are employed in the chromosome simulations; therefore, they implicitly take into account the presence of crowders, as shown in Fig. 4.3D.

Three different sets of RL polymers were used as mimics of chromosome subcompartments with different densities. As noted earlier, heterochromatin is a highly condensed domain of chromosomes with a density of ~0.2–0.4 g/ml, and euchromatin is a rather open domain with a density of ~0.1 g/ml (Bancaud et al., 2009). The volume fraction occupied by these domains in nuclei was estimated to be roughly between 0.073 and 0.29, again assuming a partial specific volume of 0.73 ml/g. By increasing the number of looping pairs in the RL polymers, we designed three sets of RL polymers whose volume fraction occupations were 0.15, 0.23, and 0.29 (Fig. 4.4).

4.2. Structural alterations under varying degrees of crowding

In earlier experiments, morphological changes in chromosomes were understood based on microscopic images of the nucleus in hypotonic and hypertonic media (Albiez et al., 2006; Richter et al., 2007). Cells in hypotonic media were swollen and the decompaction of chromosomes, which

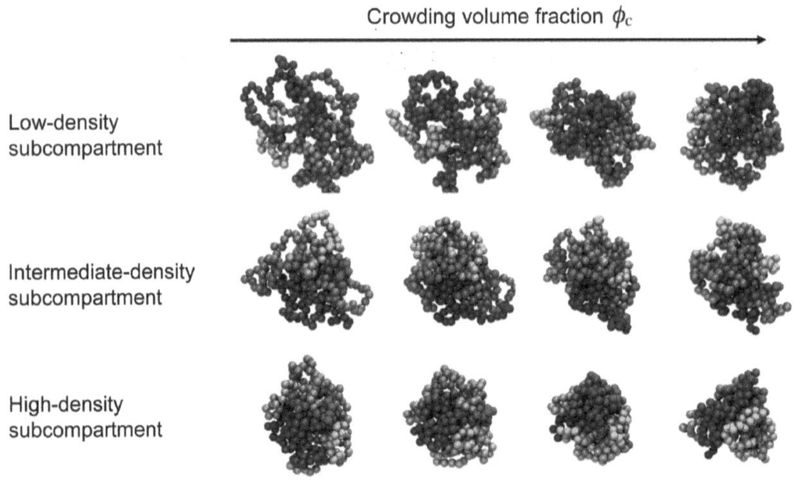

Figure 4.4 Graphical representation of the structural change of the RL polymer at three different densities. *Reprinted with permission from Kim et al. (2011b). Copyright (2011) by the American Physical Society.* (For color version of this figure, the reader is referred to the online version of this chapter.)

was attributed to the reduced level of crowding, was observed. On the other hand, in hypertonic media, cells shrank and the chromosomes became more compact because of the increased macromolecular crowding.

The graphical representation of representative simulation results for three RL models with different densities is presented in Fig. 4.4. Our simulation results were consistent with experimental findings (Albiez et al., 2006; Richter et al., 2007; Hancock, 2007, 2012), showing a clear condensation of chromosome subcompartments with increasing degree of crowding. Quantitative analysis in terms of the polymer size and volume fraction (Fig. 4.5) confirmed that the chromosomes become compacted because of crowding when the degree of crowding is moderate.

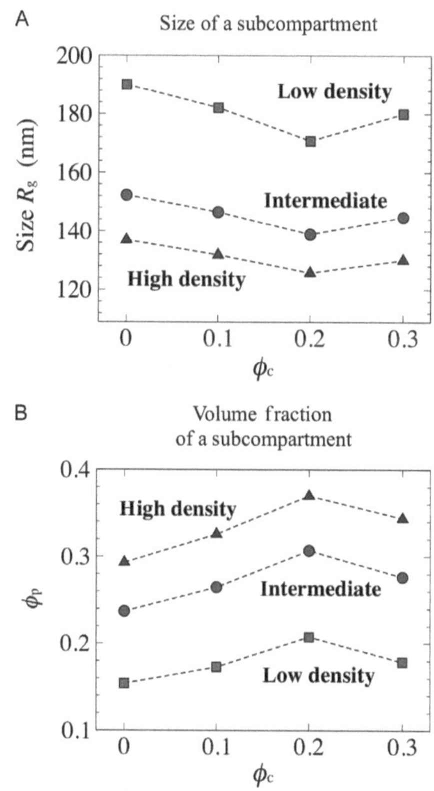

Figure 4.5 Structural alterations of the RL polymer. (A) Size in terms of the polymer radius of gyration; (B) volume fraction of the RL polymer as a function of the crowder volume fraction ϕ_c. *Reprinted with permission from Kim et al. (2011b). Copyright (2011) by the American Physical Society.* (For color version of this figure, the reader is referred to the online version of this chapter.)

Interestingly, however, the RL polymer was again decompacted with further crowding. The size of the RL polymers in terms of the radius of gyration (one measure for the size of a polymer) decreased initially, as the crowder volume fraction ϕ_c increased from 0 up to 0.2. However, the size increased again between 0.2 and 0.3. Thus, the volume fraction (which is equivalent to the density) of chromosome subcompartments increased initially with the degree of crowding, but decreased upon a further increase in the crowder volume fraction. In conclusion, the compaction of RL polymers is a nonmonotonical function of the degree of crowding; they become more condensed with an increase in crowding at small ϕ_c, and decondensed with a further increase in crowding at higher ϕ_c. To confirm that this observation was not model-dependent, we performed additional simulations with a simple linear polymer interacting with the effective depletion potential, and the nonmonotonic size dependence on ϕ_c was also predicted in simulations using this simple linear polymer. Therefore, the nonmonotonic structural alterations by crowding are a generic effect that arises from the crowding-induced effective interaction between polymer segments.

The nonmonotonic structural alteration of chromosomes with an increase in the degree of crowding has not been described in earlier experiments but is consistent with published microscopic images. Albiez et al. (2006) reported that a more apparent condensation of chromosome subcompartments did not occur at the highest level of crowding, which was in contrast to the apparent condensation observed at moderate crowding. Further, a close evaluation of the images of Richter et al. (2007) suggests that the hypercompaction of chromosomes decreased at the highest levels of crowding studied. Although more quantitative analyses are necessary, these findings are consistent with a nonmonotonic effect of crowding on chromosome structure.

Based on our general understanding of the effect of crowding on protein associations, the results just presented may seem counterintuitive. As discussed earlier, protein association is more favorable in more crowded environments because of the increase in the equilibrium constant and/or the increased effective attraction between associating proteins. As shown in Fig. 4.3C, the effective attraction between polymer segments increased with crowding at all crowding volume fractions. Therefore, compaction of chromosomes was expected to occur and decompaction at higher crowding seems counterintuitive. However, it should be noted that the repulsive barrier also becomes significantly higher as the crowding volume fraction increases (Fig. 4.3C). From ϕ_c of 0.20–0.30, the value of the repulsive maximum increases from 0.2 to 0.9 $k_B T$ while that of the attractive minimum decreases from −1.5 to −1.8 $k_B T$. Since the difference between the repulsive maximum and the attractive minimum increases from 1.7 to 2.7 $k_B T$, the polymer segments within a distance less than that of the repulsive maximum (~4.5 nm) are strongly bound to each other. However, their neighbors connected together through binding are subjected to a significant repulsive interaction. Thus while polymer segments close to each other feel the strongest attraction, their neighbors experience strong repulsions, which do not stabilize the strongly attractive segments. The balance between the maximum number of segments at the potential minimum and the minimum number at the potential maximum results in nonmonotonic behavior.

4.3. Physical consequences and biological implications

The effect of crowding on structural alterations of chromosome subcompartments may be significant for genetic processes in the nucleus. One of the most interesting consequences of the crowding effect is the change in the surface area of chromosome subcompartments that are exposed to proteins, which modulate gene activity (Beato and Eisfeld, 1997; Verschure et al., 2003). When the chromatin is highly condensed, access to these proteins is hindered, which may result in a decrease in gene activity. Therefore, we calculated the change in the accessible chromosome surface area from our simulations.

As shown in Fig. 4.6, the accessible surface area decreases as the compaction of chromosome subcompartments increases at moderate crowding, which could reduce the binding of proteins to target DNA.

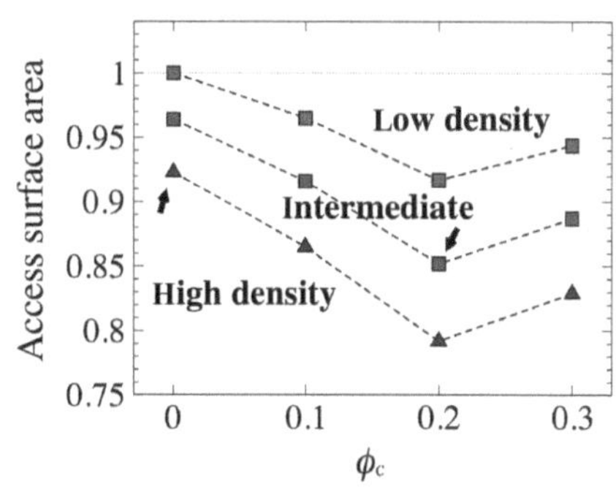

Figure 4.6 Accessible surface area normalized by the value of a low-density subcompartment. *Reprinted with permission from Kim et al. (2011b). Copyright (2011) by the American Physical Society.* (For color version of this figure, the reader is referred to the online version of this chapter.)

The low-density chromosome subcompartment lost about 8% of its surface area when the crowding volume fraction increased from 0.0 to 0.2, while the high-density subcompartment lost about 14% of its surface area. By simply assuming a uniform distribution of active genes on the whole surface, 8–14% of gene activity is predicted to be inhibited because of steric hindrance, which could have a significant effect on genome processes.

The nonmonotonic compaction of the polymers results in a nonmonotonic dependence of the accessible surface area on crowder density, and therefore it is tempting to conclude that gene activity would be restored at a very high crowding volume fraction. However, it should be noted that the accessible surface area at $\phi_c=0.30$ has a very different environment from those at $\phi_c=0.10$ and 0.20. The chromosome subcompartments at $\phi_c=0.10$ have a certain concentration of crowding particles around them whose presence exerts an attractive interaction between chromosome segments, but at $\phi_c=0.3$, an even higher concentration of crowding particles is present around the chromosome segments that exerts a repulsive interaction as well as an attraction. Some crowding particles may always be present between chromosome segments in this case, and any genome process mediated by binding of gene-regulating proteins to DNA might be inhibited because of steric hindrance from crowding particles despite the increased accessible surface area. The exact quantification of the changes in local density and surface accessible area upon genomic activity is challenging. However, the results just discussed demonstrate the importance of considering the role of crowding and, in particular, the relevance of the nonmonotonic properties that arise.

There are several possible ways to compact chromosome subcompartments: one is by expressing architectural proteins that can induce more loops; another is by posttranslationally modifying histone tails such as methylation and acetylation (Görisch et al., 2005); and a third is by increasing the effective concentration of inert crowding particles that can induce an effective attraction between chromosome segments (Lang et al., 1998). All of these effects may result in a seemingly identical degree of compaction of chromosomes, but their structure may be significantly different. In our simulations we compared the effect of an increased level of architectural protein expression as well as an increased level of crowding particles. When the high-density chromosome subcompartment at $\phi_c=0.00$ was compared with the intermediate subcompartment density at $\phi_c=0.20$, their sizes and volume fractions were found to be very similar, as shown in Fig. 4.5. However, their accessible surface areas were significantly different, which is noted by arrows in Fig. 4.6. Therefore, we conclude that the seemingly identical degrees of chromosome compaction with different origins may give rise to significantly different structures.

Recently, through simulations of a chromosome model complementary to experimental observations (Subramanian et al., 2008), we showed that minute compaction of chromosomes may occur in the early stage of carcinogenesis (Kim et al., 2011a). Therefore, we simulated decondensation of model mitotic chromosomes during interphase and compared the density change with the disorder strength observed using a backscattering technique called partial wave spectroscopy. The reason why the chromosomes are compacted in carcinogenesis remains unclear. We hypothesize that the expression of some architectural proteins increases so that chromosome decondensation from metaphase to interphase is not complete; another potential reason is that posttranslational modification is so frequent that the morphology of the chromosomes is changed. Finally, it is also possible that the level of crowding is more significant in cancer cells, resulting in more condensed chromosomes. All three hypotheses are plausible, and the question may be resolved when the detailed chromosome structure and its changes with environmental cues are revealed.

5. FORMATION AND MAINTENANCE OF NBs INFLUENCED BY MACROMOLECULAR CROWDING

5.1. Computational model for NBs

In contrast to chromosomes for which several computational models have been developed in the last two decades, no molecular model of NBs has been proposed. This is probably due to a lack of detailed information on their components and functions. Recently, we proposed a simple generic model termed Lennard-Jones (LJ) particles as the simplest suitable model for the study of NB formation in crowded environments (Cho and Kim, 2012a,b). The LJ particles used to coarsely mimic NB components (proteins and RNAs) have a spherical shape and interact with each other through an LJ potential, with volume-exclusive repulsions at short distances and moderate attractions at longer distances. Obviously, LJ particles are too simple to model various NB proteins that have different shapes, interactions, and compositions accurately; nevertheless, their known phase behavior agrees well with the physical features seen in NB formation, and application of this model may provide insightful information on the understanding of generic properties of NBs.

The most basic feature of NB proteins is their self-interaction to form assembled structures (Hebert and Matera, 2000). Recently, it was reported that any Cajal body protein and RNA can nucleate the formation of entire Cajal bodies *de novo* (Kaiser et al., 2008; Mao et al., 2011a; Shevtsov and Dundr, 2011). Among all possible model particles that self-interact, LJ particles are one of the simplest, and their phase behavior has physical similarities with NB formation. The phase separation of LJ particles is depicted in Fig. 4.7, which shows snapshots taken at different interaction strengths for a fixed density of LJ particles. The interaction strength is expressed in terms of a model parameter T^*, which is inversely proportional to the interaction strength. The stronger the attraction between LJ particles, the smaller T^* becomes and, therefore, T^* is referred to here as an inverse attraction parameter. Each system in Fig. 4.7 contains LJ particles in a cubic box, such that 5.0% of the total box volume is occupied by the LJ particles ($\phi_{\mathrm{LJ}}=0.050$), and their concentration is estimated as 0.068 g/ml assuming a partial specific volume of 0.73 ml/g (Brown et al., 2011). The phase diagram of LJ particles is well established, showing the range of ϕ_{LJ} and T^* at which phase separation occurs (Cho and Kim, 2012a,b). At a given ϕ_{LJ}, LJ particles phase-separate into condensed and dilute phases below $T^*=1.00$. The condensed

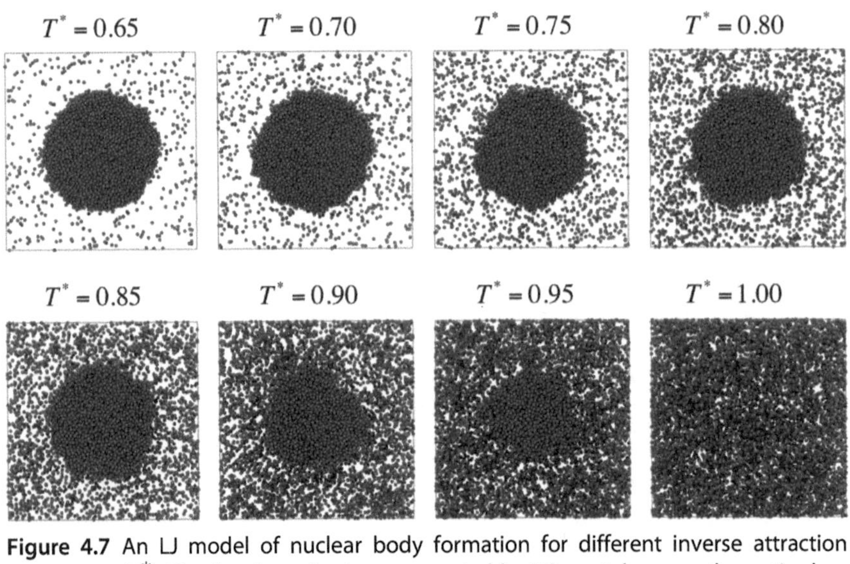

Figure 4.7 An LJ model of nuclear body formation for different inverse attraction parameters (T^*). The fraction of volume occupied by NB particles over the entire box volume ϕ_{NB} is 0.050. The lower value of T^* implies a stronger attractive interaction between NB particles. Phase separation is observed at T^* less than 1.00. (For color version of this figure, the reader is referred to the online version of this chapter.)

phase becomes denser as T^* decreases, or equivalently as the attraction becomes stronger. At a weaker attraction strength (higher T^*) the condensed phase disappears, resulting in a single phase of LJ particles. It should be noted that at a strong attraction strength where two phases coexist, there is no physical boundary between them and LJ particles keep exchanging between the two phases with no significant change in the size of the condensed phase, implying that the two phases are in dynamic equilibrium with each other.

The phase behavior of LJ particles displays common features observed in NB formation (Brangwynne, 2011; Mao et al., 2011b; Matera et al., 2009). The concentration of NB components is high in NBs, but these components also have nonzero density in nucleoplasmic space. No physical boundary, such as a membrane, is present between the NB and nucleoplasmic space and NB components exchange between these two regions. Therefore, we believe that while not perfect in terms of describing the detailed mechanisms of NB formation, LJ particles can provide an important insight into the formation and maintenance of NBs, especially in a crowded environment. Here, LJ particles are referred to as NB particles since they mimic the self-interaction of the proteins and RNAs of NBs.

To study the effect of crowding on the formation and maintenance of NBs, we first prepared a highly condensed domain of NB particles in the center of each simulation system. Then a certain number of crowding particles corresponding to a desired crowding volume fraction ϕ_c were placed in the rest of the system. The sizes of the NB particles and crowding particles were set as equal. While NB particles interact with each other through an LJ potential, crowding particles have only repulsive interactions between themselves and with NB particles. Several simulations were run at various T^* for each volume fraction of crowding particles (ϕ_c of 0.00, 0.05, 0.10, and 0.15) and the stability of the preformed condensed domain was compared at different ϕ_c values. When the condensed domain remained stable during molecular dynamics simulations, we concluded that NBs can form under the given condition of T^* and ϕ_c.

While a more realistic model of NB components would be preferred, simple LJ particles were used because of the impracticability of performing computer simulations that contained all the necessary detail at the atomistic level and the lack of structural information of key proteins in NB formation. With fast-developing computational methods and facilities, simulation studies of biomolecules have become a very powerful method to better understand the molecular details in aqueous environments. However, these studies are still limited to simulations of only a few proteins at most, and studies on aggregation, where a large number of proteins are involved, remain unachievable. Although a recent study reported a millisecond simulation of a single protein using a highly specialized computer (Shaw et al., 2010), simulations of a single protein with atomistic details can be performed for only up to a few microseconds with the supercomputers typically at hand. Studies of NB formation in the crowded nuclear environment would require simulations of several tens of thousands of NB components for a time scale longer than minutes, and such computations are not currently possible.

To overcome such difficulties in studying biomolecular aggregation using simulations, some simplified, coarse-grained models of biomolecules have been proposed. To develop successful coarse-grained models, more detailed structural information on aggregating monomers and of the aggregated structure should be available. Owing to the lack of detailed structural information, the coarse-graining approach still remains very challenging for the study of NB formation. For example, we lack structural information on important NB proteins such as survival of motor neurons and coilin, which are the key components of Cajal bodies (Hebert and Matera, 2000; Shanbhag et al., 2010). Therefore, the use of LJ particles is a reasonable first-order approach, especially when studying the physical consequences of macromolecular crowding for the assembly of NBs.

5.2. Stability of NBs under varying degrees of crowding

NB particles form either a single phase or two phases (condensed and dilute phases) depending on the inverse attraction parameter T^*. In our recent studies, we investigated the stability of self-assembled NB particles under different crowding conditions for several values of the inverse attraction parameter T^* at a fixed NB volume fraction. Two phases formed when T^* was less than a critical value, while only a single phase was observed when it was larger than this critical value (Fig. 4.8).

In Fig. 4.8, the sets of simulation point (ϕ_c, T^*) are marked as squares (with phase separation) or triangles (without phase separation), depending on the occurrence of phase separation. The critical value of T^* for phase separation increased as the volume fraction of crowding particles increased from 0.975 at $\phi_c=0.00$, 1.075 at $\phi_c=0.05$, 1.175 at $\phi_c=0.10$, and 1.325 at $\phi_c=0.15$. In the absence of crowders ($\phi_c=0.00$), phase separation of NB particles was observed when $T^*<0.975$ whereas one fluid phase was observed at larger T^*. However, in the presence of crowders ($\phi_c>0.00$), phase separation of NB particles was observed even when T^* was larger than

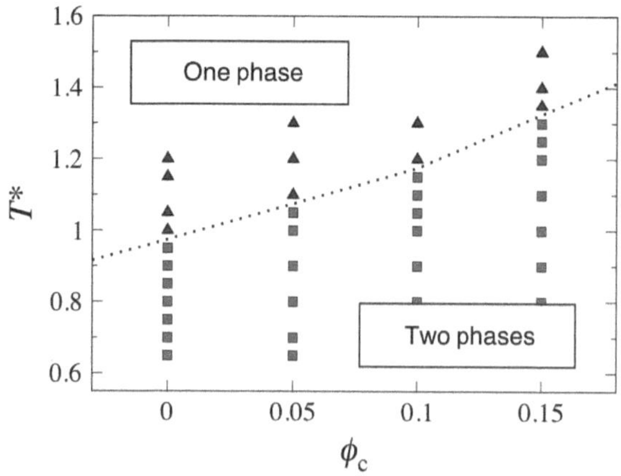

Figure 4.8 Range of T^* and ϕ_c for phase separation. *Reprinted from Cho and Kim (2012b). Copyright (2012) with permission from Elsevier.* (For color version of this figure, the reader is referred to the online version of this chapter.)

0.975. This implies that as the degree of crowding increases, phase separation into condensed and dilute phases could occur when the attraction between NB particles was weaker. In addition, the crowding-induced phase separation became more significant as ϕ_c was increased. For example, at $T^* = 1.00$, a preformed condensed domain did not remain stable in the absence of crowding particles ($\phi_c = 0.00$), but a single phase was formed. When ϕ_c was made nonzero by adding crowding particles, the condensed domain remained stable in equilibrium with the dilute phase, resulting in the formation of two phases. At $T^* > 1.20$, formation of two phases was observed only when ϕ_c exceeded 0.10. As shown in the snapshots in Fig. 4.9, phase separation of NB particles was facilitated by the presence of crowding particles: at $T^* = 1.00$, NB particles did not phase-separate in the absence of crowding particles, whereas phase separation occurred in their presence.

Our simulation results are in good agreement with the experimental observations reported by Hancock (2004). When cells were incubated in hypotonic media and became swollen, NBs disappeared and nucleolar transcription decreased by about 85%. However, when reincubated in normal nuclear buffer, NBs formed again and function was restored. These results confirmed that the disassembly and reassembly of NBs were caused by a change in the crowding condition. Interestingly, NBs were also recovered in swollen nuclei when inert crowders, such as dextran or polyethylene glycol, were added, and the nuclear environment again became crowded. These results suggest that the formation and maintenance of NBs were

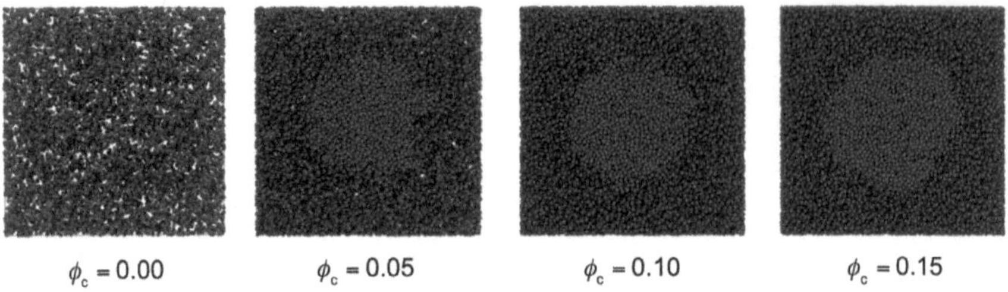

Figure 4.9 Crowding-induced nuclear body formation at a reduced temperature of $T^* = 1.00$. *Reprinted from Cho and Kim (2012b). Copyright (2012) with permission from Elsevier.* (For color version of this figure, the reader is referred to the online version of this chapter.)

largely determined by crowding-induced attractions between NB particles and not simply by their intrinsic attractive interaction. Therefore, it can be hypothesized that in the absence of crowders, the attraction strength of NB particles may not be large enough to form NBs; that is, T^* may be ≥ 1.00 so that phase separation does not occur. When crowding particles were added and the nucleus became crowded ($\phi_c > 0$), phase separation was observed at the same T^*, explaining the role of macromolecular crowding in the reassembly of NBs.

Figure 4.10A depicts the change in the volume fraction of NB particles (ϕ_{NB}) in the condensed domain with increasing crowding volume fraction (ϕ_c). The domain density increases as the degree of crowding increases, and it can be concluded that the presence of inert crowding particles confines NB particles into a smaller volume. However, the analysis of the domain size in Fig. 4.10B reveals that the size also increases with crowding and, therefore, the number of NB particles assembled into the condensed domain increases as well. Therefore, macromolecular crowding does not just confine NB particles into a smaller volume, thus increasing the density, but must also induce an effective attraction between NB particles so that more can assemble.

Direct evidence for the crowding-induced effective attraction is provided by calculating the potential of mean force, the interaction between two NB particles averaged over all degrees of freedom of the particles in the environment (Kim and Szleifer, 2010). In Fig. 4.11, only the attractive interaction between NB particles is shown as a function of their distance from each other. When the distance is less than the diameter of the NB particles (σ), the repulsive interactions dominate while at distances greater than σ, the attractive interactions dominate at short distances and approach zero at longer distances. Without crowding, the strength of the attractive

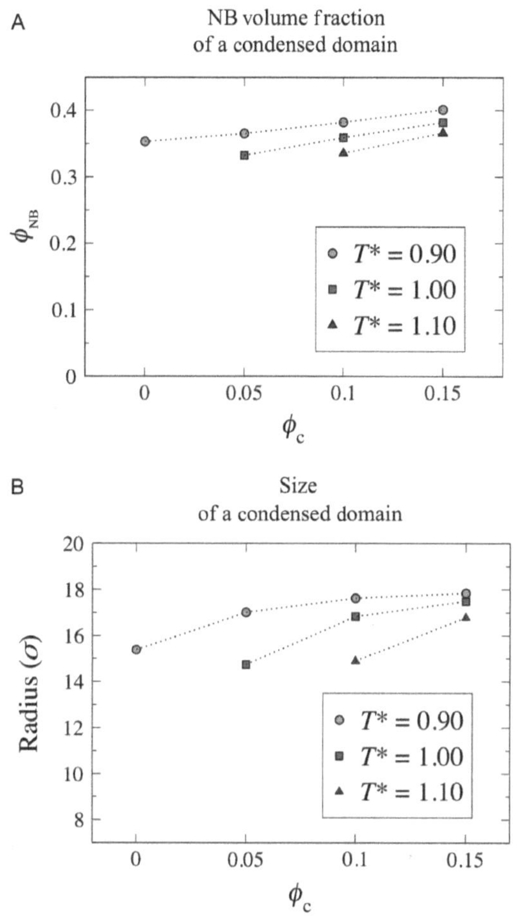

Figure 4.10 (A) Changes in NB volume fraction; (B) the size of a condensed domain with an increase in the crowding volume fraction. The radius of a condensed domain is represented in units of σ, the diameter of NB particles. *Reprinted from Cho and Kim (2012b). Copyright (2012) with permission from Elsevier.* (For color version of this figure, the reader is referred to the online version of this chapter.)

interaction was set to 1 $k_B T$ at a distance of 1.12. However, the attraction strength increased by about 0.1 or 0.2 $k_B T$ each as ϕ_c increased from 0.00 to 0.15. The increase in the effective attraction induced by crowding has been explained earlier in terms of the entropy increase. At $T^* = 1.00$, the change of attraction strength by 0.1 or 0.2 $k_B T$ was large enough to dramatically change the phase behavior. We therefore conclude that the crowding-induced phase separation was a direct result of the increased intermolecular attractions.

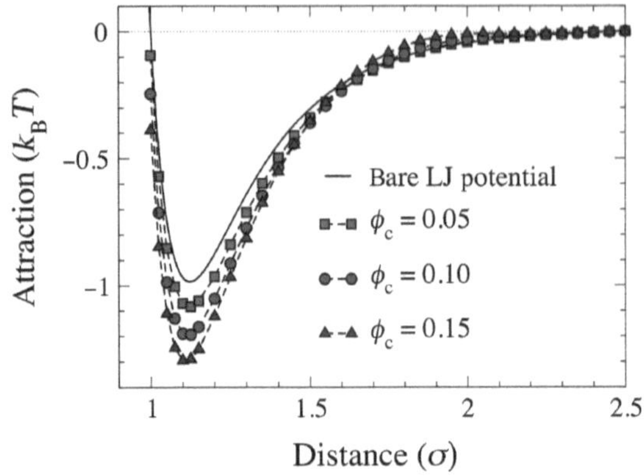

Figure 4.11 The effective potential between two NB particles at different crowder volume fractions. *Reprinted from Cho and Kim (2012b). Copyright (2012) with permission from Elsevier.* (For color version of this figure, the reader is referred to the online version of this chapter.)

5.3. Physical consequences and biological implications

In this section, we discuss the possible physical consequences and biological implications arising from a change in the crowding conditions in the nucleus. One of the major physical consequences is a change in the particle density of condensed chromatin domains. The densities in terms of volume fractions in the condensed and dilute phases at different crowding conditions are summarized in Table 4.1. As the crowding volume fraction ϕ_c increases from 0.05 to 0.15, the volume fraction of particles in the condensed domain changes from 0.333 to 0.383 and that in the dilute phase changes from 0.084 to 0.179. These increases in the volume fractions in and out of the condensed domain may have a huge impact on diffusive transport and macromolecular associations.

The transport of proteins and RNAs are crucial steps for all cellular processes, and molecules that are processed within NBs need to be exported from them for their biological roles, in many cases even out to the cytoplasm (Dundr, 2012). To investigate the effect of varying crowding conditions on the dynamics, we calculated the times required for an NB particle in the middle of the condensed domain to diffuse a given distance within or out of the domain, τ_{in} and τ_{out}. Obviously, the diffusion of NB particles in and out of the domain is delayed by the presence of crowders, as shown in Table 4.2, since diffusion becomes more difficult because of the increased particle volume fractions ϕ_{dense} and ϕ_{dilute} (Dix and Verkman, 2008). Interestingly, however, an increase in the delay caused by the same degree of

Table 4.1 Total particle volume fractions in the condensed and dilute phases, when the volume fractions of crowders ϕ_c, calculated at $T^* = 1.00$, was varied (Cho and Kim, 2012b)

ϕ_c	ϕ_{dense}	ϕ_{dilute}
0.05	0.333	0.084
0.10	0.360	0.129
0.15	0.383	0.179

Table 4.2 Increase in the normalized diffusion time to specific distances within a domain and out of a domain, τ_{in}^{normal} and τ_{out}^{normal}, calculated at $T^* = 1.00$ (Cho and Kim, 2012b)

ϕ_c	τ_{in}^{normal}	τ_{out}^{normal}
0.05	1.00	1.00
0.10	1.21	1.76
0.15	1.40	2.33

increase in crowding becomes more pronounced when the NB particles are exported out of the condensed domain than when they move within a domain (Table 4.2). The diffusion time within a domain increased by a factor of 1.40 when ϕ_c was changed from 0.05 to 0.15; on the other hand, it increased by a factor of 2.33 to reach a location out of the domain when ϕ_c was between 0.05 and 0.15. This implies that NB particles are likely to be retained for a longer time close to condensed domains under more crowded conditions.

The delay of NB particle diffusion may have both positive and negative effects on genome processes. On the positive side, delayed diffusion within an NB accelerates the association reaction between RNAs and associated proteins, as shown later. On the negative side, the delay of delivery to a functional site may slow down the overall genome process. Nature may have found optimal conditions between these extremes, and possibly this balance is perturbed under pathological conditions.

Several macromolecular interactions occur within NBs, including the assembly of U4/U6 spliceosomal di-snRNPs in Cajal bodies (Dundr, 2012; Klingauf et al., 2006). Although association also occurs in the nucleoplasmic space, Klingauf et al. (2006) suggest through their mathematical modeling approach that the rate of association increases 11-fold within Cajal bodies; they further suggest that the rate of macromolecular association events can be promoted in other types of NBs. In their study, the presence of a high content of macromolecules other than U4 and U6 RNPs was not considered. Since the particle volume fraction in and out of NBs increases because of macromolecular crowding in the nucleoplasm, we investigated the change in the rate of macromolecular association that occurs within NBs by calculating the rate constant under different crowding conditions.

The crowding effect on macromolecular associations has been studied using theoretical and computational methods with simple spherical models (Kim and Yethiraj, 2009, 2010). In our recent work, we used a theoretical expression of the rate constant from the modified Smoluchowski equation that takes into account the potential of mean force (Cho and Kim, 2012b; Zhou and Szabo, 1991). In this model, the reaction rate constant was determined based on the rate of diffusive contact between reactants k_D and the rate of product formation from the contact pair k_{prod} as follows:

$$\frac{1}{k} = \frac{1}{k_D} + \frac{1}{k_{prod}} \tag{4.1}$$

It was shown that the reaction rate can be accelerated or decelerated by crowding, varying for different types of protein associations. Such contrasting results are explained in terms of two opposing effects of crowding, the reduced k_D caused by slowed diffusion of reactants and the increased k_{prod} because of more frequent recollision between reactants caused by the increased local density. In Eq. (4.1), the inverse of the reaction rate constant k is determined by the sum of the inverse of k_D and that of k_{prod}. It can be understood from the inverse relation that between k_D and k_{prod}, the parameter with a smaller value contributes more significantly to the determination of k. That is, when the product formation from the contacting pair of reactants is slower than the rate of diffusive contact, the value of k_{prod} is much smaller than that of k_D, and then k is determined mostly by k_{prod}. In such a case, the increase in the local density of reactants under a more crowded environment results in an increase in the overall rate k as well as in k_{prod}. On the other hand, when the value of k_D is smaller than that of k_{prod}, k_D determines k. As the crowder volume fraction increases, the diffusion rate of reactants is reduced, resulting in a decrease in k as well as k_D, and macromolecular association is decelerated. In this work, a reaction model similar to that used by Klingauf et al. (2006) was employed and k_D, k_{prod}, and k values were calculated. The diffusion rate of reactants becomes slower under more crowded conditions as does k_D, which depends on the diffusion rate

Table 4.3 Prediction of the rate constant k ($M^{-1}\ s^{-1}$) in the condensed domain (NBs), at different volume fractions of crowders ϕ_c, calculated at $T^* = 1.00$ (Cho and Kim, 2012b)

ϕ_c	ϕ_{dense}	k_D	k_{prod}	k
0.05	0.333	1.77×10^7	3.48×10^5	3.42×10^5
0.10	0.360	1.48×10^7	3.88×10^5	3.78×10^5
0.15	0.383	1.24×10^7	4.27×10^5	4.13×10^5

(Table 4.3). On the other hand, the local density of reactants increases with crowding and k_{prod} also increases. It should be noted that the change in k is very similar to that of k_{prod} because it is much smaller than k_D and, thus, k_{prod} contributes more to the determination of k when the model parameters for U4/U6 spliceosomal di-snRNP assembly are used.

Macromolecular crowding in the nucleoplasm increases the particle density within NBs, which in turn is likely to increase the chance of macromolecular associations. In our simulations, when the crowder volume fraction was increased from 0.05 to 0.15, the volume fraction of the condensed domain changed from 0.333 to 0.383, and the overall reaction rate constant k increased by 21%. Albeit small, this 21% acceleration of macromolecular associations may have a significant effect on various cellular processes; if several association reactions occur in a cascade, the accelerating effect can become amplified to several powers. In conclusion, the rate of macromolecular associations was accelerated by the increase in density of the condensed domain, which was originally induced by the increase in the density of crowding particles.

The question of whether or not accelerated macromolecular association is favorable for a certain process *in vivo* cannot be answered because of the divergent roles of crowding in the stability of macromolecular complexes and the dynamics of their constituents. Genome processes are very complex, with many biochemical reactions involved, and in a genetic network, some of them may be enhanced or accelerated by crowding while others may be delayed or decelerated. We can only speculate that the regulation of the crowding condition in a timely manner during the cell cycle, sometimes more crowded and at other times less crowded, may facilitate the coordination of biological processes. To better understand the biological implications of the physical consequences induced by macromolecular crowding and their alterations, we need to quantify the concentrations of macromolecules at a specific location and time in a living cell. Such measurements have been attempted recently (Strasser et al., 2012), and these efforts should allow us to better understand the crowding effects under normal and pathological conditions, in the course of the cell cycle, or at different stages of carcinogenesis.

6. CONCLUDING REMARKS

It is widely accepted that the presence of a high content of macromolecules may induce a significant change in the structure and function of nuclear compartments. In this chapter, we first described the general effects of crowding on macromolecular association. As the degree of crowding increases, the equilibrium constant increases and an effective attraction is induced. However, an increase in the repulsive interaction at a high degree of crowding was not previously considered, which may result in dramatically different consequences. We found that the compaction of a chromosome subcompartment is induced at moderate crowding because of increased attraction, whereas the chromosome subcompartment becomes slightly decompacted at higher crowding as a result of the increased repulsion between chromosome segments (Kim et al., 2011b). This conclusion is partially supported by early experiments where no more compaction was observed at the highest crowding, and the development of a more detailed analysis of microscopic images may help to confirm this conclusion. However, it is not clear whether such a high degree of crowding is relevant to cellular environments under normal and pathological conditions. Therefore, it is of critical importance to develop quantitative tools to determine the change in crowding conditions *in vivo* with high spatial and temporal resolution.

We have also discussed the effect of crowding on the formation and maintenance of NBs using a simple spherical model whose phase behavior is similar (Cho and Kim, 2012b). Although this model may be too simplistic, it explains the experimentally observed disassembly and reassembly of NBs. Further, the effect on macromolecular transport and association reactions were also examined. The results of these simulations may help us to better understand genome processes occurring in the crowded nuclear environment, especially those associated with NBs.

Experimentally, the crowding effect of macromolecules is typically investigated by inducing volume changes in the cell nucleus and morphological changes are observed using microscopic methods. Although they have provided very useful information in understanding the effects of crowding on nuclear compartments, they are limited by the resolution of detection. Computer simulations, when combined with prior information from these experiments, can provide further physical insights into how crowding regulates the structure and function of nuclear compartments. At the moment, large scale changes in these nuclear compartments are rather limited in computer simulations and, therefore, only primitive models have been used. However, as more detailed experimental findings continue to be generated, more successful simulations can be expected in the near future.

ACKNOWLEDGMENTS

We thank Prof. Arun Yethiraj for providing simulation data for Fig. 4.1B. J. S. K. acknowledges support by the National Research Foundation of Korea (NRF) under Grant Nos. NRF-2011-0024621 and NRF-2011-220-C00030. J. S. K. also acknowledges support by the Ewha Womans University Research Grant 2011. I. S. acknowledges financial support from the National Science Foundation under Grant EFRI CBET-0937987 and Grant EAGER CBET-1249311. I. S. also acknowledges support from the National Cancer Institute of the National Institutes of Health under Award Number U54CA143869. The content is solely the responsibility of the authors and does not necessarily represent the official views of the National Institutes of Health.

REFERENCES

Albiez, J., Cremer, M., Tiberi, C., Vecchio, L., Schermeller, L., Dittrich, S., Küpper, K., Joffe, B., Thormeyer, T., von Hase, J., Yang, S., Rohr, K., Leonhardt, H., Solovei, I., Cremer, C., Fakan, S., Cremer, T., 2006. Chromatin domains and the interchromatin compartment form structurally defined and functionally interacting nuclear networks. Chromosome Res. 14, 707–733.

Asakura, S., Oosawa, F., 1958. Interactions between particles suspended in solutions of macromolecules. J. Polym. Sci. B 33, 183–192.

Bancaud, A., Huet, S., Daigle, N., Mozziconacci, J., Beaudouin, J., Ellenberg, J., 2009. Molecular crowding affects diffusion and binding of nuclear proteins in heterochromatin and reveals the fractal organization of chromatin. EMBO J. 28, 3785–3798.

Beato, M., Eisfeld, K., 1997. Transcription factor access to chromatin. Nucleic Acids Res. 25, 3559–3563.

Bickmore, W.A., Sutherland, H.G.E., 2002. Addressing protein localization within the nucleus. EMBO J. 21, 1248–1254.

Brangwynne, C.P., 2011. Soft active aggregates: mechanics, dynamics self-assembly of liquid-like intracellular protein bodies. Soft Matter 7, 3052–3059.

Brangwynne, C.P., Eckmann, C.R., Courson, D.S., Rybarska, A., Hoege, C., Gharakhani, J., Jülicher, F., Hyman, A.A., 2009. Germline P granules are liquid droplets that localize by controlled dissolution/condensation. Science 324, 1729–1732.

Brown, P.H., Balbo, A., Zhao, H., Ebel, C., Schuck, P., 2011. Density contrast sedimentation velocity for the determination of protein partial-specific volumes. PLoS One 6, e26221.

Cai, S., Lee, C.C., Kohwi-Shigematsu, T., 2006. SATB1 packages densely looped, transcriptionally active chromatin for coordinated expression of cytokine genes. Nat. Genet. 38, 1278–1288.

Carmo-Fonseca, M., Ferreira, J., Lamond, A.I., 1993. Assembly of snRNP-containing coiled bodies is regulated in interphase and mitosis: evidence that the coiled body is a kinetic nuclear structure. J. Cell Biol. 120, 841–852.
Cho, E.J., Kim, J.S., 2012a. Crowding-induced phase separation of Lennard-Jones particles: implications to nuclear structures in a biological cell. J. Phys. Chem. B 116, 3874–3879.
Cho, E.J., Kim, J.S., 2012b. Crowding effects on the formation and maintenance of nuclear bodies: insights from molecular-dynamics simulations of simple spherical model particles. Biophys. J. 103, 424–433.
Cremer, T., Cremer, C., 2001. Chromosome territories, nuclear architecture and gene regulation in mammalian cells. Nat. Rev. Genet. 2, 292–301.
Cremer, T., Cremer, C., 2006. Rise, fall, and resurrection of chromosome territories: a historical perspective. Eur. J. Histochem. 50, 223–272.
Cremer, T., Cremer, M., Dietzel, S., Müller, S., Solovei, I., Fakan, S., 2006. Chromosome territories—a functional nuclear landscape. Curr. Opin. Cell Biol. 18, 307–316.
Croft, J.A., Bridger, J.M., Boyle, S., Perry, P., Teague, P., Bickmore, W.A., 1999. Differences in the localization and morphology of chromosomes in the human nucleus. J. Cell Biol. 145, 1119–1131.
Dekker, J., Rippe, K., Dekker, M., Kleckner, N., 2002. Capturing chromosome conformation. Science 295, 1306–1311.
Dix, J.A., Verkman, A.S., 2008. Crowding effects on diffusion in solutions and cells. Annu. Rev. Biophys. 37, 247–263.
Dundr, M., 2012. Nuclear bodies: multifunctional companions of the genome. Curr. Opin. Cell Biol. 24, 415–422.
Edgar, B.A., Kim, K.J., 2009. Sizing up the cell. Science 325, 158–159.
Ellis, R.J., 2001. Macromolecular crowding: an important but neglected aspect of the intracellular environment. Curr. Opin. Struct. Biol. 11, 114–119.
Görisch, S.M., Wachsmuth, M., Tóth, K.F., Lichter, P., Rippe, K., 2005. Histone acetylation increases chromatin accessibility. J. Cell Sci. 118, 5825–5834.
Hancock, R., 2004. A role for macromolecular crowding effects in the assembly and function of compartments in the nucleus. J. Struct. Biol. 146, 281–290.
Hancock, R., 2007. Packing of the polynucleosome chain in interphase chromosomes: evidence for a contribution of crowding and entropic forces. Semin. Cell Dev. Biol. 18, 668–675.
Hancock, R., 2012. Structure of metaphase chromosomes: a role for effects of macromolecular crowding. PLoS One 7, e36045.
Handwerger, K.E., Cordero, J.A., Gall, J.G., 2005. Cajal bodies, nucleoli, and speckles in the Xenopus oocyte nucleus have a low-density, sponge-like structure. Mol. Biol. Cell 16, 202–211.
Hebert, M.D., Matera, A.G., 2000. Self-association of coilin reveals a common theme in nuclear body localization. Mol. Biol. Cell 11, 4159–4171.
Hirano, T., 2006. At the heart of the chromosome: SMC proteins in action. Nat. Rev. Mol. Cell Biol. 7, 311–322.
Kaiser, T.E., Intine, R.V., Dundr, M., 2008. De novo formation of a subnuclear body. Science 322, 1713–1717.
Kim, J.S., Szleifer, I., 2010. Depletion effect on polymers induced by small depleting spheres. J. Phys. Chem. C 114, 20864–20869.
Kim, J.S., Yethiraj, A., 2009. Effect of macromolecular crowding on reaction rates: a computational and theoretical study. Biophys. J. 96, 1333–1340.
Kim, J.S., Yethiraj, A., 2010. Crowding effects on association reactions at membranes. Biophys. J. 98, 951–958.

Kim, J.S., Yethiraj, A., 2011. Crowding effects on protein association: effect of interactions between crowding agents. J. Phys. Chem. B 115, 347–353.

Kim, J.S., Backman, V., Szleifer, I., 2011b. Crowding-induced structural alterations of random-loop chromosome model. Phys. Rev. Lett. 106, 168102.

Kim, J.S., Pradhan, P., Backman, V., Szleifer, I., 2011a. The influence of chromosome density variations on the increase in nuclear disorder strength in carcinogenesis. Phys. Biol. 8, 015004.

Klingauf, M., Stanek, D., Neugebauer, K.M., 2006. Enhancement of U4/U6 small nuclear ribonucleoprotein particle association in Cajal bodies predicted by mathematical modeling. Mol. Biol. Cell 17, 4972–4981.

Lang, F., Busch, G.L., Ritter, M., Völkl, H., Waldegger, S., Gulbins, E., Häussinger, D., 1998. Functional significance of cell volume regulatory mechanisms. Physiol. Rev. 78, 247–306.

Lieberman-Aiden, E., van Berkum, N.L., Williams, L., Imakaev, M., Ragoczy, T., Telling, A., Amit, I., Lajoie, B.R., Sabo, P.J., Dorschner, M.O., Sandstrom, R., Bernstein, B., Bender, M.A., Groudine, M., Gnirke, A., Stamatoyannopoulos, J., Mirny, L.A., Lander, E.S., Dekker, J., 2009. Comprehensive mapping of long-range interactions reveals folding principles of the human genome. Science 326, 289–293.

Maeshima, K., Eltsov, M., 2008. Packaging the genome: the structure of mitotic chromosomes. J. Biochem. 143, 145–153.

Mao, Y.S., Sunwoo, H., Zhang, B., Spector, D.L., 2011a. Direct visualization of the co-transcriptional assembly of a nuclear body by noncoding RNAs. Nat. Cell Biol. 13, 95–101.

Mao, Y.S., Zhang, B., Spector, D.L., 2011b. Biogenesis and function of nuclear bodies. Trends Genet. 27, 295–306.

Marenduzzo, D., Micheletti, C., Cook, P.R., 2006. Entropy-driven genome organization. Biophys. J. 90, 3712–3721.

Mateos-Langerak, J., Bohn, M., de Leeuw, W., Giromus, O., Manders, E.M.M., Verschure, P.J., Indemans, M.H.G., Gierman, H.J., Heermann, D.W., van Driel, R., Goetze, S., 2009. Spatially confined folding of chromatin in the interphase nucleus. Proc. Natl. Acad. Sci. U.S.A. 106, 3812–3817.

Matera, A.G., Izaguire-Sierra, M., Praveen, K., Rajendra, T.K., 2009. Nuclear bodies: random aggregates of sticky proteins or crucibles of macromolecular assembly? Dev. Cell 17, 639–647.

Minton, A.P., 1998. Molecular crowding: analysis of effects of high concentrations of inert cosolutes on biochemical equilibria and rates in terms of volume exclusion. Methods Enzymol. 295, 127–149.

Minton, A.P., 2001. The influence of macromolecular crowding and macromolecular confinement on biochemical reactions in physiological media. J. Biol. Chem. 276, 10577–10580.

Münkel, C., Eils, R., Dietzel, S., Zink, D., Mehring, C., Wedemann, G., Cremer, T., Langowski, J., 1999. Compartmentalization of interphase chromosomes observed in simulation and experiment. J. Mol. Biol. 285, 1053–1065.

Naumova, N., Dekker, J., 2010. Integrating one-dimensional and three-dimensional maps of genomes. J. Cell Sci. 123, 1979–1988.

Richter, K., Nessling, M., Lichter, P., 2007. Experimental evidence for the influence of molecular crowding on nuclear architecture. J. Cell Sci. 120, 1673–1680.

Rosa, A., Everaers, R., 2008. Structure and dynamics of interphase chromosomes. PLoS Comput. Biol. 4, e1000153.

Rosa, A., Becker, N.B., Everaers, R., 2010. Looping probabilities in model interphase chromosomes. Biophys. J. 98, 2410–2419.

Rustici, G., Mata, J., Kivinen, K., Lió, P., Penkett, C.J., Burns, G., Hayles, J., Brazma, A., Nurse, P., Bähler, J., 2004. Periodic gene expression program of the fission yeast cell cycle. Nat. Genet. 36, 809–817.

Sachs, R.K., van den Engh, G., Trask, B., Yokota, H., Hearst, J.E., 1995. A random-walk/giant-loop model for interphase chromosomes. Proc. Natl. Acad. Sci. U.S.A. 92, 2710–2714.

Shanbhag, R., Kurabi, A., Kwan, J.J., Donaldson, L.W., 2010. Solution structure of the carboxy-terminal Tudor domain from human Coilin. FEBS Lett. 584, 4351–4356.

Shaw, D.E., Maragakis, P., Lindorff-Larsen, K., Piana, S., Dror, R.O., Eastwood, M.P., Bank, J.A., Jumper, J.M., Salmon, J.K., Shan, Y., Wriggers, W., 2010. Atomic-level characterization of the structural dynamics of proteins. Science 330, 341–346.

Shevtsov, S.P., Dundr, M., 2011. Nucleation of nuclear bodies by RNA. Nat. Cell Biol. 13, 167–173.

Strasser, S.D., Shekhawat, G., Rogers, J.D., Dravid, V.P., Taflove, A., Backman, V., 2012. Near-field penetrating optical microscopy: a live cell nanoscale refractive index measurement technique for quantification of internal macromolecular density. Opt. Lett. 37, 506–508.

Subramanian, H., Pradhan, P., Liu, Y., Capoglu, I.R., Li, X., Rogers, J.D., Heifetz, A., Kunte, D., Roy, H.K., Taflove, A., Backman, V., 2008. Optical methodology for detecting histologically unapparent nanoscale consequences of genetic alterations in biological cells. Proc. Natl. Acad. Sci. U.S.A. 105, 20118–20123.

Sutherland, H.G.E., Mumford, G.K., Newton, K., Ford, L.V., Farrall, R., Dellaire, G., Cáceres, J.F., Bickmore, W.A., 2001. Large-scale identification of mammalian proteins localized to nuclear sub-compartments. Human Mol. Genet. 10, 1995–2011.

Tremethick, D.J., 2007. Higher-order structures of chromatin: the elusive 30 nm fiber. Cell 128, 651–654.

Verschure, P.J., van der Kraan, I., Manders, E.M.M., Hoogstraten, D., Houtsmuller, A.B., van Driel, R., 2003. Condensed chromatin domains in the mammalian nucleus are accessible to large macromolecules. EMBO Rep. 4, 861–866.

Zhou, H.X., Szabo, A., 1991. Comparison between molecular dynamics simulations and the Smoluchowski theory of reactions in a hard-sphere liquid. J. Chem. Phys. 95, 5948–5952.

Zhou, H.X., Rivas, G., Minton, A.P., 2008. Macromolecular crowding and confinement: biochemical, biophysical, and potential physiological consequences. Annu. Rev. Biophys. 37, 375–397.

Zimmerman, S.B., 1993. Macromolecular crowding effects on macromolecular interactions: some implications for genome structure and function. Biochim. Biophys. Acta 1216, 175–185.

CHAPTER FIVE

Phase Separation as a Possible Means of Nuclear Compartmentalization

William M. Aumiller Jr.[2], Bradley W. Davis[2], Christine D. Keating[1]
Department of Chemistry, Pennsylvania State University, University Park, Pennsylvania, USA
[1]Corresponding author: e-mail address: keating@chem.psu.edu

Contents

[2] These authors contributed equally.

International Review of Cell and Molecular Biology, Volume 307
ISSN 1937-6448
http://dx.doi.org/10.1016/B978-0-12-800046-5.00005-9

Abstract

The nucleus is perhaps the most familiar organelle within eukaryotic cells, serving as a compartment to house the genetic material. The nuclear volume is subdivided into a variety of functional and dynamic nuclear bodies not separated from the nucleoplasm by membranes. It has been hypothesized that aqueous phase separation brought about by macromolecular crowding may be in part responsible for these intranuclear compartments. This chapter discusses macromolecular solution chemistry with regard to several common types of phase separation in polymer solutions as well as to recent evidence that suggests that cytoplasmic and nuclear bodies may exist as liquid phases. We then examine the functional significance of phase separation and how it may serve as a means of compartmentalizing various nuclear activities, and describe recent studies that have used simple model systems to generate coexisting aqueous phase compartments, concentrate molecules within them, and perform localized biochemical reactions.

1. INTRODUCTION

The nucleus is organized into regions of distinct function and composition that are not delimited from the nucleoplasm by membranous boundaries. Table 5.1 provides an overview of several subnuclear structures with information on their size, number, content, and functions. The mechanisms

Table 5.1 Basic characteristics of several nuclear bodies

Body name	Number per cell	Typical size (μm)	Defining components[a]	(Putative) Functions
Nucleolus	1–4	0.5–8.0	RNA Pol I machinery	Involved in the transcription and processing of rRNA and the assembly of ribosomal subunits. Plays a role in the modification and assembly of other nuclear RNAs and RNPs. Regulates cell cycle progression by sequestering and modifying many proteins.
Cajal body	0–10	0.1–2.0	Coilin, SMN	Involved in snRNA and snoRNA modification, and assembly and trafficking of snRNPs and snoRNPs. Also plays a role in telomerase assembly and telomere length regulation.

Table 5.1 Basic characteristics of several nuclear bodies—cont'd

Body name	Number per cell	Typical size (μm)	Defining components	(Putative) Functions
Nuclear speckle	25–50	0.8–1.8	SRSF2, SRSF1, Malat1	Involved in the storage, assembly, and modification of pre-mRNA splicing factors.
Paraspeckle	10–20	0.5	PSP1, p54nrb, Men ε/β (Neat1)	Involved in nuclear retention of some A-to-I hyperedited mRNAs.
PML-nuclear body	10–30	0.3–1.0	PML	Involved in response to many forms of stress, viral defense, and genome stability by the sequestration, modification, and degradation of many partner proteins.
Nuclear stress body	2–10	0.3–3.0	HSF1, HAP	Contains satellite III ncRNAs and is a part of the general response to stress. Precise function not yet determined.
Histone locus body	2–4	0.2–1.2	NPAT, FLASH	Involved in the transcription and processing of histone pre-mRNAs.
Clastosome	0–3	0.2–1.2	19*S*, 20*S* proteasome	Contains 20*S* and 19*S* proteasomes, ubiquitin conjugates, and protein substrates of the proteasome. Forms in response to stimuli that activate proteasome-dependent proteolysis.
Perinucleolar compartment	1–4	0.2–1.0	PTB, CUGBP	Precise functions are unknown, but its prevalence positively correlates with metastatic capacity.
Polycomb body	12–16	0.3–1.0	Bmi1, Pc2	Involved in Polycomb proteins-mediated gene paring and silencing in Drosophila. Precise function in mammalian cells remains to be determined.

[a]Further information can be found in Mao et al. (2011) or Dundr and Misteli (2010).
Adapted from Mao et al. (2011), Copyright 2011, with permission from Elsevier.

by which these subnuclear compartments are formed and maintained as well as what determines their composition, size, shape, and number at various points during the cell cycle remain incompletely understood.

The most prominent nuclear compartment is the nucleolus, which is notably responsible for ribosome synthesis. Cajal bodies (CBs) and nuclear speckles are additional closely related nuclear compartments shown in Fig. 5.1. CBs are thought to be involved in posttranscriptional RNA processing and have been implicated in the biogenesis of pre-mRNA-splicing factors and the assembly of small nuclear ribonucleic proteins (Dundr, 2012; Zimber et al., 2004). Once processed, these splicing factors require further assembly and modifications within nuclear speckles, which are often found to be directly associated with CBs and are thought to serve as a storage compartment for the spliceosomal components (Spector and Lamond, 2011; Zimber et al., 2004). This colocalization of splicing factor processing and assembly with subsequent storage near active transcription sites shows a highly organized compartmentalization within the nucleoplasm. A final example, promyelocytic leukemia (PML) bodies, which are enriched in the PML tumor suppressor protein and require it to form, closely associate to chromatin and have been associated with various functions concerning cell proliferation, DNA repair, senescence, and apoptosis (Bernardi and Pandolfi, 2007). Several forms of stress known to have toxic effects on the genome induce formation of PML bodies (Dundr, 2012).

These and other nuclear bodies are characterized by the presence of a specific set of proteins, and their structure is highly dependent on protein–protein and protein–RNA interactions. The numbers of the specific compartments can range from just a few, for example, one to four nucleoli, up to several tens

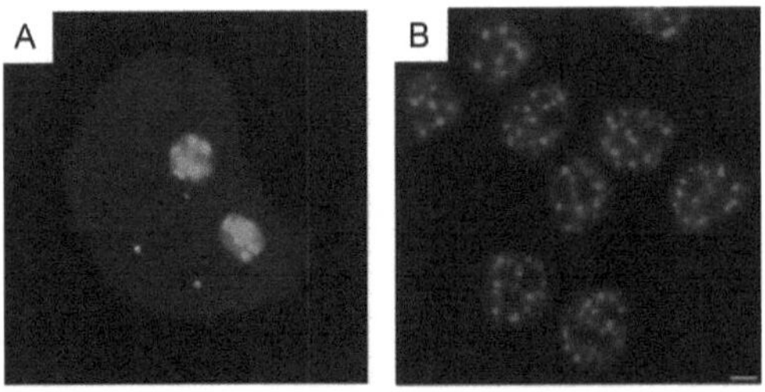

Figure 5.1 (A) Nucleoli (blue) and smaller Cajal bodies (yellow) are two prominent nuclear bodies. (B) Fluorescently labeled splicing factors localize in nuclear speckles (red). Scale bar = 5 μm. *Panel (A) From (Misteli, 2001), adapted with permission from AAAS. Panel (B): Adapted with permission from Spector and Lamond (2011). Copyright to Cold Spring Harbor Laboratory Press.* (For interpretation of the references to color in this figure legend, the reader is referred to the online version of this chapter.)

of speckles. Sizes can also vary, from hundreds of nanometers to several microns in diameter (Mao et al., 2011). The formation of nuclear compartments can be provoked by injecting known nuclear body components into the nucleoplasm. This has been done through the introduction of plasmid DNA encoding rRNA genes as well as chromatin-bound structural and functional components such as coilin into nuclei, which seeded the formation of nucleoli and CBs, respectively (Kaiser et al., 2008; Oakes et al., 1998).

Many of the subnuclear compartments are observed during specific phases of the cell cycle. For instance, upward of 30 PML bodies have been observed within nuclei and their number reaches this maximum during the G2 phase, which follows DNA replication during the S phase and precedes mitosis (Bernardi and Pandolfi, 2007). In addition, molecules in the nucleoplasm can exchange with those in the subnuclear compartments and can be targeted to them by specific peptide sequences, which illustrates that nuclear compartments are highly dynamic (Andersen et al., 2005; Fujiwara et al., 2006; Phair and Misteli, 2000). Such attributes argue against the existence of these subnuclear bodies as solid complexes and suggest instead a more open network, gel, or liquid structure. Recent observations demonstrating the liquid-like nature of nonmembranous bodies in both the cytoplasm and nucleoplasm (Brangwynne, 2011; Brangwynne et al., 2009, 2011) indicate that such structures can occur in living cells and point to the intriguing possibility that liquid–liquid phase separation could be a common mechanism for subcellular compartmentalization.

In this chapter, we discuss the physical chemistry of macromolecule solutions with an emphasis on aqueous phase separation and its possible role in nuclear organization. The nucleus harbors a myriad of biomacromolecules such as proteins and nucleic acids with an approximate total macromolecular concentration of 100 mg/mL (Daban, 2000; Hancock, 2004b). Given the concentration and molecular diversity of macromolecules in the nucleus, this environment can be expected to exhibit the effects of macromolecular crowding, for example, changes in macromolecular association equilibria and enzymatic rates. In addition, and of particular interest for this chapter, such crowded solutions can undergo phase separation to yield multiple coexisting aqueous phases.

2. MACROMOLECULE SOLUTION CHEMISTRY

To simplify interpretation and to avoid wasting precious biomacromolecular reagents, biochemicals and their reactions have typically been studied in idealized, dilute buffer solution conditions (typically less than 1 mg/mL of nucleic acids, proteins, small molecules, etc.) (Minton, 2001). This approach has been productive, and is responsible for the major part of our understanding of biomolecule structure, function, and associations (Alberts, 2008; Lehninger et al., 2013). In recent years it has become clear that the crowded intracellular environment in which these macromolecules typically act can both quantitatively and qualitatively change the outcome of important biochemical reactions ranging from conformational states to association equilibria to enzymatic activities (Zhou et al., 2008). The importance of performing studies under more realistic solution conditions is increasingly appreciated, particularly with regard to mimicking the environment of the cytoplasm (Ellis, 2001; Hall and Minton, 2003); these concerns also apply to nucleoplasm. To better understand the solution chemistry of the nucleoplasm and its compartments, it is necessary to first briefly review the physical chemistry of macromolecule solutions. The presence of macromolecules at relatively high concentrations, termed "macromolecular crowding," leads to a considerable excluded volume as well as chemically attractive and repulsive forces between solutes (Zhou et al., 2008). Consequently, such solutions are not thermodynamically ideal, and solute chemical activities can differ greatly depending on the solute concentration.

2.1. Thermodynamic nonideality

The assumption of thermodynamic ideality, while generally reasonable for dilute solutions, is invalid for macromolecularly crowded media such as the nucleoplasm. Ideal solutions are defined to have an enthalpy of mixing equal to zero. In addition, it is assumed that the total volume does not change upon mixing. This is analogous to the assumptions made when dealing with ideal gases (i.e., the volume of the individual gas molecules can be ignored and interactions between the molecules are negligible). Consider the chemical potential μ_A of a solute, A, in a solution. It can be expressed using the thermodynamic activity a_A in the following equation:

$$\mu_A = \mu_A^0 + RT \ln a_A$$

where μ_A^o is the standard state chemical potential, R is the gas constant, and T is the temperature. The chemical activity of a solute is related to its concentration by the activity coefficient,

$$a_A = \gamma_A c_A$$

where, γ_A is the activity coefficient of solute A and c_A is the concentration of the solute. The activity coefficient can be further defined as

$$\gamma_A = \exp\left(\frac{\langle f_A \rangle}{k_B T}\right)$$

where $\langle f_A \rangle$ is the free energy change associated with the equilibrium free energy of interaction between a molecule A and all other solutes, k_B is the Boltzmann constant, and T is the temperature (Atkins and De Paula, 2010; Ellis, 2001; Minton, 1983).

In ideal solutions, the activity or effective concentration is assumed to be equal to the concentration of solute A ($a_A = c_A$ and $\gamma_A = 1$). Chemical interactions between all species in solution are assumed to be identical. This assumption can introduce large errors when dealing with concentrated mixtures of macromolecules. Enthalpy of mixing is not negligible when interactions between different pairs of molecules (solute A–solute A, solute B–solute B, A–B, A–solvent, B–solvent) are substantially different because of differences in intermolecular forces between them from van der Waals, electrostatic, and hydrogen bonding. In solutions of macromolecules, the volume occupied by solutes can also be a large fraction of the total volume of solution (e.g., up to 30%). Some simple predictions can be made: If interactions are repulsive, and/or volume exclusion dominates, γ_A will be greater than 1, resulting in increased chemical activity for solute A; attractive interactions can result in γ_A being greater or less than 1, depending on the magnitude of excluded volume effects (Minton, 1983).

2.2. Macromolecular crowding

Macromolecular crowding refers to the effects of adding macromolecules to a solution, as compared to a solution containing no macromolecules. Substantial differences are observed for a wide range of biomacromolecular interactions and activities in dilute vs. crowded solutions (Zhou et al., 2008). This is interpreted as the combined outcome of both general types of interactions introduced earlier: (1) excluded volume, which is a result of inaccessibility of molecules to occupy space because of the presence of background macromolecules, and (2) chemical effects, which are caused by the attraction and repulsion between molecules. It is not necessary that any one solute should be present in high concentrations for the solution to be crowded. For example, no single protein, nucleic acid sequence, or other polymer is necessarily present in high concentration in a particular intracellular environment; rather, it is the cumulative effect of all macromolecules that renders it crowded.

A significant consequence of volume exclusion is that the thermodynamic activity, *a*, of a biological macromolecule of interest in crowded conditions can increase orders of magnitude over its value in the absence of crowding agents as illustrated in Fig. 5.2 (Ellis, 2001; Zhao et al., 2008; Zhao et al., 2011). An early example of this was the case of hemoglobin (Hb). At a concentration of 200 g/L in solution, the activity coefficient, γ, is about 10. At 300 g/L (similar to the concentration in a normal red blood cell), γ rises to 100 (Minton, 1983). At this Hb concentration, a considerable fraction of the total volume is occupied by Hb (for comparison, at 350 g/L Hb, excluded volume is 30% of the total volume; Hall and Minton, 2003). A simple geometrical model using a hard quasi-sphere that factors in the available volume was developed to account for this dependence, and it agreed well with the crystal structure of Hb (Ross and Minton, 1977). Since then, numerous studies have found substantial effects caused by crowding, and in many cases models based on volume exclusion have provided satisfactory qualitative explanations. Crowding agents and reactants are often modeled as hard spheres, which is useful for identifying trends. More elaborate models such as the direct simulation coarse-grained model by Cheung

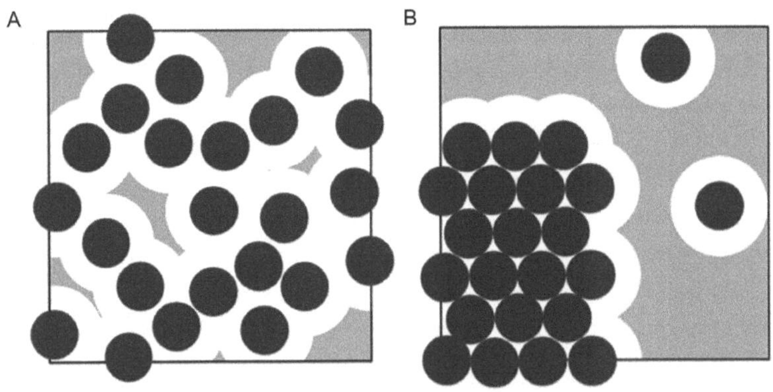

Figure 5.2 The effect of volume exclusion on chemical activity can be visualized within the volume encompassed by a square containing macromolecules (dark spheres). Another macromolecule can be added only if the core is located in the accessible volume (gray). (A) When macromolecules freely diffuse, the accessible volume is smaller and, therefore, the chemical potential is higher. (B) If the macromolecules associate with each other, for example, by assembly or aggregation, the accessible volume increases and chemical potential decreases. *Adapted with permission from Zhao et al. (2011) © IOP Publishing. Reproduced by permission of IOP Publishing. All rights reserved.*

et al. (2005) and atomistic simulations by Feig and Sugita (2012) have been developed, and future models will offer more insight into the effects by quantitatively describing chemical interactions (Elcock, 2010; Zhou, 2013).

Predictions for crowding effects caused by volume exclusion alone favor more condensed structures over less condensed forms. For example, equilibria for folding, association, polymerization, and aggregation reactions shift toward the multimeric from the monomeric forms. However, the chemical effects of crowding can either enhance stabilization of native, folded states or actually lead to unfolded states. Nonspecific repulsive interactions that occur as a result of minimizing surface exposure (such as during protein folding and association) lead to a stabilizing effect and these can enhance the effects of volume exclusion. However, nonspecific attractive interactions of a protein with the surroundings can have quite the opposite effect and result in destabilization, which can completely negate the stabilizing effects of volume exclusion (Minton, 2013; Wang et al., 2012) Miklos and coworkers found that macromolecular crowding agents lysozyme and bovine serum albumin destabilized the structure of the enzyme barley chymotrypsin inhibitor 2, the opposite of what would be expected as a result of volume exclusion alone (Miklos et al., 2011). They interpret this as a result of nonspecific interactions with the protein crowding agents. Record and coworkers examined the effects of both chemical interactions and excluded volume using polyethylene glycol (PEG) of varying chain lengths (i.e., a single ethylene glycol unit up to MW 20,000) and quantitatively describing the effects of each on duplex and hairpin formation within DNA (Knowles et al., 2011). Oligoethylene glycols, up to 200 Da, had a destabilizing effect on both hairpin and duplex formation, because of favorable interactions between the PEG and DNA bases. Increasing the molecular weight of the PEG (and therefore increasing the effect of excluded volume) caused stabilization of the duplex and less destabilization of the hairpin. Stabilization was a result of inaccessibility of the DNA to the interior PEG subunits of the PEG polymer at higher molecular weight and a decrease in volume fraction to favor DNA duplex and hairpin formations.

Because of the complexity of the combined entropic and enthalpic aspects of macromolecular crowding, which can be additive or opposing, it is not yet possible to make quantitative predictions for the impact of crowding on arbitrary reactions of interest. It is, nonetheless, important to bear these effects in mind when considering the nuclear milieu because the same types of phenomena at play in macromolecular crowded solutions *in vitro* can be expected to be important. In addition to the effects described earlier, an additional consequence of macromolecular crowding can be the formation of distinct coexisting aqueous phases. In the next section, we introduce aqueous phase separation and consider how it could contribute to the organization of the nucleus.

3. AQUEOUS PHASE SEPARATION

Solutions that contain one or more polymers at sufficient concentrations commonly phase-separate to form two or more coexisting aqueous phases (Albertsson, 1986; Wohlfarth, 2013). This is a relatively general phenomenon that occurs for a wide range of different polymeric solutes, including proteins and nucleic acids. Several types of phase separations are possible, and they can be classified based on the polymers being polyelectrolytes or otherwise, a salt being present in high concentration, and the number of phases being observed. Since the coexisting phases have different properties, biomolecule partitioning among these phases is possible and could provide a means of compartmentalization.

The nucleic acids, as polyanions, can be expected to undergo condensation and, potentially, phase separation in the presence of polycationic cosolutes (Estévez-Torres and Baigl, 2011; Zimmerman, 2006). Phase separation in concentrated protein solutions is a well-known phenomenon (Dumetz et al., 2008; Tolstoguzov, 2000; Vekilov, 2012; Wang et al., 2013). Proteins in the nucleus carry a variety of surface charges but generally a lower charge density than nucleic acids. Major classes of phase separation in aqueous polymer solutions that are most relevant for the intracellular environment are considered later. They are necessarily simpler than anything one might anticipate in the nucleus but provide a framework for understanding the types of demixing one can anticipate in crowded polymer solutions. We refer the interested reader to more comprehensive sources for additional information on types of phase separation not described here, including polymer-surfactant and polymer-alcohol systems and polymer-salt (Hatti-Kaul, 2000; Wohlfarth, 2013; Zaslavsky, 1995).

3.1. Polyelectrolyte-rich phases

Solutions in which one or more polyelectrolytes are present can, under the right conditions, phase-separate to form polyelectrolyte-rich phases, referred to as coacervates. The coacervate phase is a relatively small, dense, and polymer-rich liquid phase, and is accompanied by a dilute phase of much larger volume, which is referred to as the equilibrium phase or supernatant phase. The coacervate droplets can be differentiated from precipitates or aggregates that form in solutions of similar composition, because the latter are solids rather than liquids and hence will not exhibit spherical shapes, flow, etc. (de Kruif et al., 2004).

Simple coacervation refers to systems in which a single polyelectrolyte species phase-separates in the presence of a small molecule or salt. For example, type-B gelatin (a polyampholyte obtained from denatured collagen) forms a simple coacervate when alcohols are added, because of decreased solvation (Mohanty and Bohidar, 2003). The phase behavior of the gelatin-alcohol system depends strongly on ionic strength and pH, with a single-phase, liquid–liquid coexistence, and aggregation of all possible outcomes depending on the composition of the solution. This is typical for many different polyelectrolytes of synthetic or biological origin. Biologically important polyanions such as DNA or RNA can also form aggregates in the presence of condensing agents such as polyamines (such as spermine, spermidine), cobalt hexamine, other cationic polymers such as polylysine, highly basic proteins (histones), multivalent cations (Ca^{2+} and Mg^{2+}), neutral polymers, and alcohols (Bloomfield, 1997; Pelta et al., 1996; Tang and Szoka, 1997). These agents generally work by inducing precipitation by a few means: decreasing DNA repulsions by screening the charge of the phosphate, reordering water structure, or making DNA–solvent interactions less favorable (Bloomfield, 1997). Ethanol-induced precipitation of DNA is not unlike the alcohol-induced coacervation and, eventually, the aggregation observed for simple coacervates.

Complex coacervation involves two or more macromolecules that are oppositely charged. The charge density of the macromolecule(s) must be large enough for electrostatic interactions to keep the droplets from dissolving into the equilibrium phase to form a homogenous solution, but not large enough to induce precipitation of solid aggregates. A scheme of complex coacervate formation with a generalized representation the polyanion and polycation macromolecule is given in Fig. 5.3A. Optical microscopy images of complex coacervates with poly(L-glutamic acid) and poly(L-lysine) as the phase-forming components are given as phase droplets in Fig. 5.3B (Priftis et al., 2012). Figure 5.3C shows for comparison a solid precipitate of poly(acrylic acid) and poly(allylamine) (Chollakup et al., 2013).

The driving force for complex coacervate formation has been described as predominantly entropically driven. As the oppositely charged macromolecules interact, counter ions that were once associated with the

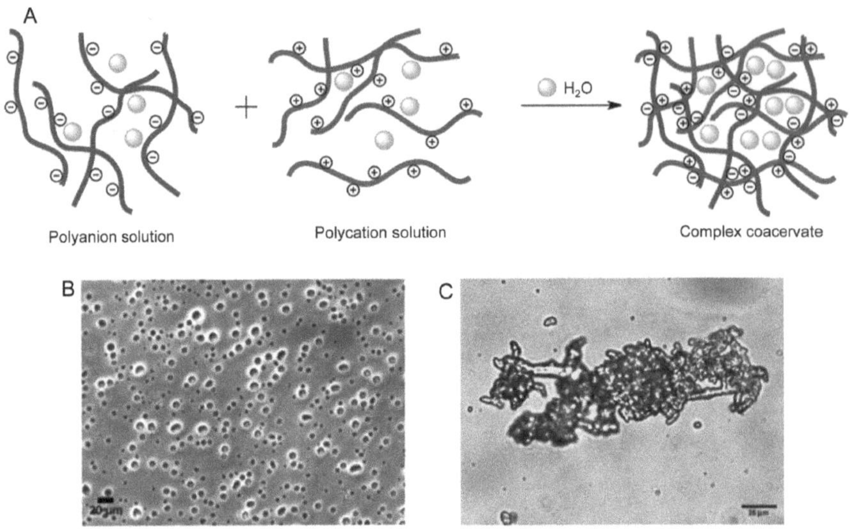

Figure 5.3 (A) General scheme of complex coacervate formation of oppositely charged polyelectrolytes. (B) Phase droplets of a poly(L-lysine) and poly(L-glutamic acid) complex coacervates. (C) Solid precipitates of poly(acrylic acid) and poly(allylamine). *Panels (A) and (B): Adapted with permission from Priftis et al. (2012). Copyright 2012, American Chemical Society. Panel (C): Adapted with permission from Chollakup et al. (2013). Copyright 2013, American Chemical Society.* (For the color version of this figure, the reader is referred to the online version of this chapter.)

macromolecule are released (de Kruif et al., 2004). Reports also suggest that it is partially enthalpically driven, because of the electrostatic free energy of the system (de Kruif and Tuinier, 2001; Turgeon et al., 2003). The phase behavior of complex coacervate systems has been studied since their original discovery, with a few key general principles that apply to coacervate systems. Mainly, the complex phase as a whole is nearly electrically neutral, yet highly polarizable (de Kruif et al., 2004). For example, Burgess and Carless used microelectrophoretic measurements on complex coacervates of gelatin and acacia and found that the optimum pH for coacervation (that which leads to the greatest volume of the coacervate phase) was equal to the electrical equivalence pH (net charge = 0) (Burgess and Carless, 1984). Other studies also suggest that there is a fixed stoichiometry determined by the charge density of the phase-forming components at a specific pH (de Kruif et al., 2004). Salt can also affect the phase formation; for every coacervate, there exists a critical salt concentration at which the two-phase system will become a homogenous solution as a result of charge screening. This value is experimentally determined and varies with the polyanion/polycation pair increasing as the polymer chain lengths and charge densities are increased (Priftis and Tirrell, 2012). For example, for a charge-matched poly(L-lysine)/poly(L-glutamic acid) mixture, the critical salt concentration increased from 300 mM NaCl for a chain length $N=30$ to ~1000 mM NaCl for $N=400$ (Priftis and Tirrell, 2012).

Given the importance of polyelectrolytes such as RNA, DNA, and cationic polypeptides/proteins in the nucleus, it is reasonable to suspect that aqueous phase separation based on macromolecular charge may be at play, but straightforward phase separation based on coacervation alone does not adequately describe the nucleus since none of the subnuclear compartments constitute a dilute, supernatant phase.

3.2. Coexisting polymer-rich phases

The entire nuclear volume is macromolecularly crowded and hence, despite the prevalence of polyelectrolytes in the nucleus, the simplified examples of coacervation described earlier—which provide one crowded and one dilute phase—do not entirely capture this milieu. Aqueous phase separation can also lead to two or more separate crowded phases, each enriched in one of the polymers (Albertsson, 1986). The process involves enthalpic interactions between polymers, changes in hydration, and entropic contributions that can dominate because of macromolecular crowding. Developing models that can predict phase behavior and solute partitioning has been an active area of research, because of interest in using these systems for bioseparations and the empirical manner in which separation protocols are generally determined. (da Silva and Loh, 2000; Guan et al., 1993a,b; Johansson et al., 1998; Mohite and Juvekar, 2009; Pessoa and Mohamed, 2004; Yu et al., 1993). One of the most widely studied aqueous phase systems of two polymers is that formed from PEG and dextran; it undergoes phase separation above a few weight percent of each polymer. An aqueous two-phase system (ATPS) of 10% (w/w) PEG 8 kDa and 10% (w/w) dextran 10 kDa is shown in Fig. 5.4. The upper phase, stained blue, is the PEG-rich phase, and the lower colorless phase is the dextran-rich phase. Mechanical agitation of the ATPS disperses droplets of one phase within the other (in this case, PEG-rich phase droplets are suspended in a continuous dextran-rich phase). Other phase systems composed of synthetic polymers, polysaccharides, or proteins have also been developed and studied, and have found uses in biotechnology for biomolecule separation (Hatti-Kaul, 2000; Zaslavsky, 1995). The process is not limited to biphasic systems; multiphase systems can be

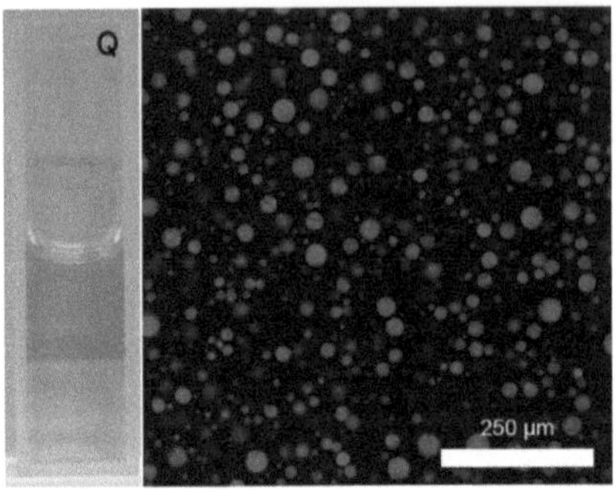

Figure 5.4 An aqueous two-phase system of polyethylene glycol and dextran with Alexa 647 labeled PEG within a cuvette (left) and confocal microscope image after agitation (right). (For the color version of this figure, the reader is referred to the online version of this chapter.)

generated by addition of more phase-forming components (Albertsson, 1986), and it should be noted that multiple types of phase separation can occur in the same solution (e.g., coacervates and nonionic polymer-rich).

4. NUCLEAR COMPARTMENTS AS CROWDED AND DYNAMIC STRUCTURES

Nuclear bodies are highly dynamic structures that exist in various numbers and sizes, and exhibit the ability to disassemble/assemble relative to the transcriptional state of the nucleus (Dundr and Misteli, 2010). In this section, we discuss how the crowded nuclear environment could induce formation of nuclear compartments and how the structure of an aqueous phase compartment is consistent with their dynamic nature.

4.1. Macromolecular crowding within nucleus

Recent experiments by Hancock suggest a role for macromolecular crowding in nuclear organization (Hancock, 2004b). When isolated nuclei were incubated in buffer, their volume increased by slightly more than two-fold, diluting the nuclear interior, which resulted in the disappearance of nucleoli and PML bodies. The process could be reversed and the nuclear bodies reformed if expanded nuclei were then incubated with either 12% PEG 8 kDa or 12% dextran 10.5 kDa, which can diffuse across the nuclear membrane and throughout the nuclear volume (Fig. 5.5). The polymeric macromolecular crowding agents led to similar numbers of nucleoli and PML bodies being formed as observed in untreated nuclei. In addition,

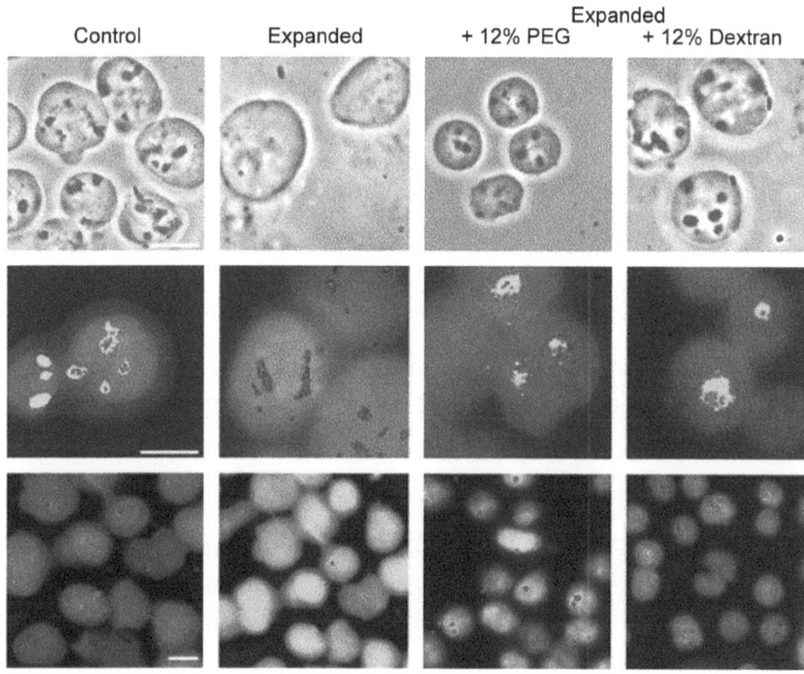

Figure 5.5 (Top) Phase-contrast images of nuclei. Nuclear compartments dissociate upon expansion in buffer, but reassemble when incubated with either 12% PEG 8 kDa or 12% dextran 10.5 kDa. (Middle) Double immunofluorescence images of two nucleolar proteins, nucleolin (green) and a subunit of RNA polymerase I (red), with surrounding DNA (blue). (Bottom) Fluorescently labeled PML protein (red) localizes to PML bodies upon incubation with crowding agents. Scale bars = 10 μm. *Reprinted from Hancock (2004a) with permissions from Elsevier.* (For interpretation of the references to color in this figure legend, the reader is referred to the online version of this chapter.)

nucleoli regained on average 50%, and as much as 80%, of their ability to incorporate radiolabeled UTP into rRNA transcripts after reformation (Hancock, 2004b).

Other researchers have investigated the accessibility of nuclear bodies to fluorescent solutes to acquire a better understanding of their structure. Handwerger and coworkers injected fluorescent dextrans of molecular weights ranging from 3–2000 kDa into the cytoplasm or directly into the nuclei of *Xenopus* oocytes, which are relatively large germinal vesicles that serve as model platforms to study nuclear bodies. The individual nuclear compartments (nucleoli, CBs, and speckles) excluded fluorescent dextrans to different degrees, with the larger molecular weight polymers increasingly excluded (Fig. 5.6). Interferometric microscopy techniques were used to

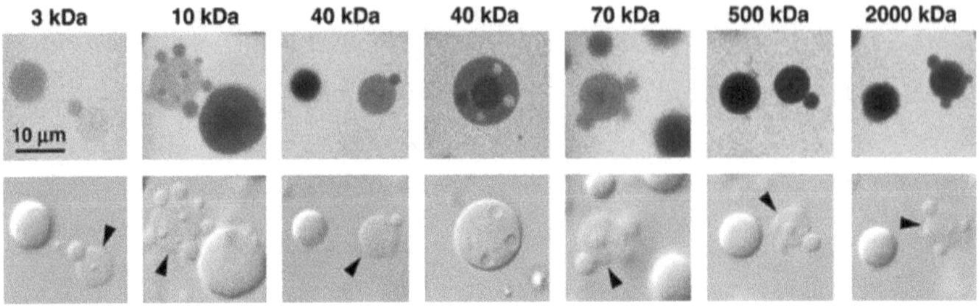

Figure 5.6 Fluorescent images (top) and optical images (bottom) of fluorescently labeled dextrans of increasing molecular weights that were incubated from a *Xenopus laevis* oocyte. Bright regions represent high concentrations of fluorescent dextran. Lower molecular weight dextrans readily diffused into nucleoli and Cajal Bodies (arrows) while larger dextrans were excluded. *Reprinted with permission from Handwerger et al. (2005)*

measure the refractive indices of nuclear bodies so that their protein densities could be estimated. Nucleoplasm was found to have an approximate protein concentration of 0.106 g/mL while the densities of speckles, CBs, and the dense fibrillar component of nucleoli were found to be 0.162, 0.136, and 0.215 g/mL, respectively (Handwerger et al., 2005). All of the described nuclear compartments had a higher protein density than the surrounding nucleoplasm, presumably because of the concentration of their associated components within these structures.

4.2. Dynamic structure of nuclear compartments

Besides the studies with fluorescent dextrans described earlier, the ability of labeled nuclear components to diffuse in and out of various nuclear compartments has been extensively studied with fluorescent microscopy, especially utilizing various photobleaching techniques such as fluorescent recovery after photobleaching (FRAP). These studies have revealed that nuclear bodies rapidly exchange some of their major components. Notable examples include fibrillarin and coilin, two major components of nucleoli and CBs, respectively (Dundr et al., 2004; Phair and Misteli, 2000). Interestingly, although fibrillarin concentrations are approximately 4.6 times higher in the nucleoli than the nucleoplasm, its diffusion coefficient is only approximately 10 times slower (Phair and Misteli, 2000). These studies along with several others underscore the dynamic nature of nuclear bodies, which exhibit a rapid flux of their major components (Dundr and Misteli, 2001, 2010; Pontvianne et al., 2013).

Brangwynne and coworkers recently demonstrated that the nucleoli of *Xenopus* oocytes behave as liquid droplets (Brangwynne et al., 2011). When nucleoli fuse, the volume is conserved and a spherical shape is maintained, as one would expect when phase droplets coalesce. The spherical nature was attributed to an apparent surface tension that acts to minimize the surface area of the nucleolus. This surface tension is approximately 10 μN/m and is roughly an order of magnitude higher than the surface tension of P granules, which are cytoplasmic germline granules of *Caenorhabditis elegans* that are also enriched in RNA and RNA-binding proteins and exhibit liquid droplet behavior as well (Brangwynne et al., 2009, 2011). Both of these nuclear and cytoplasmic RNA-protein compartments display surface tensions that fall in the expected regime for macromolecular liquids. From these surface tension estimates, they were able to conclude that nucleoli display an apparent viscosity of approximately 10^6 cP, which they relate to the viscosity of honey, and it is 1000 times more viscous than that of comparable P granules (Brangwynne et al., 2009, 2011). Figure 5.7 shows the fusion

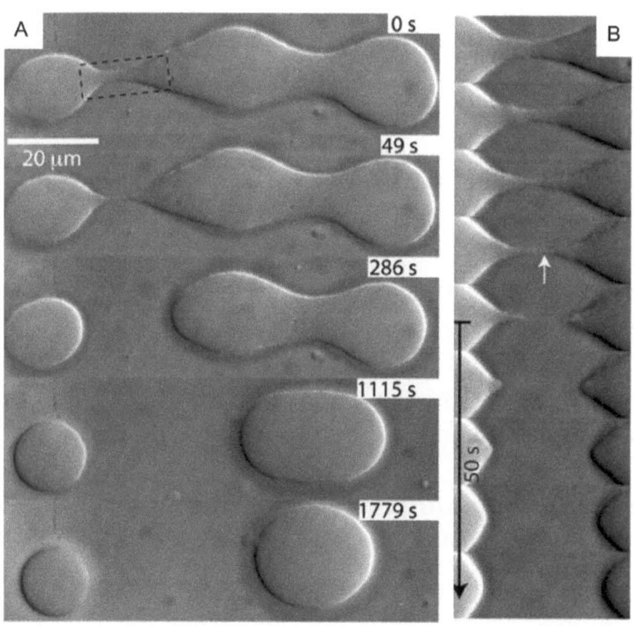

Figure 5.7 DIC images of nucleoli from a *Xenopus laevis* oocyte displaying liquid-like behavior. Three nucleoli have begun to fuse (A) in the first time frame. Two of the nucleoli coalesce while the other pinches off, which is magnified to the right (B). *Adapted with permission from Brangwynne et al. (2011). Copyright 2011, National Academy of Sciences, USA.*

and fission of nucleoli. The fluid-like behavior of nucleoli permits the coalescence of two nucleoli while a third displays a characteristic "pinch-off" behavior of viscous liquids as the nucleoli separate (Brangwynne et al., 2011).

Although not yet confirmed to display this behavior, other nuclear bodies such as CBs are known to exhibit distinct spherical morphologies that could suggest an effective surface tension with fluid-like properties as well (Brangwynne, 2011). Like nucleoli, CBs are able to undergo both fusion and fission events within the nucleoplasm (Platani et al., 2000). With mounting evidence that nuclear organization may be induced by phase separation, it is of interest to consider the possible consequences of forming intranuclear phases.

5. POTENTIAL FUNCTIONAL SIGNIFICANCE OF PHASE SEPARATION FOR NUCLEAR COMPARTMENTALIZATION

In this section, we consider the possible consequences of aqueous phase separation with respect to nuclear organization. It must be remembered that the fact that conditions seem right for phase separation does not necessarily mean that it will occur. For example, the eye lens cytoplasm contains three main proteins, the α, β, and γ-crystallins, whose phase behavior has been extensively studied because when phase separation occurs *in vivo,* it leads to a loss of lens transparency (i.e., cataract) (Dorsaz et al., 2011; Liu et al., 1996). Phase separation can readily be induced *in vitro,* but *in vivo* it occurs only in the disease state. Nonetheless, the liquid nature of the nucleolus strongly suggests phase separation for at least this subnuclear organelle and highlights the possibility that if the same experimental techniques were brought to bear on additional subnuclear compartments, we may learn that one or more of them are also liquid organelles.

5.1. Mechanism for compartmentalization by partitioning between coexisting phases

The coexistence of multiple phases with differing compositions, viscosities, densities, and so on means that solutes may accumulate more in one of the phases over the others, leading to differences in local concentration between the phases. The equilibrium distribution of a solute in an ATPS is described in terms of the partitioning coefficient, K, which is the concentration of the solute in the top phase (C_{top}) relative to that in the bottom phase (C_{bottom}):

$$K = \frac{C_{\text{top}}}{C_{\text{bottom}}}$$

Many factors influence how a solute partitions in an ATPS, including the properties of the polymer phases such as polymer identities, structures, molecular weights, concentrations, pH, ionic strength as well as the properties of the solute including the size, hydrophilicity/hydrophobicity, and any interactions with the ATPS components (Albertsson, 1986; Ruiz-Ruiz et al., 2012). Protein partitioning studies in the PEG/dextran ATPS have demonstrated that native proteins usually partition to the more hydrophilic dextran phase, while denatured proteins partition to the hydrophobic PEG-rich phase. In the denatured state, more internal hydrophobic amino acids of the protein are exposed (Dominak et al., 2010). Charge and size play a role in partitioning, as charged species are likely to interact with any charged components of the ATPS, and large molecules have larger areas of interaction with the phase-forming components (Albertsson, 1986; Walter et al., 1985). For example, RNA partitioning in a PEG/dextran ATPS is strongly length dependent, with $1/K = \sim 40$ and ~ 3000 for 15 and 159 nt RNAs, respectively (Strulson et al., 2012; Fig. 5.8). Coacervates composed of poly(diallyldimethylammonium) chloride (PDDA) and adenosine triphosphate (ATP) were also able to sequester globular proteins such as green fluorescent protein (GFP) at a 86-fold higher concentration within the coacervate phase droplet as compared to the surrounding phase (Williams et al., 2012).

Considering the possibility of coexisting phases in the nucleus, we can anticipate that a solute's partitioning behavior would be dominated by binding interactions with molecules already localized in the phase. These binding partners could be the protein and nucleic acids responsible for its formation, or another solute that partitioned strongly to the phase. The number of binding sites and strength of the interactions would lead to different degrees of localization to the phase compartment and residence times within it. Affinity interactions are routinely used to increase partitioning for bioseparations in a synthetic polymer ATPS. For example, polymers can be modified with affinity tags such as Ni-NTA or avidin (Ruiz-Ruiz et al., 2012).

Localization by partitioning into different aqueous phase compartments offers a mechanism for maintaining distinct molecular composition in the different types of nuclear bodies, while also exhibiting rapid turnover of these components and permeability to exogenous macromolecules such as

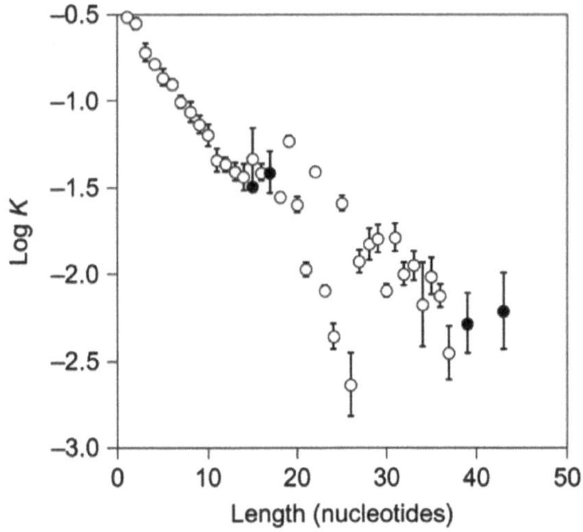

Figure 5.8 The partitioning of RNA to the dextran-rich phase of a PEG:dextran ATPS increases with RNA length. Various lengths of the hammerhead ribozyme enzyme strand were introduced to a 10% PEG 8 kDa and 16% dextran 10 kDa ATPS. *Adapted with permission from Macmillan Publishers Ltd.,* Nature Chemistry *(Strulson et al., 2012), Copyright 2012.*

the fluorescent dextrans. It has been hypothesized that colocalization of enzymatic activity may serve as a general means of metabolic regulation (Ovádi and Saks, 2004; Walter and Brooks, 1995; Wilson and Gitai, 2013). Colocalization of cellular activity reduces intermediate transient time between sequential processing events while increasing local intermediate concentrations to enhance catalytic efficiency (Ovádi and Srere, 2000; Ovádi et al., 1989). These theories have thus far been focused more on applications within the cytoplasm or organelles; however, such principles are equally valid within the nucleoplasm because of the comparable environments.

5.2. Particulates collected at aqueous/aqueous interfaces

For particulates added to an aqueous bi- or multiphasic system, partitioning occurs between not only the coexisting phases but also the interface(s) between them. For example, cell organelles, macromolecular assemblies, denatured proteins, and whole cells may accumulate at the interface of an ATPS (Albertsson, 1986; Nguyen et al., 2013). Nonbiological particulates such as polystyrene latex beads and metallic nanoparticles have also been assembled at aqueous/aqueous interfaces (Balakrishnan et al., 2012; Firoozmand et al., 2009; Helfrich et al., 2005; Tripp et al., 1996). The most straightforward explanation for this behavior is that interfacial accumulation reduces the interfacial tension. The interfacial tensions of aqueous phases are quite low, often on the order of 0.5–500 μN/m, as compared to 72 mN/m for air/H_2O or 18 mN/m for hexane/water (Zaslavsky, 1995). For the lower end of this range, stabilizations from adsorbing even a relatively large object such as a several micron diameter sphere at an aqueous/aqueous interface are small compared to $k_B T$. It is likely, therefore, that one or more additional mechanism(s) are at play in these systems, arising from macromolecular crowding effects in the aqueous phases.

Trends in interfacial tension for varying ATPS compositions can be identified. The interfacial tensions in an ATPS for nonionic polymer mixtures, for example, PEG and dextran of various molecular weights, are generally on the order of one to several hundred μN/m, with the trend that increasing polymer concentration and/or molecular weight of the polymers causes an increase in the interfacial tension (Forciniti et al., 1990; Liu et al., 2012). Polyelectrolyte-rich phases are very sensitive to changes in charge screening. For example, in coacervates of the positively charged polymer poly(2-(methacryloyloxy)-ethyl-trimethylammonium chloride and the negatively charged poly(3-sulfopropyl methacrylate), interfacial tension decreased from ~400 μN/m at 400 mM potassium chloride to ~50 μN/m at 1 M KCl (Spruijt et al., 2010).

6. EXPERIMENTAL MODEL SYSTEMS FOR CROWDED, PHASE-SEPARATED MICROCOMPARTMENTS

Thus far, we have discussed how the intrinsically crowded environment of the nucleoplasm may give rise to aqueous phase separation. Several studies have illustrated how macromolecular crowding induces the formation of nuclear bodies and how these microcompartments are accessible to a variety of macromolecules (Cho and Kim, 2012; Richter et al., 2008). Despite important work on the content, function, and physical properties of the nuclear bodies, much remains to be learned. Fundamental understanding of how these compartments form and function *in vivo* is still lacking because of the inherent complexity of studying them in living cells. On the other hand, a great deal of experimental and theory/computational work has been devoted to understanding aqueous polymer solutions and their phase behavior, including work meant to help design systems for optimized solute partitioning and to understand the phase behavior of these systems as a function of parameters such as structure, concentration, ionic strength, pH, and the presence of other solutes (Benavides and Rito-Palomares, 2008; Hatti-Kaul, 2000; Zaslavsky, 1995). Nonetheless, there remains a substantial gap between the *in vivo* studies and the polymer solution work. Later we describe several efforts made toward bridging this gap by moving toward more biologically relevant polymer model systems, studying the effect of molecular compartmentalization on biochemical reactions, or encapsulating coexisting aqueous phases inside cell-sized lipid vesicles.

6.1. Physical properties of some relevant phase systems

The idea that phase separation may be important for intracellular organization dates back many years, and has cycled into and out of favor (Luby-Phelps, 2013; Wilson, 1899). Microscopic observation of the cell and the physical chemistry of macromolecular solutions seems to have kept bringing this idea back in more sophisticated forms over the years (Welch and Clegg, 2010). Some of the earliest experimental work aimed at reconstituting some aspect of the cellular milieu through phase separation was done by Oparin, who, for example, prepared RNA/histone coacervate droplets that enzymatically produced polyadenylic acid (Oparin et al., 1963). Since then, however, the vast majority of studies on aqueous phase separation and the properties of the resulting phases have focused purely on their physical chemistry and/or their use in bioseparations. As the importance of macromolecular crowding in living cells has begun to be appreciated, interest in understanding the biological implications of coexisting aqueous phases has grown.

Our lab has used the well-characterized PEG and dextran system to form coexisting aqueous-phase compartments as a simple experimental model for intracellular crowding and compartmentalization (Keating, 2012). PEG and dextran are both nonionic polymers and will phase separate into a PEG-rich upper phase and a dextran-rich lower phase. Both phases exhibit macromolecular crowding; for example, a typical PEG 8 kDa/dextran 10 kDa ATPS may have around 30% (w/w-dextran and <5% (w/w) PEG in the dextran-rich phase and around 15% (w/w) PEG and a few weight percent dextran in the PEG-rich phase (Cacace and Keating, 2013; Strulson et al., 2012). Because the polymers are unevenly distributed between the phases, they exhibit different chemical and physical properties (hydrophobicities, viscosities, etc.). For example, in a 10% PEG 8 kDa and 10% dextran 10 kDa

ATPS, the viscosity of the dextran-rich phase is nearly twice that of the PEG-rich phase. Differences in phase composition lead to differences in biomolecular reaction rates for reactions performed in the two phases and to unequal distributions of solutes between the two phases, as described later (Cacace and Keating, 2013; Strulson et al., 2012).

Differences between the two phases of an ATPS can be quite extreme, for example, in the case of complex coacervation, where both polyelectrolytes concentrate into the droplet phase. For example, Mann and coworkers reported that when PDDA and ATP are mixed in water at 25 mM each component, the concentration of ATP was approximately 320 mM in the resulting coacervates and less than 20 mM in the supernatant phase. PDDA was also greatly concentrated in the coacervate phase, which had an apparent viscosity of 900 cP and a density of 1.1 g/L (Williams et al., 2012). The viscosity of coacervate phases increases with the strength of attraction between the macromolecules (Weinbreck et al., 2004). This can be varied with pH for polyelectrolytes that have protonatable groups; for the whey protein/gum arabic system, pH 4 gives maximal electrostatic interaction and hence the highest viscosity coacervates (~4000 cP). It should be noted that viscosity measurements in these model systems are not yet common and an effort has not been made to reach the high viscosity of specific intracellular compartments (e.g., nucleoli (10^6 cP)) (Brangwynne et al., 2009, 2011).

Mann and coworkers have investigated polycation-nucleotide and cationic peptide-nucleotide coacervates as a membrane-free protocell model. Protocells are often considered as a means of exploring the possible molecular origins of modern cells and/or for biotechnological applications that require some functional aspect of living cells but that would benefit from greater simplicity or robustness. These investigations also provide valuable insight into the potential outcomes of important physicochemical mechanisms that are known to be important for biomacromolecules and may shed light on intracellular compartmentalization such as the nuclear bodies. Koga et al. produced coacervates with nucleoside triphosphates and up to a 24 kDa poly-α-L-lysine peptide of ~115 monomer units. At 50 mM peptide and nucleotide concentrations, phase separation occurred, with compositions of 1.6 M and 400 mM in the coacervate phase, respectively, whereas the supernatant had concentrations of submillimolar and 18 mM, respectively (Koga et al., 2011). These authors also demonstrated repetitive growth and disassembly of the coacervate phase via pH cycling, as illustrated in Fig. 5.9 (Koga et al., 2011).

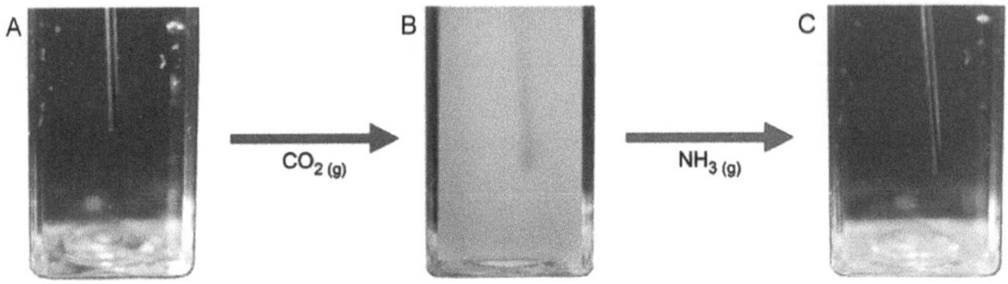

Figure 5.9 Reversible coacervation of poly(L-lysine) and ATP. (A) A solution at pH 12.5 does not form coacervates. (B) Upon treatment with gaseous CO_2, turbidity indicates coacervation. (C) Subsequent exposure to gaseous NH_3 induced the disassembly of coacervates. *Adapted with permission from Macmillan Publishers Ltd.,* Nature Chemistry *(Koga et al., 2011), Copyright 2011.*

McKnight and coworkers have studied the formation and properties of hydrogels of RNA and RNA-binding proteins that have low-complexity sequences (Han et al., 2012; Kato et al., 2012). Hydrogel droplets were formed by a phase transition from soluble forms of the low-complexity sequence containing proteins in the presence of microcrystals of a chemical, biotinylated isooxazole; the authors postulated that the crystals facilitated a conformational change in the proteins to form amyloid-like fibers. The formation of hydrogels, while not a focus of this chapter, is another possible outcome in crowded macromolecule solutions, particularly when crosslinking interactions between one or more components are possible. This work serves to remind us that not all phase separation events in biomacromolecule mixtures need lead to coexisting liquid phases, and that both liquid and gel microcompartments could well coexist as different functional bodies in the cytoplasm or nucleoplasm.

6.2. Compartmentalization induced by phase separation

Several studies have shown enhanced enzymatic activity through compartmentalization. Strulson and coworkers recently utilized an ATPS to enhance RNA cleavage via microcompartmentalization (Fig. 5.10). Variants of the hammerhead ribozyme were composed of a separate enzyme and substrate strand that induced the ribozyme to self-cleave upon association. As a result of partitioning, the RNA was localized into dextran-rich phase droplets by as much as 3000-fold over its concentration in the PEG-rich continuous phase. With a constant total volume and amount of RNA added, the local concentration of RNA depended on the volume ratio for the two phases, with smaller volumes of the dextran-rich phase leading to higher local

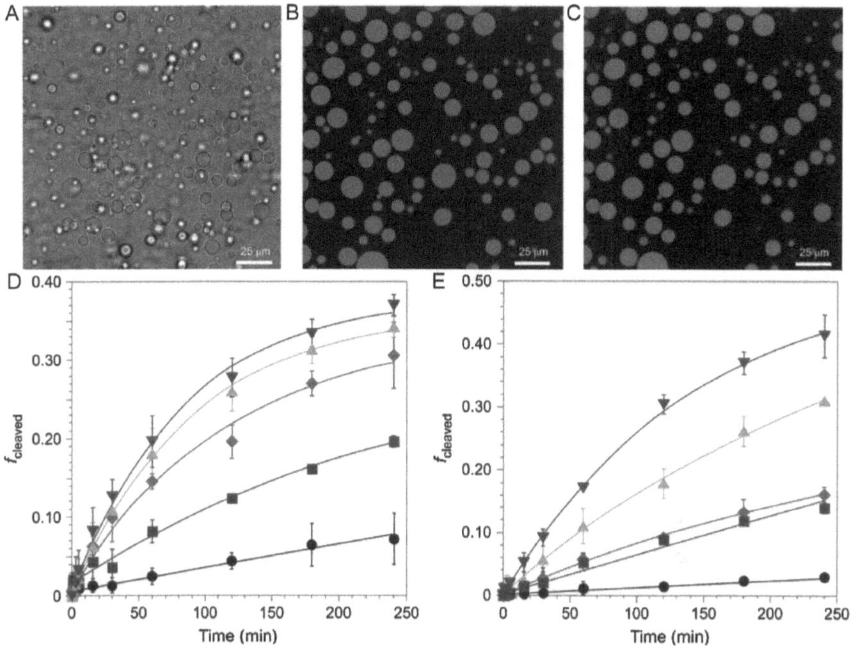

Figure 5.10 Confocal images of a PEG and dextran ATPS (A) with fluorescently labeled dextran (B). Fluorescent RNA localizes to the dextran-rich phase droplets (C). By varying the dextran:PEG volume ratio in 10 mM Mg^{2+} (D) and 0.5 mM Mg^{2+} (E), the concentration of the enzyme and substrate RNA strands within the dextran-rich phase increases, thus enhancing catalysis. Dextran:PEG volume ratios included were 1:0 (black circles), 1:5 (blue squares), 1:12.5 (red diamonds), 1:50 (blue triangles), and 1:100 (inverted green triangles). *Adapted with permission from Macmillan Publishers Ltd.,* Nature Chemistry *(Strulson et al., 2012), Copyright 2012.* (For interpretation of the references to color in this figure legend, the reader is referred to the online version of this chapter.)

RNA concentrations and, consequently, increasing the cleavage rate. Compartmentalization increased reaction rates only when the reaction was performed under subsaturating conditions where rate was proportional to the concentration of the enzyme strand; this was demonstrated using ribozyme variants with different K_M. Rate enhancements as high as 66-fold were observed, with increases tunable by the phase volume ratio (Strulson et al., 2012). This work highlights how local concentrations of enzyme, in this case the enzyme strand of a two-piece ribozyme, can have dramatic effects on rate within liquid-like phase droplets. It also points to a possible functional significance of the total volume of any given phase as a means of tuning local concentration and, hence, reaction rates without varying the partitioning preference of the molecule, which could be relevant to nuclear body sizes/numbers.

Increased rates for protein enzymes have also been observed upon compartmentalization in aqueous phases. Koga et al observed a twofold rate increase for an ATP-dependent reaction involving hexokinase when the enzyme was encapsulated within poly(L-lysine)/ATP coacervate droplets as compared to bulk solution (Koga et al., 2011). Colocalization of sequential enzymes was accomplished by incorporating the minimal complex of actinorhodin polyketide synthase (PKS), which contains three enzymatic functions, into the PDDA/ATP coacervate system. The minimal PKS complex was sequestered into the coacervate droplets, as was its substrate malonyl-CoA, leading to an 18-fold increase in product yield as compared to control assays with similar concentrations of either PDDA or ATP present (Crosby et al., 2012).

Sokolova and coworkers concentrated cell lysate droplets suspended in an oil phase by osmotic removal of water to increase the rate of transcription and translation reactions. Concentration led to phase separation within the lysate to produce a concentrated, biomolecule-rich coacervate phase. When recombinant GFP was produced in the lysate, 81% of it partitioned into the coacervate phase (Fig. 5.11). The authors then measured the transcriptional rate of the commercially available *in vitro* transcription kit used to produce GFP and observed a five to six times transcription rate constant enhancement as well as an increase in translation compared to single-phase controls. Interestingly, this observed rate is similar to that of the in vivo rate for

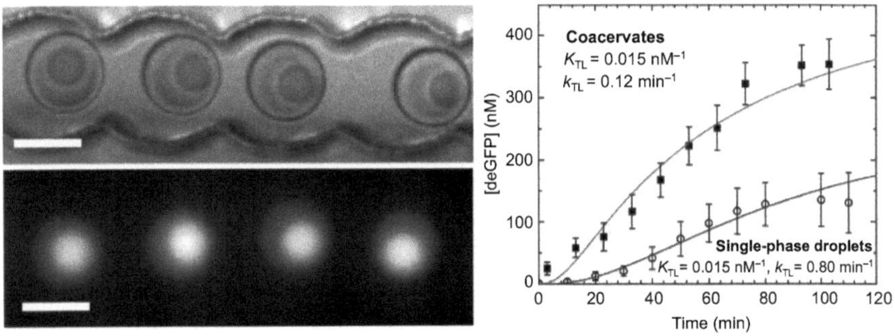

Figure 5.11 (Left) Optical microscope images of concentrated droplets that form a coacervate phase (top) with corresponding fluorescence image (bottom) of produced GFP that partitions to the coacervate phase. (Right) GFP expression over time in coacervates droplets (solid squares) compared to single-phase controls (open circles). *Adapted with permission from Sokolova et al. (2013), Copyright 2013, National Academy of Sciences, USA.*

E. coli, which further supports the idea that such crowded conditions may be necessary for efficient transcription (Sokolova et al., 2013).

Another possible advantage of compartmentalization is control over the location of a reaction. Cacace and Keating demonstrated that even a weakly-partitioned enzyme ($K=0.12$) could lead to nearly quantitative product formation only in the dextran-rich phase of a PEG/dextran ATPS (Cacace and Keating, 2013). This work used urease to catalyze formation of the mineral calcium carbonate by locally producing carbonate anion in the dextran-rich phase. Under optimized conditions of enzyme concentration and volume ratio, <1 mol% mineral could be detected in the PEG-rich phase despite the only 8.3-fold higher concentration of urease in the dextran-rich phase and the fact that Ca^{2+} (*aq*) was initially present in equal concentrations in both phases (Cacace and Keating, 2013). Mineral formation is not relevant to the nuclear bodies; however, the ability to restrict product formation to a compartment that lacks a membranous barrier is potentially as important as increasing local reaction rates.

Phase transitions can be readily initiated in an ATPS by changes in dilution, temperature, ionic strength, or pH; however, it is unlikely that these stimuli are responsible for the assembly/dissolution of intracellular or intranuclear phases. More reasonable options could include changes in binding affinity, conformation, or local concentration of a key component of the phase compartment. Phosphorylation states of transcription factors within the nucleolus correlate with its assembly/disassembly (Hernandez-Verdun, 2006), suggesting a possible trigger for a phase transition. The formation of CBs has also been tied to the phosphorylation state of coilin, one of its major protein components known to self-associate (Dundr, 2012). It is of particular interest, then, that a recent study by Li and coworkers has shown that the degree of phosphorylation is a viable means of controlling phase separation through multivalent interactions between signaling proteins. The researchers engineered two types of proteins, one of which was composed of multiple SRC homology 3 (SH3) domain repeats and the other contained repeats of its known proline-rich motif (PRM) binding domain. They were chosen because of their roles in the nephrin–NCK–N-WASP system in which three interacting components are responsible for actin-regulation. The SH3 domain of nephrin contains several phosphorylation sites that bind to the SH2 domain of NCK, which contains multiple SH3 domains that may then bind to the PRMs of N-WASP. The multivalent engineered proteins readily formed phase droplets in Fig. 5.12 that on the macroscale exhibited liquid-like properties, yet on the microscale

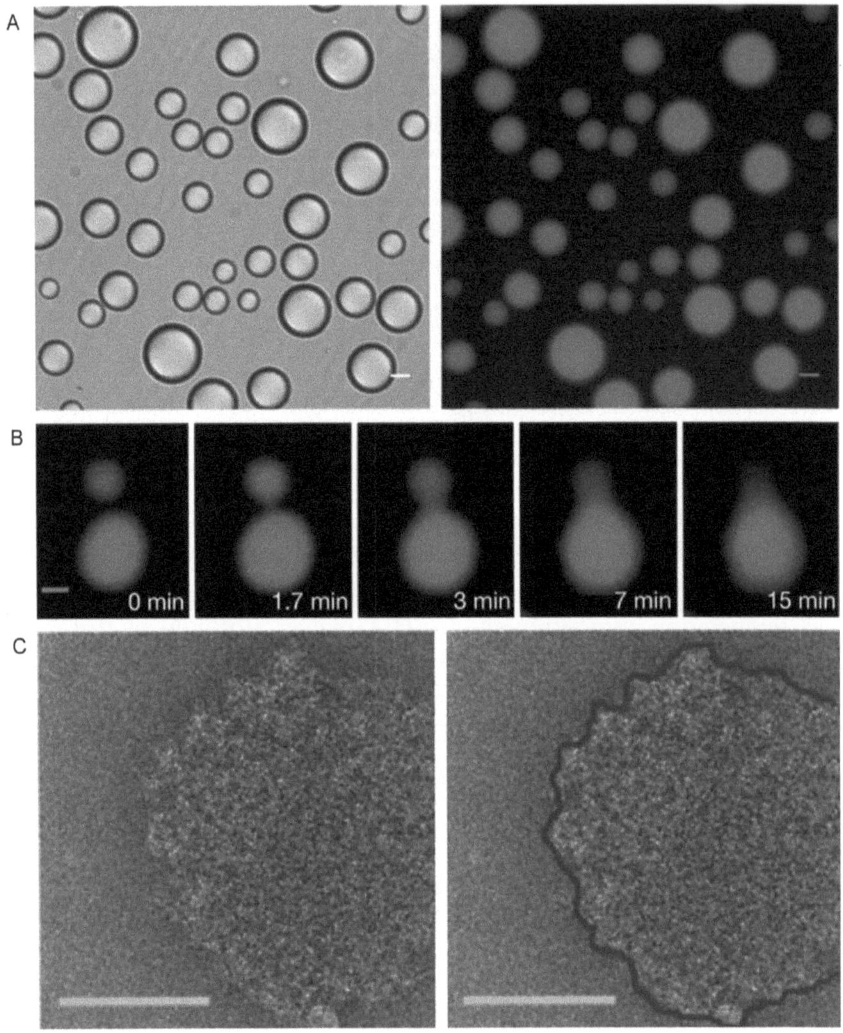

Figure 5.12 (A) DIC images of phase droplets formed with multivalent protein domains (left) with corresponding wide-field fluorescence images of Oregon Green-labeled protein (right). Scale bar = 20 μm. (B) Two-phase droplets fuse over time, thus displaying macroscopic liquid properties. Scale bar = 10 μm (C) Cryo-electron microscope images of a phase droplet displaying microscale structural characteristics of a sol-gel (left) with boundary outlined in red (right). Scale bar = 100 nm. *Adapted with permission from Macmillan Publishers Ltd.,* Nature *(Li et al., 2012), Copyright 2012.* (For interpretation of the references to color in this figure legend, the reader is referred to the online version of this chapter.)

displayed a structure similar to that of sol-gels (Li et al., 2012). Therefore, altering the phosphorylation state of a few components within a nuclear body may present a viable means of regulating phase transitions to permit rapid reorganization of nuclear compartments.

6.3. Encapsulating ATPS within lipid vesicles

PEG/dextran ATPS have been encapsulated within giant lipid vesicles (GVs, generally >10 μm in diameter) to generate microcompartments into which solutes can be localized by partitioning (Fig. 5.13). Thus far, the focus of this work has been to think of the encapsulated ATPS more as a model for cytoplasm than nucleoplasm; however, several observations for these GV-encapsulated ATPS are also instructive with respect to nuclear bodies. First, although in bulk studies phase droplets can be easily maintained by agitation of the sample, inside of the GVs, they ultimately coalesce such that there are only two phase compartments (one for each phase) rather than one dispersed in the other. This differs from the nuclear bodies and may result from their anchoring to or nucleation from specific sites within the nucleus, which could prevent their coalescence with other bodies of the

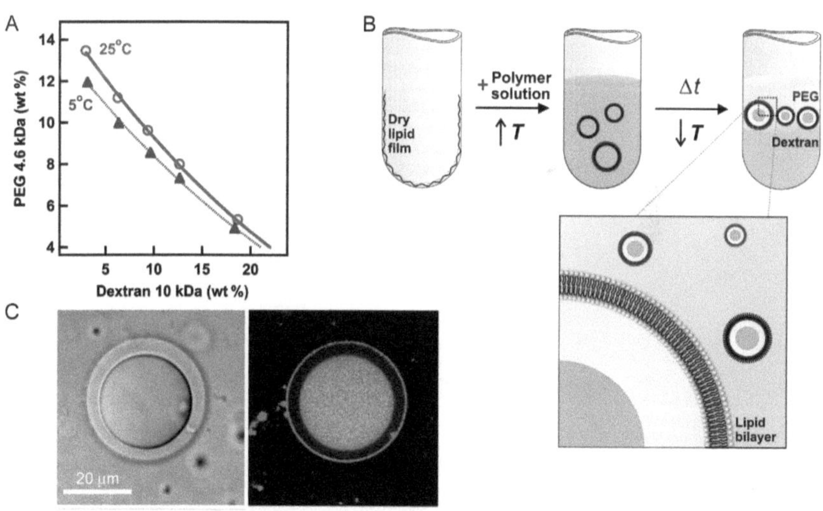

Figure 5.13 Aqueous two-phase systems encapsulated within lipid membranes. (A) Phase diagram showing temperature dependence of an ATPS. (B) Lipid films can be hydrated with a warm polymer solution that phase-separates within vesicles upon cooling. (C) Fluorescently labeled dextran (blue) partitions to the dextran-rich phase within the vesicle. *Panels (A) and (B): Adapted with permission from Long et al. (2005). Copyright 2005, National Academy of Sciences, USA. Panel (C): Reprinted with permission from Keating (2012). Copyright 2012, American Chemical Society.* (For interpretation of the references to color in this figure legend, the reader is referred to the online version of this chapter.)

same type (Matera et al., 2009). Surface affinity anchoring has been shown to stabilize nonspherical patterns of PEG/dextran ATPS on planar solid surfaces (Kojima and Takayama, 2013).

A second important factor for a GV-encapsulated ATPS is the possibility of interactions with the membrane, the strength of which will differ for the two aqueous phases. When the membrane itself also contains coexisting phase domains (i.e., liquid-ordered, L_o, and liquid-disordered, L_d, regions of a ternary DOPC/DPPC/cholesterol system), the aqueous interior phases are preferentially coated by the more favorable lipid phase domains (Cans et al., 2008). Figure 5.14 illustrates this for a budded ATPS structure in

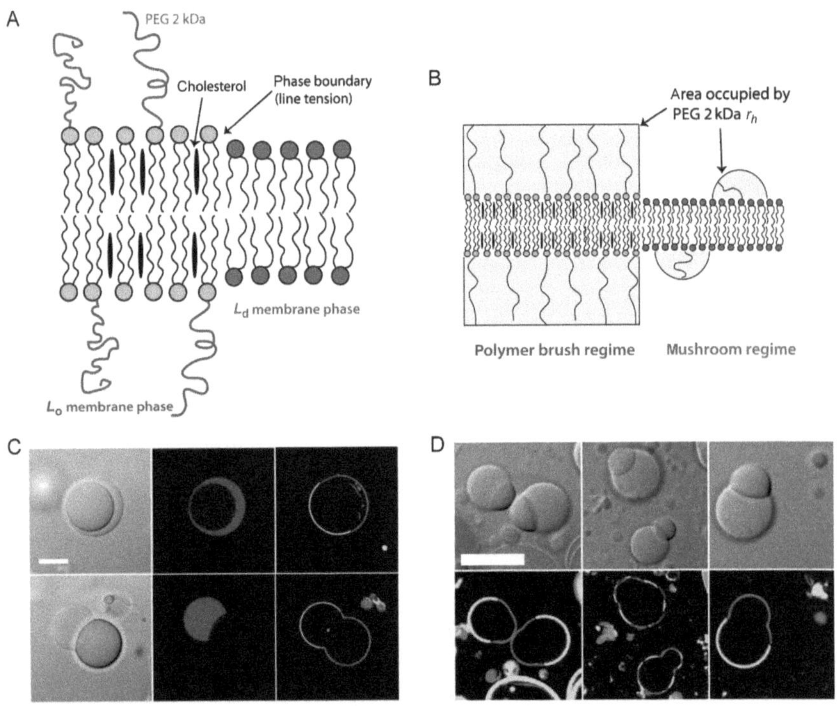

Figure 5.14 (A) Coexisting phase domains within a lipid membrane. (B) PEGylated lipids are more densely packed within the L_o domain (green) than the L_d domain (red). (C) Optical microscope images of vesicles containing ATPS before and after the budding transition. The PEG-rich phase (blue) wets the more PEGylated L_o domain of the membrane. Scale bar = 10 μm. (D) Effect on membrane domains when exposed to a temperature increase and subsequent cooling. Scale bar = 50 μm. *Reprinted with permission from Cans et al. (2008). Copyright 2008, American Chemical Society. Reprinted with permission from Keating (2012). Copyright 2012, American Chemical Society.* (For interpretation of the references to color in this figure legend, the reader is referred to the online version of this chapter.)

which the PEGylated L_o lipid domain preferentially wets the PEG-rich phase, while the L_d domain wets the dextran-rich phase. No specific binding partners were needed to drive this interaction between the aqueous polymer-rich phases on the inside of the GV and the membrane domains on the outside. We can expect aqueous-phase droplets to nucleate on, and associate with, other structures for which they have affinity, perhaps including the nuclear envelope or other intranuclear assemblies.

A third observation for the GV-encapsulated ATPS is that reorganization in the form of phase transitions or protein repartitioning is relatively rapid, occurring on a timescale of minutes or less. An example of a temperature-induced phase transition can be seen in Fig 5.15A (Helfrich et al., 2002). Encapsulated ATPS were reversibly converted between a single phase and numerous microcompartments that coalesce over time. This process also dynamically controls biomolecule localization through partitioning in these phase-separated environments as highlighted in Fig. 5.15B for the protein concanavalin A, which localizes into the dextran-rich droplets (Long et al., 2005). The phase transition in aqueous phases shows a striking resemblance

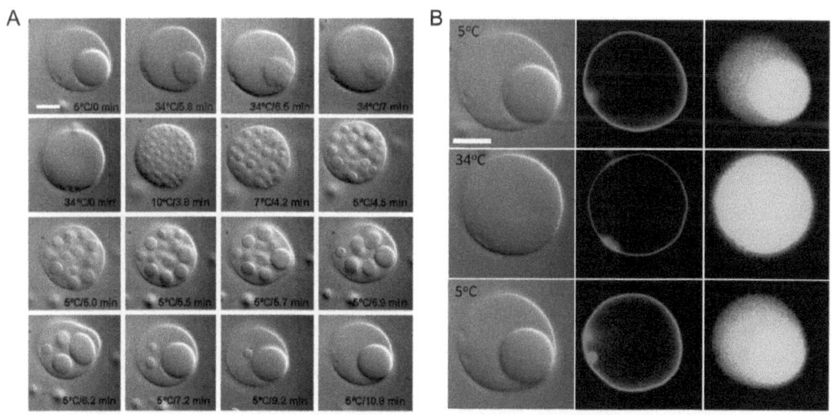

Figure 5.15 (A) Optical microscopy images illustrating reversible compartmentalization within vesicles. Upon heating, the ATPS forms a homogeneous mixture of polymers and after subsequent cooling, separate phases reform and several dextran-rich phase droplets coalesce. (B) A fluorescently labeled protein, concanavalin A, partitions to the dextran-rich phase. Upon the heat-induced phase transition, the protein distributes evenly throughout the vesicle before repartitioning after cooling and phase separation. Scale bars = 10 μm. *Adapted with permission from Long et al. (2005). Copyright 2005, National Academy of Sciences, USA. Reprinted with permission from Keating (2012). Copyright 2012, American Chemical Society.* (For the color version of this figure, the reader is referred to the online version of this chapter.)

to the dissociation and fusion of nuclear bodies as observed throughout the cell cycle within a semipermeable membrane. Temperature is not required to induce phase transitions; osmotic dilution/concentration has also been demonstrated (Andes-Koback and Keating, 2011; Dimova and Lipowsky, 2012; Long et al., 2008). Protein relocalization has also been demonstrated in response to a change in protein conformation. This did not require any disassembly of the phase compartments but rather took advantage of pH-induced protein conformational changes to alter the partitioning preference of proteins. For example, denatured human serum albumin partitioned to the PEG-rich phase; however, upon renaturation at pH 6.5, it partitions instead to the dextran-rich phase, as shown in Fig. 5.16 (Dominak et al., 2010). Accumulation of protein aggregates (at low pH) and lipid material (as the protein disperses) can also be seen in Fig. 5.16. This type of accumulation of material is not uncommon for the encapsulated ATPS, and depends on the composition of the phases. Dimova and coworkers have shown that the ATPS interface can serve as a reservoir for lipid membrane material. Upon osmotic deflation, lipid nanotubes were formed from excess membrane at the interface of a GV-encapsulated PEG/dextran ATPS. This process was reversible, with nanotubes retracted back to the membrane by increasing the interfacial tension (Li et al., 2011).

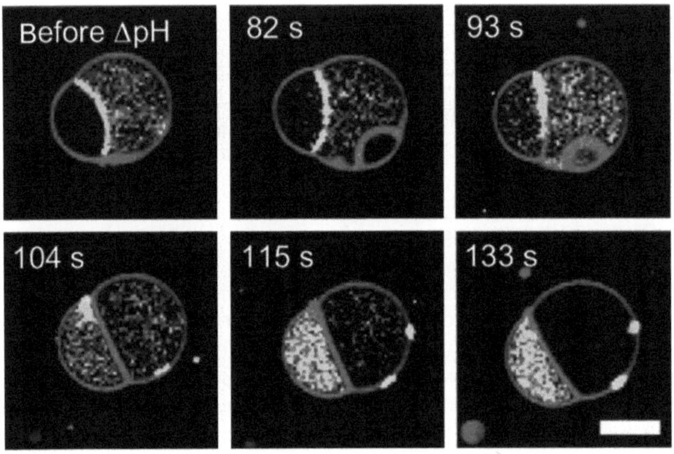

Figure 5.16 At pH 4.1, fluorescently labeled human serum albumin (HSA) partitions to the PEG-rich phase and interface of a PEG and dextran ATPS encapsulated vesicle. Upon a change in pH to 6.5, the protein migrates to the dextran-rich phase compartment. Scale bar = 5 μm. *Adapted with permission from Dominak et al. (2010). Copyright 2009, American Chemical Society.* (For the color version of this figure, the reader is referred to the online version of this chapter.)

6.4. Generation of aqueous-phase compartments inside living cells by expressing nonnative proteins

The intracellular environment is of course much more complex than the experimental model systems described earlier. Although in studies using synthetic polymers, generally adding complexity to the solution in the form of additional polymers, salts, or small molecules is likely to increase the tendency to phase-separate, it is, nonetheless, of interest to test whether systems that lead to aqueous phase separation in test tubes also do so in cells. A few studies have taken advantage of the ability to express phase-forming, nonnative proteins inside cells using molecular biology techniques to generate protein-rich phase compartments inside living cells. In the nephrin–NCK–N-WASP multivalent system, the engineered proteins were expressed in HeLa cells as fluorescent fusion proteins. When the mCherry-SH3 domain protein was coexpressed with the proline binding motif eGFP fusion protein, the two proteins were observed to form cytoplasmic phases with diameters of 0.5–2 μm. FRAP studies verified that these existed as dynamic phases that permitted rapid recovery within 10 s of photobleaching. When expressed separately, no such phases formed (Li et al., 2012).

Ge and coworkers produced elastin-like polypeptides (ELP) as GFP fusions in *E. coli* and tobacco cells. Fluorescent imaging and FRAP studies showed that the GFP–ELP was localized within aqueous phase compartments that occupied a relatively large fraction of the cell's internal volume but that excluded its genetic material (Fig. 5.17). The authors also hypothesized that aqueous phases *in vivo* may serve as storage compartments for proteins that may otherwise be degraded by proteases. They evaluated this possibility for the GFP–ELP phase systems by generating a thioredoxin–ELP fusion with a thrombin cleavage site. When this fusion was incubated with thrombin above its phase transition temperature (where it exists as two phases), it was protected from cleavage, while at lower temperatures, where phase separation does not occur, more than 95% was cleaved (Ge et al., 2009). The protective effect of the phase compartment in this case depends upon exclusion of other proteins from the ELP-rich phase, which is a known property of ELP fusions that has led to their use in protein purification. Nuclear bodies are unlikely to exclude so many other different proteins as effectively as these ELP constructs and, indeed, would probably not be able to function effectively if they did; however, exclusion of certain proteins has been observed (Kumari et al., 2007). This complements the idea that nuclear speckles serve as a storage compartment (Spector and Lamond, 2011; Zimber et al., 2004).

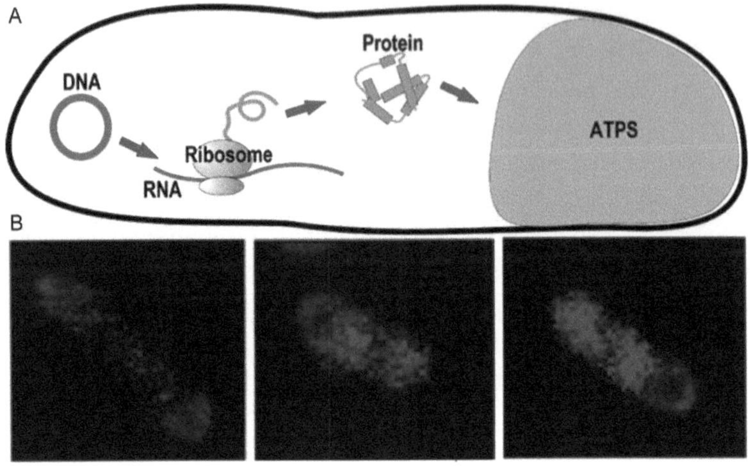

Figure 5.17 (A) Scheme depicting GFP–ELP expression within *E. coli*. Upon inducing phase separation, the ELP-rich phase is isolated within the cell. (B) Fluorescent images of the separate GFP–ELP phase and nucleic acids stained with SYTO 62. *Adapted with permission from Ge et al. (2009). Copyright 2009, American Chemical Society.* (For the color version of this figure, the reader is referred to the online version of this chapter.)

The formation of intentional, nonnative protein-rich phases inside living cells has potential for increasing our understanding of the cellular environment and how it is impacted by phase separation. One could imagine intentionally forming liquid compartments for desired enzymes or nucleic acid sequences, and using such systems to ask questions about the role of that molecule's spatial distribution and/or to change its function inside the cell by, for example, protecting it from degradation, or colocalizing it with particular reaction partners.

7. LOOKING FORWARD

Experimental work has shown that nuclear compartments are highly dynamic structures that rely, in part on the crowded, phase-separated nucleoplasm to exist. Although various model systems have begun to replicate some of the compartmentalized features of the nucleoplasm, not all aspects can be captured by existing models. Thus far, work has shown that nucleotide-peptide coacervates readily form and can have incorporated enzymatic activity while the PEG and dextran studies provide two distinct, yet crowded, phases in which biomolecules may partition to increase activity. However, a combination of these phenomena exists within the nucleoplasm in addition to a variety of other conditions that promote nuclear body formation and function. Ionic strength, for example, is known to affect nuclear organization, specifically divalent cation concentration such as that of Mg^{2+} that has been shown to alter the distribution of chromatin (Aaronson and Woo, 1981; Brasch et al., 1971). In addition, specific binding interactions are important in targeting different macromolecular components to the various intranuclear compartments, and have been underrepresented in current model systems. It will be of particular interest to gain a greater understanding of the interplay between specific binding interactions and the more general physicochemical effects that lead to macromolecular crowding and phase separation. Model systems that begin to incorporate a larger range of these nuclear conditions and specific interactions, in addition to providing heterogeneous and crowded media, are desired and will further the understanding of microcompartmentalization within the nucleus.

ACKNOWLEDGMENT

The authors were supported by the National Science Foundation grant MCB-1244180 during the writing of this chapter.

REFERENCES

Aaronson, R.P., Woo, E., 1981. Organization in the cell-nucleus: divalent-cations modulate the distribution of condensed and diffuse chromatin. J. Cell Biol. 90, 181–186.
Alberts, B., 2008. Molecular Biology of the Cell, fifth ed. Garland Science, New York, NY.
Albertsson, P.-Å., 1986. Partition of Cell Particles and Macromolecules. Wiley, New York, NY.
Andersen, J.S., Lam, Y.W., Leung, A.K.L., Ong, S.-E., Lyon, C.E., Lamond, A.I., Mann, M., 2005. Nucleolar proteome dynamics. Nature 433, 77–83.
Andes-Koback, M., Keating, C.D., 2011. Complete budding and asymmetric division of primitive model cells to produce daughter vesicles with different interior and membrane compositions. J. Am. Chem. Soc. 133, 9545–9555.
Atkins, P.W., De Paula, J., 2010. Physical Chemistry, ninth ed. W.H. Freeman, New York, NY.
Balakrishnan, G., Nicolai, T., Benyahia, L., Durand, D., 2012. Particles trapped at the droplet interface in water-in-water emulsions. Langmuir 28, 5921–5926.
Benavides, J., Rito-Palomares, M., 2008. Practical experiences from the development of aqueous two-phase processes for the recovery of high value biological products. J. Chem. Technol. Biotechnol. 83, 133–142.
Bernardi, R., Pandolfi, P.P., 2007. Structure, dynamics and functions of promyelocytic leukaemia nuclear bodies. Nat. Rev. Mol. Cell Biol. 8, 1006–1016.
Bloomfield, V.A., 1997. DNA condensation by multivalent cations. Biopolymers 44, 269–282.
Brangwynne, C.P., 2011. Soft active aggregates: mechanics, dynamics and self-assembly of liquid-like intracellular protein bodies. Soft Matter 7, 3052–3059.

Brangwynne, C.P., Eckmann, C.R., Courson, D.S., Rybarska, A., Hoege, C., Gharakhani, J., Jülicher, F., Hyman, A.A., 2009. Germline P granules are liquid droplets that localize by controlled dissolution/condensation. Science 324, 1729–1732.

Brangwynne, C.P., Mitchison, T.J., Hyman, A.A., 2011. Active liquid-like behavior of nucleoli determines their size and shape in *Xenopus laevis* oocytes. Proc. Natl. Acad. Sci. U. S. A. 108, 4334–4339.

Brasch, K., Seligy, V.L., Setterfi, G., 1971. Effects of low salt concentration on structural organization and template activity of chromatin in chicken erythrocyte nuclei. Exp. Cell Res. 65, 61–72.

Burgess, D.J., Carless, J.E., 1984. Microelectrophoretic studies of gelatin and acacia for the prediction of complex coacervation. J. Colloid Interface Sci. 98, 1–8.

Cacace, D.N., Keating, C.D., 2013. Biocatalyzed mineralization in an aqueous two-phase system: effect of background polymers and enzyme partitioning. J. Mater. Chem. B 1, 1794–1803.

Cans, A.S., Andes-Koback, M., Keating, C.D., 2008. Positioning lipid membrane domains in giant vesicles by micro-organization of aqueous cytoplasm mimic. J. Am. Chem. Soc. 130, 7400–7406.

Cheung, M.S., Klimov, D., Thirumalai, D., 2005. Molecular crowding enhances native state stability and refolding rates of globular proteins. Proc. Natl. Acad. Sci. U. S. A. 102, 4753–4758.

Cho, E.J., Kim, J.S., 2012. Crowding-induced phase separation of Lennard-Jones particles: implications to nuclear structures in a biological cell. J. Phys. Chem. B 116, 3874–3879.

Chollakup, R., Beck, J.B., Dirnberger, K., Tirrell, M., Eisenbach, C.D., 2013. Polyelectrolyte molecular weight and salt effects on the phase behavior and coacervation of aqueous solutions of poly(acrylic acid) sodium salt and poly(allylamine) hydrochloride. Macromolecules 46, 2376–2390.

Crosby, J., Treadwell, T., Hammerton, M., Vasilakis, K., Crump, M.P., Williams, D.S., Mann, S., 2012. Stabilization and enhanced reactivity of actinorhodin polyketide synthase minimal complex in polymer-nucleotide coacervate droplets. Chem. Commun. 48, 11832–11834.

Daban, J.R., 2000. Physical constraints in the condensation of eukaryotic chromosomes. Local concentration of DNA versus linear packing ratio in higher order chromatin structures. Biochemistry 39, 3861–3866.

da Silva, L.H.M., Loh, W., 2000. Calorimetric investigation of the formation of aqueous two-phase systems in ternary mixtures of water, poly(ethylene oxide) and electrolytes (or dextran). J. Phys. Chem. B 104, 10069–10073.

de Kruif, C.G., Tuinier, R., 2001. Polysaccharide protein interactions. Food hydrocolloids 15, 555–563.

de Kruif, C.G., Weinbreck, F., de Vries, R., 2004. Complex coacervation of proteins and anionic polysaccharides. Curr. Opin. Colloid Interface Sci. 9, 340–349.

Dimova, R., Lipowsky, R., 2012. Lipid membranes in contact with aqueous phases of polymer solutions. Soft Matter 8, 6409–6415.

Dominak, L.M., Gundermann, E.L., Keating, C.D., 2010. Microcompartmentation in artificial cells: pH-induced conformational changes alter protein localization. Langmuir 26, 5697–5705.

Dorsaz, N., Thurston, G.M., Stradner, A., Schurtenberger, P., Foffi, G., 2011. Phase separation in binary eye lens protein mixtures. Soft Matter 7, 1763–1776.

Dumetz, A.C., Chockla, A.M., Kaler, E.W., Lenhoff, A.M., 2008. Protein phase behavior in aqueous solutions: crystallization, liquid-liquid phase separation, gels, and aggregates. Biophys. J. 94, 570–583.

Dundr, M., 2012. Nuclear bodies: multifunctional companions of the genome. Curr. Opin. Cell Biol. 24, 415–422.

Dundr, M., Misteli, T., 2001. Functional architecture in the cell nucleus. Biochem. J. 356, 297–310.
Dundr, M., Misteli, T., 2010. Biogenesis of nuclear bodies. Cold Spring Harb. Perspect. Biol. 2, a000711.
Dundr, M., Hebert, M.D., Karpova, T.S., Stanek, D., Xu, H., Shpargel, K.B., Meier, U.T., Neugebauer, K.M., Matera, A.G., Misteli, T., 2004. In vivo kinetics of Cajal body components. J. Cell Biol. 164, 831–842.
Elcock, A.H., 2010. Models of macromolecular crowding effects and the need for quantitative comparisons with experiment. Curr. Opin. Struct. Biol. 20, 196–206.
Ellis, R.J., 2001. Macromolecular crowding: obvious but underappreciated. Trends Biochem. Sci. 26, 597–604.
Estévez-Torres, A., Baigl, D., 2011. DNA compaction: fundamentals and applications. Soft Matter 7, 6746–6756.
Feig, M., Sugita, Y., 2012. Variable interactions between protein crowders and biomolecular solutes are important in understanding cellular crowding. J. Phys. Chem. B 116, 599–605.
Firoozmand, H., Murray, B.S., Dickinson, E., 2009. Interfacial structuring in a phase-separating mixed biopolymer solution containing colloidal particles. Langmuir 25, 1300–1305.
Forciniti, D., Hall, C.K., Kula, M.R., 1990. Interfacial-tension of polyethyleneglycol-dextran-water systems—influence of temperature and polymer molecular-weight. J. Biotechnol. 16, 279–296.
Fujiwara, T., Suzuki, S., Kanno, M., Sugiyama, H., Takahashi, H., Tanaka, J., 2006. Mapping a nucleolar targeting sequence of an RNA binding nucleolar protein, Nop25. Exp. Cell Res. 312, 1703–1712.
Ge, X., Conley, A.J., Brandle, J.E., Truant, R., Filipe, C.D.M., 2009. *In Vivo* formation of protein based aqueous microcompartments. J. Am. Chem. Soc. 131, 9094–9099.
Guan, Y., Lilley, T.H., Treffry, T.E., 1993a. A new excluded-volume theory and its application to the coexistence curves of aqueous polymer two-phase systems. Macromolecules 26, 3971–3979.
Guan, Y., Lilley, T.H., Treffry, T.E., 1993b. Theory of phase-equilibria for multicomponent aqueous-solutions—applications to aqueous polymer two-phase systems. J. Chem. Soc. Faraday Trans. 89, 4283–4298.
Hall, D., Minton, A.P., 2003. Macromolecular crowding: qualitative and semiquantitative successes, quantitative challenges. Biochim. Biophys. Acta 1649, 127–139.
Han, T.W., Kato, M., Xie, S., Wu, L.C., Mirzaei, H., Pei, J., Chen, M., Xie, Y., Allen, J., Xiao, G., McKnight, S.L., 2012. Cell-free formation of RNA granules: bound RNAs identify features and components of cellular assemblies. Cell 149, 768–779.
Hancock, R., 2004a. Internal organisation of the nucleus: assembly of compartments by macromolecular crowding and the nuclear matrix model. Biol. Cell. 96, 595–601.
Hancock, R., 2004b. A role for macromolecular crowding effects in the assembly and function of compartments in the nucleus. J. Struct. Biol. 146, 281–290.
Handwerger, K.E., Cordero, J.A., Gall, J.G., 2005. Cajal bodies, nucleoli, and speckles in the *Xenopus* oocyte nucleus have a low-density, sponge-like structure. Mol. Biol. Cell 16, 202–211.
Hatti-Kaul, R., 2000. Aqueous Two-Phase Systems: Methods and Protocols. Humana Press, Totowa, NJ.
Helfrich, M.R., Mangeney-Slavin, L.K., Long, M.S., Djoko, K.Y., Keating, C.D., 2002. Aqueous phase separation in giant vesicles. J. Am. Chem. Soc. 124, 13374–13375.
Helfrich, M.R., El-Kouedi, M., Etherton, M.R., Keating, C.D., 2005. Partitioning and assembly of metal particles and their bioconjugates in aqueous two-phase systems. Langmuir 21, 8478–8486.

Hernandez-Verdun, D., 2006. Nucleolus: from structure to dynamics. Histochem. Cell Biol. 125, 127–137.
Johansson, H.O., Karlstrom, G., Tjerneld, F., Haynes, C.A., 1998. Driving forces for phase separation and partitioning in aqueous two-phase systems. J. Chromatogr. B 711, 3–17.
Kaiser, T.E., Intine, R.V., Dundr, M., 2008. De novo formation of a subnuclear body. Science 322, 1713–1717.
Kato, M., Han, T.W., Xie, S., Shi, K., Du, X., Wu, L.C., Mirzaei, H., Goldsmith, E.J., Longgood, J., Pei, J., Grishin, N.V., Frantz, D.E., Schneider, J.W., Chen, S., Li, L., Sawaya, M.R., Eisenberg, D., Tycko, R., McKnight, S.L., 2012. Cell-free formation of RNA granules: low complexity sequence domains form dynamic fibers within hydrogels. Cell 149, 753–767.
Keating, C.D., 2012. Aqueous phase separation as a possible route to compartmentalization of biological molecules. Acc. Chem. Res. 45, 2114–2124.
Knowles, D.B., LaCroix, A.S., Deines, N.F., Shkel, I., Record, M.T., 2011. Separation of preferential interaction and excluded volume effects on DNA duplex and hairpin stability. Proc. Natl. Acad. Sci. U. S. A. 108, 12699–12704.
Koga, S., Williams, D.S., Perriman, A.W., Mann, S., 2011. Peptide-nucleotide microdroplets as a step towards a membrane-free protocell model. Nat. Chem. 3, 720–724.
Kojima, T., Takayama, S., 2013. Patchy surfaces stabilize dextran-polyethylene glycol aqueous two-phase system liquid patterns. Langmuir 29, 5508–5514.
Kumari, G., Singhal, P.K., Rao, M.R.K.S., Mahalingam, S., 2007. Nuclear transport of Ras-associated tumor suppressor proteins: different transport receptor binding specificities for arginine-rich nuclear targeting signals. J. Mol. Biol. 367, 1294–1311.
Lehninger, A.L., Nelson, D.L., Cox, M.M., 2013. Lehninger Principles of Biochemistry, sixth ed. W.H. Freeman, New York.
Li, Y.H., Lipowsky, R., Dimova, R., 2011. Membrane nanotubes induced by aqueous phase separation and stabilized by spontaneous curvature. Proc. Natl. Acad. Sci. U. S. A. 108, 4731–4736.
Li, P., Banjade, S., Cheng, H.-C., Kim, S., Chen, B., Guo, L., Llaguno, M., Hollingsworth, J.V., King, D.S., Banani, S.F., Russo, P.S., Jiang, Q.-X., Nixon, B.T., Rosen, M.K., 2012. Phase transitions in the assembly of multivalent signalling proteins. Nature 483, 336–341.
Liu, C.W., Asherie, N., Lomakin, A., Pande, J., Ogun, O., Benedek, G.B., 1996. Phase separation in aqueous solutions of lens gamma-crystallins: special role of gamma s. Proc. Natl. Acad. Sci. U. S. A. 93, 377–382.
Liu, Y., Lipowsky, R., Dimova, R., 2012. Concentration dependence of the interfacial tension for aqueous two-phase polymer solutions of dextran and polyethylene glycol. Langmuir 28, 3831–3839.
Long, M.S., Jones, C.D., Helfrich, M.R., Mangeney-Slavin, L.K., Keating, C.D., 2005. Dynamic microcompartmentation in synthetic cells. Proc. Natl. Acad. Sci. U. S. A. 102, 5920–5925.
Long, M.S., Cans, A.-S., Keating, C.D., 2008. Budding and asymmetric protein microcompartmentation in giant vesicles containing two aqueous phases. J. Am. Chem. Soc. 130, 756–762.
Luby-Phelps, K., 2013. The physical chemistry of cytoplasm and its influence on cell function: an update. Mol. Biol. Cell 24, 2593–2596.
Mao, Y.T.S., Zhang, B., Spector, D.L., 2011. Biogenesis and function of nuclear bodies. Trends Genet. 27 (8), 295–306.
Matera, A.G., Izaguire-Sierra, M., Praveen, K., Rajendra, T.K., 2009. Nuclear bodies: random aggregates of sticky proteins or crucibles of macromolecular assembly? Dev. Cell 17, 639–647.

Miklos, A.C., Sarkar, M., Wang, Y., Pielak, G.J., 2011. Protein crowding tunes protein stability. J. Am. Chem. Soc. 133, 7116–7120.
Minton, A.P., 1983. The effect of volume occupancy upon the thermodynamic activity of proteins—some biochemical consequences. Mol. Cell. Biochem. 55, 119–140.
Minton, A.P., 2001. The influence of macromolecular crowding and macromolecular confinement on biochemical reactions in physiological media. J. Biol. Chem. 276, 10577–10580.
Minton, A.P., 2013. Quantitative assessment of the relative contributions of steric repulsion and chemical interactions to macromolecular crowding. Biopolymers 99, 239–244.
Misteli, T., 2001. Nuclear structure—protein dynamics: implications for nuclear architecture and gene expression. Science 291, 843–847.
Mohanty, B., Bohidar, H.B., 2003. Systematic of alcohol-induced simple coacervation in aqueous gelatin solutions. Biomacromolecules 4, 1080–1086.
Mohite, L.V., Juvekar, V.A., 2009. Quantification of thermodynamics of aqueous solutions of poly(ethylene glycols): role of calorimetry. Fluid Phase Equilib. 278, 41–53.
Nguyen, B.T., Nicolai, T., Benyahia, L., 2013. Stabilization of water-in-water emulsions by addition of protein particles. Langmuir 29, 10658–10664.
Oakes, M., Aris, J.P., Brockenbrough, J.S., Wai, H., Vu, L., Nomura, M., 1998. Mutational analysis of the structure and localization of the nucleolus in the yeast *Saccharomyces cerevisiae*. J. Cell Biol. 143, 23–34.
Oparin, A.I., Serebrovskaia, K.B., Pantskhava, S.N., Vasileva, N.V., 1963. Enzymatic synthesis of polyadenylic acid in coacervate drops. Biokhimiia 28, 671–675.
Ovádi, J., Saks, V., 2004. On the origin of intracellular compartmentation and organized metabolic systems. Mol. Cell. Biochem. 256/257, 5–12.
Ovádi, J., Srere, P.A., 2000. Macromolecular compartmentation and channeling. Int. Rev. Cytol. 192, 255–280.
Ovádi, J., Tompa, P., Vértessy, B., Orosz, F., Keleti, T., Welch, G.R., 1989. Transient-time analysis of substrate-channelling in interacting enzyme-systems. Biochem. J. 257, 187–190.
Pelta, J., Livolant, F., Sikorav, J.-L., 1996. DNA aggregation induced by polyamines and cobalthexamine. J. Biol. Chem. 271, 5656–5662.
Pessoa, P.D., Mohamed, R.S., 2004. A hydration shell-based thermodynamic model for aqueous two-phase systems. Can. J. Chem. Eng. 82, 530–538.
Phair, R.D., Misteli, T., 2000. High mobility of proteins in the mammalian cell nucleus. Nature 404, 604–609.
Platani, M., Goldberg, I., Swedlow, J.R., Lamond, A.I., 2000. In vivo analysis of Cajal body movement, separation, and joining in live human cells. J. Cell Biol. 151, 1561–1574.
Pontvianne, F., Blevins, T., Chandrasekhara, C., Mozgová, I., Hassel, C., Pontes, O.M., Tucker, S., Mokroš, P., Muchová, V., Fajkus, J., Pikaard, C.S., 2013. Subnuclear partitioning of rRNA genes between the nucleolus and nucleoplasm reflects alternative epiallelic states. Genes Dev. 27, 1545–1550.
Priftis, D., Tirrell, M., 2012. Phase behaviour and complex coacervation of aqueous polypeptide solutions. Soft Matter 8, 9396–9405.
Priftis, D., Farina, R., Tirrell, M., 2012. Interfacial energy of polypeptide complex coacervates measured via capillary adhesion. Langmuir 28, 8721–8729.
Richter, K., Nessling, M., Lichter, P., 2008. Macromolecular crowding and its potential impact on nuclear function. BBA Mol. Cell Res. 1783, 2100–2107.
Ross, P.D., Minton, A.P., 1977. Analysis of non-ideal behavior in concentrated hemoglobin solutions. J. Mol. Biol. 112, 437–452.
Ruiz-Ruiz, F., Benavides, J., Aguilar, O., Rito-Palomares, M., 2012. Aqueous two-phase affinity partitioning systems: current applications and trends. J. Chromatogr. 1244, 1–13.

Sokolova, E., Spruijt, E., Hansen, M.M.K., Dubuc, E., Groen, J., Chokkalingam, V., Piruska, A., Heus, H.A., Huck, W.T.S., 2013. Enhanced transcription rates in membrane-free protocells formed by coacervation of cell lysate. Proc. Natl. Acad. Sci. U. S. A. 110, 11692–11697.

Spector, D.L., Lamond, A.I., 2011. Nuclear speckles. Cold Spring Harb. Perspect. Biol. 3, a000646.

Spruijt, E., Sprakel, J., Stuart, M.A.C., van der Gucht, J., 2010. Interfacial tension between a complex coacervate phase and its coexisting aqueous phase. Soft Matter 6, 172–178.

Strulson, C.A., Molden, R.C., Keating, C.D., Bevilacqua, P.C., 2012. RNA catalysis through compartmentalization. Nat. Chem. 4, 941–946.

Tang, M.X., Szoka, F.C., 1997. The influence of polymer structure on the interactions of cationic polymers with DNA and morphology of the resulting complexes. Gene Ther. 4, 823–832.

Tolstoguzov, V., 2000. Compositions and phase diagrams for aqueous systems based on proteins and polysaccharides. Int. Rev. Cytol. 192, 3–31.

Tripp, B.C., Kaler, E.W., Sandler, S.I., Kopatsis, A., 1996. Effect of ionic strength and ionic species on partitioning behavior of hydrophobic and hydrophilic polystyrene latex beads in aqueous two-phase polymer systems. Colloids Surf., B 6, 151–164.

Turgeon, S.L., Beaulieu, M., Schmitt, C., Sanchez, C., 2003. Protein-polysaccharide interactions: phase-ordering kinetics, thermodynamic and structural aspects. Curr. Opin. Colloid Interface Sci. 8, 401–414.

Vekilov, P.G., 2012. Phase diagrams and kinetics of phase transitions in protein solutions. J. Phys. Condens. Matter 24, 193101.

Walter, H., Brooks, D.E., 1995. Phase-separation in cytoplasm, due to macromolecular crowding, is the basis for microcompartmentation. FEBS Lett. 361, 135–139.

Walter, H., Brooks, D.E., Fisher, D., 1985. Partitioning in Aqueous Two-Phase Systems: Theory, Methods, Uses, and Application to Biotechnology. Academic Press, Orlando.

Wang, Y., Sarkar, M., Smith, A.E., Krois, A.S., Pielak, G.J., 2012. Macromolecular crowding and protein stability. J. Am. Chem. Soc. 134, 16614–16618.

Wang, Y., Lomakin, A., Latypov, R.F., Laubach, J.P., Hideshima, T., Richardson, P.G., Munshi, N.C., Anderson, K.C., Benedek, G.B., 2013. Phase transitions in human IgG solutions. J. Chem. Phys. 139, 1219041–1219049.

Weinbreck, F., Wientjes, R.H.W., Nieuwenhujse, H., Robijn, G.W., de Kruif, C.G., 2004. Rheological properties of whey protein/gum arabic coacervates. J. Rheol. 48, 1215–1228.

Welch, G.R., Clegg, J.S., 2010. From protoplasmic theory to cellular systems biology: a 150-year reflection. Am. J. Physiol. Cell Physiol. 298, C1280–C1290.

Williams, D.S., Koga, S., Hak, C.R.C., Majrekar, A., Patil, A.J., Perriman, A.W., Mann, S., 2012. Polymer/nucleotide droplets as bio-inspired functional micro-compartments. Soft Matter 8, 6004–6014.

Wilson, E.B., 1899. The structure of protoplasm. Science 10, 33–45.

Wilson, M.Z., Gitai, Z., 2013. Beyond the cytoskeleton: mesoscale assemblies and their function in spatial organization. Curr. Opin. Microbiol. 16, 177–183.

Wohlfarth, C., 2013. CRC Handbook of Phase Equilibria and Thermodynamic Data of Aqueous Polymer Solutions. Taylor & Francis, Boca Raton, FL.

Yu, M., Nishiumi, H., Arons, J.D., 1993. Thermodynamics of phase-separation in aqueous-solutions of polymers. Fluid Phase Equilib. 83, 357–364.

Zaslavsky, B.Y., 1995. Aqueous Two-Phase Partitioning: Physical Chemistry and Bioanalytical Applications. M. Dekker, New York, NY.

Zhao, H., Magone, M.T., Schuck, P., 2011. The role of macromolecular crowding in the evolution of lens crystallins with high molecular refractive index. Phys. Biol. 8, 1–10.

Zhou, H.-X., 2013. Influence of crowded cellular environments on protein folding, binding, and oligomerization: biological consequences and potentials of atomistic modeling. FEBS Lett. 587, 1053–1061.

Zhou, H.-X., Rivas, G.N., Minton, A.P., 2008. Macromolecular crowding and confinement: biochemical, biophysical, and potential physiological consequences. Annu. Rev. Biophys. 37, 375–397.

Zimber, A., Nguyen, Q.-D., Gespach, C., 2004. Nuclear bodies and compartments: functional roles and cellular signalling in health and disease. Cell. Signal. 16, 1085–1104.

Zimmerman, S.B., 2006. Shape and compaction of *Escherichia coli* nucleoids. J. Struct. Biol. 156, 255–261.

CHAPTER SIX

Formation of Multiprotein Assemblies in the Nucleus: The Spindle Assembly Checkpoint

Victor M. Bolanos-Garcia[1]
Faculty of Health and Life Sciences, Department of Biological and Medical Sciences, Oxford Brookes University, Oxford, United Kingdom
[1]Corresponding author: e-mail address: vbolanos-garcia@brookes.ac.uk

Contents

Abstract

Specific interactions within the cell must occur in a crowded environment and often in a narrow time-space framework to ensure cell survival. In the light that up to 10% of individual protein molecules present at one time in mammalian cells mediate signal transduction, the establishment of productive, specific interactions is a remarkable achievement. The spindle assembly checkpoint (SAC) is an evolutionarily conserved and essential self-monitoring system of the eukaryotic cell cycle that ensures the high fidelity of chromosome segregation by delaying the onset of anaphase until all chromosomes are properly bi-oriented on the mitotic spindle. The function of the SAC involves communication with the kinetochore, an essential multiprotein complex crucial for chromosome segregation that assembles on mitotic or meiotic centromeres to link centromeric DNA with microtubules. Interactions in the SAC and kinetochore–microtubule network often involve the reversible assembly of large multiprotein complexes in which regions of the polypeptide chain that exhibit low structure complexity undergo a disorder-to-order transition. The confinement and high density of protein molecules in the cell has a profound effect on the stability, folding rate, and biological functions of individual proteins and protein assemblies. Here, I discuss the role of large and highly flexible surfaces that mediate productive intermolecular interactions in SAC

International Review of Cell and Molecular Biology, Volume 307
ISSN 1937-6448
http://dx.doi.org/10.1016/B978-0-12-800046-5.00006-0

signaling and postulate that macromolecular crowding contributes to the exquisite regulation that is required for the timely and accurate segregation of chromosomes in higher organisms.

1. INTRODUCTION

The communication between macromolecules often involves the highly specific and reversible assembly of large multiprotein complexes to generate a signal that overcomes background noise. This is a remarkable achievement considering that up to 10% of individual protein molecules that are present in mammalian cells at one time mediate cell-signaling events, and that in such a busy environment numerous specific interactions must occur in a narrow time-space framework to ensure cell survival. Indeed, the high density of macromolecules inside the cell gives place to a cumulative excluded volume, a phenomenon commonly referred to as macromolecular crowding. Such confinement and crowding of macromolecules in the cellular space can have a profound effect on the stability, folding rate, and biological functions of macromolecules (Elcock, 2010; Ellis, 2001; Hancock, 2012; Minton, 1997, 2000; Zhou et al., 2008; Zimmerman and Minton, 1993; Zimmerman and Trach, 1991).

One experimental strategy to study the forces that promote macromolecular crowding is the use of certain polymers to recreate crowding conditions. For instance, the addition of Ficoll 70 to an aqueous solution of phosphoglycerate kinase stimulated the catalytic activity of this protein kinase *in vitro* (Dhar et al., 2010), whereas the addition of dextran to an acidic solution of cytochrome *c* promoted a transition from the unfolded state into a near-native molten globule state (Sasahara et al., 2003). An independent study on a family of electron-transfer proteins, the flavodoxins that combined an experimental approach with *in silico* simulations showed that addition of Ficoll 70 increased the thermal stability and secondary structure content of the proteins in the native state but had no observable effect on the proteins in the denatured state (Stagg et al., 2007). Perhaps, a more dramatic example of the effect of macromolecular crowding on protein function is the formation of amyloid fibrils of human and bovine prion proteins, which is significantly enhanced by addition of Ficoll 70 (150–200 g/L) (Batra et al., 2009; Huang et al., 2010; Zhou et al., 2011). An intriguing exception occurs in rabbits, where the prion protein amyloid fibril formation seems to be inhibited by the presence of crowding agents (Ma et al., 2012; Zhou et al., 2011). In any case, a common theme in all of the aforementioned studies is that the addition of compounds that recreate macromolecular crowding conditions can induce long-range conformational changes. Since macromolecular crowding can play an important role in the etiology of human diseases, definition of the molecular details of the conditions that promote macromolecular crowding in nuclei should provide valuable new insights into the understanding of the stability, regulation, structure, and function of macromolecules that reside in this organelle. Because one dramatic example of the exquisite regulation of specific interactions in a complex signaling system is the spindle assembly checkpoint (SAC), also commonly referred to as mitotic checkpoint, the implications of macromolecular crowding in the nucleus will be discussed in the context of this signaling system.

2. SAC SIGNALING

Inside cells, the concentration of macromolecules can reach up to 400 g/L. Considering that in mammalian cells, the average diameter of the nucleus is approximately 6 μm, which represents approximately 10% of the total cell volume (Zimmerman and Minton, 1993; Zimmerman and Trach, 1991), the establishment of specific, productive interactions in the nucleus is a remarkable achievement. One dramatic example of the exquisite regulation of specific interactions in a complex signaling system is the SAC, also commonly referred to as the mitotic checkpoint. In a nutshell, the SAC is an essential, evolutionary conserved, self-monitoring regulatory mechanism of the cell cycle that ensures the maintenance of genomic stability in higher organisms by delaying the onset of anaphase until all chromosomes are properly bi-oriented and attached to the mitotic spindle (Foley and Kapoor, 2013; Hardwick et al., 2000; Jia et al., 2013; Morrow et al., 2005; Warren et al., 2002; Yao and Dai, 2012). A brief description of the function and structure features of individual central protein components of the SAC signaling system is presented as follows.

2.1. Central protein components of the SAC

Three serine/threonine protein kinases, Bub1, BubR1, and Mps1 play key roles in SAC signaling: Bub1 mediates the recruitment of other checkpoint components in cells that have the checkpoint incompleted and is important for assembly of the inner centromere; BubR1 is required for the establishment of proper kinetochore–microtubule attachment and chromosome alignment and together with Bub3, Mad2, and Cdc20 forms part of the mitotic checkpoint complex that inhibits the E3 ubiquitin ligase activity of the anaphase-promoting complex (also known as the cyclosome, APC/C) toward Securin and Cyclin B1 (Bolanos-Garcia and Blundell, 2011; Boyarchuk et al., 2007; Elowe, 2011; Tang et al., 2004; Vanoosthuyse and Hardwick, 2005). The multidomain protein kinase Mps1 is also essential for proper SAC signaling (Abrieu et al., 2001; Maciejowski et al., 2010; Weiss and Winey, 1996) and 1 of the top 25 genes associated with cancer (Carter et al., 2006; Janssen et al., 2009). Mad1, Mad2, and Cdc20 are other central components of the SAC. In humans, Mad1 depletion severely affects the SAC (Luo et al., 2000; Maciejowski et al., 2010; Meyer et al., 2013). Mad1 is a coiled-coil protein of ~79 kDa (718 residues) that forms a stable complex with Mad2 *in vitro* (Luo et al., 2002), whereas Cdc20 acts as coactivator of the APC/C, the macromolecular assembly that is responsible for targeting proteins for ubiquitin-mediated degradation during mitosis (Izawa and Pines, 2012; Nilsson et al., 2008; Sedgwick et al., 2013).

The N-terminal regions of Bub1, BubR1, and Mps1 are organized as a single domain consisting of a triple tandem arrangement of the tetratricopeptide (TPR) motif, a protein motif defined by a consensus of 34 amino acids that are organized in a helix-loop-helix. The three TPR units of Bub1, BubR1, and Mps1 share features typical of other TPR motifs such as the presence of small and large hydrophobic residues located at specific positions within the helix-loop-helix, and the assembly of the TPR units into a relatively extended structure forms a regular series of antiparallel α-helices rotated relative to one another by a constant 24° (Fig. 6.1).

The uniform arrangement of neighboring α-helices gives rise to the formation of a right-handed superhelical structure with a continuous concave surface on the one side and a contrasting convex surface on the other. Essential for the stability of TPR tandem arrays are short-range and long-range interactions (Cliff et al., 2006; D'Arcy et al., 2010; Zeytuni and Zarivach, 2012), the disruption of which largely accounts for the instability of N-terminal truncated mutants of human Bub1, BubR1, Mad3 (the BubR1 homologue in yeast, which lacks the catalytic kinase domain), and Mps1 (Bolanos-Garcia and Blundell, 2011; D'Arcy et al., 2010; Kadura et al., 2005; Lee et al., 2012; Thebault et al., 2012). Bub3 is a central mitotic checkpoint protein that binds to Bub1 and BubR1 and exhibits a WD40-repeat fold. Bub3 is organized in a single globular domain (Larsen and Harrison, 2004; Wilson et al., 2005), whereas Mad2 adopts a distinctive

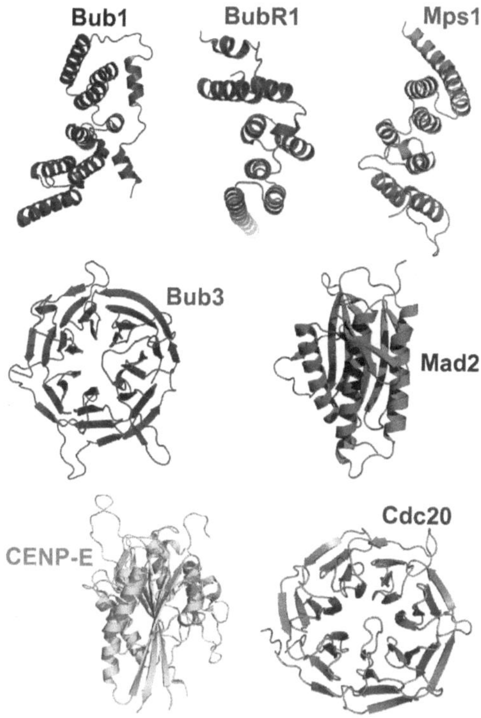

Figure 6.1 Three-dimensional structures of SAC protein components. The N-terminal regions of Bub1, BubR1, and Mps1 are organized as a triple tandem of the TPR motif (pdb 3ESL, 2WVI, 4H7X, and 4H7Y, respectively); Bub3 and Cdc20 both adopt a seven-blades, WD40-fold (pdb 1UAC and 4GGA, respectively); the architecture of Mad2 defines the HORMA domain (pdb 1DUJ); the crystal structure of the motor domain and linker region of human CENP-E with MgADP bound in the active site revealed that this CENP-E fragment is organized as a canonical kinesin motor domain (pdb 1T5C). Figures are generated with PyMOL (DeLano, 2002). (For color version of this figure, the reader is referred to the online version of this chapter.)

architecture, the HORMA (for Hop1, Rev7, and Mad2) domain (Luo et al., 2000) (Fig. 6.1). Cdc20 is also organized as a WD40-repeat fold (Fig. 6.1) and in mammals contain two independent degradation signals: the KEN box (Pfleger and Kirschner, 2000) and the CRY box (Reis et al., 2006). At least in budding yeast, APC/C-Cdh1-dependent degradation of Cdc20 is mediated by one of its two amino-terminal destruction (D) boxes (Huang et al., 2001). Mad1 is a predominantly coiled-coil protein that in mammalian cells encompasses 718 amino acid residues (Bolanos-Garcia, 2007). Mad1 depletion severely affects the SAC, thus evidencing the essential role of this protein in the process (Luo et al., 2002).

Centromere-associated protein E (CENP-E) is a member of the kinesin family that regulates the SAC by acting on BubR1 (Mao et al., 2003). Bipolar kinetochores lacking CENP-E are unable to generate sufficient poleward force to achieve normal levels of tension between sister kinetochores, demonstrating that this protein is essential for some aspects of kinetochore–microtubule attachments (Weaver et al., 2003). Human CENP-E is a 316 kDa protein organized in three distinct segments: the N-terminal motor domain (residues M1–K327) which includes binding sites for ATP and microtubules (Fig. 6.1), a long discontinuous α-helix/coiled region (residues 336–2471), and a C-terminal ATP-independent microtubule-binding domain (residues 2472–2663). Two regions with homology to PEST sequences (residues 459–489 and 2480–2488) seem to be responsible for the rapid degradation of human CENP-E at the end of mitosis (Brown et al., 1994). The kinetochore-binding region is located in the C-terminal end, residues 2126–2476, whereas phosphorylation of the C-terminal, ATP-independent microtubule-binding site (residues 2565–2663) inhibits the microtubule-binding activity of CENP-E. However, the association of CENP-E with mitogen-activated protein kinases in mitotic cells suggests that the interaction between microtubules and chromosomes to control mitotic progression involves an additional layer of regulation (Mayes et al., 2013; Zecevic et al., 1998).

2.2. SAC and the kinetochore

The function of the SAC involves communication with the kinetochore, an essential multiprotein complex crucial for chromosome segregation that assembles on mitotic or meiotic centromeres to link centromeric DNA with microtubules (Funabiki and Wynne, 2013; Westhorpe and Straight, 2013). The kinetochore is highly conserved in structure across species, despite the amino acid sequence divergence in most of the kinetochore protein components (Przewloka and Glover, 2009; Tanaka, 2013; Westhorpe and Straight, 2013). The structural core of the kinetochore is formed by the kinetochore–microtubule network (KMN), which comprises the protein KNL-1/Blinkin/Spc105 and the Mis12/Mtw1/MIND and Ndc80/HEC1 protein complexes (Fig. 6.2A). For simplicity, they are commonly referred to as KNL1, Mis12 complex, and Ndc80 complex, respectively. The KMN acts as docking platform for the recruitment of SAC proteins into the kinetochore. Also, the KMN physically links the centromere with microtubules and regulates microtubule capturing and plus end dynamics (Cheeseman et al., 2006; Kiyomitsu et al., 2007; Przewloka and Glover,

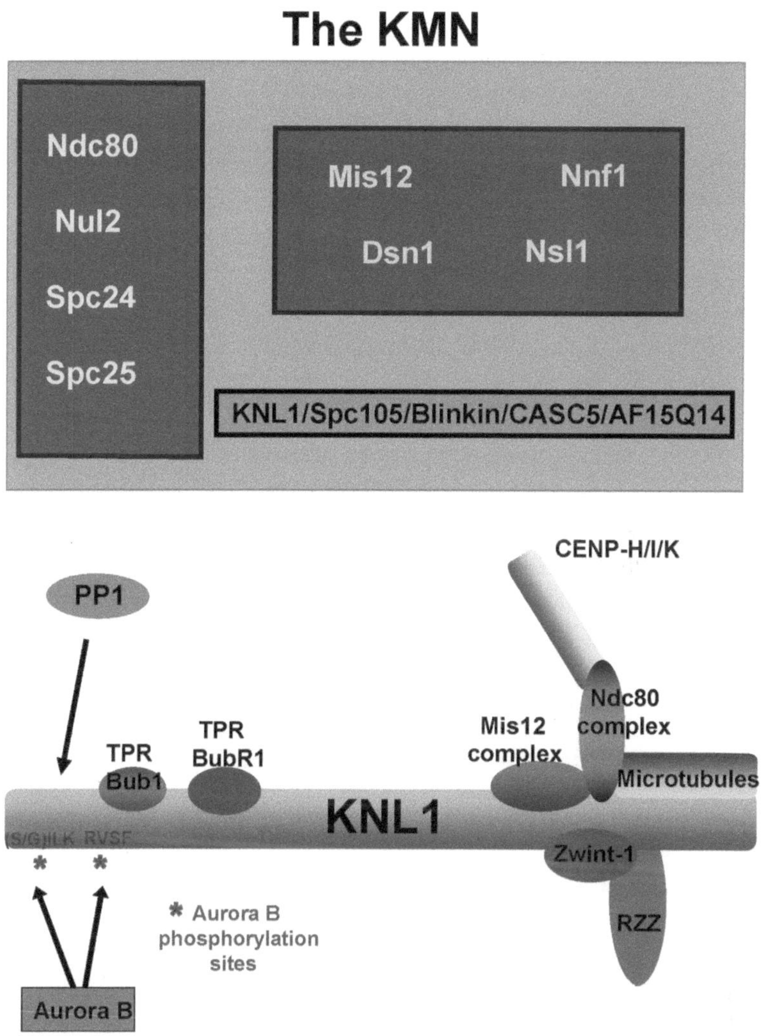

Figure 6.2 Composition of the KMN network. (A) The outer kinetochore (salmon box) is defined by the KMN network: KNL1 (gray box), the Mis12 complex, which is composed of Mis12, Dsn1, Nsl1, and Nnf1 (blue box), and the Ndc80 complex, constituted by Ndc80, Nuf2, Spc24, and Spc25 (red box). (B) Kinetochore localization of Bub1 and BubR1 is mediated by Blinkin, a central component of the kinetochore–microtubule network (KMN). (For interpretation of the references to color in this figure legend, the reader is referred to the online version of this chapter.)

2009; Przewloka et al., 2011). The kinetochore protein KNL1 (also often referred to as Blinkin, Spc105, AF15Q14, and CASC5) (Bolanos-Garcia et al., 2009; Kiyomitsu et al., 2007, 2011) is localized to kinetochores throughout mitosis and is required for the localization of a number of outer kinetochore proteins. Depletion of KNL1 in higher organisms by RNAi causes severe chromosome segregation defects that closely resemble the phenotypes associated with depletion of Bub1 and BubR1 (Cheeseman et al., 2006, 2008; Kiyomitsu et al., 2007).

KNL1 is predicted to be a predominantly intrinsically disordered protein and known to function as a multisubstrate docking platform. For instance, the KNL1 C-terminal region interacts with the Nsl1 and Dsn1 components of the Mis12 complex (Cheeseman et al., 2006; Kiyomitsu et al., 2007), whereas its N-terminal region binds to TPR Bub1 and TPR BubR1 (Bolanos-Garcia et al., 2009; Bolanos-Garcia and Blundell, 2011) and recruits protein phosphatase 1 (PP1) to kinetochores, an interaction that is required to silence the SAC (Fig. 6.2B) (Funabiki and Wynne, 2013; London et al., 2012; Rosenberg et al., 2011; Shepperd et al., 2012). The fact that Mis12, Nnf1a/b, Nsl1, and Spc105 are interdependent for their mitotic recruitment to the nascent kinetochores of *Drosophila* (Venkei et al., 2012) suggests these proteins interact in a coordinated manner. The physical interaction of KNL1 with proteins that are essential for proper chromosome segregation, including PP1, Bub1, and BubR1, strongly suggest a critical regulatory role of KNL1 in the assembly of the KMN and in SAC signaling.

3. DISORDER-TO-ORDER TRANSITIONS

Although kinetochore assembly is a defining aspect of mitotic progression, the exact sequence of events whereby the kinetochore is assembled upon entry into mitosis remains to be established. Knowledge of the precise timing and kinetics of recruitment of individual KMN components to centromeres is essential if we are to understand the mechanism of kinetochore assembly. We know that SAC signaling involves the timely assembly of protein subcomplexes in which at least one of the components often shows low structural complexity. The 3D structures of several binary complexes of SAC: Bub3 with individual fragments of Bub1 and Mad3, the individual TPR motifs of Bub1 and Bubr1 in a complex with KNL1, and of Mad2 with Mad1 and Cdc20 mimic peptides have provided molecular details of how the interactions are established. For instance, Bub1 and BubR1 (Mad3 in yeast) have a conserved stretch of about 40 amino acid residues downstream of the N-terminal TPR domain that is predicted to be mainly disordered and to contain a Bub3 binding motif that is commonly referred to as the GLE2p-binding sequence (GLEBS) motif. However, the crystal structures of two independent complexes formed between peptides that mimic the GLEBS motifs of Mad3 and yeast Bub1 with yeast Bub3 show that the peptides form an extensive interface along the top surface of Bub3 (Larsen and Harrison, 2007) (Fig. 6.3) in a process that seems to imply a large conformational transition of the peptides from a disordered to an ordered state. In a similar fashion, the crystal structure of a Mad1 fragment (residues

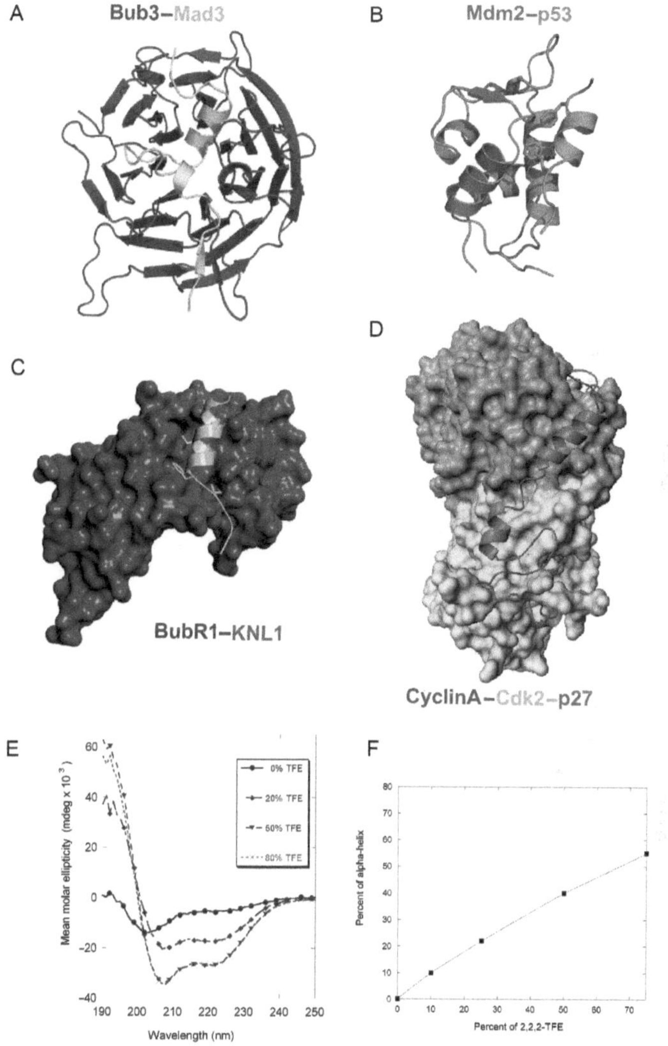

Figure 6.3 Disorder-to-order transitions in SAC signaling. (A) Crystal structure of Mad3 in complex with Bub3 (pdb 2I3T). (B) Structure of a p53 fragment bound to Mdm2 (pdb 1T4F). (C) Structure of the BubR1–KNL1 complex (pdb 3SI5). (D) p27^{Kip1} bound to Cdk2–Cyclin A (pdb 1JSU). (E and F) Far-UV circular dichroism studies of peptides mimicking the BubR1 binding region of human KNL1 show that the peptides undergo a disorder-to-order transition upon titration with TFE. (For color version of this figure, the reader is referred to the online version of this chapter.)

485–584) in complex with Mad2 revealed that the Mad1 fragment adopts a predominantly α-helix conformation upon complex formation (Luo et al., 2000) (Fig. 6.3). Furthermore, binding studies *in vitro* suggest a conformational mechanism in which Mad1 primes the Mad2 binding site for the interaction with Cdc20 (Luo et al., 2002). However, whether these interactions are also established *in vivo* is an aspect that remains to be confirmed.

The interaction of SAC kinases Bub1 and BubR1 with the protein KNL1 (also known as Blinkin and Spc105) physically links SAC signaling with the kinetochore (Bolanos-Garcia and Blundell, 2011; Kiyomitsu et al., 2007, 2011). The crystal structure of a chimeric protein in which the TPR-containing region of human BubR1 was fused to the N-terminal KNL1 region that directly interacts with BubR1 has shown that TPR BubR1 undergoes little conformational change upon KNL1 binding, an observation further confirmed by NMR methods coupled to peptide binding assays. The interaction of N-terminal KNL1 with TPR BubR1 defines an extensive hydrophobic interface implicating KNL1 residues I213, F215, F218, I219, and L222. Because interfering with the KNL1–BubR1 interaction results in premature exit from mitosis, chromosome segregation defects, and the impairment of BubR1 binding to Cdc20 whereas other BubR1 substitutions such as L128A/L131A and Y141A/L142A still localize to the kinetochore, suggest that a productive BubR1–KNL1 interaction involves multiple sites of contact or that BubR1 has several independent binding sites on the kinetochore. We have favored a mechanistic zipper mode of binding in which KNL1 residues I213, F215, F218, and I219 dock into BubR1 pockets in a sequential manner. An important implication of such cooperative, Velcro-like type of interaction is the achievement of high specificity and sensitive regulation. Comparison of the crystal structure of the TPR BubR1-KNL1 binary complex with free KNL1 peptides titrations using 2,2,2-trifluoroethanol and monitored by far-UV circular dichroism revealed that also in this case a disorder-to-order transition (that of N-terminal KNL1 upon binding BubR1) takes place (Fig. 6.3). It is most likely that a similar disorder-to-order transition occurs in N-terminal KNL1 upon binding TPR Bub1. Such a mode of interaction should induce local conformational changes that modulate the binding of KNL1 to other interacting partners. For instance, the disorder-to-order transition of KNL1 upon binding Bub1/BubR1 may assist the exposure of unbound flexible regions of KNL1 to specific kinases and/or phosphatases, a process that is important for a proper SAC response. For instance, recognition sites for PP1 and Aurora kinase B have been mapped onto N-terminal KNL1 and are in close proximity to the Bub1 and BubR1 binding regions (Liu et al., 2010; Rosenberg et al., 2011). Furthermore, the architecture of the KNL1–BubR1 complex shows unique features when compared to the mode of ligand binding of other TPRs of high structural similarity, suggesting that TPR Bub1 and/or TPR BubR1 may present additional protein binding sites (Bolanos-Garcia and Blundell, 2011). In this regard, one possibility is the interaction of a highly conserved motif of N-terminal BubR1 (the KEN box motif) with Cdc20, an interaction that is required for inhibition of the APC/C complex (Burton and Solomon, 2007; Davenport et al., 2006; King et al., 2007).

A transition from a disordered state to a more organized state upon ligand binding is not exclusive of the SAC signaling pathway. A similar disorder-to-order transition has been described in a number of nuclear protein complexes that regulate the eukaryotic cell cycle. One example is the interaction of the inhibitor of cyclin-dependent kinases $p27^{KIP1}$, with the tumor suppressor p53. Limited proteolysis, circular dichroism, and nuclear magnetic resonance spectroscopy analyses have shown that a large fraction of the $p27^{KIP1}$ polypeptide chain is intrinsically unstructured with a marginal content of α-helix (Bienkiewicz et al., 2002; Galea et al., 2008; Kriwacki et al., 1996). The conformational flexibility of these unstructured segments facilitates the phosphorylation of $p27^{KIP1}$ by protein kinase CK2 (Tapia et al., 2004). Phosphorylation of $p27^{KIP1}$ induces a large conformational change that primes $p27^{KIP1}$ for subsequent ubiquitylation and degradation by the proteosome, a critical event that is required for progression through the cell cycle. $p27^{KIP1}$ also undergoes a disorder-to-order transition upon binding to Cyclin A–Cdk2 (Fig. 6.3), an interaction that is important for the regulation of Cdk2 catalytic function (Russo et al., 1996). Another example of disorder-to-order transition occurs during the interaction of N-terminal p53 with murine double minute (Mdm2) protein (Davis et al., 2013; ElSawy et al., 2013; Kussie et al., 1996) (Fig. 6.3). p53 binding to Mdm2 results in the loss of transcriptional activity and stimulates p53 ubiquitination and eventual degradation of this protein by the proteasome (Davis et al., 2013; Shi and Gu, 2012). In summary, disorder-to-order transitions are an important determinant of the function of nuclear proteins that show low structure complexity and high flexibility in the unbound state.

4. MACROMOLECULAR CROWDING OF NUCLEAR PROTEINS

The organization of the polypeptide chain in regions that exhibit low structure complexity is a common feature of protein molecules regardless of the specific subcell compartment in which they are confined (Babu et al., 2012; Dunker et al, 1998; Dyson and Wright, 2002, 2005; Gsponer and Babu, 2009). For instance, extensive bioinformatics analyses have shown that 35–51% of eukaryotic proteins have at least one disordered region encompassing 50 or more amino acid residues (Dunker et al., 2002). Consistent with this notion, DisProt, a curated Protein Disorder Database (http://www.disprot.org), lists 1375 disorder regions in a total of 643 proteins (Sickmeier et al., 2007).

Therefore, it may not come as a surprise that large polypeptide segments of low structure complexity seem to be a common feature of hub proteins in interactome networks (Dosztanyi et al., 2006; Dunker et al., 2005; Haynes et al., 2006; Uversky, 2013). One important implication of such a structural feature is that the rate of interconversion between ensembles of protein conformers can dictate productive interactions with different interacting partners.

Importantly, intrinsic local disorder can accelerate the search for specific targets in the crowded environment of the cell and increase the conformational entropy of the polypeptide chain after complex formation (Meszaros et al., 2007; Sigalov et al., 2007), a feature that may be important for the exquisite regulation of SAC signaling and the communication of this pathway with the KMN. Furthermore, the establishment of large and highly flexible surfaces that mediate productive intermolecular interactions can be the most critical requirement for the establishment of macromolecular assemblies (Dunker et al., 2008; Dyson and Wright, 2005; Kim et al., 2006a,b; Schlessinger et al., 2007). Consequently, suppression or impairment of the stability of hub proteins can have a dramatic impact on the function of an entire interaction network (Albert, 2005; Albert et al., 2000).

Moreover, macromolecular crowding is a parameter used to study the microcompartmentalization of the cell nucleus (Richter et al., 2007). Furthermore, the association of proteins to form macromolecular assemblies through the interaction of regions of low structure complexity to create a crowded environment within the cell (Fig. 6.4) exerts an important influence on protein stability, diffusion of protein complexes, intracellular transport, rate of protein folding, and rate of protein association with other molecules (Banks and Fradin, 2005; Cino et al., 2012; McGuffee and Elcock, 2010; Miermont et al., 2013; Wang et al., 2010, 2012).

Kinesin motor proteins are a good example of this phenomenon, because these proteins process distinct molecular signals in order to operate effectively under the crowded conditions of the cell (Leduc et al., 2012).

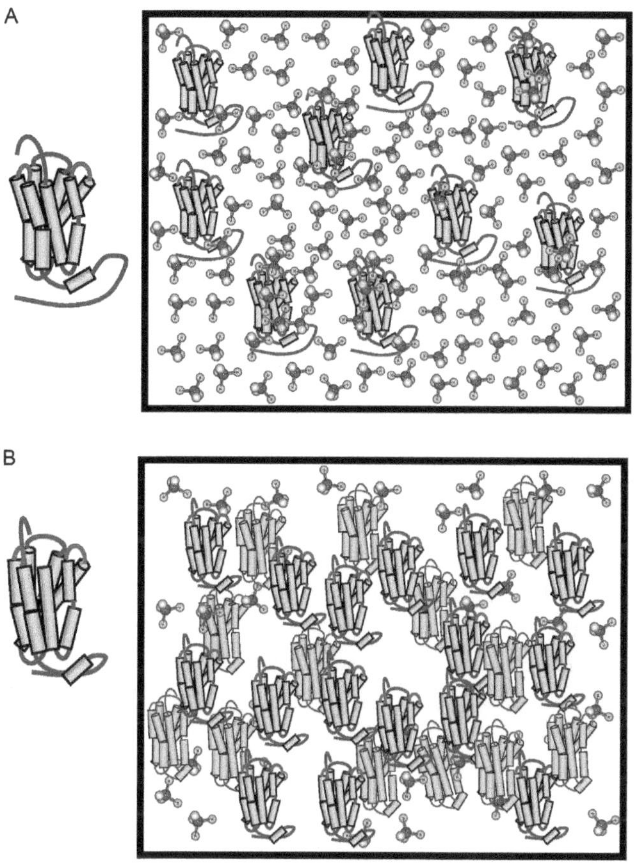

Figure 6.4 Macromolecular crowding in the nucleus. (A) Large macromolecules coexist alongside a high concentration of comparatively smaller molecules. (B) Macromolecular crowding caused by the high density of macromolecules inside the cell gives place to the exclusion of solvent molecules (show in ball and stick representation). (For color version of this figure, the reader is referred to the online version of this chapter.)

These distinctive properties of kinesin motor proteins can be very important for the regulation of protein function, as observed for the protein kinase ERK (Aoki et al., 2011). This class of regulation may also hold true for protein components of the KMN that exhibit low structure complexity, such as KNL1 and members of the CENP protein family such as CENP-C and CENP-E (Perpelescu and Fukagawa, 2011). For instance, recent studies in *Drosophila* have shown that the nuclear import of Spc105 (the fly homologue of KNL1) and its immediate association with the Mis12 complex is required for the onset of kinetochore assembly and that Spc105 nuclear import is mediated by its C-terminal region (Venkei et al., 2012). Therefore, molecular crowding of dynamic structural ensembles of Spc105 in the cellular environment may determine the correct conformational orientation of a certain region of the Spc105 polypeptide chain that is required for Spc105 to act as a licensing factor for the onset of kinetochore assembly (Venkei et al., 2012).

Compared to the study of macromolecular crowding on globular proteins, how macromolecular crowding conditions affect the behavior of intrinsically disordered proteins is less known. What is clear is that this class of proteins often shows a wide range of conformations under nondenaturing conditions. Therefore, in intrinsically disordered proteins, the quick sampling of a dynamic conformational space should have a profound impact on the recognition of their interacting molecules, a notion supported by recent experimental data. For instance, an NMR spin relaxation study of the effect of macromolecular crowding on three proteins, ProTa, TC1, and α-synuclein, revealed a different extent of disorder in each case and that despite the high concentration of other macromolecules present in the system, ProTa, TC1, and α-synuclein remained at least partially disordered (Cino et al., 2012). The same study also revealed that macromolecular crowding exerts a differential effect upon the conformational propensity of distinct regions of low structure complexity. Such a differential effect may stabilize certain ligand-binding motifs without affecting the conformation of large fragments of intrinsically disordered proteins under crowded conditions (Cino et al., 2012). These findings suggest that intrinsically disordered proteins can behave as highly dynamic structural and/or regulatory ensembles in cellular environments. For instance, inhibition of the function of a protein domain can be achieved through interactions with an autoinhibitory module present in the same polypeptide chain. Such autoinhibition can in turn tune the cell to respond only to appropriate signals, thus enhancing the signal-to-noise ratio. This type of regulatory mechanism has been observed in DNA binding to the transcription factors Ets-1 (Lee et al., 2005), NF-kB (Stöven et al., 2000), p53 (Ko and Prives, 1996), and s70 (Dombroski et al., 1993) and in the autoinhibition of the catalytic activity of a number of protein kinases (Hubbard, 2004; Trudeau et al., 2013). Several independent studies have examined in detail the effects of macromolecular crowding on the structure of intrinsically disordered proteins. For instance, it has been shown that FlgM is disordered in dilute buffer solutions and that its C-terminal adopts a regular secondary structure within the cell and in solutions containing a high concentration of glucose

(Dedmon et al., 2002). In sharp contrast, the disordered C-terminal activation domain of c-Fos and the kinase-inhibition domain of p27^{KIP1} behave quite differently: these domains do not undergo important conformational changes in the presence of dextran or Ficoll (Flaugh and Lumb, 2001). Interestingly, α-casein, MAP2c, and p21^{Cip1} have also been reported to experience minor structural changes under molecular crowding conditions (Szasz et al., 2011), thus suggesting that the maintenance of a highly dynamic structure is an important requirement for the function of these proteins.

5. COOPERATIVE INTERACTIONS OF NUCLEAR MULTIPROTEIN COMPLEXES

The cooperative assembly of higher order signaling complexes resulting from specific, low-affinity binary complexes should be advantageous for cell signaling because multiprotein complexes that form cooperatively would less likely be formed by chance (Blundell et al., 2002; Bolanos-Garcia et al., 2012). In agreement with this notion, the cooperative assembly of higher order signaling complexes has been described for the KMN subcomplexes Mis12 and Ndc80, which play an essential role in SAC signaling. The Ndc80 subcomplex is composed of four subunits Ndc80 (the subunit that gives its name to the entire subcomplex), Nuf2, Spc24, and Spc25 that define a dumbbell-shaped molecule (Fig. 6.5) (Ciferri et al., 2005, 2008; Wan et al., 2009; Wei et al., 2005, 2007). The Spc24–Spc25 and Nuf2–Ndc80 subcomplexes are located in opposite ends of the molecule (Fig. 6.5) (Ciferri et al., 2005; Wei et al., 2005). Association of the Nuf2–Ndc80 subunits mediates the binding of the Ndc80 complex to microtubules, while the association of the Spc24–Spc25 heterodimer is required for binding KNL1 and the Mis12 complex (Cheeseman et al., 2006; Ciferri et al., 2008; Joglekar and DeLuca, 2009; Kiyomitsu et al., 2007; Wan et al., 2009; Wei et al., 2007).

A critical aspect of SAC signaling is that the link made by the KMN to connect the centromere to microtubules of the mitotic spindle must be strong enough to sustain the pulling forces during anaphase, whereas at the same time it must be sufficiently dynamic to enable microtubule polymerization–depolymerization, thus ensuring proper chromosome alignment at the metaphase plate. The exquisite regulation of cell division is a fine example of how the remodeling of nuclear macromolecular assemblies in time and space has evolved as a successful strategy that allows sequential interactions and the increase of selectivity with a minimal margin for errors. At the same time, the highly versatile and dynamic remodeling of

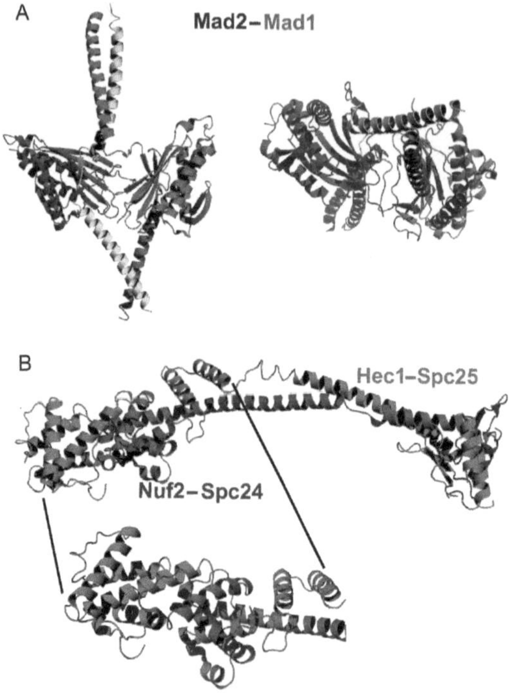

Figure 6.5 The assembly of a range of subcomplexes mediates SAC signaling. (A) The crystal structure of the Mad1–Mad2 complex shows that the two chains of Mad1 interact with Mad2 through the N-terminal coiled-coil region (pdb 1GO4). (B) Crystal structure of a chimeric (bonsai) Ndc80 complex (pdb 2VE7). Spc24 and Spc25 have N-terminal coiled-coils that mediate intersubunit interactions, while Hec1 and Nuf2 contain N-terminal Calponin homology domains followed by a coiled-coil region that is engaged in intersubunit interactions. (For color version of this figure, the reader is referred to the online version of this chapter.)

macromolecular assemblies constitutes a great challenge for their functional, biochemical, and structural characterization temporally as well as spatially. An additional level of complexity is the fact that a wide range of posttranslational modifications such as acetylation, phosphorylation, ubiquitylation, sumoylation, etc. can have a significant impact on protein stability, turnover, reversibility, subcellular localization, and the hierarchical order of assembly/disassembly (Kim et al., 2006a,b; Mao et al., 2011; Pawson and Nash, 2003; Seet et al., 2006; Simorellis and Flynn, 2006; Wan et al., 2012). Importantly, the specific roles of a large number of proteins that have recently been associated with the assembly and regulation of the kinetochore remain to be established. For instance, a study of intact chromosomes using large-scale quantitative mass spectrometry combined with stable-isotope labeling by amino acids in cell culture and bioinformatics techniques identified 4029 mitotic chromosome-associated proteins, of which 562 were previously uncharacterized (Ohta et al., 2010, 2011). Knowledge of the specific role(s) of such a large number of chromosome-associated proteins and the hierarchy of their recruitment to form kinetochore subcomplexes should provide important clues of the mechanism mediating kinetochore assembly/disassembly and molecular details of how this complex choreography of macromolecular interactions regulates the SAC.

6. CONCLUDING REMARKS

The function and regulation of eukaryotic cells depend upon a hierarchical organization of macromolecular assemblies in time and space. In the nucleus, such organization is a structural theme crucial for the regulation of SAC signaling to ensure the accurate and timely transmission of the genetic material to descendants. Intrinsically disordered proteins frequently associate with binding partners through low affinity but highly specific interactions to mediate an effective response in cell cycle regulation. This often involves multiple linear motifs that mediate interaction with one or more ligands, thus increasing the signal to noise ratio. The intrinsic flexibility of intrinsically disordered proteins in the nucleus should be important for the establishment of an effective, polyvalent mode of interaction in a crowded environment. Protein regions of low structural complexity of central components of the SAC and the KMN play essential roles in this process, as greater selectivity is gained by the involvement of multiple components. It can be anticipated that the study of macromolecular crowding of SAC protein assemblies will reveal novel molecular details of the control of chromosome segregation in higher organisms.

REFERENCES

Abrieu, A., Magnaghi-Jaulin, L., Kahana, J.A., Peter, M., Castro, A., Vigneron, S., Lorca, T., Cleveland, D.W., Labbé, J.C., 2001. Mps1 is a kinetochore-associated kinase essential for the vertebrate mitotic checkpoint. Cell 106, 83–93.

Albert, R., 2005. Scale-free networks in cell biology. J. Cell Sci. 118, 4947–4957.

Albert, R., Jeong, H., Barabasi, A.L., 2000. Error and attack tolerance of complex networks. Nature 406, 378–382.

Aoki, K., Yamada, M., Kunida, K., Yasuda, S., Matsuda, M., 2011. Processive phosphorylation of ERK MAP kinase in mammalian cells. Proc. Natl. Acad. Sci. U. S. A. 108, 12675–12680.

Babu, M.M., Kriwacki, R.W., Pappu, R.V., 2012. Structural biology. Versatility from protein disorder. Science 337, 1460–1461.

Banks, D.S., Fradin, C., 2005. Anomalous diffusion of proteins due to molecular crowding. Biophys. J. 89, 2960–2971.

Batra, J., Xu, K., Qin, S., Zhou, H.X., 2009. Effect of macromolecular crowding on protein binding stability: modest stabilization and significant biological consequences. Biophys. J. 97, 906–911.

Bienkiewicz, E.A., Adkins, J.N., Lumb, K.J., 2002. Functional consequences of preorganized helical structure in the intrinsically disordered cell-cycle inhibitor p27 (Kip1). Biochemistry 41, 752–759.

Blundell, T.L., Bolanos-Garcia, V., Chirgadze, D.Y., Harmer, N.J., Lo, T., Pellegrini, L., Sibanda, B.L., 2002. Asymmetry in the multiprotein systems of molecular biology. Struct. Chem. 13, 405–412.

Bolanos-Garcia, V.M., 2007. BUB, mad and the mitotic spindle checkpoint control: new trends in checkpoint signalling. In: Yanson, B.R. (Ed.), New Research on Signal Transduction. Nova Science Publishers, New York, pp. 1–36.

Bolanos-Garcia, V.M., Blundell, T.L., 2011. BUB1 and BUBR1: multifaceted kinases of the cell cycle. Trends Biochem. Sci. 36, 141–150.

Bolanos-Garcia, V.M., Kiyomitsu, T., D'Arcy, S., Chirgadze, D.Y., Grossmann, J.G., Matak-Vinkovic, D., Venkitaraman, A.R., Yanagida, M., Robinson, C.V., Blundell, T.L., 2009. The crystal structure of the N-terminal region of BUB1 provides insight into the mechanism of BUB1 recruitment to kinetochores. Structure 17, 105–116.

Bolanos-Garcia, V.M., Wu, Q., Ochi, T., Chirgadze, D.Y., Sibanda, B.L., Blundell, T.L., 2012. Spatial and temporal organisation of multiprotein assemblies: achieving sensitive control in information-rich cell regulatory systems. Philos. Trans. A Math. Phys. Eng. Sci. 370, 3023–3039.

Boyarchuk, Y., Salic, A., Dasso, M., Arnaoutov, A., 2007. Bub1 is essential for assembly of the functional inner centromere. J. Cell Biol. 176, 919–928.

Brown, K.D., Coulson, R.M., Yen, T.J., Cleveland, D.W., 1994. Cyclin-like accumulation and loss of the putative kinetochore motor CENP-E results from coupling continuous synthesis with specific degradation at the end of mitosis. J. Cell Biol. 125, 1303–1312.

Burton, J.L., Solomon, M.J., 2007. Mad3p, a pseudosubstrate inhibitor of APCCdc20 in the spindle assembly checkpoint. Genes Dev. 21, 655–667.

Carter, S.L., Eklund, A.C., Kohane, I.S., Harris, L.N., Szallasi, Z., 2006. A signature of chromosomal instability inferred from gene expression profiles predicts clinical outcome in multiple human cancers. Nat. Genet. 38, 1043–1048.

Cheeseman, I.M., Chappie, J.S., Wilson-Kubalek, E.M., Desai, A., 2006. The conserved KMN network constitutes the core microtubule-binding site of the kinetochore. Cell 127, 983–997.

Cheeseman, I.M., Hori, T., Fukagawa, T., Desai, A., 2008. KNL1 and the CENP-H/I/K complex coordinately direct kinetochore assembly in vertebrates. Mol. Biol. Cell 19, 587–594.

Ciferri, C., De Luca, J., Monzani, S., Ferrari, K.J., Ristic, D., Wyman, C., Stark, H., Kilmartin, J., Salmon, E.D., Musacchio, A., 2005. Architecture of the human Ndc80-Hec1 complex, a critical constituent of the outer kinetochore. J. Biol. Chem. 280, 29088–29095.

Ciferri, C., Pasqualato, S., Screpanti, E., Varetti, G., Santaguida, S., Dos Reis, G., Maiolica, A., Polka, J., De Luca, J.G., De Wulf, P., Salek, M., Rappsilber, J., Moores, C.A., Salmon, E.D., Musacchio, A., 2008. Implications for kinetochore-microtubule attachment from the structure of an engineered Ndc80 complex. Cell 133, 427–439.

Cino, E.A., Karttunen, M., Choy, W.-Y., 2012. Effects of molecular crowding on the dynamics of intrinsically disordered proteins. PLoS ONE 7, e49876.

Cliff, M.J., Harris, R., Barford, D., Ladbury, J.E., Williams, M.A., 2006. Conformational diversity in the TPR domain-mediated interaction of protein phosphatase 5 with Hsp90. Structure 14, 415–426.
D'Arcy, S., Davies, O.R., Blundell, T.L., Bolanos-Garcia, V.M., 2010. Defining the molecular basis of BubR1 kinetochore interactions and APC/C-CDC20 inhibition. J. Biol. Chem. 285, 14764–14776.
Davenport, J., Harris, L.D., Goorha, R., 2006. Spindle checkpoint function requires Mad2-dependent Cdc20 binding to the Mad3 homology domain of BubR1. Exp. Cell Res. 312, 1831–1842.
Davis, J.R., Mossalam, M., Lim, C.S., 2013. Controlled access of p53 to the nucleus regulates its proteasomal degradation by MDM2. Mol. Pharm. 10, 1340–1349.
Dedmon, M.M., Patel, C.N., Young, G.B., Pielak, G.J., 2002. FlgM gains structure in living cells. Proc. Natl. Acad. Sci. U. S. A. 99, 12681–12684.
DeLano, W.L., 2002. The PyMOL Molecular Graphics System. DeLano Scientific, San Carlos, CA.
Dhar, A., Samiotakis, A., Ebbinghaus, S., Nienhaus, L., Homouz, D., Gruebele, M., Cheung, M.S., 2010. Structure, function, and folding of phosphoglycerate kinase are strongly perturbed by macromolecular crowding. Proc. Natl. Acad. Sci. U. S. A. 107, 17586–17591.
Dombroski, A.J., Walter, W.A., Gross, C.A., 1993. Amino-terminal amino acids modulate sigma-factor DNA-binding activity. Genes Dev. 7, 2446–2455.
Dosztanyi, Z., Chen, J., Dunker, A.K., Simon, I., Tompa, P., 2006. Disorder and sequence repeats in hub proteins and their implications for network evolution. J. Proteome Res. 5, 2985–2995.
Dunker, A.K., Garner, E., Guilliot, S., Romero, P., Albrecht, K., Hart, J., Obradovic, Z., Kissinger, C., Villafranca, J.E., 1998. Protein disorder and the evolution of molecular recognition: theory, predictions and observations. Pac. Symp. Biocomput, 473–484.
Dunker, A.K., Brown, C.J., Lawson, J.D., Iakoucheva, L.M., Obradoviæ, Z., 2002. Intrinsic disorder and protein function. Biochemistry 41, 6573–6582.
Dunker, A.K., Cortese, M.S., Romero, P., Iakoucheva, L.M., Uversky, V.N., 2005. Flexible nets: the roles of intrinsic disorder in protein interaction networks. FEBS J. 272, 5129–5148.
Dunker, A.K., Oldfield, C.J., Meng, J., Romero, P., Yang, J.Y., Chen, J.W., Vacic, V., Obradovic, Z., Uversky, V.N., 2008. The unfoldomics decade: an update on intrinsically disordered proteins. BMC Genomics 9 (Suppl. 2), S1.
Dyson, H.J., Wright, P.E., 2002. Coupling of folding and binding for unstructured proteins. Curr. Opin. Struct. Biol. 12, 54–60.
Dyson, H.J., Wright, P.E., 2005. Intrinsically unstructured proteins and their functions. Nat. Rev. Mol. Cell Biol. 6, 197–208.
Elcock, A.H., 2010. Models of macromolecular crowding effects and the need for quantitative comparisons with experiment. Curr. Opin. Struct. Biol. 20, 196–206.
Ellis, R.J., 2001. Macromolecular crowding: obvious but underappreciated. Trends Biochem. Sci. 26, 597–604.
Elowe, S., 2011. Bub1 and BubR1: at the interface between chromosome attachment and the spindle checkpoint. Mol. Cell. Biol. 31, 3085–3093.
ElSawy, K.M., Verma, C.S., Joseph, T.L., Lane, D.P., Twarock, R., Caves, L.S., 2013. On the interaction mechanisms of a p53 peptide and nutlin with the MDM2 and MDMX proteins: a Brownian dynamics study. Cell Cycle 12, 394–404.
Flaugh, S.L., Lumb, K.J., 2001. Effects of macromolecular crowding on the intrinsically disordered proteins c-Fos and p27(Kip1). Biomacromolecules 2, 538–540.
Foley, E.A., Kapoor, T.M., 2013. Microtubule attachment and spindle assembly checkpoint signalling at the kinetochore. Nat. Rev. Mol. Cell Biol. 14, 25–37.

Funabiki, H., Wynne, D.J., 2013. Making an effective switch at the kinetochore by phosphorylation and dephosphorylation. Chromosoma 122, 135–158.

Galea, C.A., Wang, Y., Sivakolundu, S.G., Kriwacki, R.W., 2008. Regulation of cell division by intrinsically unstructured proteins: intrinsic flexibility, modularity, and signaling conduits. Biochemistry 47, 7598–7609.

Gsponer, J., Babu, M.M., 2009. The rules of disorder or why disorder rules. Prog. Biophys. Mol. Biol. 99, 94–103.

Hancock, R., 2012. Structure of metaphase chromosomes: a role for effects of macromolecular crowding. PLoS One 7, e36045.

Hardwick, K.G., Johnston, R.C., Smith, D.L., Murray, A.W., 2000. MAD3 encodes a novel component of the spindle checkpoint which interacts with Bub3p, CDC20p, and MAD2p. J. Cell Biol. 148, 871–882.

Haynes, C., Oldfield, C.J., Ji, F., Klitgord, N., Cusick, M.E., Radivojac, P., Uversky, V.N., Vidal, M., Iakoucheva, L.M., 2006. Intrinsic disorder is a common feature of hub proteins from four eukaryotic interactomes. PLoS Comput. Biol. 2, e100.

Huang, J.N., Park, I., Ellingson, E., Littlepage, L.E., Pellman, D., 2001. Activity of the APC(Cdh1) form of the anaphase-promoting complex persists until S phase and prevents the premature expression of Cdc20p. J. Cell Biol. 154, 85–94.

Huang, L., Jin, R., Li, J., Luo, K., Huang, T., Wu, D., Wang, W., Chen, R., Xiao, G., 2010. Macromolecular crowding converts the human recombinant PrPC to the soluble neurotoxic beta-oligomers. FASEB J. 24, 3536–3543.

Hubbard, S.R., 2004. Juxtamembrane autoinhibition in receptor tyrosine kinases. Nat. Rev. Mol. Cell Biol. 5, 464–471.

Izawa, D., Pines, J., 2012. Mad2 and the APC/C compete for the same site on Cdc20 to ensure proper chromosome segregation. J. Cell Biol. 199, 27–37.

Janssen, A., Kops, G.J., Medema, R.H., 2009. Elevating the frequency of chromosome missegregation as a strategy to kill tumor cells. Proc. Natl. Acad. Sci. U. S. A. 106, 19108–19113.

Jia, L., Kim, S., Yu, H., 2013. Tracking spindle checkpoint signals from kinetochores to APC/C. Trends Biochem. Sci. 38, 302–311.

Joglekar, A.P., DeLuca, J.G., 2009. Chromosome segregation: Ndc80 can carry the load. Curr. Biol. 19, R404–R407.

Kadura, S., He, X., Vanoosthuyse, V., Hardwick, K.G., Sazer, S., 2005. The A78V mutation in the Mad3-like domain of Schizosaccharomyces pombe Bub1p perturbs nuclear accumulation and kinetochore targeting of Bub1p, Bub3p, and Mad3p and spindle assembly checkpoint function. Mol. Biol. Cell 16, 385–395.

Kim, P.M., Lu, L.J., Xia, Y., Gerstein, M.B., 2006a. Relating three-dimensional structures to protein networks provides evolutionary insights. Science 314, 1938–1941.

Kim, S.C., Sprung, R., Chen, Y., Xu, Y., Ball, H., Pei, J., Cheng, T., Kho, Y., Xiao, H., Xiao, L., Grishin, N.V., White, M., Yang, X.J., Zhao, Y., 2006b. Substrate and functional diversity of lysine acetylation revealed by a proteomics survey. Mol. Cell 23, 607–618.

King, E.M., van der Sar, S.J., Hardwick, K.G., 2007. Mad3 KEN boxes mediate both Cdc20 and Mad3 turnover, and are critical for the spindle checkpoint. PLoS One 2, e342.

Kiyomitsu, T., Obuse, C., Yanagida, M., 2007. Human Blinkin/AF15q14 is required for chromosome alignment and the mitotic checkpoint through direct interaction with Bub1 and BubR1. Dev. Cell 13, 663–676.

Kiyomitsu, T., Murakami, H., Yanagida, M., 2011. Protein interaction domain mapping of human kinetochore protein Blinkin reveals a consensus motif for binding of spindle assembly checkpoint proteins Bub1 and BubR1. Mol. Cell. Biol. 31, 998–1011.

Ko, L.J., Prives, C., 1996. p53: puzzle and paradigm. Genes Dev. 10, 1054–1072.

Kriwacki, R.W., Hengst, L., Tennant, L., Reed, S.I., Wright, P.E., 1996. Structural studies of p21Waf1/Cip1/Sdi1 in the free and Cdk2-bound state: conformational disorder mediates binding diversity. Proc. Natl. Acad. Sci. U. S. A. 93, 11504–11509.

Kussie, P.H., Gorina, S., Marechal, V., Elenbaas, B., Moreau, J., Levine, A.J., Pavletich, N.P., 1996. Structure of the MDM2 oncoprotein bound to the p53 tumor suppressor transactivation domain. Science 274, 948–953.

Larsen, N.A., Harrison, S.C., 2004. Crystal structure of the spindle assembly checkpoint protein Bub3. J. Mol. Biol. 344, 885–892.

Larsen, N.A., Harrison, S.C., 2007. Structural analysis of Bub3 interactions in the mitotic spindle checkpoint. Proc. Natl. Acad. Sci. U. S. A. 104, 1201–1206.

Leduc, C., Padberg-Gehle, K., Varga, V., Helbing, D., Diez, S., Howard, J., 2012. Molecular crowding creates traffic jams of kinesin motors on microtubules. Proc. Natl. Acad. Sci. U. S. A. 109, 6100–6105.

Lee, G.M., Donaldson, L.W., Pufall, M.A., Kang, H.-S., Pot, I., Graves, B.J., McIntosh, L.P., 2005. The structural and dynamic basis of Ets-1 DNA binding auto-inhibition. J. Biol. Chem. 280, 7088–7099.

Lee, S., Thebault, P., Freschi, L., Beaufils, S., Blundell, T.L., Landry, C.R., Bolanos-Garcia, V.M., Elowe, S., 2012. Characterization of spindle checkpoint kinase Mps1 reveals domain with functional and structural similarities to tetratricopeptide repeat motifs of Bub1 and BubR1 checkpoint kinases. J. Biol. Chem. 287, 5988–6001.

Liu, D., Vleugel, M., Backer, C.B., Hori, T., Fukagawa, T., Cheeseman, I.M., Lampson, L.A., 2010. Regulated targeting of protein phosphatase 1 to the outer kinetochore by KNL1 opposes Aurora B kinase. J. Cell Biol. 188, 809–920.

London, N., Ceto, S., Ranish, J.A., Biggins, S., 2012. Phosphoregulation of Spc105 by Mps1 and PP1 regulates Bub1 localization to kinetochores. Curr. Biol. 22, 900–906.

Luo, X., Fang, G., Coldiron, M., Lin, Y., Yu, H., Kirschner, M.W., Wagner, G., 2000. Structure of the MAD2 spindle assembly checkpoint protein and its interaction with CDC20. Nat. Struct. Biol. 7, 224–229.

Luo, X., Tang, Z., Rizo, J., Yu, H., 2002. The MAD2 spindle checkpoint protein undergoes similar major conformational changes upon binding to either MAD1 or CDC20. Mol. Cell 9, 59–71.

Ma, Q., Fan, J.B., Zhou, Z., Zhou, B.R., Meng, S.R., Hu, J.Y., Chen, J., Liang, Y., 2012. The contrasting effect of macromolecular crowding on amyloid fibril formation. PLoS One 7, e36288.

Maciejowski, J., George, K.A., Terret, M.E., Zhang, C., Shokat, K.M., Jallepalli, P.V., 2010. Mps1 directs the assembly of Cdc20 inhibitory complexes during interphase and mitosis to control M phase timing and spindle checkpoint signaling. J. Cell Biol. 190, 89–100.

Mao, Y., Abrieu, A., Cleveland, D.W., 2003. Activating and silencing the mitotic checkpoint through CENP-E-dependent activation/inactivation of BubR1. Cell 114, 87–98.

Mao, Y.S., Zhang, B., Spector, D.L., 2011. Biogenesis and function of nuclear bodies. Trends Genet. 27, 295–306.

Mayes, P.A., Degenhardt, Y.Y., Wood, A., Toporovskya, Y., Diskin, S.J., Haglund, E., Moy, C., Wooster, R., Maris, J.M., 2013. Mitogen-activated protein kinase (MEK/ERK) inhibition sensitizes cancer cells to centromere-associated protein E inhibition. Int. J. Cancer 132, E149–E157.

McGuffee, S.R., Elcock, A.H., 2010. Diffusion, crowding & protein stability in a dynamic molecular model of the bacterial cytoplasm. PLoS Comput. Biol. 6, e1000694.

Meszaros, B., Tompa, P., Simon, I., Dosztanyi, Z., 2007. Molecular principles of the interactions of disordered proteins. J. Mol. Biol. 372, 549–561.

Meyer, R.E., Kim, S., Obeso, D., Straight, P.D., Winey, M., Dawson, D.S., 2013. Mps1 and Ipl1/Aurora B act sequentially to correctly orient chromosomes on the meiotic spindle of budding yeast. Science 339, 1071–1074.

Miermont, A., Waharte, F., Hu, S., McClean, M.N., Bottani, S., Léon, S., Hersen, P., 2013. Severe osmotic compression triggers a slowdown of intracellular signaling, which can be explained by molecular crowding. Proc. Natl. Acad. Sci. U. S. A. 110, 5725–5730.
Minton, A.P., 1997. Influence of excluded volume upon macromolecular structure and associations in 'crowded' media. Curr. Opin. Biotechnol. 8, 65–69.
Minton, A.P., 2000. Implications of macromolecular crowding for protein assembly. Curr. Opin. Struct. Biol. 10, 34–39.
Morrow, C.J., Tighe, A., Johnson, V.L., Scott, M.I., Ditchfield, C., Taylor, S.S., 2005. BUB1 and aurora B cooperate to maintain BubR1-mediated inhibition of APC/CCDC20. J. Cell Sci. 118, 3639–3652.
Nilsson, J., Yekezare, M., Minshull, J., Pines, J., 2008. The APC/C maintains the spindle assembly checkpoint by targeting Cdc20 for destruction. Nat. Cell Biol. 10, 1411–1420.
Ohta, S., Bukowski-Wills, J.C., Sanchez-Pulido, L., Alves, FdeL, Wood, L., Chen, Z.A., Platani, M., Fischer, L., Hudson, D.F., Ponting, C.P., Fukagawa, T., Earnshaw, W.C., Rappsilber, J., 2010. The protein composition of mitotic chromosomes determined using multiclassifier combinatorial proteomics. Cell 142, 810–821.
Ohta, S., Wood, L., Bukowski-Wills, J.C., Rappsilber, J., Earnshaw, W.C., 2011. Building mitotic chromosomes. Curr. Opin. Cell Biol. 23, 114–121.
Pawson, T., Nash, P., 2003. Assembly of cell regulatory systems through protein interaction domains. Science 300, 445–452.
Perpelescu, M., Fukagawa, T., 2011. The ABCs of CENPs. Chromosoma 120, 425–446.
Pfleger, C.M., Kirschner, M.W., 2000. The KEN box: an APC recognition signal distinct from the D box targeted by Cdh1. Genes Dev. 14, 655–665.
Przewloka, M.R., Glover, D.M., 2009. The kinetochore and the centromere: a working long distance relationship. Annu. Rev. Genet. 43, 439–465.
Przewloka, M.R., Venkei, Z., Bolanos-Garcia, V.M., Debski, J., Dadlez, M., Glover, D.M., 2011. CENP-C is a structural platform for kinetochore assembly. Curr. Biol. 21, 399–405.
Reis, A., Levasseur, M., Chang, H.Y., Elliott, D.J., Jones, K.T., 2006. The CRY box: a second APCcdh1-dependent degron in mammalian cdc20. EMBO Rep. 7, 1040–1045.
Richter, K., Nessling, M., Lichter, P., 2007. Experimental evidence for the influence of molecular crowding on nuclear architecture. J. Cell Sci. 20, 1673–1680.
Rosenberg, J.S., Cross, F.R., Funabiki, H., 2011. KNL1/Spc105 recruits PP1 to silence the spindle assembly checkpoint. Curr. Biol. 21, 942–947.
Russo, A.A., Jeffrey, P.D., Patten, A.K., Massagué, J., Pavletich, N.P., 1996. Crystal structure of the p27Kip1 cyclin-dependent-kinase inhibitor bound to the cyclin A-Cdk2 complex. Nature 382, 325–331.
Sasahara, K., McPhie, P., Minton, A.P., 2003. Effect of dextran on protein stability and conformation attributed to macromolecular crowding. J. Mol. Biol. 326, 1227–1237.
Schlessinger, A., Liu, J., Rost, B., 2007. Natively unstructured loops differ from other loops. PLoS Comput. Biol. 3, e140.
Sedgwick, G.G., Hayward, D.G., Di Fiore, B., Pardo, M., Yu, L., Pines, J., Nilsson, J., 2013. Mechanisms controlling the temporal degradation of Nek2A and Kif18A by the APC/C-Cdc20 complex. EMBO J. 32, 303–314.
Seet, B.T., Dikic, I., Zhou, M.M., Pawson, T., 2006. Reading protein modifications with interaction domains. Nat. Rev. Mol. Cell Biol. 7, 473–483.
Shepperd, L.A., Meadows, J.C., Sochaj, A.M., Lancaster, T.C., Zou, J., Buttrick, G.J., Rappsilber, J., Hardwick, K.G., Millar, J.B., 2012. Phosphodependent recruitment of Bub1 and Bub3 to Spc7/KNL1 by Mph1 kinase maintains the spindle checkpoint. Curr. Biol. 22, 891–899.
Shi, D., Gu, W., 2012. Dual roles of MDM2 in the regulation of p53: ubiquitination dependent and ubiquitination independent mechanisms of MDM2 repression of p53 activity. Genes Cancer 3, 240–248.

Sickmeier, M., Hamilton, J.A., LeGall, T., Vacic, V., Cortese, M.S., Tantos, A., Szabo, B., Tompa, P., Chen, J., Uversky, V.N., Obradovic, Z., Dunker, A.K., 2007. DisProt: the database of disordered proteins. Nucleic Acids Res. 35, D786–D793.

Sigalov, A.B., Zhuravleva, A.V., Orekhov, V.Y., 2007. Binding of intrinsically disordered proteins is not necessarily accompanied by a structural transition to a folded form. Biochimie 89, 419–421.

Simorellis, A.K., Flynn, P.F., 2006. Fast local backbone dynamics of encapsulated ubiquitin. J. Am. Chem. Soc. 128, 9580–9581.

Stagg, L., Zhang, S.Q., Cheung, M.S., Wittung-Stafshede, P., 2007. Molecular crowding enhances native structure and stability of alpha/beta protein flavodoxin. Proc. Natl. Acad. Sci. U. S. A. 104, 18976–18981.

Stöven, S., Ando, I., Kadalayil, L., Engström, Y., Hultmark, D., 2000. Activation of the Drosophila NF-kappaB factor Relish by rapid endoproteolytic cleavage. EMBO Rep. 1, 347–352.

Szasz, C.S., Alexa, A., Toth, K., Rakacs, M., Langowski, J., Tompa, P., 2011. Protein disorder prevails under crowded conditions. Biochemistry 50, 5834–5844.

Tanaka, K., 2013. Regulatory mechanisms of kinetochore-microtubule interaction in mitosis. Cell. Mol. Life Sci. 70, 559–579.

Tang, Z., Shu, H., Oncel, D., Chen, S., Yu, H., 2004. Phosphorylation of CDC20 by Bub1 provides a catalytic mechanism for APC/C inhibition by the spindle checkpoint. Mol. Cell 16, 387–397.

Tapia, J.C., Bolanos-Garcia, V.M., Sayed, M., Allende, C.C., Allende, J.E., 2004. Cell cycle regulatory protein p27KIP1 is a substrate and interacts with the protein kinase CK2. J. Cell. Biochem. 91, 865–879.

Thebault, P., Chirgadze, D.Y., Dou, Z., Blundell, T.L., Elowe, S., Bolanos-Garcia, V.M., 2012. Structural and functional insights into the role of the N-terminal Mps1 TPR domain in the SAC (spindle assembly checkpoint). Biochem. J. 448, 321–328.

Trudeau, T., Nassar, R., Cumberworth, A., Wong, E.T.C., Woollard, G., Gsponer, J., 2013. Structure and intrinsic disorder in protein autoinhibition. Structure 21, 332–341.

Uversky VN. 2013. Intrinsic disorder-based protein interactions and their modulators. Curr Pharm Des. 19, 4191-4213.

Vanoosthuyse, V., Hardwick, K.G., 2005. Bub1 and the multilayered inhibition of CDC20-APC/C in mitosis. Trends Cell Biol. 15, 231–233.

Venkei, Z., Przewloka, M.R., Ladak, Y., Albadri, S., Sossick, A., Juhasz, G., Novák, B., Glover, D.M., 2012. Spatiotemporal dynamics of Spc105 regulates the assembly of the Drosophila kinetochore. Open Biol. 2, 110032.

Wan, X., O'Quinn, R.P., Pierce, H.L., Joglekar, A.P., Gall, W.E., DeLuca, J.G., Carroll, C.W., Liu, S.T., Yen, T.J., McEwen, B.F., Stukenberg, P.T., Desai, A., Salmon, E.D., 2009. Protein architecture of the human kinetochore microtubule attachment site. Cell 137, 672–684.

Wan, J., Subramonian, D., Zhang, X.D., 2012. SUMOylation in control of accurate chromosome segregation during mitosis. Curr. Protein Pept. Sci. 13, 467–481.

Wang, Y., Li, C., Pielak, G.J., 2010. Effects of proteins on protein diffusion. J. Am. Chem. Soc. 132, 9392–9397.

Wang, Y., Sarkar, M., Smith, A.E., Krois, A.S., Pielak, G.J., 2012. Macromolecular crowding and protein stability. J. Am. Chem. Soc. 134, 16614–16618.

Warren, C.D., Brady, D.M., Johnston, R.C., Hanna, J.S., Hardwick, K.G., Spencer, F.A., 2002. Distinct chromosome segregation roles for spindle checkpoint proteins. Mol. Biol. Cell 13, 3029–3041.

Weaver, B.A., Bonday, Z.Q., Putkey, F.R., Kops, G.J., Silk, A.D., Cleveland, D.W., 2003. Centromere-associated protein-E is essential for the mammalian mitotic checkpoint to prevent aneuploidy due to single chromosome loss. J. Cell Biol. 162, 551–563.

Wei, R.R., Sorger, P.K., Harrison, S.C., 2005. Molecular organization of the Ndc80 complex, an essential kinetochore component. Proc. Natl. Acad. Sci. U. S. A. 102, 5363–5367.

Wei, R.R., Al-Bassam, J., Harrison, S.C., 2007. The Ndc80/HEC1 complex is a contact point for kinetochore-microtubule attachment. Nat. Struct. Mol. Biol. 14, 54–55.

Weiss, E., Winey, M., 1996. The *Saccharomyces cerevisiae* spindle pole body duplication gene *MPS1* is part of a mitotic checkpoint. J. Cell Biol. 132, 111–123.

Westhorpe, F.G., Straight, A.F., 2013. Functions of the centromere and kinetochore in chromosome segregation. Curr. Opin. Cell Biol. 25, 334–340.

Wilson, D.K., Cerna, D., Chew, E., 2005. The 1.1-angstrom structure of the spindle checkpoint protein Bub3p reveals functional regions. J. Biol. Chem. 280, 13944–13951.

Yao, Y., Dai, W., 2012. Mitotic checkpoint control and chromatin remodeling. Front. Biosci. 17, 976–983.

Zecevic, M., Catling, A.D., Eblen, S.T., Renzi, L., Hittle, J.C., Yen, T.J., Gorbsky, G.J., Weber, M.J., 1998. Active MAP kinase in mitosis: localization at kinetochores and association with the motor protein CENP-E. J. Cell Biol. 142, 1547–1558.

Zeytuni, N., Zarivach, R., 2012. Structural and functional discussion of the tetra-tricopeptide repeat, a protein interaction module. Structure 20, 397–405.

Zhou, H.X., Rivas, G., Minton, A.P., 2008. Macromolecular crowding and confinement: biochemical, biophysical, and potential physiological consequences. Annu. Rev. Biophys. 37, 375–397.

Zhou, Z., Yan, X., Pan, K., Chen, J., Xie, Z.S., Xiao, G.F., Yang, F.Q., Liang, Y., 2011. Fibril formation of the rabbit/human/bovine prion proteins. Biophys. J. 101, 1483–1492.

Zimmerman, S.B., Minton, A.P., 1993. Macromolecular crowding: biochemical, biophysical, and physiological consequences. Annu. Rev. Biophys. Biomol. Struct. 22, 27–65.

Zimmerman, S.B., Trach, S.O., 1991. Estimation of macromolecule concentrations and excluded volume effects for the cytoplasm of Escherichia coli. J. Mol. Biol. 222, 599–620.

CHAPTER SEVEN

Characteristic Behavior of Crowding Macromolecules Confined in Cell-Sized Droplets

Miho Yanagisawa*, Takahiro Sakaue*, Kenichi Yoshikawa†,1
*Department of Physics, Graduate School of Sciences, Kyushu University, Fukuoka, Japan
†Faculty of Life and Medical Sciences, Doshisha University, Kyotanabe, Kyoto, Japan
1Corresponding author: e-mail address: keyoshik@mail.doshisha.ac.jp

Contents

Abstract

We describe the unique behavior of long DNA molecules in a cell-sized confinement by using cell-sized water-in-oil droplets coated with a lipid layer. It is shown that DNA molecules exhibit specific localization on lipid membranes accompanied by conformational transitions depending on the spatial size of droplets as well as the chemical structure of lipids coating the droplets. Interestingly, such specific behavior of DNA molecules causes on/off switching of transcriptional activity. We demonstrate the effect of crowding

International Review of Cell and Molecular Biology, Volume 307
ISSN 1937-6448
http://dx.doi.org/10.1016/B978-0-12-800046-5.00007-2

by adapting various kinds of macromolecules with different electronic charges and flexibility. As recently reported, it is shown that gene expression is remarkably accelerated in cell-sized confinement. We theoretically discuss the abovementioned unique behavior of DNA molecules in a micro-confinement. Further study in comparison with the actual behavior of DNA molecules in a living cellular system is expected to be fruitful.

1. INTRODUCTION

1.1. Cell as highly crowded system

Cell cytoplasm is highly concentrated and crowded with macromolecules such as proteins, DNA, RNA, and polysaccharides, the total mass density of which exceeds hundreds of grams per liter (Konopka et al., 2006; Zimmerman and Minton, 1993; Zimmerman and Trach, 1991). In contrast to such a cytoplasmic condition, most of the biochemical and molecular biology experiments have been carried out under relatively simple conditions of purified macromolecules. Therefore, it is important to unveil the effect of macromolecular crowding in cells (Ando and Skolnick, 2010; Eisenhoffer et al., 2012; Elcock, 2010; Zhou et al., 2008). In fact, it has been indicated that crowding affects replication and transcription of DNA (Ellis, 2001; Li et al., 2009) and protein functions such as folding (Fujiwara and Nomura, 2013; Schlesinger et al., 2011). Among the crowding effects on macromolecules, it is known that the so-called depletion effect describes the conditions for which the increase in entropy is accompanied by de-mixing or phase separation in polymer systems (Adams et al., 1998; Albiez et al., 2006; Asakura and Oosawa, 1958; de Vries, 2006; Emanuel et al., 2009; Tuinier et al., 2003). When various particles with different sizes are encapsulated in a small space, the depletion interaction exists between the larger particles and the surface of the confined space (Dijkstra and van Roij, 2002; Dinsmore et al., 1996; Zhou et al., 2008). In addition, the confinement itself directly affects the structure of the encapsulated molecules when the size of the confined space is on the same order of magnitude of the characteristic length of the molecule, such as persistence length, radius of gyration, and correlation length (Claessens et al., 2006; Limozin and Sackmann, 2002; Limozin et al., 2005). Furthermore, in the case of electrically charged molecules, the electrostatic interaction with the surface of the confined space plays a significant role especially in smaller spaces having a large surface-to-volume ratio.

These phenomena should be observed even in model cells such as liposomes under the highly concentrated and cell-sized confined conditions.

However, the experiment verification using a lipid bilayer liposome has been difficult due to following reasons: (i) size control in the micrometer scale is difficult, (ii) liposomes are covered with a soft lipid membrane and thus fragile, (iii) flexible and semiflexible macromolecules change their conformation according to their surrounding conditions, (iv) a highly concentrated solution has a high osmotic pressure, and (v) almost all biomacromolecules are electrically charged. Considering such difficult situations linked to liposomes usage, we developed a simple easy-to-handle experimental system consisting in water-in-oil (W/O) droplets coated with a phospholipid layer as a simple cell model to encapsulate highly concentrated biomacromolecules. The large surface tension at the oil/water interface (>1 mN/m) can stabilize the droplet system against the high osmotic pressure difference between the inner and outer media. Indeed, it is possible to examine the crowding and confinement effects by encapsulating macromolecules together with their substrates with at almost desired physiological concentrations in the droplets with different sizes.

Here, we first describe the behavior of long DNA molecules encapsulated in droplets, where DNA shows different localizations and conformational transitions according to the droplet size and the types of lipids coating the droplet surface (Kato et al., 2009, 2010). The unfold–fold transition of DNA triggers the on/off switching of the transcription (Tsuji and Yoshikawa, 2010a). Second, we describe phase separation in binary systems: constituted by DNA and other macromolecule such as poly-alginate, actin filaments, and poly(ethylene glycol) (PEG) (Biswas et al., 2012; Negishi et al., 2010, 2011). The observed phenomena are explained theoretically according to the difference of the physicochemical properties of those macromolecules. Third, we show the remarkable acceleration of gene expression in smaller droplets (Kato et al., 2012). Fourth, we introduce a method to prepare liposomes from cell-sized droplets to mimic more realistic cell models, where gene expression systems and membrane proteins with the appropriate orientation are reconstituted (Kato et al., 2012; Nomura et al., 2003; Saito et al., 2009; Yamada et al., 2006; Yanagisawa et al., 2011). Finally, we summarize these characteristic behaviors of crowding macromolecules in a droplet from a physical point of view.

1.2. Preparation of cell-sized droplets covered with phospholipid layer

Preparation of W/O droplets coated with a phospholipid layer is rather simple (Fig. 7.1) (Kato et al., 2009; Yamada et al., 2006), First, we dissolve lipids

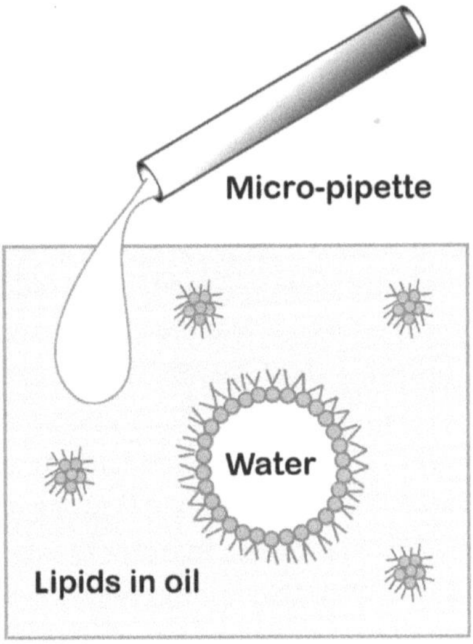

Figure 7.1 Schematic illustration of the experimental procedure used to obtain water-in-oil (W/O) droplets. Lipids spontaneously cover the surface of the W/O droplets.

in chloroform or chloroform/methanol (1:1 molar ratio) solution to be approximately 10 mM in a glass tube. To prepare the lipid membrane, we usually use zwitterionic phosphatidylethanolamine (PE) and phosphatidylcholine (PC). Evaporation of the organic solvent under a nitrogen gas flow brings dry films of lipids at the bottom of the glass tube. Oil such as mineral oil (e.g., liquid paraffin) or hexadecane is added to the lipid films prior to sonication for about 90 min at above lipid gel–liquid crystal transition, T_t. For example, for dimyristoylphosphatidylcholine (14:0) and dioleoylphosphatidylcholine (DOPC, 18:1), the T_t values are about 23 and −20 °C, respectively. Just after the sonication, vortex mixing is performed to prevent the aggregation of lipids. The final lipid concentration in oil is approximately 1 mM. This procedure results in dispersed lipid micelles in oil phase. To obtain W/O droplets, less than 10% (v/v) of aqueous solution is added to the oil phase and emulsification is performed by pipetting. The radius of the resulting droplets ranges from 5 to 200 μm. An aliquot of droplets is placed on a silicone-coated glass cover to avoid adhesion of droplets on the glass plate.

The W/O droplet system is stable against encapsulation of highly concentrated solutions due to the large surface tension at the oil/water interface. The surface tension of DOPC droplets in mineral oil or hexadecane is about 1 mN/m (Yanagisawa et al., 2013). PE droplets can be obtained, although the conical-shaped PE lipids cannot form a lipid bilayer in water, unlike PC lipids. In addition, we can easily control the droplet size in a wide range from 5 to 200 μm. These advantages enable us to analyze the interactions between the lipid membranes and encapsulated molecules, according to the types of lipids and the droplet sizes. Thus, we utilized the W/O droplet coated with a lipid layer as a simple cell model, and encapsulated various types of biomacromolecules, such as DNA and actin.

1.3. Preparation of DNA and actin solution

Here, bacteriophage T4 DNA is used as a long DNA molecule. To label the T4 DNA with YOYO-1 iodide (excitation 491 nm, emission 509 nm), we add an equal volume of 1.5 mM YOYO-1 in water to 10 ng/mL of T4 DNA (molar ratio of base pairs to YOYO-1 is 10), and allow the mixture to remain overnight at 4 °C. For the fluorescence microscope observation, T4 DNA/YOYO-1 solution is diluted 10- to 100-fold in 5–10 mM Tris–HCl, pH 7–8, 30–100 mM KCl, depending on the samples. Cations such as Mg^{2+} and spermine are added according to each system. Actin is obtained from rabbit skeletal muscles as reported previously (Ebashi and Ebashi, 1965; Higashi-Fujime, 1983) and polymerized in a buffer containing 2 mM Tris–HCl, pH 8.0, 30 mM KCl, and 0.2 mM ATP. The mother solution is diluted to the desired concentration by using 5 mM Tris–HCl, pH 8.0, 30 mM KCl, and 0.2 mM ATP buffer. Actin filaments are previously stained by Rhodamine phalloidin (excitation 550 nm, emission 575 nm) at 1/40 mol% against actin monomer (Negishi et al., 2010; Takiguchi et al., 2009).

2. CONFINEMENT OF LONG DNA MOLECULES IN DROPLETS

2.1. DNA adsorption on lipid membrane accompanied by structural transition

DNA is a negatively charged and semiflexible polymer. A long DNA, larger than several tens of kilo base pairs (kb), exhibits a conformational change from a coiled state to a highly folded state in aqueous solution, depending on the presence of condensing agents such as polyamines and polyphosphates (Yoshikawa, 2001). This conformational transition is discrete at the single-molecule level, and can be described in terms of first-order phase transition under the criterion of Landau (Koltover et al., 1998; Melnikov et al., 1995; Yoshikawa et al., 1996). Such a conformational transition is expected to be different from that in cells, where the spatial size is restricted to a micrometer scale and surrounded by a lipid membrane.

To reveal how the cell-sized confinement affects the conformation of DNA molecules, we encapsulate long DNA molecules (bacteriophage T4 DNA 166 kb) in cell-sized W/O droplets coated with a lipid layer. The buffer conditions are 10 mM Tris–HCl, pH 7.4, 100 mM KCl, with various concentrations of $MgCl_2$ and spermine (tetravalent amine). To avoid the intermolecular aggregation of T4 DNA molecules due to the addition of spermine, we perform experiments using a very low concentration of DNA (0.1–1.0 ng/mL). Consequently, encapsulated DNA molecules vary their conformations and spatial arrangements according to the concentrations of Mg^{2+} and spermine, droplet size, and the types of lipid coating the droplet (Kato et al., 2009, 2010).

Figure 7.2 shows the distributions of T4 DNA labeled with the fluorescent dye YOYO-1 inside droplets with diameter <60 μm covered with two types of lipids: DOPE (dioleoylphosphatidylethanolamine) and DOPC. The results clearly demonstrate that the DNA molecules localize on the droplet surface covered with a DOPE membrane in the presence of Mg^{2+} ions, but not on a DOPC membrane. Although these lipids are both zwitterionic, such a localization of DNA on the membrane is only observed for DOPE droplets. In addition, a single molecular observation reveals that the DNA adsorption triggers the unfolding of its structure into an extended coil conformation under high concentrations of Mg^{2+} (Fig. 7.3). The DNA molecule attaches to the membrane at several (e.g., up to 10) points within the molecule, while other parts of the molecule between the attachment points are free. Therefore, DNA localizes close to the membrane surface. Such localization of DNA molecules on a DOPE membrane is confirmed using

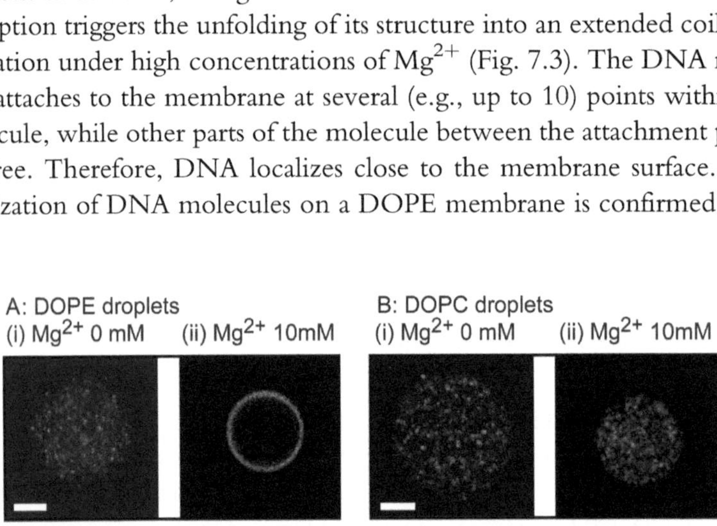

Figure 7.2 Distribution of tagged DNA molecules in a droplet coated with (A) DOPE and (B) DOPC lipid layer. The buffer condition is (i) Mg^{2+} 0 mM, spermine 0 mM, and (ii) Mg^{2+} 10 mM, spermine 0 mM. Scale bar is 20 μm. *Reproduced with permission from Kato et al. (2009).*

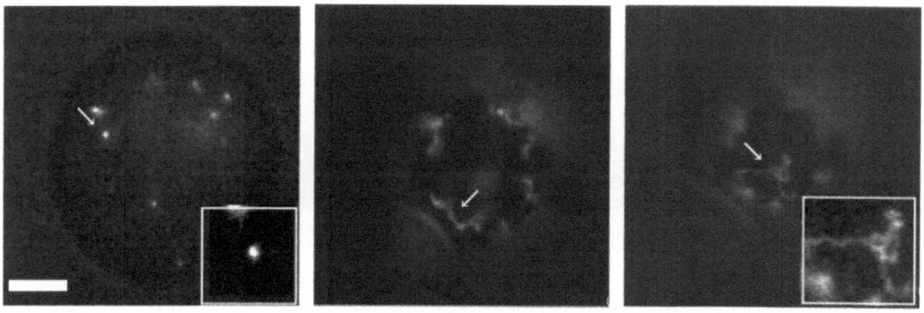

Figure 7.3 Conformation of DNA molecules in a DOPE droplet. The arrow indicates a single DNA molecule, and the magnified view is shown in each insets. The buffer condition is (left) Mg^{2+} 0 mM, spermine 1.5 mM, and (middle and right) Mg^{2+} 10 mM, spermine 1.5 mM. Scale bar is 10 μm. The left image shows an equatorial section of a droplet, and the middle and right images are focused on the top. *Reproduced with permission from Kato et al. (2009).*

a lipid bilayer liposome composed of DOPC, DPPE, and cholesterol (Chol). With a decrease in temperature, the ternary liposome separates into a DPPE–Chol-rich phase and a DOPC-rich phase. When we add DNA molecules exhibiting a compact conformation in aqueous solution, they are adsorbed onto the PE-rich phase in presence of the Mg^{2+} by assuming an elongated conformation (Kato et al., 2010).

In addition, the droplet size affects the distribution of DNA molecules. We analyze this from DOPE droplets containing DNA molecules with 10 mM Mg^{2+} and 1.5 mM spermine, and give a summary in Fig. 7.4. In the case of smaller droplets with a diameter $D < 60$ μm, DNA adsorbs onto the membrane surface (Fig. 7.2). We confirm that DNA molecules in bulk maintain the folded conformation under these experimental conditions. On the contrary, in the case of larger droplets ($D > 100$ μm), DNA molecules are not adsorbed on the DOPE membrane and thus conformational transition is not observed, similar to those in the bulk.

It has been shown that a zwitterionic phospholipid, such as PE or PC, does not interact with DNA directly; however, the interaction can be mediated by divalent cations such as Mg^{2+} and Ca^{2+} (Gromelski and Brezesinski, 2006; Mengistu et al., 2009; Tresset et al., 2007). This is attributed to the fact that divalent cations bridge the negative phosphate moieties of the lipid head groups, and the lipid layer becomes positively charged (Herbette et al., 1984; Lis et al., 1981). Our real-time observation reveals that the interaction between the DNA molecule and the PE membrane surface is rather weak because, although the DNA molecule in the coil state attaches to the

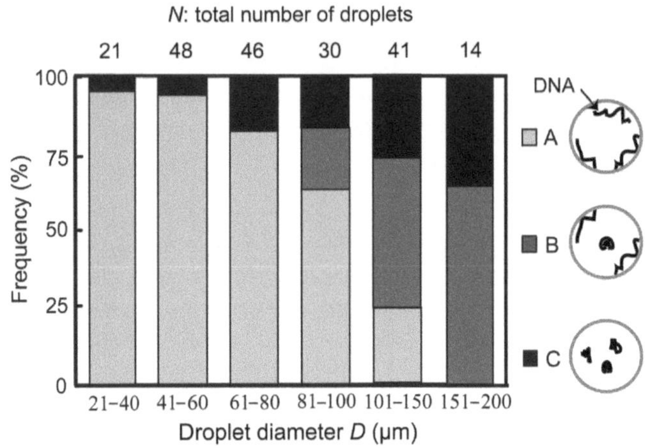

Figure 7.4 Conformation and location of DNA molecules dependent on the droplet size covered with a DOPE layer, as observed by fluorescence microscopy as in Fig. 7.3. (A) Almost all of the DNAs are located on the DOPE membrane surface. (B) Some DNAs are on the membrane surface and others are in the aqueous phase. (C) Almost all of the DNAs are in the aqueous phase. *N* indicates the number of droplets analyzed. *Reproduced with permission from Kato et al. (2009).*

membrane at several points, most parts of the molecule freely move close to the membrane (Fig. 7.3). In addition, DNA molecules are not adsorbed onto a PC membrane even in the presence of 10 mM Mg^{2+} (Fig. 7.2). Although both PC and PE membranes have a net positive charge in the presence of Mg^{2+}, the electrostatic properties of these membranes would be different from each other. We expect that the different effect in absorbing DNA onto the membrane between PC and PE is due to the different abilities of Mg^{2+} to bridge the phosphate head-group of the phospholipids and the phosphate moiety of DNA. It may be natural to expect that the smaller head-group of PE prefers the stronger bridging effect than the bulky head-group of PC. Moreover, we explain the droplet-size dependence of the DNA distribution with the thermodynamic analysis in Section 2.3 (Kato et al., 2009).

2.2. On–off switching of DNA transcription

As described above, the encapsulation of long DNA molecules into DOPE droplets leads the conformational transition from a folded state in aqueous phase to an extended coil state on the membrane surface. This conformational change of DNA strongly relates with its transcription, which can be evaluated at the individual DNA molecule level using fluorescence resonance energy transfer (Tsuji and Yoshikawa, 2010b). Transcription proceeds on almost all the DNA molecules with a coil conformation in the aqueous phase. However, in the presence of tetravalent spermine, the DNA has a folded state conformation and its transcription is completely inhibited. When the Mg^{2+} concentration is increased, DNA is adsorbed onto the inner surface of the PE membrane, exhibiting an extended conformation (Fig. 7.3). Accordingly, the transcription experiments show that this conformational transition recovers the transcriptional activity; in fact, transcription occurs on DNA molecules that are on the membrane (Tsuji and Yoshikawa, 2010a).

2.3. Thermodynamic considerations on DNA conformational transition in cell-sized space

It has been well established that a long DNA molecule exhibits a conformational transition between a swollen coiled state and a compactly folded state in aqueous solution. However, it is important to know how the above bulk scenario is altered in cell-sized spaces. Upon encapsulation of DNA in a droplet, one may immediately notice the following two factors: (i) the reduction of the free space available to DNA and (ii) the emergence of the surface. The surface provides a field where the distinct structural organizations of biopolymers may take place. Such a phenomenon should become more and more prominent with an increase in the surface-to-volume ratio. We now discuss the qualitative consequences of these two factors based on a simple free energy analysis in comparison with the experimental results.

A single DNA molecule is assumed to adopt one of the following conformations in a droplet: (1) a free coil conformation in the inner solution space; (2) a quasi-two-dimensional coil conformation that is adsorbed onto the phospholipid membrane surface; and (3) a folded conformation in the inner solution space. By setting the state (1) as a reference, we can write the free energies of the respective states as follows:

$$\left.\begin{aligned} F_1 &= k_B T \ln \varphi_V \\ F_2 &= k_B T \ln \varphi_A - \varepsilon_a N \\ F_3 &= k_B T \ln \varphi_V - \varepsilon_f N \end{aligned}\right\} \tag{7.1}$$

where k_B and T are the Boltzmann constant and temperature, respectively, and N is the number of segments in a single DNA. In these equations, the first terms represent the contribution from the translational freedom of the DNA. The concentration of the DNA in the droplet of size D is $\varphi_V \approx 1/D^3$.

This expression should be replaced by the two-dimensional concentration $\varphi_A \approx 1/D^2$ when the DNA is restricted to the surface region only (i.e., state 2). Upon adsorption onto the membrane surface, there is a net free energy change of $-\varepsilon_a$ per DNA segment. This phenomenological parameter includes the conformational entropy change of DNA as well as the contribution from the counter ion and solvent degrees of freedom, which is why it is regarded as a function of the environmental factors such as concentrations of the small ions and property of the membrane. The same considerations are applied to the parameter ε_f, which accounts for the net free energy change upon the folding induced by the addition of the condensing agent. From these free energies, one can construct a phase diagram (Fig. 7.5). The adsorption and the folding are, respectively, primarily driven by the binding of Mg^{2+} and spermine ions to DNA from the bulk solution. Therefore, in the crudest approximation, one may expect that the parameters ε_a and ε_f depend on the logarithm of the Mg^{2+} and spermine ions, respectively. In reality, these two cations (Mg^{2+} and spermine) compete in the binding to DNA. This competition in the binding equilibrium makes the above parameter mapping more involved, as can be observed from the comparison of Figs. 7.5 and 7.6. For example, spermine also has the ability to induce

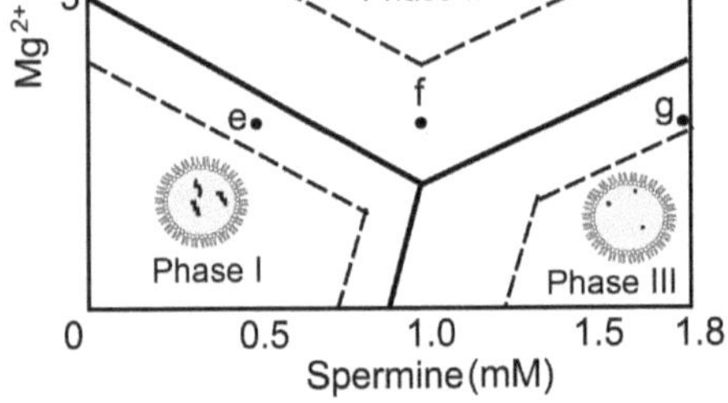

Figure 7.5 Phase diagram for the Mg^{2+} and spermine concentrations. Phase I: DNA has a coil conformation in the aqueous phase. Phase II: DNA is adsorbed onto the membrane surface. Phase III: DNA has a highly compact conformation in the aqueous phase. The field surrounded with broken lines is the intermediate region in which different states of DNA molecules (coil and compact, in the aqueous phase, and on the membrane surface) coexist within a single droplet. *Reproduced with permission from Kato et al. (2009).*

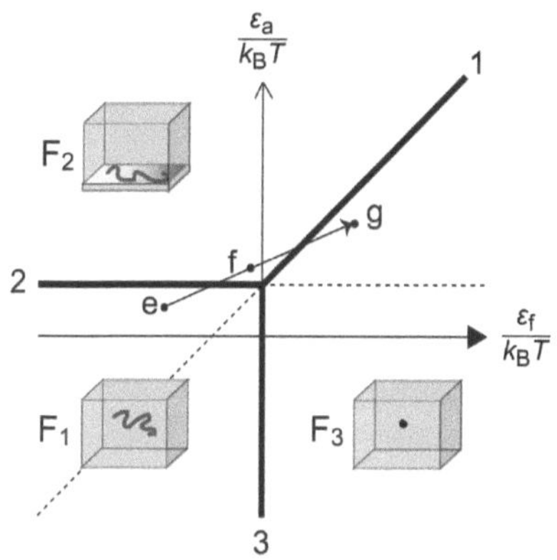

Figure 7.6 Schematic representation of the state of DNA molecules in a microdroplet. Three states are assumed: F_1, free coil conformation in the aqueous solution phase; F_2, quasi-two-dimensional coil conformation adsorbed onto the surface of the phospholipid membrane; and F_3, compactly folded conformation in the aqueous solution phase. On axis 1, where $\varepsilon_a/k_BT \sim \varepsilon_f/k_BT + 1/N \cdot \ln(R/\lambda)$, the free energy of the F_2 state is equal to that of the F_3 state. On axis 2, where $\varepsilon_a/k_BT \sim 1/N \cdot \ln(R/\lambda)$, the free energy of the F_1 state is equal to that of the F_2 state. On axis 3, where $\varepsilon_f/k_BT \sim 0$, the free energy of the F_1 state is equal to that of the F_3 state. *Reproduced with permission from Kato et al. (2009).*

DNA adsorption in cooperation with Mg^{2+} under the current experimental conditions; thus, the sequential transitions from the phase I to phase III, via phase II, along with the increase of spermine concentration could be interpreted as the route e→f→g in the phase diagram in the $(\varepsilon_a - \varepsilon_f)$ plane.

The boundary between phases I and II is given by $\varepsilon_a = k_B T \ln D$. This indicates that the adsorption is enhanced in smaller droplets in qualitative agreement with the experimental result (Fig. 7.4). With the decrease in the surface-to-volume ratio, the presence of the surface and its role becomes less and less relevant; hence, the adsorbed state becomes improbable. More quantitative analysis requires the explicit inclusion of small ions taking part in the binding equilibrium with DNA and surface through electrostatic interactions.

3. CROSS-TALK OF DNA WITH OTHER SEMIFLEXIBLE POLYMERS

To study the crowding effects of concentrated biomacromolecules in cells, here we report about three different systems of binary macromolecule mixtures in cell-sized W/O droplets: a long DNA with poly-alginate, actin filament, or PEG (Biswas et al., 2012; Negishi et al., 2010, 2011). Except for

PEG, these macromolecules (DNA, actin, and alginate) are negatively charged. The persistence lengths of DNA, alginate, and PEG are about 50, 5, and 0.9 nm, respectively, whereas that of actin is much larger comparable to droplet size (i.e., the micrometer scale). In addition, the characteristic size of PEG is also expressed with a radius of gyration R_g corresponding to $R_g \sim 0.01078 \cdot M^{0.635}$ (nm), where M is the molecular weight of PEG. In the case of PEG with $M = 6000$, R_g is about 3 nm. Consequently, the differences among the lengths cause unique aqueous phase separations inside the micro-droplets according to the droplet size and concentration of encapsulated solution. We describe the phase behaviors of each system in the following subsections.

3.1. DNA with anionic flexible polymer: Poly-alginate

The negatively charged DNA is mixed with the negatively charged alginate, which has a smaller size. We describe the encapsulation in micro-droplets with the phase diagram shown in Fig. 7.7A, which represents the phase behavior in the bulk solution after 60 min at 20 °C. In region I, the mixture is in a homogeneous condition. In region II, DNA is phase-segregated from the alginate-rich phase, and DNA exhibits micrometer scale domain structures. Figure 7.7B shows the change in conformational of 0.34 mg/mL DNA when it is mixed with various concentrations of alginate (1.0, 1.5,

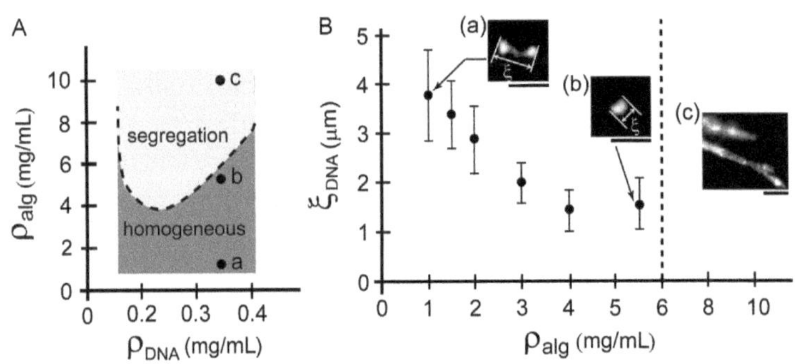

Figure 7.7 DNA conformation in a crowding alginate solution under confinement. (A) Phase diagram of DNA and alginate mixture in the bulk phase. The labels (a–c) correspond to those in (B). (B) Change in the conformation of DNA when 0.34 mg/mL DNA is mixed with alginate. The average long-axis length of single DNA molecules (ξ_{DNA}) is indicated with the standard deviation. Insets (a) and (b) show representative fluorescence images of coil or meta-stable coil DNA molecules with 1.0 and 5.5 mg/mL alginate, respectively. The scale bars are 5 μm. Inset (c) shows a fluorescence image of DNA domains with 10.0 mg/mL alginate. The scale bar is 50 μm. *Reproduced with permission from Negishi et al. (2011).*

2.0, 3.0, 4.0, or 5.5 mg/mL). The average long-axis length of single DNA molecules (n_{an}) shows that DNA is in the coil state with 1.0, 1.5, or 2.0 mg/mL alginate (Fig. 7.7B, inset a) and in the coil state with 3.0, 4.0, or 5.5 mg/mL alginate (Fig. 7.7B, inset b). We compared the phase behavior after preparation (<10 min) with that after 5 days at 4 °C, and found that the results were similar to those in Fig. 7.7B. Therefore, we conclude that the obtained phase diagram of DNA and alginate in the bulk phase represents the thermodynamic equilibrium state.

Next, we show the phase diagram of the DNA distribution within PE droplets after standing for 60 min at 20 °C (Fig. 7.8), in which DNA and various concentrations of alginate (1.0, 1.5, 2.0, 3.0, 4.0, or 5.5 mg/mL) are confined within the droplets. In region I, where DNA molecules with a relatively high concentration of alginate were confined within droplets with small diameters, DNA molecules were completely localized on the

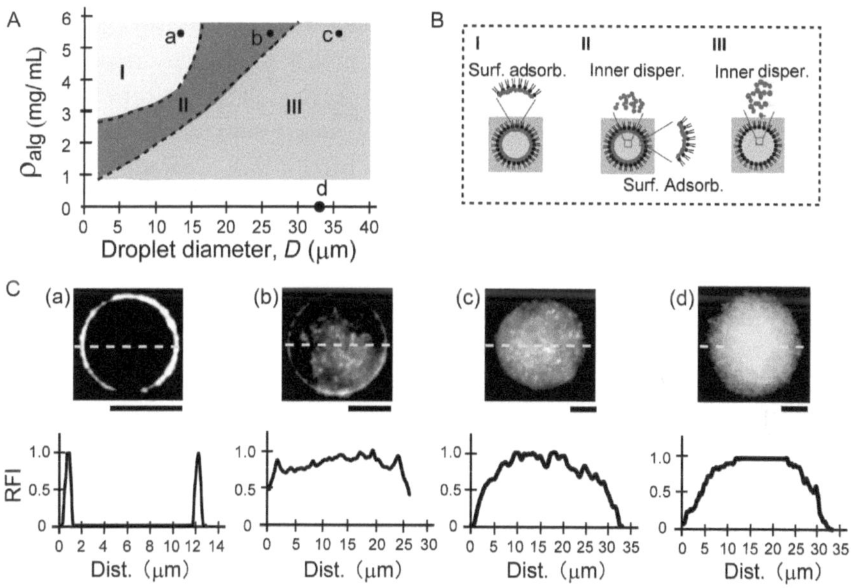

Figure 7.8 Confinement effect on DNA with alginate. (A) Phase diagram of the DNA distribution in droplets in a pseudo-equilibrium state, where 0.34 mg/mL DNA with alginate is confined in droplets. (B) Schematic illustrations of the DNA distribution are in the dotted-line rectangle. (C) Fluorescence images of DNA in the droplets and their relative fluorescence intensity profiles along the dotted lines in the fluorescence images. The labels (a–d) are the same as those in the phase diagram. RFI and Dist. denote relative fluorescence intensity and distance, respectively. Scale bars are 10 μm. *Reproduced with permission from Negishi et al. (2011).*

surface of the droplet. A representative fluorescent microscopy image is shown in Fig. 7.8A (the diameter of the droplet, D, is 12.5 μm at 0.34 mg/mL DNA with 5.5 mg/mL alginate). In region II, some of the DNA molecules were localized on the surface of the droplet, and the rest were suspended homogeneously within the droplet (Fig. 7.8B: $D=23.8$ μm at 0.34 mg/mL DNA with 5.5 mg/mL alginate). In region III, where the concentration of alginate is relatively low or where the droplet diameter D is relatively small, all the DNA molecules were suspended homogeneously within the droplet (Fig. 7.8C: $D=32.8$ μm at 0.34 mg/mL DNA with 5.5 mg/mL alginate). In "control" droplets, the presence of individual DNA existing independently implies that there is no significant interaction between individual DNA and the surface of the PE droplet under our experimental conditions (Fig. 7.8D) (Negishi et al., 2011).

As shown in Figs. 7.7 and 7.8, the droplet encapsulation of DNA and alginate mixture confers on them a unique phase behavior depending on the droplet size, whereas the mixture remains homogeneous in bulk solution. When the droplet is small, DNA with a characteristic length of ~50 nm (longer than that of alginate, ~5 nm) is completely depleted on the droplet surface. When the droplet is medium sized, a portion of the DNA is depleted on the surface and the remainder stays within the sphere. These results indicate that the depletion-induced phase separation is accelerated in smaller droplets with a large area-to-volume ratio.

3.2. DNA with anionic stiff polymer: Actin

Instead of alginate, which is smaller than DNA, the much larger actin filaments are now mixed with DNA molecules. The basic properties of the actin filament are as follows: persistent length $l_{actin}=10$–17 μm, diameter 7.5 nm, charge density e−/0.25 nm, and the critical concentration 8 μg/mL. It is reported that actin filaments reconstituted in model cells, such as liposomes and droplets, are localized on the lipid membrane when actin exists with Mg^{2+} and/or cross-linker proteins are present (Claessens et al., 2006; Hase and Yoshikawa, 2006; Limozin and Sackmann, 2002; Limozin et al., 2005). Figure 7.9 shows the structural transition of actin filaments in PE droplets with a diameter ~50 μm under several concentrations of Mg^{2+}. With an increase of Mg^{2+} from 0 to 12 mM, the actin filament was localized on the PE membrane surface of the droplets. Increase of Mg^{2+} to >12 mM led to a bundling of the actin filaments in the aqueous phase of the droplets, and decreased the number of actin filaments

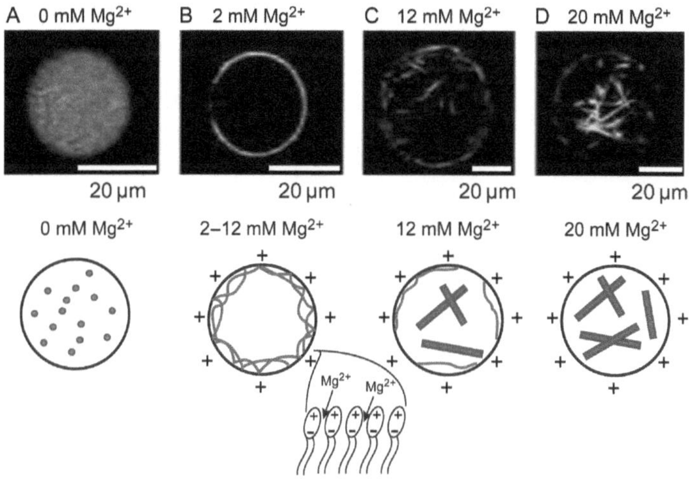

Figure 7.9 Characteristic action mechanism under confinement. Confocal images of actin filaments in droplets coated with a PE layer at different Mg^{2+} concentrations and 350 μg/mL actin. The Mg^{2+} concentration contained in the droplets are (A) 0 mM, (B) 2 mM, (C) 12 mM, and (D) 20 mM, respectively. *Reproduced with permission from Hase and Yoshikawa (2006).*

adsorbing to the membrane. This phenomenon is explained by the formation of a bridge of the divalent cations between the negatively charged actin filaments and PE membranes, similarly to the DNA adsorption on the PE droplet surface (Section 2.1) (Hase and Yoshikawa, 2006).

When we encapsulate only actin filaments in PE droplets with different sizes, the structure of the actin filaments also varies according to the droplet size. In the case of smaller PE droplets with diameter <15 μm, the actin filaments are localized on the PE membrane surface even without the Mg^{2+}. This critical size of droplets (i.e., 15 μm) is almost the same as the persistence length of the actin filament ($l_{actin}=10$–17 μm). In other words, the rigidity of the actin filament determines the structure of the droplet in confined spaces comparable to cells.

Accordingly, a binary mixture of DNA and actin filaments is encapsulated in the PE droplets. Figure 7.10 shows the phase diagrams obtained and schematic illustrations of the structures of the DNA and actin filament mixture in PE droplets and in the bulk solution, categorized according to the concentration of actin and the droplet size. The concentration of DNA is fixed at 0.17 mg/mL for all the experiments. In region I, the size of PE droplets, D, is smaller than l_{actin}; thus, actins tend to form a fuzzy cortex at the droplet surface, similarly to the small droplets containing only actins. In region II, the size of the droplet is larger (i.e., $l_{actin}<D<45$ μm), and actins

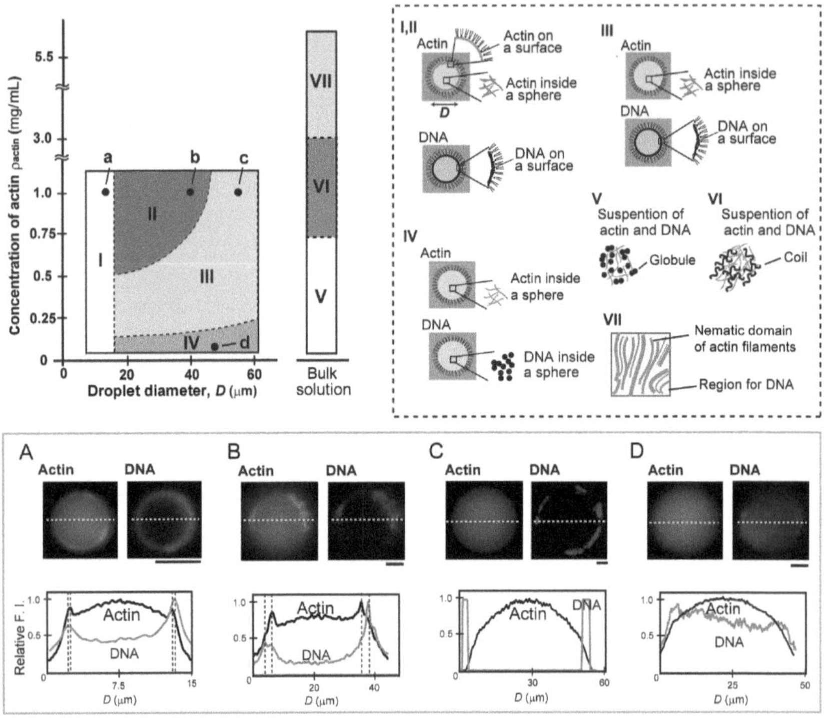

Figure 7.10 Characteristic change of the conformation and location of the mixture of DNA and actin; (upper) phase diagrams and schematic illustrations of the structures of DNA and actin filaments. (Lower) Fluorescence images of the mixture in the droplets and relative fluorescence intensity profiles along the dotted lines in the images. The labels (A–D) correspond to those in the phase diagram. Black and gray lines indicate the distribution of actin and DNA, respectively. Scale bars are 10 μm. *Reproduced with permission from Negishi et al. (2010).*

tend to form fuzzy cortex at the droplet surface, although this is not observed in similarly sized droplets containing only actins. In regions I and II, the DNA molecules are localized on the droplet surface. Representative microscopic images of regions I and II are shown in Fig. 7.10A ($D=14.3$ μm at 1.05 mg/mL actin with 0.17 mg/mL DNA) and 7.10B ($D=41.2$ μm at 1.05 mg/mL actin with 0.17 mg/mL DNA). In Fig. 7.10A and B, DNA molecules are located just on the membrane surface, whereas actin filaments are slightly apart from the surface. In some droplets in region II, these accumulations exist at the center of the droplets or penetrate through the droplet to connect to the neighboring droplets (Fig. 7.10B). In region III, phase separation occurs: DNA molecules are all localized on the membrane surface of the droplet, whereas actin filaments are suspended within the droplet

(Fig. 7.10C; $D = 54.1$ μm at 1.05 mg/mL actin with 0.17 mg/mL DNA). At low actin filament concentration (region IV), both actin filaments and DNA are suspended within the droplet without local phase separation (Fig. 7.10D; $D = 48.7$ μm at 1.05 mg/mL actin with 0.17 mg/mL DNA).

To clarify the confinement effect, we also report the phase behavior of the mixture in the bulk phase. In regions V and VI, both actin filaments and DNA molecules are suspended without local phase separation. A distinction between these regions is revealed through single DNA molecule observation. DNA is in a compact conformation (globule) in region V (at 0.53 mg/mL actin with 0.17 mg/mL DNA), while DNA is in coil conformation in region VI (at 1.05 mg/mL actin with 0.17 mg/mL DNA). Such conformational behavior is reminiscent of DNA folding and unfolding upon the addition of PEG due to the depletion effect, which is described precisely in the following section (Biswas et al., 2012). In region VII, local phase separation is observed, in which entangled percolating networks of elongated actin filament domains through a mesh of DNA coils are present. A similar structure has been reported under slightly different experimental conditions (Lai et al., 2008). The average long-axis length of the domains reaches 50 μm for a mixture of 0.53 mg/mL actin filaments and 0.17 mg/mL DNA molecules (Negishi et al., 2010).

Thus, DNA and actin mixture entrapped in PE droplets vary their structures and spatial distributions according to the droplet size. When the droplet diameter, D, is larger than the persistent length of the actin filament, $l_{actin} = 10–17$ μm, these are suspended within the droplet, and the DNA molecules are depleted on the droplet surface, like in the DNA and alginate system (Section 3.1). On the contrary, in the case of smaller droplets with, e.g., $D < l_{actin}$, both DNA molecules and actin filaments are localized on the droplet surface, and the actin filaments form a fuzzy cortex at the droplet surface (region I), because the high rigidity of the actin prevents deformation.

3.3. DNA with neutral polymer: PEG

It is known that DNA condenses in the presence of hydrophilic polymers and salts such as NaCl and KCl, a phenomenon that is which is called polymer salt-induced (psi) condensation (Zinchenko and Yoshikawa, 2005). The hydrophilic solvable polymer PEG has been frequently used in the study of psi condensation. Additionally, single DNA molecules undergo discrete transitions between the elongated coil and compact globule state in the presence of PEG and salt (Vasilevskaya et al., 1995). Similarly to concentrated

DNA solutions with PEG, DNA chains tend to adopt elongated conformation by aligning parallel to each other, similar to liquid crystal structures (Kojima et al., 2006). Here, we focus our attention on the change of the state of DNA molecules under crowded conditions in the micrometer-length scales. We adapted the experimental system of a cell-sized droplet entrapping a DNA and PEG mixture.

Initially, DNA molecules are dispersed uniformly within the droplet, and the absence of birefringence under crossed-Nicol condition is indicative of an isotropic phase in the interior of the droplet (Fig. 7.11A and B). After 2.25 h (Fig. 7.11C), a void appeared in the inner part of the droplet accompanied with DNA localization near the periphery, suggesting the onset of phase segregation. After 3 h, almost all the DNA molecules are condensed at the boundary (Fig. 7.11D). Figure 7.11D (bottom) shows that birefringence also appears at the glass–oil–water contact line. This situation remains for a long time (Fig. 7.11E), implying that the DNA segregation is a thermodynamically stable phenomenon. The whole process is thus characterized as the transition from an isotropic to an ordered phase, while the in-plane size of the droplets remains constant.

Figure 7.12 illustrates the effect of droplet size on the phase separation process at 2.5 h after the droplet preparation. The smallest droplet (with a diameter = 50 μm) already shows DNA agglomeration near the periphery and the agglomerated area has a uniform distribution of DNA (Fig. 7.12A).

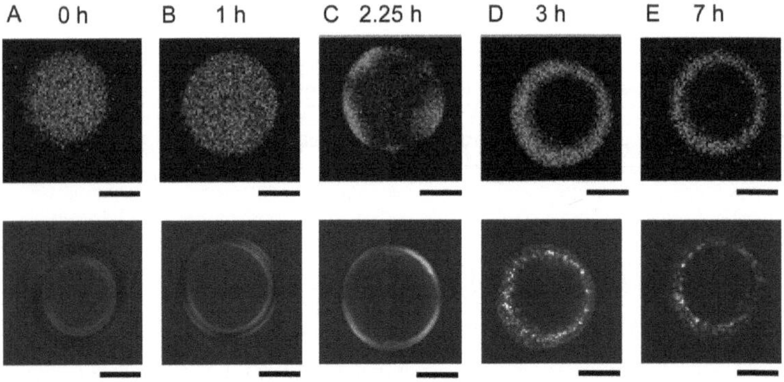

Figure 7.11 The time after the encapsulation is (A) 0 h, (B) 1 h, (C) 2.25 h, (D) 3 h, and (E) 7 h, respectively. Time evolution of phase separation in a droplet containing DNA and PEG (DNA 52 mg/mL, PEG 114 mg/mL, NaCl 71 mM). Top panels: fluorescence microscope images of tagged DNA molecules in a droplet. Bottom panels: polarization microscope images under crossed-Nicol conditions. Scale bars are 20 μm. *Reproduced with permission from Biswas et al. (2012).*

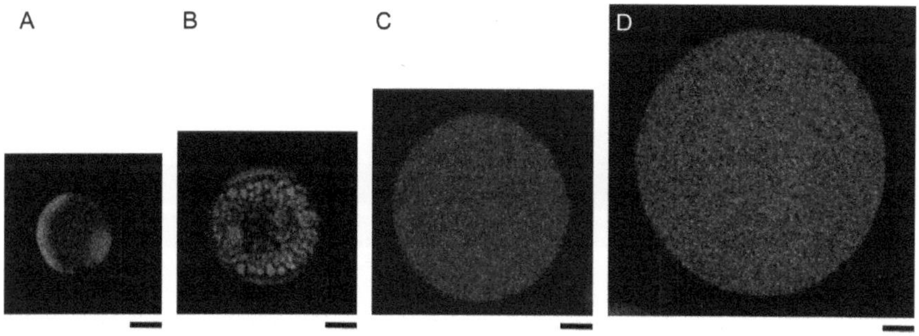

Figure 7.12 Effect of size on the segregation between DNA and PEG. Phase separation of tagged DNA (52 mg/mL), PEG (114 mg/mL), and NaCl (71 mM) mixture 2.5 h after the sample preparation. The diameter of the droplets is (A) 50 μm, (B) 78 μm, (C) 122 μm, and (D) 170 μm, respectively. Scale bars are 40 μm. *Reproduced with permission from Biswas et al. (2012).*

The larger droplet with a diameter of 78 μm shows formation of small aggregates of DNA molecules along the periphery (Fig. 7.12B). In the case of even larger droplets with $D = 122$ or 170 μm (Fig. 7.12C, D), the phase of the entrapped mixture seems to be homogeneous. Hence, we found that the phase separation is strongly dependent on the droplet size. However, we also found that the formation of the DNA aggregates initiates near the periphery or interface, and then extends to the body of the droplet. In fact, a closer look at Fig. 7.12B demonstrates the linearly aligned DNA at the upper and lower peripheries of the droplet. These aggregates of DNA gradually coalesce together and finally move to the interface (Figs. 7.11 and 7.12A). These results lead us to suggest an interface-mediated mechanism for this phase separation of DNA and PEG mixture. Such interface-mediated phase separation may be accelerated in smaller droplets with a large area-to-volume ratio (Biswas et al., 2012).

4. GENE EXPRESSION IN CELL-SIZED DROPLETS

4.1. Cell-sized confinement accelerates gene expression

Green fluorescent protein (GFP) gene expression has been also tested within the liposomes (Ishikawa et al., 2004; Murtas et al., 2007; Noireaux and Libchaber, 2004; Nomura et al., 2003; Oberholzer et al., 1999). Interestingly, gene expression is reported to be enhanced in small W/O droplets coated by a lipid layer (Fiordemondo and Stano, 2007), or by adding liposomes outside (Bui et al., 2008). Using the size-controllable droplet system,

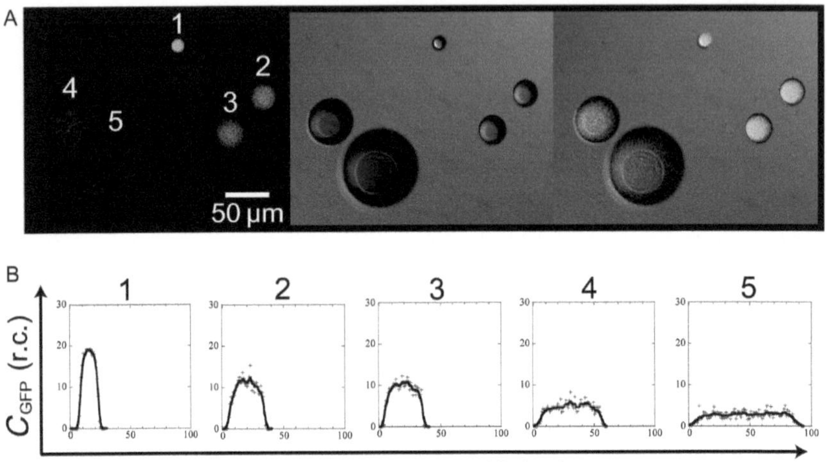

Figure 7.13 Confinement to smaller space exhibits higher production of GFP. (A) Distribution of GFP fluorescent in droplets coated with a DOPG layer. (B) Profiles of fluorescent intensities along the diameter C_{GFP} (Kato et al., 2012).

we report a remarkable acceleration of gene expression in smaller droplets entrapping a cell-free translation system to synthesize GFP. Figure 7.13A shows confocal images of DOPG droplets entrapping a GFP expression system 3 h after encapsulation. The fluorescence image (Fig. 7.13A, left) shows the distribution of GFP fluorescence. The merged image (Fig. 7.13A, right) shows that GFP fluorescence was homogeneously distributed in each droplet. Interestingly, the intensity of GFP fluorescence in the droplets strongly depended on the droplet size. We calibrated the GFP intensity per volume and obtained the GFP concentration, C_{GFP}, based on normalization by the value of the bulk solution after 3 h. As shown in Fig. 7.13B, C_{GFP} decreased with an increase in the droplet size. In the case of a small droplets with radius $R \sim 10$ μm (Fig. 7.13B, 1), the value of C_{GFP} was approximately four times larger than that for a large droplet with a radius ~45 μm (Fig. 7.13B, 5). On the other hand, C_{GFP} values in droplets with almost the same R ~15 μm were similar (Fig. 7.13B, 2, 3). We have confirmed that there was no size dependence of C_{GFP} among the droplets after encapsulation of GFP already expressed in the bulk solution. Furthermore, this result indicates that the membrane surface and/or the total volume significantly affect the GFP expression, since the surface-to-volume ratio is greater in a smaller droplet. If the lipid membrane contributed to the size dependence of GFP expression, the lipid species used should affect the relation between the droplet size and expression level. Thus, in the following experiments, we examined the effect of different lipids on the confinement effect.

Three different species of lipids, anionic DOPG, zwitterionic DOPC, and zwitterionic DOPE, were used to encapsulate the GFP gene expression system within droplets. The time-courses of GFP concentration per unit volume, C_{GFP}, were monitored for 45 h beginning soon after encapsulation. Identical droplets with a radius $R = 10$–100 μm were monitored. As shown in Fig. 7.14, the value of C_{GFP} increased linearly with time up to 20 h, and reached a saturation point at ~40 h. Among all the three types of lipids, C_{GFP} in a small droplet with R ~20 μm was higher than that in a large droplet with R ~50 μm. Additionally, C_{GFP} at a given time and the production rate ($V_{GFP} = dC_{GFP}/dt$) were both dependent on the type of lipid. V_{GFP} in each droplet remained almost constant for at least 1 day, which implies 0th-order reaction kinetics. Interestingly, V_{GFP} was inversely proportional to the radius of the droplets (R) when R was under 50 μm, with V_{GFP} in droplets with $R = 10$ μm being more than 10 times higher than that in the bulk. As shown in Fig. 7.14, the acceleration rates of GFP production in cell-sized droplets strongly depend on the lipid types. These results demonstrate that the membrane surface has the significant effect to facilitate protein production, especially when the scale of confinement is in the order of the cell size. Additionally, such time development of GFP expression can be analyzed within liposomes by using the droplet-transfer method (Section 5.1) (Saito et al., 2009).

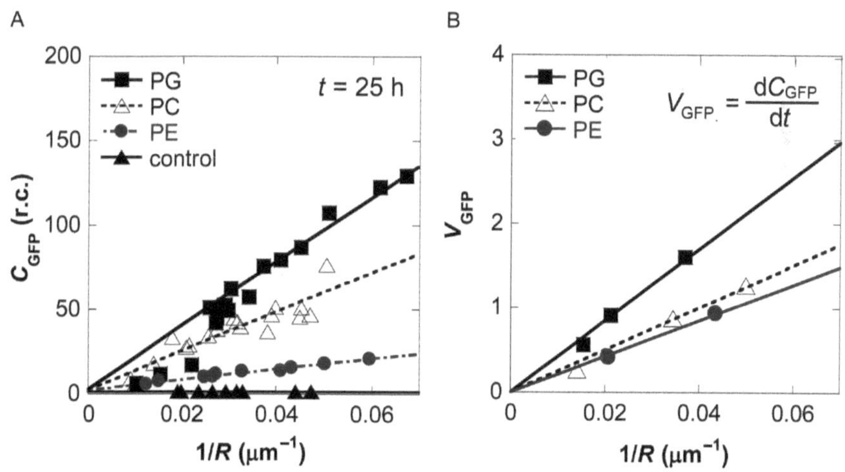

Figure 7.14 Size dependence of the GFP concentration per unit volume, C_{GFP}, at 25 h and the production rate ($V_{GFP} = dC_{GFP}/dt$) from 0 to 20 h (Kato et al., 2012).

4.2. Surface-assisted reaction

To get some insight into the above experiments, let us consider a simple enzymatic reaction process taking place in a small confined space, where the enzyme (E) catalyzes transformation of the substrate (S) into the product (P):

$$S+E \underset{k'}{\overset{k}{\leftrightarrow}} SE \overset{k^*}{\rightarrow} P+E \tag{7.2}$$

In the enzymatic reaction, the total concentration of the enzyme is kept constant; $[E]+[SE]=[E_0]$, then the production rate is described by the following Michaelis–Menten kinetics form:

$$\frac{d[P]}{dt}=\frac{G[S]}{K+[S]} \tag{7.3}$$

where $G=k^*[E_0]$ and $K=(k'+k^*)/k=[S][E]/[SE]$.

In a small droplet of spatial size R, the reactions may take place either in the interior (volume) or on the surface. The total number N_X of X molecules is the sum of the number of molecules in the volume and the surface regions ($N_X=N_{X,v}+N_{X,s}$). Therefore, the entire reaction scheme is:

$$S_V+E_V \underset{k'_V}{\overset{k_V}{\leftrightarrow}} (SE)_V \overset{k_V{}^*}{\rightarrow} P_V+E_V \text{ (reaction in the interior)} \tag{7.4}$$

$$S_S+E_S \underset{k'_S}{\overset{k_S}{\leftrightarrow}} (SE)_S \overset{k_S{}^*}{\rightarrow} P_S+E_S \text{ (reaction on the surface)} \tag{7.5}$$

$$S_V \underset{W'_S}{\overset{W_S}{\leftrightarrow}} S_S \text{ (binding equilibrium)} \tag{7.6}$$

$$E_V \underset{w'_E}{\overset{w_E}{\leftrightarrow}} E_S \text{ (binding equilibrium)} \tag{7.7}$$

$$SE_V \underset{w'_{ES}}{\overset{w_{ES}}{\leftrightarrow}} SE_S \text{ (binding equilibrium)} \tag{7.8}$$

Accordingly, the production rates in the respective parts are given by:

$$\frac{d[P_V]}{dt}=\frac{G_V[S_V]}{K_V+[S_V]} \tag{7.9}$$

and

$$\frac{d[P_S]}{dt}=\frac{G_S[S_S]}{K_S+[S_S]} \tag{7.10}$$

where $[X_v] \sim N_{X,v}/V$ and $[X_s] \cong N_{X,s}/S$ are the interior and surface concentrations of the molecule X with the interior volume $V \cong R^3$ and the surface area $S \cong R^2$. The parameters in the above set of equations in the interior/surface regions are the functions of the reaction constants and the available enzyme concentrations in the respective regions: $K_v = (k_v^* + k_v')/k_v$, $K_s = (k_s^* + k_s')/k_s$, $G_v = k_v^*[E_0]_v$, $G_v = k_v^*[E_0]_v$, and $G_s = k_s^*[E_0]_s$. The production rate $\kappa = d[(N_{P,v} + N_{P,s})/V]/dt$, defined as the growth rate of the total produced molecule per the system size, can thus be written as:

$$\kappa(R) = \kappa_V\left(1 + \frac{\alpha}{R}\right) \tag{7.11}$$

where κ_v is the production rate of the interior region, which is given by Eq. (7.3) with the subscript "v" appropriately added. There is an additional surface term in the small space, which is inversely proportional to the spatial size with a factor

$$a = \left(\frac{G_S[S_S]}{G_V[S_V]}\right)\left(\frac{K_V + [S_V]}{K_S + [S_S]}\right) \tag{7.12}$$

The factor α depends on various system parameters, but in particular on the nature of the interaction between the species and the surface. In fact, assuming that the concentrations of the substrate and the enzyme are not too high, the binding equilibrium indicates:

$$k_B T \ln \frac{N_{X,v} \cdot a_0^3}{V} = k_B T \ln \frac{N_{X,s} \cdot a_0^2}{S} - \varepsilon_x \tag{7.13}$$

$$\Leftrightarrow N_{X,S} \cong N_{X,v}\left(\frac{a_0}{R}\right) e^{\varepsilon_X/k_B T} \tag{7.14}$$

where ε_x is the energetic stabilization upon absorption onto the surface, and a_0 is the size of individual molecules (supposed to be of the same order of magnitude). In terms of the reaction rates given in Eqs. (7.6)–(7.8) this corresponds to $wX/w'X = a_0 e^{\varepsilon_X/k_B T}$. To obtain a rough estimate, we set $\varepsilon_E \cong \varepsilon_{EP} = \varepsilon$ and then $G_S/G_V \cong a_0\left(k_S^*/k_V^*\right)e^{\varepsilon/k_B T}$. On the other hand, the reaction constant k in the diffusion-limited reaction depends on the diffusivity of the species: $k_V \cong a_0\ (D_{S,v} + D_{E,v})$ and $k_S \cong D_{S,s} + D_{E,s}$, where $D_{X,v}$ and $D_{X,s}$ are the 3-D (in the interior part) and 2-D (on the surface) diffusion coefficients of the species X, respectively. Here, the strong absorption is expected to suppress the surface diffusion, and thus, $K_s \sim (1/k_s)$ will become larger with the increase in ε.

To summarize, our simple analysis indicates that, in the cell-sized small space, the efficiency of the chemical reaction may be substantially affected by the surface. This surface-assisted reaction gives rise to an additional production rate generally proportional to the inverse of the spatial size. Its relative importance, thus, becomes larger in smaller spaces, and may primarily depend on the nature of the interaction between the species and the surface. These points are in qualitative agreement with the experimental trends summarized in Section 4.1. The stronger attraction accommodates more substrates and enzymes on the surface, but at the same time could suppress their lateral diffusivity. These two competing effects suggest the nonmonotonic dependence of α on ε. As Eq. (7.11) shows, the spatial size is an important factor to determine the reaction rate in cell-sized space, where the surface-assisted effect may be optimized at suitable interaction strengths between molecules and the surface.

5. FROM CELL-SIZED DROPLET IN OIL PHASE TO LIPOSOME IN AQUEOUS PHASE

In the above sections, we introduced the W/O droplet systems as a cellular model. These systems are quite stable against the osmotic pressure. However, they are isolated from the surrounding environment. Here, we introduce a method to prepare liposomes from W/O droplets, which enables us to construct more realistic cellular models with highly concentrated inner solutions of proteins and harboring membrane proteins (Hamada et al., 2008; Hase et al., 2007; Pautot et al., 2003a; Stachowiak et al., 2008; Takiguchi et al., 2009; Yamada et al., 2006).

5.1. Droplet-transfer method to prepare liposomes

Lipid bilayer liposomes have been the most popular model of cellular membranes. A liposome can be formed by transferring a W/O droplet coated with a lipid monolayer to an oil/water interface having a preformed monolayer (Fig. 7.15) (Pautot et al., 2003a,b; Saito et al., 2009; Yamada et al., 2006). A thin layer in water phase is introduced at the bottom of a cylindrical hole (4 mm in diameter) in a polydimethylsiloxane sheet (5 mm thick) and covered with a lipid/oil solution, which is prepared as described above. After 1 h, a lipid monolayer is stably formed at the oil/water interface. Then, the oil phase of W/O droplets is added immediately after the emulsification above the lipid layer. To transfer the droplets easily, we add about 100 mM sucrose solution to the internal media and an

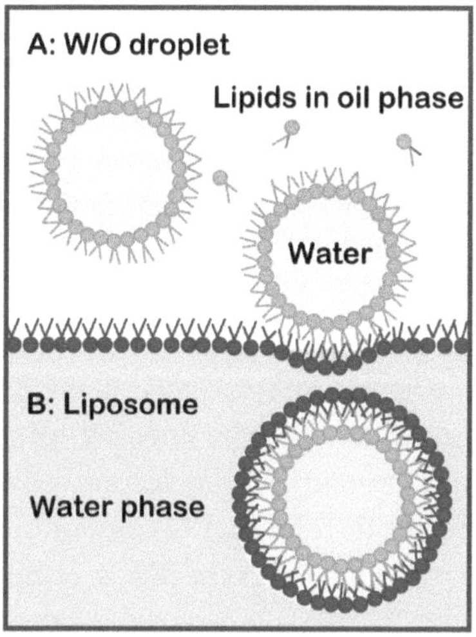

Figure 7.15 Schematic illustration of the droplet-transfer method to prepare asymmetric liposomes. (A) W/O droplets coated with a lipid monolayer. (B) A liposome formed by transferring the droplet through an oil/water interface with another lipid monolayer.

equimolar glucose solution to the external media. The droplets in the oil phase then spontaneously cross the lipid monolayer, driven by the difference in molar density (i.e., the gravity difference) between sucrose and glucose. During the transfer, the droplets are converted to liposomes where the monolayer covering the droplets becomes the inner leaflet of the liposome, and the monolayer at the oil/water interface becomes the outer leaflet. The average diameter of the liposomes obtained is about 20 μm. Additionally, asymmetric liposomes can be obtained by using chemically different types of lipids to cover the droplet and to prepare the lipid monolayer at the oil/water interface. Although some of the liposomes remain anchored to the interface, this issue is circumvented by either the microfluid-jetting method (Stachowiak et al., 2008) or the use of a centrifuge (Pautot et al., 2003b).

5.2. Reconstitution of membrane protein in liposomes with desired orientation

By using the droplet-transfer technique, membrane proteins can be reconstituted into liposomes with the appropriate orientation. Here, we introduce a rather simple method employing the potassium channel KcsA. First, we prepared W/O droplets lined with a lipid monolayer (Fig. 7.13). The droplet was then transferred through an oil/water interface having a preformed monolayer. The interface monolayer covered the droplet to generate a bilayer liposome. By creating chemically different lipid monolayers at the droplet and oil/water interface, we can obtain liposomes with asymmetric lipid compositions in the outer and inner leaflets. The solubilized KcsA, the concentration of which should be below the critical micelle concentration, was spontaneously inserted into liposomes from the inside or outside, and this phenomenon was accelerated in liposomes that contained PE or PG. Integrated insertion into the liposome membrane and the KcsA orientation was examined by a functional assay, exploiting the pH sensitivity of the opening of the KcsA when the pH-sensitive cytoplasmic domain (CPD) faces toward the acidic medium. KcsA loaded from the inside of the PG-containing liposomes becomes permeable only when the intravesicular pH is acidic, while KcsA loaded from the outside becomes permeable when the extravesicular pH is acidic. Therefore, the internal or external insertion of KcsA leads to an outside-out or inside-out configuration to retain its hydrophilic CPD on the added aqueous side. The CPD-truncated KcsA exhibited a random orientation, supporting the idea that the CPD determines the orientation of the protein. Furthermore, the application of the droplet-transfer method is promising for the reconstitution of other types of membrane proteins with a desired orientation into cell-sized liposomes (Yanagisawa et al., 2011) and also actin filaments (Takiguchi et al., 2008; Walde et al., 2010). Additionally, the use of multi-component liposomes with various shapes may reveal the complex interactions between the reconstituted protein and the lipid membrane upon phase separation of lipids (Yanagisawa et al., 2008, 2010, 2012).

6. CONCLUDING REMARKS

We have shown several experimental evidences about the markedly different behavior of biopolymers under cell-sized confinement. Interestingly, in micro-confinement biopolymers exhibit characteristic conformational transitions accompanied by specific localization. It has been also demonstrated that key biological reactions, such as transcription and expression, are also significantly influenced by the conformational transitions under confinement. These results suggest the scientific importance of the actual effect of confinement and crowding in living cells.

We have interpreted the application of cell-sized droplets to explore the effect of cell-sized confinement as an easy-to-do experimental system. It has been also stressed that cell-sized liposomes in an aqueous phase can be easily prepared using the transfer method of the W/O droplet. We believe that active application of these experimental systems would be promising toward the development of more complex studies on cell-sized confinement.

Finally, we would like to note the possible relationship of the cell-sized droplet with the subject of the "Origin of Life." Currently, studies have been performed to unveil the origin of life. For example, the hypothesis of RNA world attracts much interest as a plausible scenario. However, one should note that under usual aqueous solution conditions, RNA tends to be hydrolyzed into monomers because the free energy of the polymeric state is significantly higher than the monomers. Thus, we may propose the working hypothesis that primitive life on the Earth would have evolved through micro-droplets existing in a natural oily environment. Thus, it is also tempting to stress the scientific importance of the cell-sized droplets in relation to the possible scenario of the "Origin of Life."

REFERENCES

Adams, M., Dogic, Z., Keller, S.L., Fraden, S., 1998. Entropically driven microphase transitions in mixtures of colloidal rods and spheres. Nature 393, 349–352.

Albiez, H., Cremer, M., Tiberi, C., Vecchio, L., Schermelleh, L., Dittrich, S., Kupper, K., Joffe, B., Thormeyer, T., von Hase, J., Yang, S., Rohr, K., Leonhardt, H., Solovei, I., Cremer, C., Fakan, S., Cremer, T., 2006. Chromatin domains and the interchromatin compartment form structurally defined and functionally interacting nuclear networks. Chrom. Res. 14, 707–733.

Ando, T., Skolnick, J., 2010. Crowding and hydrodynamic interactions likely dominate *in vivo* macromolecular motion. Proc. Natl. Acad. Sci. U. S. A. 107, 18457–18462.

Asakura, S., Oosawa, F., 1958. Interaction between particles suspended in solutions of macromolecules. J. Polym. Sci. 33, 183–192.

Biswas, N., Ichikawa, M., Datta, A., Sato, Y.T., Yanagisawa, M., Yoshikawa, K., 2012. Phase separation in crowded micro-spheroids: DNA–PEG system. Chem. Phys. Lett. 539, 157–162.

Bui, H.T., Umakoshi, H., Ngo, K.X., Nishida, M., Shimanouchi, T., Kuboi, R., 2008. Liposome membrane itself can affect gene expression in the *Escherichia coli* cell-free translation system. Langmuir 24, 10537–10542.

Claessens, M.M.A.E., Tharmann, R., Kroy, K., Bausch, A.R., 2006. Microstructure and viscoelasticity of confined semiflexible polymer networks. Nat. Phys. 2, 186–189.

de Vries, R., 2006. Depletion-induced instability in protein-DNA mixtures: influence of protein charge and size. J. Chem. Phys. 125, 014905.

Dijkstra, M., van Roij, R., 2002. Entropic wetting and many-body induced layering in a model colloid–polymer mixture. Phys. Rev. Lett. 89, 208303.

Dinsmore, A.D., Yodh, A.G., Pine, D.J., 1996. Entropic control of particle motion using passive surface microstructures. Nature 383, 239–242.

Ebashi, S., Ebashi, F., 1965. Alpha-actinin, a new structural protein from striated muscle. I. Preparation and action on actomyosi-ATP interaction. J. Biochem. 58, 7–12.

Eisenhoffer, G.T., Loftus, P.D., Yoshigi, M., Otsuna, H., Chien, C.B., Morcos, P.A., Rosenblatt, J., 2012. Crowding induces live cell extrusion to maintain homeostatic cell numbers in epithelia. Nature 484, 546–549.

Elcock, A.H., 2010. Models of macromolecular crowding effects and the need for quantitative comparisons with experiment. Curr. Opin. Struct. Biol. 20, 196–206.

Ellis, R.J., 2001. Macromolecular crowding: obvious but underappreciated. Trends Biochem. Sci. 26, 597–604.

Emanuel, M., Radja, N.H., Henriksson, A., Schiessel, H., 2009. The physics behind the larger scale organization of DNA in eukaryotes. Phys. Biol. 6, 025008.

Fiordemondo, D., Stano, P., 2007. Lecithin-based water-in-oil compartments as dividing bioreactors. ChemBioChem 8, 1965–1973.

Fujiwara, K., Nomura, S.M., 2013. Condensation of an additive-free cell extract to mimic the conditions of live cells. PLoS One 8, e54155.

Gromelski, S., Brezesinski, G., 2006. DNA condensation and interaction with zwitterionic phospholipids mediated by divalent cations. Langmuir 22, 6293–6301.

Hamada, T., Miura, Y., Komatsu, Y., Kishimoto, Y., Vestergaard, M., Takagi, M., 2008. Construction of asymmetric cell-sized lipid vesicles from lipid-coated water-in-oil microdroplets. J. Phys. Chem. B 112, 14678–14681.

Hase, M., Yoshikawa, K., 2006. Structural transition of actin filament in a cell-sized water droplet with a phospholipid membrane. J. Chem. Phys. 124, 104903.

Hase, M., Yamada, A., Hamada, T., Baigl, D., Yoshikawa, K., 2007. Manipulation of cell-sized phospholipid-coated microdroplets and their use as biochemical microreactors. Langmuir 23, 348–352.

Herbette, L., Napolitano, C.A., Mcdaniel, R.V., 1984. Direct determination of the calcium profile structure for dipalmitoyllecithin multilayers using neutron diffraction. Biophys. J. 46, 677–685.

Higashi-Fujime, S., 1983. Phosphorylation of myosin light chain modulates the *in vitro* movement of fibrils composed of actin and myosin filaments from skeletal muscle. J. Biochem. 94, 1539–1545.

Ishikawa, K., Sato, K., Shima, Y., Urabe, I., Yomo, T., 2004. Expression of a cascading genetic network within liposomes. FEBS Lett. 576, 387–390.

Kato, A., Shindo, E., Sakaue, T., Tsuji, A., Yoshikawa, K., 2009. Conformational transition of giant DNA in a confined space surrounded by a phospholipid membrane. Biophys. J. 97, 1678–1686.

Kato, A., Tsuji, A., Yanagisawa, M., Saeki, D., Juni, K., Morimoto, Y., Yoshikawa, K., 2010. Phase separation on a phospholipid membrane inducing a characteristic localization of DNA accompanied by its structural transition. J. Phys. Chem. Lett. 1, 3391–3395.

Kato, A., Yanagisawa, M., Sato, Y.T., Fujiwara, K., Yoshikawa, K., 2012. Cell-sized confinement in microspheres accelerates the reaction of gene expression. Sci. Rep. 2, 283.

Kojima, M., Kubo, K., Yoshikawa, K., 2006. Elongation/compaction of giant DNA caused by depletion interaction with a flexible polymer. J. Chem. Phys. 124, 024902.

Koltover, I., Salditt, T., Radler, J.O., Safinya, C.R., 1998. An inverted hexagonal phase of cationic liposome–DNA complexes related to DNA release and delivery. Science 281, 78–81.

Konopka, M.C., Shkel, I.A., Cayley, S., Record, M.T., Weisshaar, J.C., 2006. Crowding and confinement effects on protein diffusion *in vivo*. J. Bacteriol. 188, 6115–6123.

Lai, G.H., Butler, J.C., Zribi, O.V., Smalyukh, I.I., Angelini, T.E., Purdy, K.R., Golestanian, R., Wong, G.C.L., 2008. Self-organized gels in DNA/F-actin mixtures without crosslinkers: networks of induced nematic domains with tunable density. Phys. Rev. Lett. 101, 218303.

Li, G.-W., Berg, O.G., Elf, J., 2009. Effects of macromolecular crowding and DNA looping on gene regulation kinetics. Nat. Phys. 5, 294–297.

Limozin, L., Sackmann, E., 2002. Polymorphism of crosslinked actin networks in giant vesicles. Phys. Rev. Lett. 89, 168103.
Limozin, L., Roth, A., Sackmann, E., 2005. Microviscoelastic moduli of biomimetic cell envelopes. Phys. Rev. Lett. 95, 178101.
Lis, L.J., Lis, W.T., Parsegian, V.A., Rand, R.P., 1981. Biochemistry 20, 1771–1777.
Melnikov, S.M., Sergeyev, V.G., Yoshikawa, K., 1995. Binding of divalent cations to dipalmitoylphosphatidylcholine bilayers and its effect on bilayer interaction. J. Am. Chem. Soc. 117, 2401–2408.
Mengistu, D.H., Bohinc, K., May, S., 2009. Binding of DNA to zwitterionic lipid layers mediated by divalent cations. J. Phys. Chem. B 113, 12277–12282.
Murtas, G., Kuruma, Y., Bianchini, P., Diaspro, A., Luisi, P.L., 2007. Protein synthesis in liposomes with a minimal set of enzymes. Biochem. Biophys. Res. Commun. 363, 12–17.
Negishi, M., Sakaue, T., Kingo, T., Yoshikawa, K., 2010. Cooperation between giant DNA molecules and actin filaments in a microsphere. Phys. Rev. E 81, 051921.
Negishi, M., Ichikawa, M., Nakajima, M., Kojima, M., Fukuda, T., Yoshikawa, K., 2011. Phase behavior of crowded like-charged mixed polyelectrolytes in a cell-sized sphere. Phys. Rev. E 83, 061921.
Noireaux, V., Libchaber, A., 2004. A vesicle bioreactor as a step toward an artificial cell assembly. Proc. Natl. Acad. Sci. U. S. A. 101, 17669–17674.
Nomura, S., Tsumoto, K., Hamada, T., Akiyoshi, K., Nakatani, Y., Yoshikawa, K., 2003. Gene expression within cell-sized lipid vesicles. ChemBioChem 4, 1172–1175.
Oberholzer, T., Nierhaus, K.H., Luisi, P.L., 1999. Protein expression in liposomes. Biochem. Biophys. Res. Commun. 261, 238–241.
Pautot, S., Frisken, B.J., Cheng, J.X., Xie, X.S., Weitz, D.A., 2003a. Spontaneous formation of lipid structures at oil/water/lipid interfaces. Langmuir 19, 10281–10287.
Pautot, S., Frisken, B.J., Weitz, D.A., 2003b. Engineering asymmetric vesicles. Proc. Natl. Acad. Sci. U. S. A. 100, 10718–10721.
Saito, H., Kato, Y., Le Berre, M., Yamada, A., Inoue, T., Yosikawa, K., Baigl, D., 2009. Time-resolved tracking of a minimum gene expression system reconstituted in giant liposomes. ChemBioChem 10, 1640–1643.
Schlesinger, A.P., Wang, Y., Tadeo, X., Millet, O., Pielak, G.J., 2011. Macromolecular crowding fails to fold a globular protein in cells. J. Am. Chem. Soc. 133, 8082–8085.
Stachowiak, J.C., Richmond, D.L., Li, T.H., Liu, A.P., Parekh, S.H., Fletcher, D.A., 2008. Unilamellar vesicle formation and encapsulation by microfluidic jetting. Proc. Natl. Acad. Sci. U. S. A. 105, 4697–4702.
Takiguchi, K., Yamada, A., Negishi, M., Tanaka-Takiguchi, Y., Yoshikawa, K., 2008. Entrapping desired amounts of actin filaments and molecular motor proteins in giant liposomes. Langmuir 24, 11323–11326.
Takiguchi, K., Yamada, A., Negishi, M., Honda, M., Tanaka-Takiguchi, Y., Yoshikawa, K., 2009. Construction of cell-sized liposomes encapsulating actin and actin-cross-linking proteins. Methods Enzymol. 464, 31–53.
Tresset, G., Cheong, W.C.D., Lam, Y.M., 2007. Role of multivalent cations in the self-assembly of phospholipid–DNA complexes. J. Phys. Chem. B 111, 14233–14238.
Tsuji, A., Yoshikawa, K., 2010a. ON–OFF switching of transcriptional activity of large DNA through a conformational transition in cooperation with phospholipid membrane. J. Am. Chem. Soc. 132, 12464–12471.
Tsuji, A., Yoshikawa, K., 2010b. Real-time monitoring of RNA synthesis in a phospholipid-coated microdroplet as a live-cell model. ChemBioChem 11, 351–357.
Tuinier, R., Rieger, J., de Kruif, C.G., 2003. Depletion-induced phase separation in colloid–polymer mixtures. Adv. Colloid Interface Sci. 103, 1–31.
Vasilevskaya, V.V., Khokhlov, A.R., Matsuzawa, Y., Yoshikawa, K., 1995. Collapse of single DNA in poly(ethylene glycol) solutions. J. Chem. Phys. 102, 6595–6602.

Walde, P., Cosentino, K., Engel, H., Stano, P., 2010. Giant vesicles: preparations and applications. ChemBioChem 11, 848–865.
Yamada, A., Yamanaka, T., Hamada, T., Hase, M., Yoshikawa, K., Baigl, D., 2006. Spontaneous transfer of phospholipid-coated oil-in-oil and water-in-oil micro-droplets through an oil/water interface. Langmuir 22, 9824–9828.
Yanagisawa, M., Imai, M., Taniguchi, T., 2008. Shape deformation of ternary vesicles coupled with phase separation. Phys. Rev. Lett. 100, 148102.
Yanagisawa, M., Imai, M., Taniguchi, T., 2010. Periodic modulation of tubular vesicles induced by phase separation. Phys. Rev. E 82, 051928.
Yanagisawa, M., Iwamoto, M., Kato, A., Yoshikawa, K., Oiki, S., 2011. Oriented reconstitution of a membrane protein in a giant unilamellar vesicle: experimental verification with the potassium channel KcsA. J. Am. Chem. Soc. 133, 11774–11779.
Yanagisawa, M., Shimokawa, N., Ichikawa, M., Yoshikawa, K., 2012. Micro-segregation induced by bulky-head lipids: formation of characteristic patterns in a giant vesicle. Soft Matter 8, 488–495.
Yanagisawa, M., Yoshida, T., Furuta, M., Nakata, S., Tokita, M., 2013. Adhesive force between paired microdroplets coated with lipid monolayers. Soft Matter 9, 5891–5897.
Yoshikawa, K., 2001. Controlling the higher-order structure of giant DNA molecules. Adv. Drug Deliv. Rev. 52, 235–244.
Yoshikawa, K., Takahashi, M., Vasilevskaya, V.V., Khokhlov, A.R., 1996. Large discrete transition in a single DNA molecule appears continuous in the ensemble. Phys. Rev. Lett. 76, 3029–3031.
Zhou, H.X., Rivas, G., Minton, A.P., 2008. Macromolecular crowding and confinement: biochemical, biophysical, and potential physiological consequences. Annu. Rev. Biophys. 37, 375–397.
Zimmerman, S.B., Minton, A.P., 1993. Macromolecular crowding: biochemical, biophysical, and physiological consequences. Annu. Rev. Biophys. Biomol. Struct. 22, 27–65.
Zimmerman, S.B., Trach, S.O., 1991. Estimation of macromolecule concentrations and excluded volume effects for the cytoplasm of *Escherichia coli*. J. Mol. Biol. 222, 599–620.
Zinchenko, A.A., Yoshikawa, K., 2005. Na^+ shows a markedly higher potential than K^+ in DNA compaction in a crowded environment. Biophys. J. 88, 4118–4123.

CHAPTER EIGHT

Noncanonical Structures and Their Thermodynamics of DNA and RNA Under Molecular Crowding: Beyond the Watson–Crick Double Helix

Naoki Sugimoto[1]
Frontier Institute for Biomolecular Engineering Research (FIBER) and Faculty of Frontiers of Innovative Research in Science and Technology (FIRST), Konan University, Kobe, Japan
[1]Corresponding author: e-mail address: sugimoto@konan-u.ac.jp

Contents

International Review of Cell and Molecular Biology, Volume 307
ISSN 1937-6448
http://dx.doi.org/10.1016/B978-0-12-800046-5.00008-4

Abstract

How does molecular crowding affect the stability of nucleic acid structures inside cells? Water is the major solvent component in living cells, and the properties of water in the highly crowded media inside cells differ from that in buffered solution. As it is difficult to measure the thermodynamic behavior of nucleic acids in cells directly and quantitatively, we recently developed a cell-mimicking system using cosolutes as crowding reagents. The influences of molecular crowding on the structures and thermodynamics of various nucleic acid sequences have been reported. In this chapter, we discuss how the structures and thermodynamic properties of nucleic acids differ under various conditions such as highly crowded environments, compartment environments, and in the presence of ionic liquids, and the major determinants of the crowding effects on nucleic acids are discussed. The effects of molecular crowding on the activities of ribozymes and riboswitches on noncanonical structures of DNA- and RNA-like quadruplexes that play important roles in transcription and translation are also described.

1. INTRODUCTION

Living cells contain a large number of macromolecules including nucleic acids, proteins, carbohydrates, and others. The total concentration of biomolecules is estimated to be 300–400 g L^{-1} in *E. coli*, 50–400 g L^{-1} in the eukaryotic cytoplasm, 100–400 g L^{-1} in nuclei, 100–200 g L^{-1} in nuclear organelles, and 270–560 g L^{-1} in the mitochondrial matrix (Cheung et al., 2005; Ellis, 2001; Fulton, 1982; Goodsell, 1991; Hancock, 2007; Handwerger et al., 2005; Partikian et al., 1998; Srere, 1980, 1981; Zimmerman and Trach, 1991). These concentrations are much higher than those used for *in vitro* experiments, which are typically below 10 g L^{-1}. Studies *in vitro* using molecular crowding agents are helping to bridge the gaps in our understanding of nucleic acid reactions and their functions in *in vivo* environments.

In principle, the space excluded by macromolecules is inaccessible to other molecules due to steric effects, and thus effective concentrations of biomolecules are increased inside cells by the excluded volume effect (Minton, 1998). Furthermore, reaction volumes are orders of magnitude smaller and surface areas more substantial in cytoplasmic organelles and cell membranes than in the dilute solution conditions of *in vitro* studies. A large number of macromolecules and organelles in a small cellular volume restrict molecular movements and dynamics (Dix and Verkman, 2008; Misteli, 2008). The mobility of water molecules proximal to intracellular components and cell membranes is also highly restricted, resulting in differences in the solvent parameters, including viscosity, water activity, and dielectric property, in cells relative to dilute solution. Thus, the intracellular environment likely results in molecular interactions that cannot be anticipated based on our understanding of thermodynamic and kinetic properties measured in dilute solution.

How does molecular crowding affect nucleic acids inside cells? It is known that molecular crowding promotes the organization of the nuclear architecture (Hancock, 2004; Richter et al., 2007, 2008). Experiments in which penetrating polymers of polyethylene glycol (PEG) and dextran were injected into the nucleus show that the degree of molecular crowding plays an essential role in driving the formation and maintaining the functions of nuclear compartments (Hancock, 2004; Handwerger et al., 2005). Intracellular organelles, the cytoskeleton, and macromolecules significantly hinder the diffusion of nucleic acids and cause an anomalous rate of diffusion-mediated searching for interaction partners (Guigas and Weiss, 2008). A spot photobleaching technique was used to show that the diffusion rate of DNA fragments in HeLa cell cytoplasm is much slower than that in an aqueous solution, and the diffusion coefficients relative to that in water were 0.19 for a 100-base pair (bp) DNA, 0.06 for a 250-bp DNA, and <0.01 for >2000-bp DNA fragments (Lukacs et al., 2000). The DNA size-dependence of diffusion rates suggests that cytoplasmic organelles are barriers for nucleic acid transport and gene delivery.

There seems to be considerable diffusional heterogeneity within cells and intracellular organelles. Eukaryotic nuclei are highly heterogeneous and crowded. The nucleus contains a variety of subnuclear structures and large protein complexes including the nucleolus, splicing-factor compartments, Cajal bodies, promyelocytic leukemia bodies, replication factors, and transcription factors (Lewis and Tollervey, 2000; O'Brien et al., 2003). In the eukaryote nucleus, DNA is highly condensed and wrapped around histone proteins, forming the chromosome structure (Luger et al., 1997; Richmond and Davey, 2003). Histones regulate nucleosome assembly by causing a dense compaction of DNA. These structures are flexible, however, and structural modulations are reversible (Minsky et al., 1997; Ringrose et al., 1999; Zimmerman and Murphy, 1996). Changes of DNA compaction occur during the cell cycle and lead to considerable alterations in the condensation of metal ions and the degree of hydration. DNA hybridization kinetics in a single HeLa cell measured using microscopy showed accelerated and decelerated kinetics in comparison with data obtained in dilute solution that were strongly dependent on the length of the hybridization probe

(Schoen et al., 2009). Moreover, molecular crowding effects on gene interactions with transcription factors play a key role in transcriptional bursts and effectively make messenger RNA (mRNA) and protein concentrations much higher than those predicted by first-order kinetics (Meyer et al., 2012).

Water is the major solvent in living cells and typically makes up around 70% of the total weight of a cell. The average mobility of intracellular free water is considered to be similar to that of pure water. There are significant layers of ordered water that hydrate intracellular components, however, and these layers are highly ordered and osmotically inactive (Garlid, 2000; Luby-Phelps, 2000; Parsegian and Rau, 1984; Srere, 1981). Thus, the properties of water in highly crowded media are quite unique. The amounts of intracellular free and ordered water molecules change depending on the status of the cell. For example, the fraction of ordered water in a eukaryotic cell changes during the cell cycle (DuPre and Hempling, 1978), and fewer water molecules can move freely in tumor cells than in normal cells (Damadian, 1971). It has also been suggested that the amount of cytoplasmic water is the primary determinant of the growth rate of osmotically stressed *E. coli*; presumably cytoplasmic molecular interactions are perturbed at an increased osmolality (Cayley and Record, 2003). *E. coli* can adapt to changes in osmolality of at least 100-fold, and DNA–protein interactions are relatively insensitive to changes in the salt concentration *in vivo*, whereas they are sensitive *in vitro* (Record et al., 1998a).

The buffering system to relieve the perturbations in an intracellular environment arises from changes in the amounts of water and metabolites such as inorganic ions, polyols, amino acids, and their derivatives, and methylamines including sarcosine and betaine (Yancey et al., 1982). Small hydrophilic molecules are abundant in cells and tightly bind water molecules and restrict their mobility, thus acting as osmolytes. An extreme example in mammals is the kidney medulla cell; the urea concentrations in these cells are as high as 5.4 M, corresponding to 30% by mass (MacMillen and Lee, 1967; Rösgen et al., 2005). In addition to water activity, the dielectric properties in cells are quite different from those in aqueous solutions. The dielectric constant of yeast cells is estimated to be around 50 or even less, which is significantly lower than the value of pure water at about 80 (Asami et al., 1976; Tanizaki et al., 2008). The decrease of the dielectric constant results in increased electrostatic attractions and repulsions. Therefore, not only macromolecular crowding but also the influences of small organic molecules have an impact on the molecular environment in a living cell (Lang et al., 1998; Yancey et al., 1982).

It is difficult to directly measure the thermodynamics of oligonucleotide hybridization or folding in cells, but their structures can be evaluated. Dimethyl sulfate (DMS) provides a methylation pattern that reflects the RNA secondary structure and protein associations. Chemical modification experiments have revealed that the ribosomal RNA (rRNA) intron in *Tetrahymena thermophila*, the U3A small nuclear RNA in *S. cerevisiae*, the thrS leader sequence in *Bacillus subtilis*, and MFA2 mRNA in a yeast strain adopt structures nearly equivalent to those determined *in vitro* (Doktycz et al., 1998; Luo et al., 1998; Mereau et al., 1997; Zaug and Cech, 1995). In contrast, the 18S rRNA in soybean seedling leaves studied with DMS has a structure that differs from that predicted based on *in vitro* analyses (Senecoff and Meagher, 1992). It remains difficult to investigate systematically and quantitatively individual structures and reactions of intracellular nucleic acids.

In this chapter, we discuss how the structures and thermodynamic properties of nucleic acids differ in various conditions such as highly crowded environments, compartment environments, and ionic liquids (ILs). We also describe the major determinants of the crowding effects on nucleic acids. The impacts of molecular crowding on functions of nucleic acids such as ribozymes and riboswitches and noncanonical structures of DNA and RNA-like quadruplexes are also discussed.

2. THERMODYNAMIC STUDIES ON NUCLEIC ACIDS

The thermodynamics of nucleic acid interactions are responsible for the energetic aspects of hybridization and the folding of oligonucleotides. Quantitative data regarding interaction energies are quite useful for designing sequences to maximize the effects of oligonucleotide technologies including antisense, antigene, ribozyme, deoxyribozyme, nucleic acid decoy, riboswitches, and RNA interference methods (Matveeva et al., 2010; Praseuth et al., 1999; Rosi et al., 2006; Yen et al., 2004). Two stabilizing forces in Watson–Crick base pairing are interbase hydrogen bonding and base stacking (Fig. 8.1A). The strengths of these interactions primarily determine the efficiencies of sequence-specific hybridization and the structure adopted. Counterion binding and hydration are also important determinants of thermodynamic stability, because cations and water molecules associate with the polar purine and pyrimidine bases and with the negatively charged sugar-phosphate backbone (Fig. 8.1B). In most cases, the association or dissociation of ions and waters is energetically coupled to nucleic acid folding (Fig. 8.1C) (Marky and Kupke, 2000; Nakano et al., 2008a; Record et al., 1998b; Westhof, 1988).

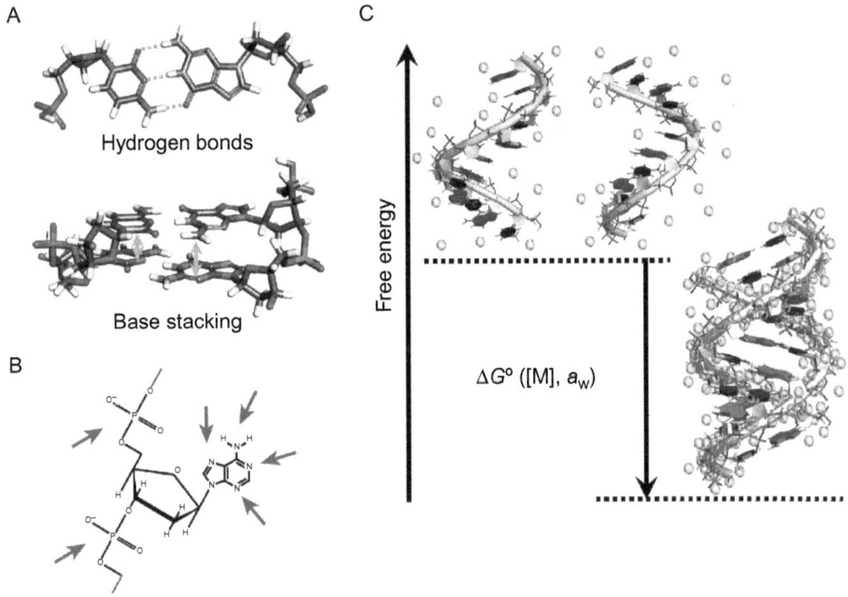

Figure 8.1 Nucleotide interactions in hybridization and nucleic acid folding. (A) Interbase hydrogen bonds and base stacking in Watson–Crick base pairs. (B) Potential binding sites of metal ions and water molecules on a DNA nucleotide unit indicated by the arrows. (C) Free energy diagram for the Watson–Crick duplex formation as a function of the cation concentration [M] and the water activity a_w. (For color version of this figure, the reader is referred to the online version of this chapter.)

Investigation of thermodynamic aspects of base-pairing interactions is essential to understanding of secondary and tertiary structures of DNA and RNA. In early studies, it was assumed that the number of interbase hydrogen bonds determines the stability of Watson–Crick base pairs; that is, the three hydrogen bonds of a C-G base pair make it more stable than the A-T base pair with only two hydrogen bonds (Saenger, 1984). Now, the nearest-neighbor model is widely used to account for the thermodynamic properties of Watson–Crick duplexes (Turner et al., 1988). This model assumes that the formation of base pairing is mostly affected by the nearest-neighbor (adjacent) base pair already formed. Thus, this model places significant weight on the contributions of base stacking interactions, as well as hydrogen bonds, to thermodynamic stability of nucleic acid structures. Experimentally, the thermal stability of a nucleotide (nt) duplex is dependent on sequence, and duplex stability is well accounted for by the sum of thermodynamics of individual nearest-neighbor interactions.

The nearest-neighbor parameters of ΔH°, ΔS°, and ΔG°_{37} (changes in the enthalpy, entropy, and Gibbs free energy at 37 °C, respectively) for Watson–Crick base pair formation can be used to reliably predict the thermodynamic parameters of Watson–Crick duplex formation between any DNA or RNA sequences.

Since oligonucleotides are commercially available, they are often used for therapeutic and diagnostic purposes. Short oligonucleotides of DNA and RNA can be synthesized chemically. A sequence can be designed for the purpose of forming continuous base pairs with a target sequence without unpaired sites, providing a large negative free energy change, ΔG°, for duplex formation at physiological temperatures. The thermodynamic stability of oligonucleotide base pairing has been studied extensively in aqueous solutions containing salts (Bloomfield et al., 2000). These studies have revealed that the thermodynamic parameters for the formation of oligonucleotide duplexes can be simply explained by the sum of nearest-neighbor base pair interactions as mentioned above, and context-dependent values of ΔH°, ΔS°, and ΔG° for the formation of Watson–Crick base pairs have been estimated (Mathews et al., 1999; Sugimoto et al., 1995, 1996; Xia et al., 1998). The thermodynamic parameters allow for accurate predictions of the hybridization energy and secondary structures of a given oligonucleotide sequence (Mathews, 2004; Zuker, 2003). Predictions for long RNA structures are greatly improved when considering chemical modification data (Low and Weeks, 2010). Consequently, the energetic aspects of nt interactions are useful for the rational design of oligonucleotide sequences. As an example of this applied to the antisense method of gene silencing, the prediction parameters determined for RNA/DNA hybrid duplexes (Sugimoto et al., 1995, 2000; Watkins et al., 2011) can be used as a general guideline for designing antisense DNA sequences. It has also been proposed that the thermodynamic properties of base pairing between RNA and DNA sequences have a strong correlation with the level of mRNA transcription in a cell (Kraeva et al., 2007).

Most thermodynamic studies have been performed using aqueous solutions of dilute and homogenous nucleic acid, and it is unclear whether those thermodynamic data can be used to predict interaction energies under non-aqueous conditions. Molecular conditions in cells and on material surfaces are remarkably different from those in aqueous solutions. One of the most distinguishing features of the molecular conditions in cells is an environment crowded with proteins and cellular organelles. These crowded conditions

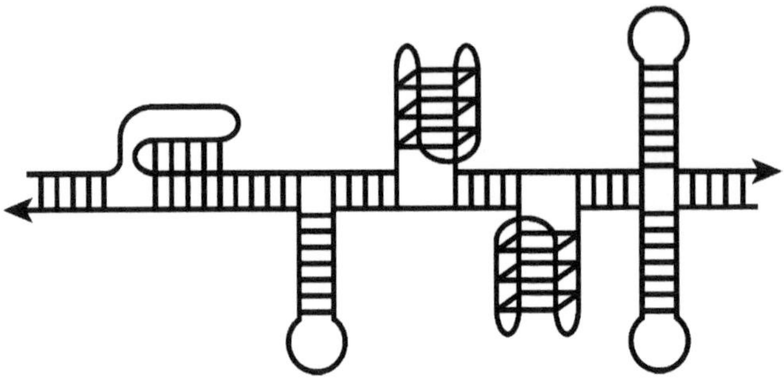

Figure 8.2 Schematic illustration of the secondary structures of DNA or RNA. A triplex, a hairpin loop, quadruplexes, and a junction (from left to right).

affect diffusion rates, molecular activity, conformations of biomolecules, and the properties of solvent water, which in turn will affect equilibria and rates of molecular reactions. In addition to macromolecules, small hydrophilic molecules such as metabolites and osmolytes are present in cells, resulting in the properties of intracellular water being quite different from those of a dilute solution. It has been proposed that regulation of cell volume can play an essential role in cell growth and proliferation by modulating the extent of intracellular crowding or osmolality (Al-Habori, 2001). The intracellular environment is possibly a key determinant of the efficiency of cellular processes including tumor generation, susceptibility to diseases with aging, intracellular phase separation, molecular compartmentation, and glucose metabolism (Johansson and Walter, 2000; Ovádi and Srere, 2000; Vazquez and Oltvai, 2011).

It is important to note that DNA and RNA fold into distinct structures that are stabilized by noncanonical interactions as well as Watson–Crick base pairs, and species-specific sequence regions often adopt uniquely folded structures, such as hairpin loops, left-handed helices, triplexes, quadruplexes, and branched junctions (Fig. 8.2). Since these noncanonical structures may be binding sites for proteins that regulate nucleic acid functions, these structural sites are of interest as promising drug targets (Blount and Breaker, 2006; Oganesian and Bryan, 2007; Wilson and Li, 2000).

3. MOLECULAR CROWDING EFFECTS ON THE CANONICAL STRUCTURES OF NUCLEIC ACIDS

3.1. Polynucleotide duplexes

The double helical structure is arguably the major conformation of nucleic acids and, with two strands of complementary sequence, exists stably under physiological conditions. Cosolute effects on the thermal stability of DNA and RNA duplexes have been well characterized. The thermal melting temperature T_m, at which one-half of the folded structure is dissociated, is useful for evaluation of the strength of base-pairing interactions (Puglisi and Tinoco, 1989), although certain polymer nucleic acids do not show complete dissociation of nt strands even at high temperatures.

Relatively inert polymers, like PEG, dextran, bovine serum albumin, and Ficoll, that are used to mimic the crowded environment of cells, have dramatic effects on duplex stability. The T_m of the RNA polymer duplex poly(I)/poly(C) (polyinosine associated with polycytidine) increases by 1.8 °C upon addition of PEG of molecular weight (MW) 2×10^4 (PEG 20000), 2.0 °C with PEG 4000, 0.7 °C with dextran of MW 7×10^4, and 0.7 °C with dextran of MW 1×10^4 at 10 wt% (Woolley and Wills, 1985). The T_m of the DNA polymer duplex poly(dA)/poly(dT) also increases as the PEG concentration or MW increases (e.g., ca. 5 °C at 19 wt% PEG 8000) (Spink and Chaires, 1995). The changes in the thermal stability of the duplexes can be accounted for by the excluded volume effect that favors the association of complementary strands.

In contrast to the effect of polymers, small cosolutes such as ethylene glycol, glycerol, acetamide, and sucrose decrease the T_m of *E. coli* DNA and poly(dA)/poly(dT) (Spink and Chaires, 1995, 1999). Since the formation of Watson–Crick base pairs is coupled with water binding, the duplex stability depends on water activity, which is substantially decreased by small cosolutes (Marky and Kupke, 2000). Calorimetric data show that, compared to dilute solutions, polyols, ethylene glycol, glycerol, urea, or PEG 200 at a high concentrations decrease the enthalpy changes of the melting of poly(dA-dT)/poly(dA-dT) and poly(dG-dC)/poly(dG-dC), whereas betaine and sucrose increase the melting enthalpy, perhaps due to preferential interactions with DNA base pairs (Spink et al., 2007). Small polyols such as glycerol and several monosaccharides (*i*-erythritol, D-sorbitol, D-mannitol, myo-inositol, D-arabitol, and L-arabitol) also decrease the T_m of calf thymus DNA by 1.1–8.3 °C (Del Vecchio et al., 1999).

3.2. Oligonucleotide duplexes

More insights into the cosolute effect have been gained by the studies of short oligonucleotides. The influence of PEG on the stability of DNA duplexes depends on the length of the duplex and on the MW of PEG (Nakano et al., 2004). Increases in the equilibrium constants for DNA duplex formation are more significant for longer duplexes than shorter duplexes; for shorter duplexes, the destabilizing effect due to decreased water activity has more of an impact.

Investigations using short oligonucleotides have several benefits in quantitative studies. A short double helix (e.g., less than a 15-mer) tends to melt into an unfolded single-stranded state in a two-state transition, and this allows for accurate analysis of the thermodynamic parameters from thermal transition curves (Puglisi and Tinoco, 1989). Short oligonucleotides also have a great advantage in solvating polymer cosolutes. PEG and alcohols precipitate long DNA fragments, but no precipitation occurs with short oligonucleotides. PEG is highly soluble in water and has a relatively low vapor pressure, both useful properties for experiments in which melting of oligonucleotide structure is thermally induced. Accordingly, studies using short oligonucleotides are convenient for investigating the structures, thermal stabilities, and kinetic properties of nucleic acids under conditions enriched in polymer molecules.

Using a large amount of a neutral cosolute such as PEG tends to induce longer polynucleotide duplexes to adopt a compact structure and gives rise to precipitation (Bloomfield, 1996). On the other hand, short oligonucleotides are considerably more resistant to precipitation, allowing for experiments using high concentrations of cosolutes. The overall helical conformation of short DNA and RNA duplexes is changed only slightly by the addition of cosolutes (Nakano et al., 2004). Structural studies using circular dichroism (CD) and electron paramagnetic resonance spectra suggest that a high-molecular weight PEG slightly changes the conformation of oligonucleotides in the single-stranded state (Nakano et al., 2008b). It is assumed that the helical conformation of single strands, which are relatively flexible compared to those of base-paired structures, undergo conformational restrictions upon the addition of cosolutes.

Influences of small neutral cosolutes (e.g., methanol, ethanol, ethylene glycol, 1-propanol, 2-propanol, formamide, dimethyl sulfoxide, *N,N*-dimethylformamide, urea, and dioxane) on the thermodynamics of short DNA and RNA duplexes have been reported (Nakano et al., 2013). These cosolutes destabilize short duplexes with the degree of destabilization differing depending on the cosolute type. There is no good correlation between duplex stability in 10 mol% cosolute solutions and viscosity, dipole moment, surface tension, or the solubility of the nucleotide.

As mentioned above, base pair formation is accompanied with the association of water molecules and creates a new ordered hydration layer that counteracts effects of dehydration of polar purine and pyrimidine bases. The number of water molecules that bind during DNA base pair formation

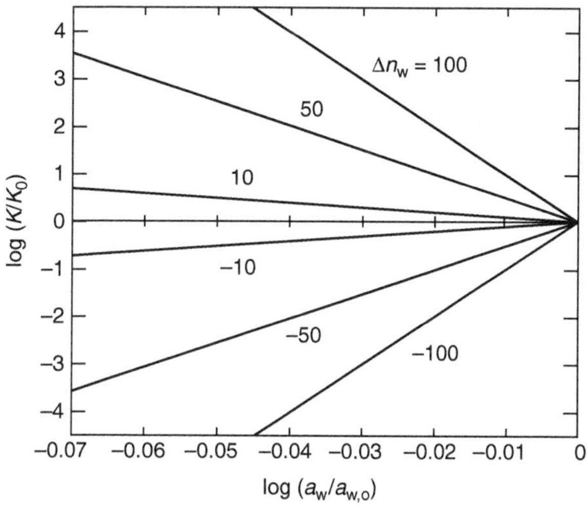

Figure 8.3 Small cosolutes influence biomolecular reactions involving the binding or release of water molecules. Dependence of the water activity, a_w, on the association equilibrium constant K of the reaction $2B \Leftrightarrow B_2 + \Delta n_w H_2O$. The parameters $a_{w,0}$ and K_0 indicate the water activity and the association equilibrium constant without cosolute, respectively. The linear slope is approximately $-\Delta n_w$.

determines the degree of destabilization of nucleic acid structures due to any reduction in water activity by cosolutes. According to the equation $(\partial \log K/\partial \log a_w) = -\Delta n_w$, the plot of the log of the association equilibrium constant (K) versus the log of water activity (a_w) gives a linear slope that represents the number of waters released via duplex formation, Δn_w (Fig. 8.3). This plot for all oligonucleotide duplex and hairpin structures evaluated is linear, demonstrating water uptake upon structure formation (Nakano et al., 2004, 2012). This linear relationship supports the hypothesis that water activity is the primary determinant of base pair stability.

Solution properties further affect the rates of association and the folding process. When an association reaction is limited by diffusional encounter of molecules, cosolutes may decrease the association rate as the increased viscosity may interfere with the movements and dynamics of molecules. On the other hand, the excluded volume effect may increase the association rates as this effect results in an increase in the effective concentration of reactant molecules. For duplex formation of short oligonucleotides, ethylene glycol and large and small PEGs at 20 wt% increase the dissociation rate constant of DNA strands with little impact on association rate (Gu et al., 2007). This observation suggests that neither the excluded volume effect nor the viscosity significantly contribute to kinetic parameters of duplex formation.

3.3. Hydration of nucleic acids

Osmotic stress experiments using a cosolute as an osmolyte provide valuable insights regarding nucleic acid interactions in aqueous solutions. The plot of ln K for duplex formation against ln a_w shows a linear relationship and gives a negative slope representing the hydration change. This hydration analysis reveals a difference in the hydration numbers between folded ($n_{w,folded}$) and unfolded states ($n_{w,unfolded}$) of oligonucleotides. The degree of hydration change, $-\Delta n_w$ ($=n_{w,folded}-n_{w,unfolded}$), is correlated with the effect of osmotic pressure on the stability of oligonucleotide structures. This number differs depending on the cosolute used, at least partly reflecting different interactions with DNA strands. Specifically, DNA duplexes are partly dehydrated due to competitive solvation with hydrophilic cosolutes such as polyols that replace hydrating waters and thus affect the hydration number. The value of $-\Delta n_w$ determined from the slope of the linear plot may not represent the true hydration number if cosolutes directly interact with nucleic acids. It has been proposed that PEG may interact with exposed DNA bases because of their hydrophobic nature (Knowles et al., 2011); however, whether or not PEG stably interacts with oligonucleotides is still debated (Nakano et al., 2012).

A study of DNA hairpins of the same nearest-neighbor compositions showed similar thermodynamic parameters and similar $-\Delta n_w$ values in the presence of various types of cosolutes, in support of the nearest-neighbor interaction model (Nakano et al., 2012). In addition, osmotic stress experiments using small PEGs showed that the values of $-\Delta n_w$ for flexible loops, unstable mismatches, and an abasic site upon DNA folding were negligible, whereas the binding to stable mismatch pairs was significant (Nakano et al., 2012). It appears that the $-\Delta n_w$ value, representing the number of associated water molecules, increases as the length of the oligonucleotide and the stability of a duplex increase. The cosolute effect that decreases the duplex stability can be reversed when dehydration is more significant than hydration; this can occur upon formation of noncanonical structures, as discussed next.

4. MOLECULAR CROWDING EFFECTS ON NONCANONICAL STRUCTURES OF NUCLEIC ACIDS

Nucleic acids can adopt many types of noncanonical structures, such as a left-handed double helix, a triple helix, and the quadruplex, depending on

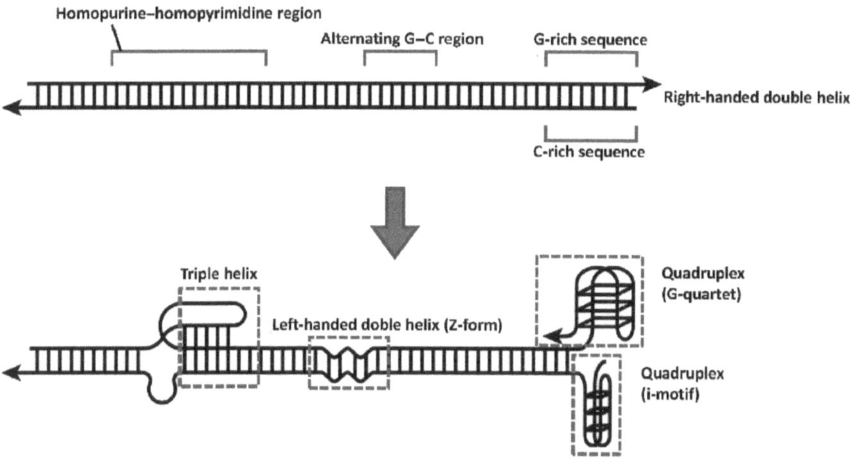

Figure 8.4 Noncanonical structures in genomic DNA. Triple helices (homopurine–homopyrimidine regions), left-handed double helices (alternating G–C), G-quadruplexes (G-rich), and i-motifs (C-rich) have the potential to form in regions with the indicated sequence types. (For color version of this figure, the reader is referred to the online version of this chapter.)

the sequence and solution conditions including pH, ion concentration, and temperature. Sequences with the potential to form these structures are found in the control regions and regulatory protein binding sites in the genome (Fig. 8.4). Satellite DNAs consisting of repeat sequences at the centromeric region of chromosomes tend to adopt noncanonical structures. These structures are relevant not only to gene regulation through their association with protein factors, but also because mutations in these regions are correlated with certain diseases (Wang and Vasquez, 2006; Wells, 2007). Such uniquely folded structures have been validated as promising drug targets (Oganesian and Bryan, 2007; Wilson and Li, 2000). Since Watson–Crick duplexes are relatively stable, the transition from duplexes to noncanonical structures in genomic DNA requires a high energy cost. However, it seems that the energy barrier is lowered in media containing background molecules, and that the conformations of certain sequence regions determined from *in vitro* and *in vivo* experiments may be different.

4.1. Triple helixes

A triple helix is formed after the binding of a third strand to the major groove of a duplex DNA through Hoogsteen base pairing. Pyrimidine-rich and purine-rich sequences can form stable triplex structures as a consequence

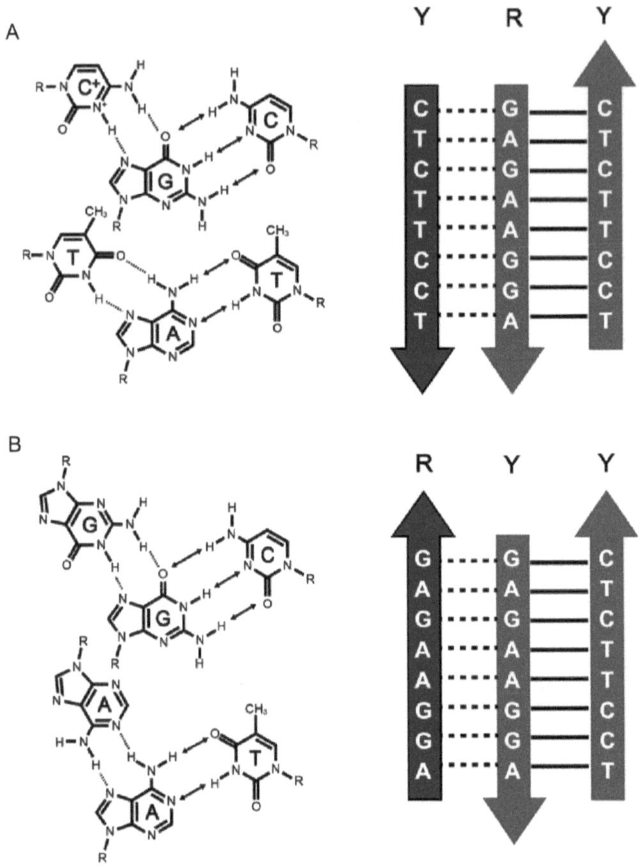

Figure 8.5 Chemical structure of base pairs in DNA triplex structures. (A) A parallel triplex structure. (B) An antiparallel triplex structure. Dotted and solid lines indicate Hoogsteen and Watson–Crick base pairs, respectively. Y and R indicate pyrimidine and purine stands, respectively. (For color version of this figure, the reader is referred to the online version of this chapter.)

of the formation of A–T–A and C^{+}–G–C triplets (Fig. 8.5). Unlike the duplex of poly(dA)/poly(dT) (consisting of Watson–Crick A–T pairs) with a T_m that is not significantly changed by cosolutes, the triplex of poly(dA)/poly(dT) • poly(dA) (consisting of Watson–Crick A–T and Hoogsteen T–A pairs) has an increased T_m in the presence of PEG (e.g., an increase of 14–21 °C in 19 wt% PEGs of an average MW of 400–8000 compared to that without PEG), but the T_m is decreased in the presence of small cosolutes (e.g., −1 to −3 °C by ethylene glycol or glycerol) (Spink and Chaires, 1995, 1999).

PEG also stabilizes the Hoogsteen base pairs of oligomer DNA triplexes (Goobes and Minsky, 2001; Goobes et al., 2002, 2003). The degree of stabilization increases as the MW of PEG is increased from 200 to 8000 (8–16 °C increase in the T_m of Hoogsteen base pairs at 15 wt% PEGs), whereas the stability of a Watson–Crick duplex is decreased or remains constant (−5 to 1 °C) (Goobes et al., 2003). This dissimilarity is due to differences in the involvement of water molecules in the formation of Hoogsteen base pairs, which is a water releasing reaction, and in the formation of Watson–Crick base pairs, which is a water uptake reaction. It is assumed that the third strand associates with a Watson–Crick duplex by replacing water molecules that are accommodated in the major groove of the double helix. The opposing effects of cosolutes on Hoogsteen base pairing and Watson–Crick base pairing are caused by the different contributions of water molecules in the pair formations (Miyoshi et al., 2009).

4.2. Quadruplexes

The most frequently observed repetitive DNA sequences throughout the genomes of most organisms are G-rich sequences and their complementary C-rich regions (Du et al., 2008; Qin and Hurley, 2008). G-rich and C-rich sequences are found at the ends of eukaryotic chromosomes in telomeres. The G-rich and C-rich regions can fold into G-quadruplexes and i-motifs, respectively. The G-quadruplex is formed by intermolecular or intramolecular association of Hoogsteen-paired coplanar guanines called a G-quartet (Fig. 8.6) (Gellert et al., 1962).

The G-quadruplex, which can be formed by intermolecular or intramolecular association of G-rich strands in antiparallel or parallel orientations, shows high structural polymorphism depending on conditions (Davis, 2004). Interestingly, G-quadruplex polymorphism is critically affected by molecular crowding. Molecular crowding can induce a structural transition from antiparallel to parallel form in the oligonucleotides $d(G_4T_4G_4)$ and $d(G_4T_4)_3G_4$, sequences derived from the *Oxytricha nova* telomere (Fig. 8.7A) (Miyoshi et al., 2002). Chaires and coworkers further found that PEG 400 induces dramatic changes in the G-quadruplex structure of human telomere DNA in the presence of K^+ (Li et al., 2005), indicating that the crystal structure, which revealed a parallel-stranded G-quadruplex, may not be the predominant structure in solution. On the other hand, it has also been reported that human telomere DNA forms a parallel-stranded G-quadruplex in a K^+ solution under molecular crowding conditions induced by PEG 200 (Xue et al., 2007).

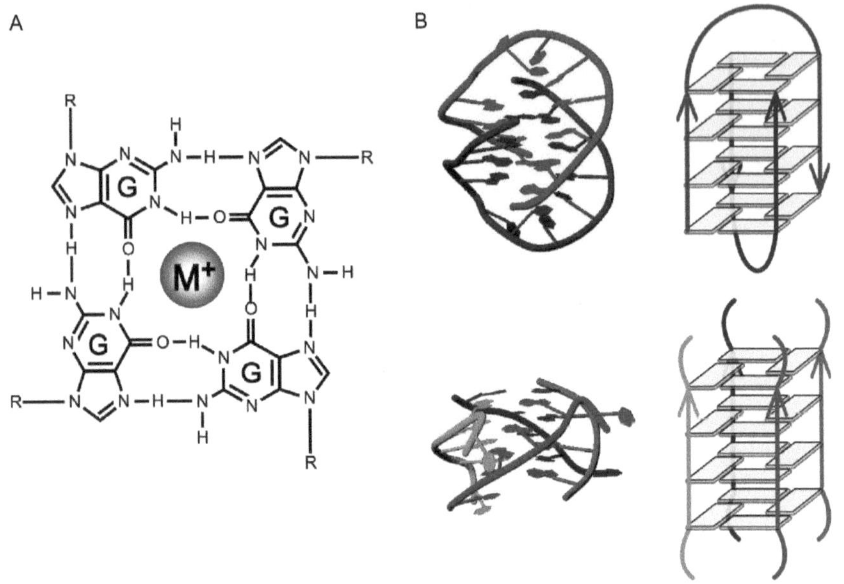

Figure 8.6 (A) Chemical structure of a G-quartet. (B) Structure of two-stranded antiparallel (top, PDB ID: 1JPQ) and four-stranded parallel (bottom, PDB ID: 352D) G-quadruplexes. (For color version of this figure, the reader is referred to the online version of this chapter.)

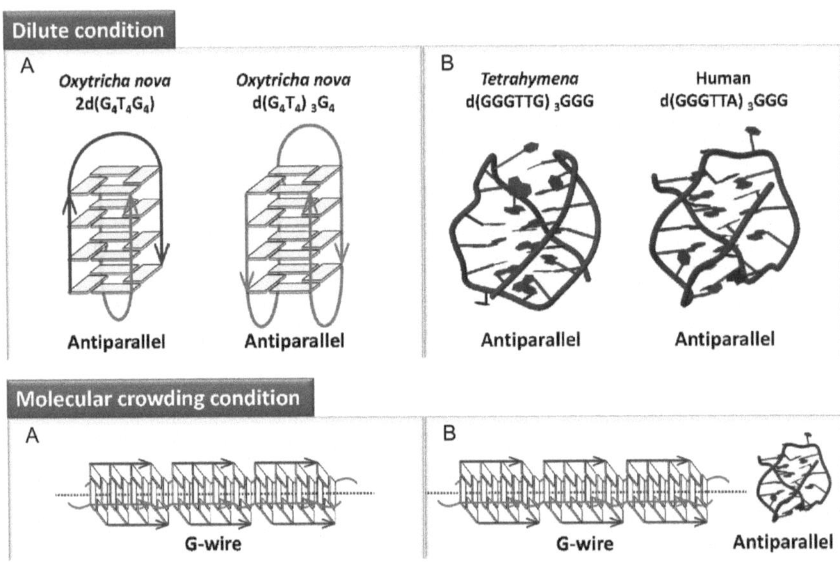

Figure 8.7 Molecular crowding effects on the conformations of telomeric DNA. (A) *Oxytricha nova* telomere DNA and (B) *Tetrahymena* and human telomeric DNAs (PDB ID: 186D and 143D). Molecular crowding induces a parallel-stranded G-quadruplex (G-wire) conformation in all of the telomeric DNA sequences except for human telomeric DNA. *Tetrahymena* and human telomeric structures are modified from the solution structure (PDB ID: 143D). (For color version of this figure, the reader is referred to the online version of this chapter.)

The intermolecular and intramolecular G-quadruplex structures adopted by the *Tetrahymena* telomere sequence d($G_3TTG)_3G_3$ and the human telomere sequence d($G_3TTA)_3G_3$ have been studied in the absence and presence of various cosolutes (Fig. 8.7B) (Miyoshi et al., 2005). The oligonucleotide based on the *Tetrahymena* telomere forms a very long, well-ordered nanostructure, called a G-wire, in the presence of high concentrations of cosolutes such as PEG. Although this structure is adopted by many parallel-stranded telomere sequences, the intermolecular and intramolecular human telomere sequences remain compact antiparallel G-quadruplexes in the presence of cosolutes. These results demonstrate that a single G to A base replacement in a repetitive sequence leads to a drastically different structure under molecular crowding conditions. The important roles of hydration in the structures and stabilities of G-quadruplexes have been demonstrated by the use of acetonitrile as a cosolute (Miller et al., 2010).

It is thought that most telomere DNAs form into a double helical structure through Watson–Crick base pairing between the G-rich and C-rich strands. However, duplexes formed with G-rich DNA d($G_4T_4G_4$) and C-rich DNA d($C_4A_4C_4$) in dilute conditions are dissociated upon molecular crowding, and the G- and C-rich DNAs fold into G-quadruplex and i-motif structures, respectively (Fig. 8.8) (Miyoshi et al., 2004).

It has also been reported that molecular crowding induces telomere G-quadruplex formation under salt-deficient conditions and enhances its competition with duplex formation (Kan et al., 2006). Kinetic and thermodynamic studies of G-quadruplex formation from duplexes showed that the formation rate of a G-quadruplex depends on PEG, K^+, and G-quadruplex-ligand concentrations (Kumar and Maiti, 2005; Zhou et al., 2009). As shown using these short oligonucleotides, molecular crowding makes it possible for a G-rich sequence to form a stable G-quadruplex, presumably even in the context of chromosomal DNA (Zheng et al., 2010). These results indicate that molecular crowding is essential for G-quadruplex formation in a G-rich/C-rich double-stranded region. G-quadruplexes are polymorphic and the structure adopted depends on solution conditions. These results also illustrate the difficulty of correctly extrapolating the structures of not only telomeric DNA sequences but also many other biomolecules to those *in vivo* based on *in vitro* analyses. Thus, molecular crowding as well as other cellular environmental factors (Kumar and Maiti, 2004; Li et al., 2003; Miyoshi et al., 2003, 2006a) play critical roles in the structure of biomolecules, and analyzing their effects is useful for understanding biomolecular behaviors *in vivo* and for efficient drug and ligand targeting.

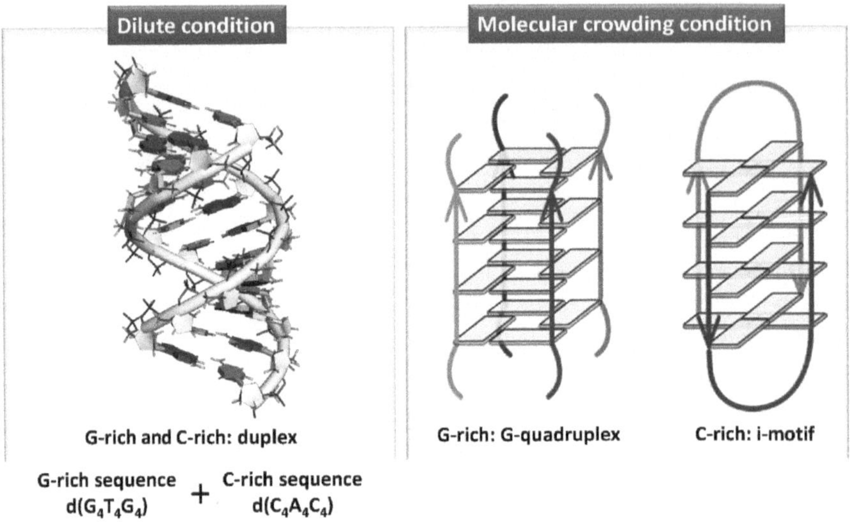

Figure 8.8 Structures of telomeric DNAs under dilute and molecular crowding conditions. Under dilute conditions, a 1:1 mixture of G-rich and C-rich strands forms a duplex (left), whereas under molecular crowding conditions the strands form a G-quadruplex and an i-motif, respectively (right). (For color version of this figure, the reader is referred to the online version of this chapter.)

Repeat sequences are frequently found in vertebrate genomes. More than 200 tracts of CCT repeats and their complementary AGG repeats have been identified in the human genome (Bacolla et al., 2008). These cytosine-rich regions, like those in telomeres, can form a four-stranded helix called an i-motif (Fig. 8.9), and this structure has been suggested to be involved in telomere function.

Because protonation of cytosine residues (the pK_a of *N*3 of a cytosine base is around 4) is required for the formation of the i-motif, a CCT repeat sequence forms an i-motif structure only at acidic pH. In PEG-containing solutions, however, C-rich sequences can adopt the i-motif structure at neutral pH. In the presence of PEG, the structure is stabilized due to an apparent pK_a shift of the cytosine base to near neutrality; this is more significant with high-molecular-weight PEGs than with those of lower MW (Rajendran et al., 2010). In summary, water-soluble cosolutes stabilize the triplex helix and the quadruplexes of G-rich and C-rich sequences. The formation of these noncanonical structures appears to be relevant to gene regulation and diseases.

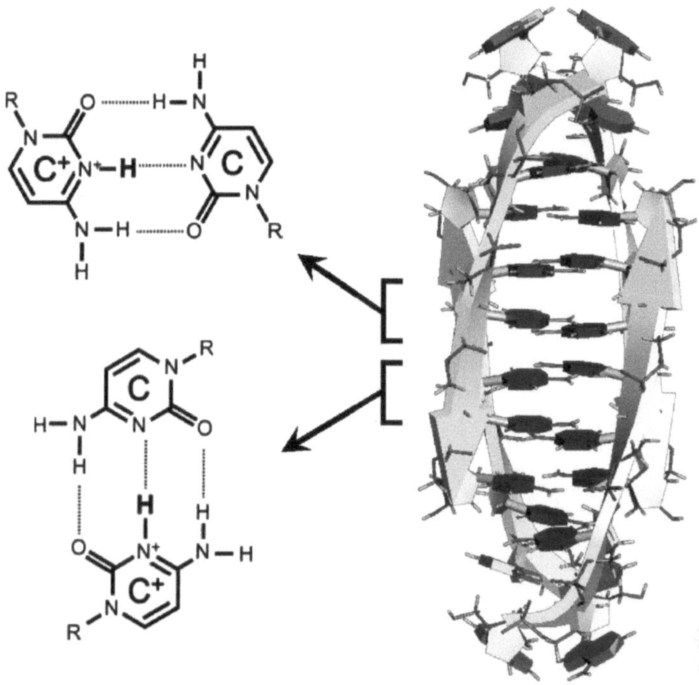

Figure 8.9 The C–C interaction formed in the i-motif DNA at acidic pH (modification of PDB ID: 1YBL). (For color version of this figure, the reader is referred to the online version of this chapter.)

4.3. High-order structures and left-handed double helix

Junctions in nucleic acid structures arise when three or more helices meet at a single point (Fig. 8.10A). These branched junctions are important intermediates in many biological processes (Kitts and Nash, 1987). Three-way junctions (TWJs) are the simplest type of junction and consist of three double helical arms connected at a junction point. TWJs occur both in DNA and RNA. In RNA they are involved in splicing (Guthrie and Patterson, 1988) and translation (Wimberly et al., 2000), whereas in DNA they are formed transiently during DNA replication (the replication fork) (Singleton et al., 2001). TWJs in DNA also arise during recombination involving phages (Hutchins et al., 1986). Furthermore, TWJs in DNA have been proposed to occur in the expansion of triplet repeats found in regions of genetically unstable genomic DNA associated with human diseases such as Huntington's (Jensch and Kemper, 1986).

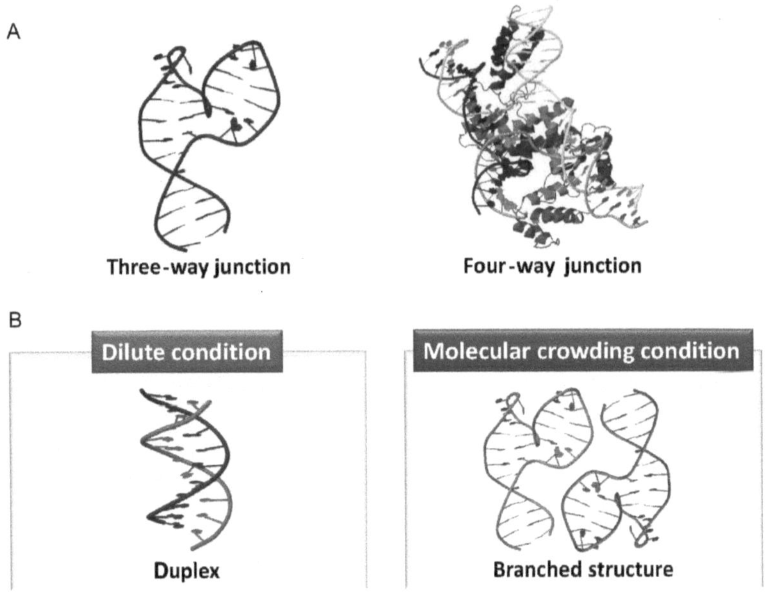

Figure 8.10 Three-dimensional structures of three- and four-way junctions. (A) A three-way junction (PDB ID: 1SNJ) and a four-way junction (2QNC) with T4 endonuclease VII. (B) Schematic illustration of the molecular crowding effect on the equilibrium between an intermolecular duplex and an intramolecular three-way junction. (For color version of this figure, the reader is referred to the online version of this chapter.)

Because TWJs are the smallest type of junction structures, oligonucleotides that adopt this structure have been used as model systems to gain insight into other complex and multibranched junctions (Lu et al., 1992; Rosen and Patel, 1993). Coaxial stacking is one of the most important conformational features of TWJs, and a divalent cation such as Mg^{2+} is critical for coaxial stacking (Lu et al., 1992; Rosen and Patel, 1993). Moreover, TWJs create a unique electrostatic environment near the junction point due to the close proximity of the opposing charges. Thus, there should be larger scale changes in charge neutralization and hydration during junction formation than during formation of a simple duplex (Zhong et al., 1997).

The effects of molecular crowding have been studied systematically on the formation of TWJs in DNA (Muhuri et al., 2009). TWJs are stabilized considerably with the addition of 5 mM Mg^{2+} in dilute conditions and destabilized by molecular crowding agents both in the absence and presence of 5 mM Mg^{2+}. Moreover, the destabilizing effect caused by molecular crowding impacts the entire TWJ structure, including the junction point and three helical duplex arms, demonstrating that hydration is important. On the other hand, comparison of the numbers of water molecules bound upon TWJ formation and by each helical arm individually suggests that water molecules are released from the nts at the junction point, leading to the stabilization of the junction point by molecular crowding. In fact, molecular crowding conditions are able to induce a structural transition from a bimolecular duplex to a TWJ, even in the absence of Mg^{2+} (Fig. 8.10B). This finding indicates that intracellular conditions, in which water activity is decreased and hydration is less favorable than in dilute solution, likely facilitates the formation of TWJ structures.

There have been several studies on the molecular crowding effects on other high-order RNA structures and their catalytic activity. It has been reported that methanol stabilizes the tertiary structure of a 58-nt RNA but not basic RNA secondary structure (Shiman and Draper, 2000). The addition of trimethylamine-oxide (TMAO) as an osmolyte enhances the efficiency of ribosome reconstitution by up to 100-fold, providing a substantially improved system for the *in vitro* analysis of mutant ribosomes (Semrad and Green, 2002). This effect may be due to the stabilizing effect of TMAO on the RNA tertiary structure. Indeed, TMAO can counteract the denaturing effects of urea on tRNA tertiary structure (Gluick and Yadav, 2003). In order to understand the basis of this effect, the TMAO-induced stabilization of RNAs has been studied. Results suggest that the formation of RNA tertiary structures is accompanied by substantial dehydration of the phosphate groups and that TMAO affects this process by reducing the energetic penalty associated with this dehydration (Lambert et al., 2010). These results are further supported by molecular dynamics simulations of a 22-nt RNA hairpin in aqueous TMAO solution (Pincus et al., 2008). Another molecular simulation study indicated that molecular crowding enhances the thermodynamic stability of a pseudoknot in human telomerase RNA, leading to a structural switch from a hairpin to the pseudoknot (Denesyuk and Thirumalai, 2011). These findings demonstrate that cosolutes play an important role in stabilizing high-order RNA structures and their interactions with other biomolecules.

Genomic DNA rich in alternating G–C sequence regions may adopt a left-handed double helix, called a Z-form (Fig. 8.11). The left-handed structure may be formed during gene transcription mediated by proteins, which would increase the potential for recombination, deletion, and translocation events (Wang and Vasquez, 2007). There are several proteins that bind and induce formation of the Z-form conformation (Kim et al., 2010;

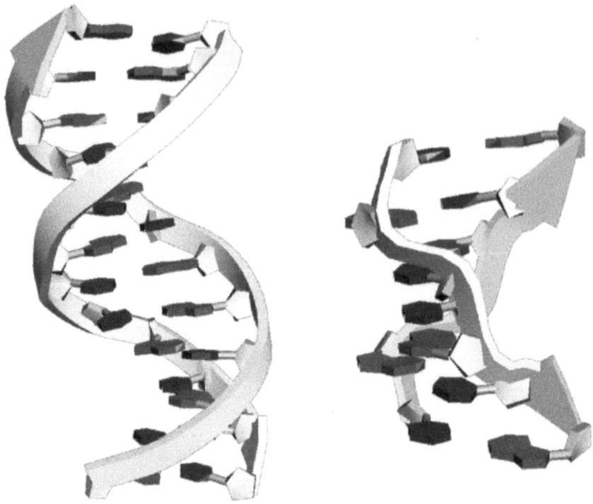

Figure 8.11 Structures of B-form (left, PDB ID: 2L8Q) and Z-form (right, PDB ID: 2ELG) DNA duplexes. (For color version of this figure, the reader is referred to the online version of this chapter.)

Wong et al., 2007). The structure is stabilized in the presence of high salt concentrations *in vitro,* whereas the canonical right-handed B-form duplex is formed at moderate and low salt concentrations (Mao et al., 1999). The structural transformation from the B-form to Z-form is induced by adding neutral cosolutes such as methanol, ethanol, glycerol, ethylene glycol, glucose, sucrose, and PEG, even with a moderate concentration of salts (Nakano et al., 2008b; Preisler et al., 1995). The Z-form is also favored under high hydrostatic pressure (Krzyzaniak et al., 1991, 1994). Formation of the Z-form with small cosolutes or under hydrostatic pressure is consistent with the fact that the left-handed duplex is less hydrated than the canonical duplex structure (Westhof, 1988).

5. MOLECULAR CROWDING EFFECTS ON FUNCTIONAL RNAs

5.1. Ribozymes

Ribozymes are RNA enzymes that catalyze site-specific phosphodiester bond cleavage of a target RNA sequence. The ribozyme reaction is facilitated by divalent metal ions that stabilize the catalytically active conformation and that may also play a direct role in catalysis (Bevilacqua et al.,

2004; Doudna and Lorsch, 2005; Lilley, 2005; Rangan and Woodson, 2003; Sugimoto et al., 1988, 1989; Toor et al., 2008). The hammerhead ribozyme is one of the smallest ribozymes, forming an active structure with a three-stem junction surrounding the catalytic core, which hybridizes and cleaves a substrate RNA in the presence of high or moderate Mg^{2+} concentrations (Fig. 8.12) (Forster and Symons, 1987; Haseloff and Gerlach, 1988; Uhlenbeck, 1987). Crystallography and molecular dynamics simulations have shown that Mg^{2+} stabilizes the active structure but does not play a catalytic role (Lee et al., 2008; Martick and Scott, 2006). Because the hammerhead ribozyme is relatively small and rapidly cleaves target RNAs *in vitro*, the *trans*-acting ribozyme sequence can be used for *in vivo* suppression of a target gene (Citti and Rainaldi, 2005; Khvorova et al., 2003; Li et al., 2008; Yen et al., 2004). PEG and other cosolutes at 20 wt% increase the RNA hydrolysis rate; however, there is no substantial change in the reaction rate or the product yield when using 1 wt% PEG, poly(dA)/poly(dT), tRNA, or bovine serum albumin (Nakano et al., 2009). Although the cosolutes decrease the stability of the base pairs of the stem helixes of the ribozyme/substrate complex, the thermal inactivation temperature of the ribozyme is inversely increased by the presence of cosolute. It is also remarkable that a large PEG (MW = 8000) decreases the metal ion concentration required for the reaction, facilitating catalytic turnover and activating catalytically inactive ribozymes. This ribozyme shows efficient activities with limiting Mg^{2+} concentrations in the presence of crowding agents, implying that the hammerhead ribozyme can work efficiency in living cells.

The hairpin ribozyme is composed of two domains, the substrate-binding domain loop A and the catalytic loop B domain. These domains

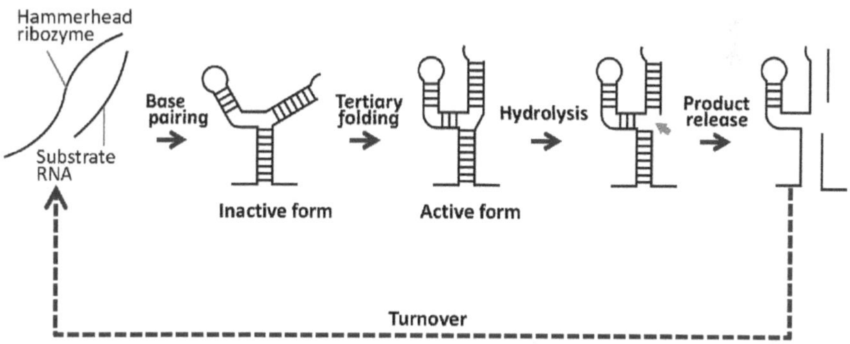

Figure 8.12 The reaction mechanism of targeting RNA hydrolysis by the hammerhead ribozyme. (For color version of this figure, the reader is referred to the online version of this chapter.)

are docked into a compact active structure through extensive interactions between the two loops (Butcher et al., 1999). The RNA cleavage activity of this ribozyme is enhanced in the presence of PEG 400 or dextran ($MW = 1 \times 10^4$), and RNA compaction accompanied by water release has been proposed to be responsible for the enhanced activity under osmotic pressure induced by cosolutes (Herve et al., 2006; Tobe et al., 2005).

In large group I ribozymes, the formation of an RNA tertiary structure between a GAAA tetraloop and a receptor RNA tertiary motif is favored in the presence of PEG or dextran but not of sucrose or glycerol (Downey et al., 2007). This observation is explained by folding which yields a change in the partial molar volume associated with the release of water molecules. Small angle X-ray scattering experiments have been performed on a large, bacterial group I ribozyme in the presence of PEG 1000. It was found that PEG favors more compact RNA structures and folding at low Mg^{2+} concentrations, in agreement with the dominant influence of the excluded volume effects (Kilburn et al., 2010). Thus, the cosolute effect differs among the ribozymes examined likely due to different RNA sizes, different contributions of water molecules to their folding, different structures, or different rate-determining steps.

Recent discoveries of self-cleaving RNA sequences functioning as regulatory cellular ribozymes demonstrate the versatility of RNA enzymes in biology (Martick et al., 2008; Salehi-Ashtiani et al., 2006; Teixeira et al., 2004; Winkler et al., 2004). The catalytic activities of these ribozymes are probably strongly influenced by intracellular conditions. The cosolute effects provided by very simple hydrophilic molecules imply that RNA could have had enhanced catalytic activity in primordial cells relative to what we observe under typical *in vitro* conditions.

5.2. Riboswitches

A riboswitch is a region in an mRNA that controls gene expression in response to specific metabolites (Montange and Batey, 2008). Riboswitches are generally located in the 5′ untranslated regions (UTRs) of mRNAs related to biosynthesis and transport of the target metabolites (Kim and Breaker, 2008; Mandal et al., 2003). The aptamer domain of the riboswitch undergoes an RNA conformational transition upon ligand binding to activate or suppress gene expression. The conformational transition is thus a trigger that controls the level of protein synthesis in response to levels of a metabolite. The dynamics of riboswitch conformational changes upon ligand binding have been the focus of study but are not well understood at present. Among riboswitches, those that are specific for purine (adenine or guanine) are the simplest and best characterized. Purine riboswitches are ubiquitous among prokaryotic species (Barrick and Breaker, 2007; Mandal et al., 2003). The secondary structures of their aptamer domains consist of a TWJ formed by RNA helices P1, P2, and P3 and involve two loops (L2 and L3) (Fig. 8.13) (Serganov et al., 2004).

Conserved bases in the riboswitches are primarily in the junction and loop regions (Lemay and Lafontaine, 2007). The bases in the junction region are essential for direct interaction with ligands. Although the loop regions do not interact directly with ligands, loop–loop interactions form when the riboswitch binds to a ligand (Lemay et al., 2006; Rieder et al., 2007; Serganov et al., 2004). Mutation of the bases in L2 and L3 of a *pbuE* adenine-specific riboswitch derived from *B. subtilis* cause the loss of the ability to form the L2–L3 interaction and significantly decrease binding affinity for the ligand (Lemay et al., 2006).

We investigated the conformational dynamics and ligand binding properties of the aptamer domain of an adenine-specific riboswitch (*add-A* riboswitch) (Mandal and Breaker, 2004; Serganov et al., 2004) derived from the mRNA of *Vibrio vulnificus* adenine deaminase under the molecular crowding conditions induced by PEG 200 (Kumar et al., 2012). Ligand binding affinity was evaluated by monitoring the fluorescence of 2-aminopurine (2-AP). 2-AP was used as a model ligand for the *add-A* riboswitch because 2-AP binds to adenine-specific riboswitches with an

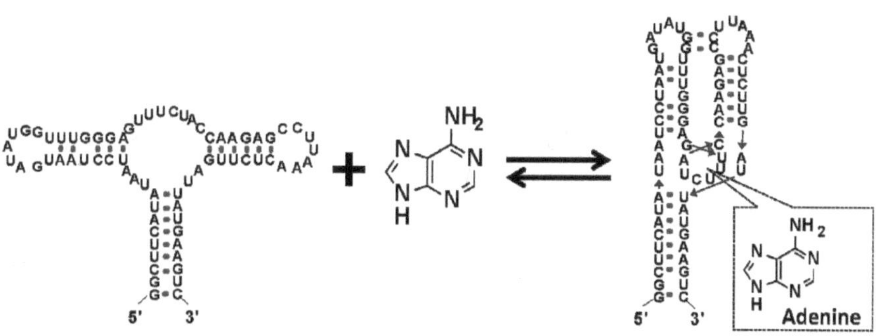

Figure 8.13 Conformational transition of the adenine-specific riboswitch (*add-A* riboswitch) derived from the mRNA of *V. vulnificus* adenine deaminase in response to adenine binding. (For color version of this figure, the reader is referred to the online version of this chapter.)

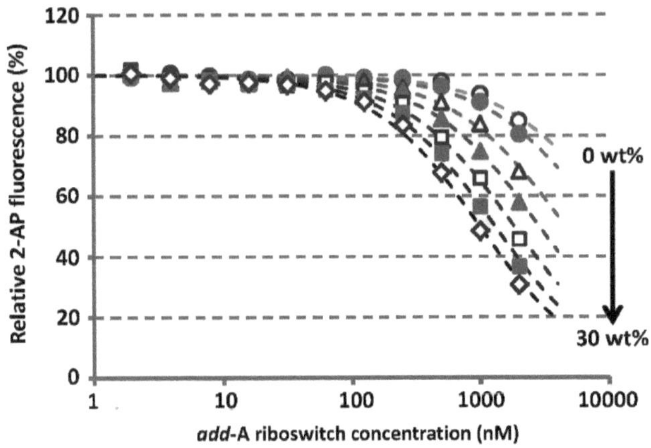

Figure 8.14 Relative fluorescence intensities of 2-AP (200 nM) in the presence of the *add-A* riboswitch and 0 (open circle), 5 (filled circle), 10 (open triangle), 15 (filled triangle), 20 (open square), 25 (filled square), or 30 (open diamond) wt% of PEG 200 in a buffer containing 50 mM sodium MOPS (pH 7.3) and 100 mM KC1 at 25 °C. Fluorescence of 2-AP was measured at 300 nm excitation and 380 nm emission.

affinity similar to the natural ligand (adenine) and binding results in a decrease in its fluorescence (Lemay and Lafontaine, 2007; Lemay et al., 2006). A reduction in fluorescence that depended on the concentration of the *add-A* riboswitch was observed at all concentrations of PEG 200 (Fig. 8.14). The binding between the *add-A* riboswitch and 2-AP was assumed to occur in a simple 1:1 stoichiometry as revealed by the crystal structure. The binding constant, K_{obs}, at 25 °C increased from 0.08×10^6 M^{-1} to 1.13×10^6 M^{-1} as the concentration of PEG 200 was increased from 0 to 30 wt%, indicating that the ligand binding affinity of the riboswitch increased under conditions of molecular crowding.

To investigate the effect of PEG 200 on the tertiary structure of the *add-A* riboswitch, we performed a partial digestion using RNase T1 in the absence and presence of 300 mM 2-AP with and without 30 wt% PEG 200. Without PEG 200, cleavage of the *add-A* riboswitch at G37 and G38 was significantly decreased by the addition of 2-AP. These results correspond well to previous reports showing that G37 and G38 form Watson–Crick base pairs with C61 and C60, respectively, in the ligand-bound structure (Fig. 8.13) (Rieder et al., 2007; Serganov et al., 2004). In the presence of PEG 200, the signals due to cleavage at G37 and G38 were decreased in the presence of 2-AP. The ratio of the cleaved products in the absence and presence of 2-AP at positions G37 and G38 was similar with and without 30 wt% PEG 200. These results suggest that PEG 200 does not induce the formation of the tertiary structure of the *add-A* riboswitch but rather that the addition of PEG 200 increases the ligand binding affinity.

Nucleic acid conformations are highly sensitive to hydration (Ball, 2008). Thus, if the hydration state of the *add-A* riboswitch changes upon ligand binding, the PEG 200, which decreases the water activity, would affect the ligand binding affinity. To determine the contribution of water molecules under the molecular crowding conditions, we determined the correlation between K_{obs} and water activity. In the presence of Mg^{2+}, there is no water release or uptake upon the binding of 2-AP to the *add-A* riboswitch; this is to be expected as the tertiary structure forms in the absence of ligand binding in 20 mM Mg^{2+} (Brenner et al., 2010; Lemay et al., 2006; Noeske et al., 2007). In contrast, 54.2 water molecules were released upon 2-AP binding in the absence of Mg^{2+}. In the absence of Mg^{2+} and both in the absence and presence of PEG 200, the *add-A* riboswitch did not form the tertiary structure prior to ligand binding, and 2-AP binding caused the formation of the loop–loop interaction. Thus, water molecules were released during the RNA conformational transition.

In solution, water molecules form a hydration shell around an RNA, and the major grooves of RNA are more hydrated than the minor grooves. The RNA conformational transition of the *add-A* riboswitch caused by ligand binding juxtaposes the P2 and P3 stems, bringing the major grooves close together as revealed in the crystal structure (Serganov et al., 2004). Thus, it is likely that the dehydration observed resulted from the P2 and P3 stem interaction. In addition, although PEG 200 did not affect K_{obs} at 20 mM Mg^{2+}, significant increases of K_{obs} were observed at more moderate and physiologically relevant concentrations of Mg^{2+} of less than 1 mM. Synergistic effects of dehydration caused by PEG 200 and neutralization of the negatively charged RNA caused by Mg^{2+} possibly stabilized the tertiary structure and facilitated the 2-AP binding.

Although the binding affinity might be different, the facilitation of the binding affinity through the dehydration of the stem regions should be observed with natural ligand adenine, because adenine induces formation of the same tertiary structure of the *add-A* riboswitch as does 2-AP (Rieder et al., 2007). A similar effect likely occurs in the case of the hammerhead ribozyme, whose catalytic activity was increased in the presence of PEG 200 (Nakano et al., 2009). Various natural riboswitches with affinities for ligands other than purines, such as S-adenosyl methionine, flavin mononucleotide, and thiamine triphosphate, also have functional structures characterized by interactions between stem regions and by loop–loop interactions (Serganov et al., 2009; Stoddard et al., 2010). Thus, the intracellular crowding conditions might favorably affect the *in vivo* functions of such riboswitches through dehydration during the RNA conformational transitions that are fundamental for the regulation of gene expression.

6. MOLECULAR CROWDING EFFECTS UNDER EXTREME ENVIRONMENTS

6.1. Compartment environments

Macromolecular surfaces confine biomolecules to nanometer-scale volumes within cells. Confinement on this scale can modify the fundamental properties of biomolecules and create a heterogeneous intracellular environment that is different from bulk solution (Ball, 2008; Chaplin, 2006; Zhou et al., 2008). Unfortunately, the complexity of biological systems makes it very difficult to obtain information about biomolecular behavior in locally confined spaces. A surfactant-based reverse micellar interior is an excellent biophysical model to study the confinement effect (Abel et al., 2006; Levinger, 2002; Nucci et al., 2011; Valentine et al., 2010).

Reverse micelles (RMs) are simple artificial systems that mimic in-cell conditions (Fig. 8.15A) (Chang et al., 2000; Van Horn et al., 2009). Their structures are made up of amphiphilic molecules similar to the amphiphilic phospholipids of a biomembrane. Moreover, the water pool of the RM

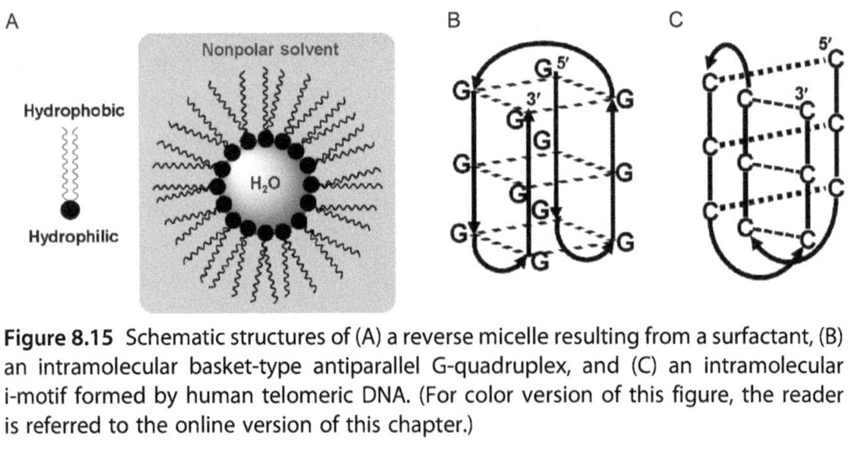

Figure 8.15 Schematic structures of (A) a reverse micelle resulting from a surfactant, (B) an intramolecular basket-type antiparallel G-quadruplex, and (C) an intramolecular i-motif formed by human telomeric DNA. (For color version of this figure, the reader is referred to the online version of this chapter.)

provides a compartmentalized environment with properties considered to be similar to those found at polar/apolar interfaces *in vivo* (Luisi et al., 1988). RMs tend to be spherical in shape at low concentrations of surfactant and polar solvent. Micellar size in a suspension is characterized by the parameter ϖ, which is equal to $[H_2O]/[\text{surfactant}]$ and which is directly proportional to the micellar radius (Bridge and Fletcher, 1983; Kotlarchyk et al., 1985). The radial dimensions of RMs are similar to the confined aqueous compartments in cells and tissues (McIntosh et al., 2005). Thus, RMs have been used to study the structures and biophysical properties of proteins and nucleic acids (Babu et al., 2001; Park et al., 2003; Wand et al., 1998; Workman and Flynn, 2009; Zhou et al., 2010).

To study the effect of localization in cellular compartments on structures formed by telomeric DNA, we have analyzed G-quadruplex- and i-motif-forming oligonucleotides in RMs. Telomere sequences are highly conserved (Blackburn, 1991) and telomeres play an important role in chromatin organization in the cell nucleus (Arthanari and Bolton, 2001). In humans the telomeric repeat motif is (TTAGGG):(CCCTAA) (Moyzis et al., 1988). As discussed above, the G-rich strands can fold into G-quadruplex structures *in vitro* (Fig. 8.15B) (Simonsson, 2001), whereas the C-rich strands can adopt an i-motif structure under acidic conditions (Fig. 8.15C) (Gehring et al., 1993). At physiological pH, temperature and salt concentrations, equimolar mixtures of G-rich and C-rich telomeric strands predominantly adopt a canonical Watson–Crick double helix form. The competition between tetraplex and duplex structures is only observed under nonphysiological conditions (e.g., at lower pH and high temperature) (Phan and Mergny, 2002).

We investigated the formation of tetraplex structure from an equimolar mixture of the oligonucleotides based on the human telomeric sequence, d[AGGG(TTAGGG)$_3$] (22AG) and d[(CCCTAA)$_3$CCCT] (22CT), in bis(2-ethylhexyl)sulfosuccinate (AOT) RMs (Pramanik et al., 2012). The structures of the 22AG and 22CT were studied by CD spectroscopy both in dilute aqueous buffer and in RMs. The CD spectrum of 4 mM 22AG in dilute solution had a positive peak at around 295 nm and an intense negative peak at around 260 nm, characteristic of a basket-type antiparallel G-quadruplex structure (Fig. 8.16A).

In AOT RMs at $\varpi = 10$, the CD signature of 22AG was unaltered compared to that in the buffer alone, indicating no structural modification (Fig. 8.16B), although there were malformed CD signatures below 240 nm in AOT RMs due to the significant CD contribution of the medium

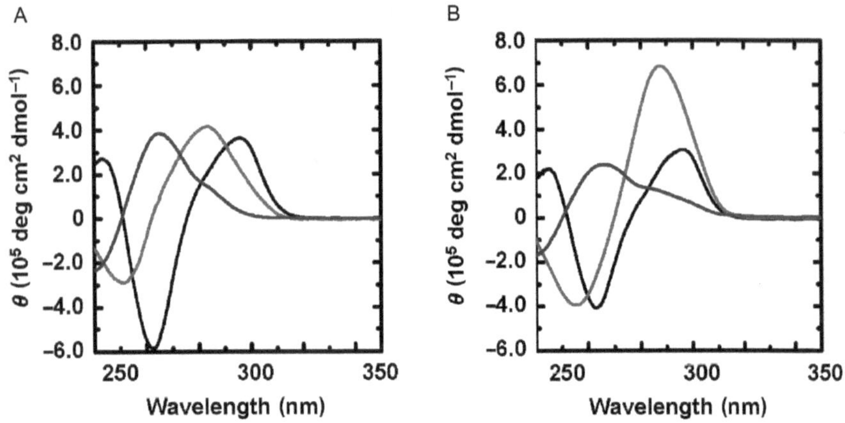

Figure 8.16 CD spectra of 4 mM 22AG (black), 4 mM 22CT (red), and a mixture of 22AG and 22CT (blue) in (A) dilute solution and (B) AOT RM at $\varpi = 10$ and 5 °C. (For interpretation of the references to color in this figure legend, the reader is referred to the online version of this chapter.)

in that region. Moreover, with increasing water content, the CD signature of the G-quadruplex remained unaltered, indicating no structural modification of 22AG at higher ϖ. In contrast, the CD spectrum of 4 mM 22CT in AOT RMs at $\varpi = 10$ was different from that in dilute solution and was typical of an i-motif signature with a positive peak at around 287 nm and a relatively weak negative peak at around 255 nm (Fig. 8.16A, B). Unlike 22AG, the positive and negative peak positions and peak intensities of 22CT decreased with increasing water content. Thus, 22CT underwent a structural transition upon changing the water content. The equimolar mixture of 2 mM each of 22AG and 22CT showed a positive peak at 265 nm in the dilute solution (Fig. 8.16A). On the other hand, in micellar solution at $\varpi = 10$, the CD signature of the equimolar mixture of the 22AG and 22CT had decreased positive peak intensity relative to that in buffer alone and a broad positive peak from 310 to 250 nm (Fig. 8.16B). This complex CD signature is likely due to the existence of mixed conformations. With increasing ϖ, the overall CD signature was not altered significantly.

We then studied the effects of temperature on the CD spectra in RMs. For 22AG at $\varpi = 10$, increase of temperature from 5 to 95 °C resulted in decreases in the positive peak intensities and increases in negative peak intensities with a distinct dichroic point. With increasing temperature, the spectrum of 22CT also changed; the positive peak intensity at

287 nm decreased with a concomitant increase in negative peak intensity at 255 nm. However, the spectra of 22CT as a function of temperature do not have a dichroic point. This indicates that 22AG undergoes a two-state transition in AOT RM but that 22CT does not. At higher ϖ (20 and 30), as well as in dilute solution, 22AG undergoes a two-state transition as temperature increased; the transition for 22CT is not two-state. The peak intensities of the equimolar mixture of the two complementary strands, 22AG and 22CT, in AOT RM at $\varpi = 10$, were irregularly altered as a function of temperature (Fig. 8.17A). The transition appeared to be non-two-state, indicating the presence of more than one structure. At higher ϖ (20 and 30), as well as in the dilute solution, the change in positive peak intensities was regular. The G-rich strand, 22AG, had a high thermal stability in AOT RM. The structure did not melt completely at $\varpi = 10$. With increasing ϖ, the thermal stability of 22AG decreased. In contrast, the i-motif structure formed by 22CT is relatively unstable. With increasing ϖ the 22CT structure became less stable, resulting in decreased CD intensities of the positive and negative peaks. In addition, the normalized CD melting curves at 295 and 265 nm of the equimolar mixture of 22AG and 22CT (2 mM each strand) in AOT RM at $\varpi = 10$ showed that the 295 nm profile of the mixture is more complex than that at 265 nm (Fig. 8.17B). At 295 nm, three transitions were observed. CD intensity decreased to some extent upon heating from 5 to 60 °C, increased rapidly from 65 to 80 °C, and decreased upon heating above 85 °C. It should be noted that G-quadruplex formation by 22AG

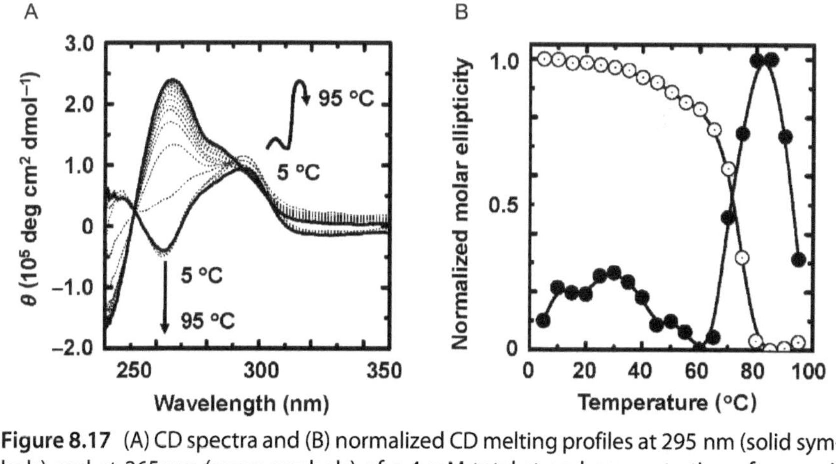

Figure 8.17 (A) CD spectra and (B) normalized CD melting profiles at 295 nm (solid symbols) and at 265 nm (open symbols) of a 4 mM total strand concentration of an equimolar mixture of 22AG and 22CT in AOT RM at $\varpi = 10$.

leads to an increase in CD intensity at 295 nm. As the mixture is slowly heated from 5 °C, the initial decrease in CD intensity at 295 nm corresponds to melting of the i-motif structure. Above 60 °C, the increase in CD intensity at 295 nm results from the melting of duplex structure and formation of the G-quadruplex structure. Above 85 °C, the G-quadruplex structure begins to melt as indicated by the decrease in CD intensity at 295 nm. At higher ϖ (20 and 30), three transitions were not observed. Irrespective of water content (ϖ), melting curves at 265 nm showed two-state transitions, corresponding to the melting of duplex.

At physiological temperature (37 °C), the CD spectrum of the equimolar mixture of 22AG and 22CT had a sharp positive peak at 265 nm in dilute solution (Fig. 8.18A). In AOT RMs at $\varpi = 10$ and 37 °C, the CD spectrum of the equimolar mixture had a broad positive peak in the range of 310–250 nm (Fig. 8.18A) and decreased peak intensity at 265 nm relative to the spectrum in buffer alone. Thus, in a cell-mimicking environment at physiological temperature a significant fraction of the DNA formed the G-quadruplex structure. With increasing ϖ, the intensity of the shoulder at 295 nm decreased and the intensity at 265 nm increased (Fig. 8.18A); thus, in AOT RM the duplex–quadruplex conversion depends on ϖ.

G-Quadruplex structures are stabilized by inter-guanine hydrogen bonding, stacking interactions between two tetrads, and specific and

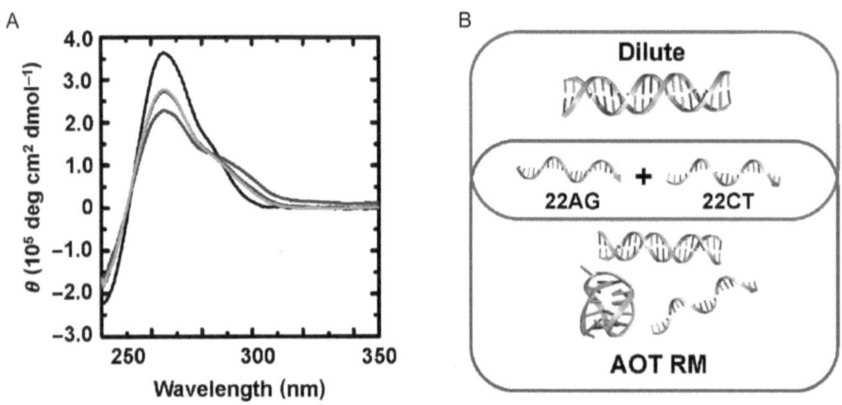

Figure 8.18 (A) CD spectra of a 4 mM total strand concentration of an equimolar mixture of 22AG and 22CT at 37 °C in dilute solution (black) and in AOT RM at $\varpi = 10$ (red), $\varpi = 20$ (blue), and $\varpi = 30$ (green). (B) Schematic representation of the equimolar mixture of 22AG and 22CT in dilute solution and in AOT RM. (For interpretation of the references to color in this figure legend, the reader is referred to the online version of this chapter.)

nonspecific ion–DNA electrostatic interactions. On the other hand, the stability of i-motif structures mainly depends on the pH of the medium, and slightly acidic pH favors their formation. Encapsulation of DNA within nano-sized water pools of RMs reduces the available volume; the impact of this restriction on biomolecular structure has been called the excluded volume effect (Van Horn et al., 2009). The G-quadruplex structure is stabilized by the excluded volume effect (Zhou et al., 2008). Recently, we reported that the i-motif DNA structure can also be formed under conditions of molecular crowding at neutral and slightly basic pH due to a cytosine pK_a shift (Rajendran et al., 2010). The water activity inside the water pool of the RM differs from that of bulk water and depends on the water pool size (Politi and Chaimovich, 1986). Previously, we reported that the osmotic environment influences the thermodynamics of water molecules associated with a G-quadruplex structure, and thus osmolarity could determine the stability or structure formed (Miyoshi et al., 2006b). Therefore, the stability of a DNA G-quadruplex differs in dilute solution from that in the water pool of the RM that mimics the confined cellular space. Several groups have reported that osmolytes shift the tetraplex–duplex equilibria toward the tetraplex structures (Kan et al., 2007; Kumar and Maiti, 2005; Zhou et al., 2009). Consequently, a significant fraction of an equimolar mixture of complementary strands with the human telomere sequence adopts the noncanonical tetraplex structure in AOT RMs (Fig. 8.18B).

There is evidence for the existence and importance of noncanonical DNA structures in various chromosomal events. Chromatin structure is dynamic and exerts profound control over gene expression. Condensed regions of chromosomes are called heterochromatin, and more loosely packed regions are called euchromatin (Luger and Hansen, 2005). In other words, a heterochromatin state is more confined than a euchromatin state. We proposed previously that a histone H3 tail can induce noncanonical DNA structure formation in heterochromatin regions (Pramanik et al., 2012). It is possible that nuclear confinement can also induce noncanonical DNA structures in telomere regions that are in the heterochromatin state (Schoeftner and Blasco, 2009). In conclusion, fundamental biological functions might be regulated by the degrees of confinement.

6.2. Ionic liquids

The canonical DNA structure is a B-form duplex consisting of Watson–Crick base pairs (Bloomfield et al., 2000; Saenger, 1984). In physiologically relevant buffered solutions, G–C base pairs are more stable than A–T base pairs (Bloomfield et al., 2000; Saenger, 1984). This stability difference affects local and long-range DNA structure formation. DNAs have enormous potential in the fields of nanobiotechnology and biomedical technology due to the ability of single strands of DNA to recognize and hybridize with their complementary sequences through highly specific base-pairing interactions (Balasubramanian et al., 2011; Liu and Lu, 2007; Miyoshi et al., 2006a, 2007; Ohmichi et al., 2005; Seeman, 2003a; Taton et al., 2000; Venkatesan et al., 2008). DNAs can undergo structural transitions in response to environmental stimuli (Liu and Lu, 2007; Miyoshi et al., 2006a, 2007; Ohmichi et al., 2005; Seeman, 2003b; Taton et al., 2000; Venkatesan et al., 2008). Such stimuli include pH (Miyoshi et al., 2006a; Ohmichi et al., 2005), small molecules (e.g., cosolutes and ions) (Liu and Lu, 2007; Miyoshi et al., 2007; Seeman, 2003b), macromolecules (e.g., oligonucleotides and proteins) (Venkatesan et al., 2008), and electrical signals (Taton et al., 2000). Certain ions are known to preferentially stabilize certain base pairs. For example, alkylammonium ions bind to A–T base pairs in the minor groove of a DNA duplex (Dong et al., 2009; Hud et al., 1999; Melchior and Von Hippel, 1973; Stellwagen et al., 2007), stabilizing regions of A–T base pairs (Stellwagen et al., 2007). Some osmolytes such as glycine betaine, which has alkylammonium-derivative ions, destabilize G–C rich DNA duplexes by binding to unpaired guanines at high salt concentrations (Felitsky et al., 2004; Hong et al., 2004; Nordstrom et al., 2006; Tateishi-Karimata and Sugimoto, 2012; Vasudevamurthy et al., 2009).

We evaluated the effect of an IL on base pair stability (Tateishi-Karimata and Sugimoto, 2012). ILs provide favorable environments for a wide range of chemical reactions (Ding et al., 2010; Earle and Seddon, 2000; Nishimura and Ohno, 2002; Seddon, 1997; Welton, 1999). A representative IL, choline dihydrogen phosphate (choline dhp) (Attri and Venkatesu, 2011; Byrne and Angell, 2008; Fujita et al., 2005; Vijayaraghavan et al., 2010), dissolved in a small amount of water ensures long-term stability of biomolecules like DNA (Vijayaraghavan et al., 2010) and proteins (Byrne and Angell, 2008; Fujita et al., 2005). As choline dhp contains alkylammonium ions, we reasoned that this IL might impact DNA duplex stability. In addition, we have demonstrated that water activity is an important factor in DNA stability (Miyoshi et al., 2006b, 2009; Nakano et al., 2009), and the water activity in the hydrated ILs is low (Attri and Venkatesu, 2011). It is also known that solute behavior can be regulated by the combination of anions and cations in an IL (Seddon, 1997; Welton, 1999).

Table 8.1 Sequences and melting temperatures for duplex formation in the presence of 4 M NaCl or 4 M choline dhp (IL)

		T_m (°C)[b]	
Sequence name[a]	**Sequence**	**NaCl**	**IL**
ODN1 (10)	5′-AAAAAAAAAA-3′/5′-TTTTTTTTTT-3′	30.7	53.3
ODN2 (9)	5′-AAAAAAAAAC-3′/5′-GTTTTTTTTT-3′	31.5	48.6
ODN3 (8)	5′-AAAAAAAACC-3′/5′-GGTTTTTTTT-3′	36.0	46.6
ODN4 (7)	5′-AAAAAAACCC-3′/5′-GGGTTTTTTT-3′	38.7	42.7
ODN5 (6)	5′-AAAAAACCCC-3′/5′-GGGGTTTTTT-3′	40.0	36.3
ODN6 (5)	5′-AAAAACCCCC-3′/5′-GGGGGTTTTT-3′	45.3	33.3
ODN7 (10)	5′-AAATATATTT-3′/5′-AAATATATTT-3′	17.4	37.7
ODN8 (0)	5′-GGGCGCGCCC-3′/5′-GGGCGCGCCC-3′	61.0	33.1
ODN9 (8)	5′-TTATAACCTA-3′/5′-TAGGTTATAA-3′	26.7	34.8
ODN10 (2)	5′-CGGCAAGCGC-3′/5′-GCGCTTGCCG-3′	47.2	30.5

[a]The number of A–T base pairs in the DNA duplex is shown in parentheses.
[b]Melting temperature was calculated at a strand concentration of 5 μM.

The T_ms of 10-mer DNA duplexes with different A–T base pair contents were measured in a solution containing 4 M choline dhp (80 wt%) and in solution containing 4 M NaCl (Table 8.1).

In the NaCl-containing solution, the T_m values of the DNA duplexes decreased from 45.3 to 30.7 °C as A–T content increased. In contrast, in choline dhp a reverse trend for the stabilities of DNA duplexes was observed. The T_m values of the DNA duplexes in the choline dhp-containing solution increased from 33.3 to 53.3 °C as A–T content increased. To confirm the effect of choline dhp, we also measured the T_m values of four additional DNA duplexes (Table 8.1). The DNA duplexes with highest A–T contents had the highest T_ms in the solution containing choline dhp. Intriguingly, ODN7, a duplex composed only of A and T, was more stable in the choline dhp-containing solution than in NaCl-containing solution, whereas ODN8 with only G–C base pairs was destabilized in choline dhp relative to the NaCl solution. The all A–T duplex was more stable than the all G–C duplex in the choline dhp-containing solution with T_m values of 37.7 and 33.1 °C, respectively. The ΔT_m was calculated by subtracting the T_m value obtained in NaCl-containing solution from that in choline dhp solution. The correlation between ΔT_m values and A–T content of the duplexes is linear

(Fig. 8.19). The DNA duplexes with more than 61% A–T content were stabilized in the choline dhp-containing solution, whereas the DNA duplexes with <61% A–T content were destabilized. A similar trend was also seen for 12-mer DNA duplexes of different sequences.

To gain insight into how choline dhp altered the stabilities of DNA duplexes, we measured the thermodynamic parameters for ODN9 and ODN10 (Table 8.2).

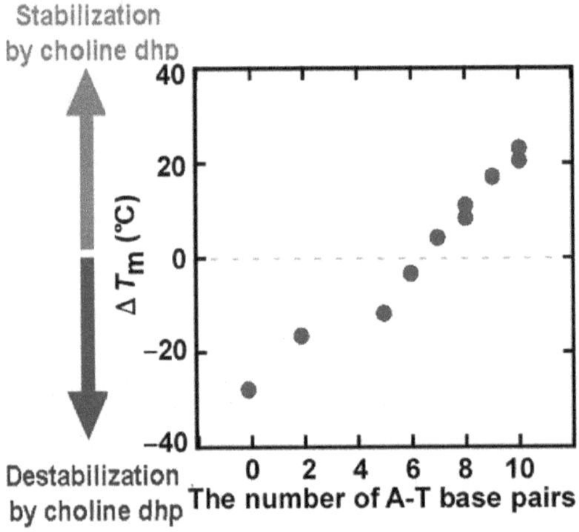

Figure 8.19 The correlation between ΔT_m values and A–T content of the duplexes. The ΔT_m was calculated by subtracting the T_m value obtained in NaCl-containing solution from that in choline dhp solution. (For color version of this figure, the reader is referred to the online version of this chapter.)

Table 8.2 Thermodynamic parameters for DNA duplex formation measured in NaCl or choline dhp[a]

		$\Delta H°$ (kcal mol^{-1})	$T\Delta S°$ (kcal mol^{-1})	ΔG_{25} (kcal mol^{-1})	T_m (°C)[b]
NaCl	ODN9	−46.6 ± 2.8	−38.2 ± 2.0	−8.4 ± 0.3	38.6
	ODN10	−54.7 ± 3.3	−42.0 ± 2.9	−12.7 ± 0.4	63.2
Choline dhp	ODN9	−66.8 ± 3.4	−56.7 ± 2.7	−10.1 ± 0.4	43.6
	ODN10	−47.8 ± 3.7	−39.4 ± 3.4	−8.4 ± 0.5	38.2

[a]All the experiments were carried out in a buffer containing 50 mM MES (pH 6.0), 1 mM Na_2EDTA, and 4 M NaCl or 4 M choline dhp. Thermodynamic parameters were evaluated from the average values obtained from curve fitting and T_m^{-1} versus log(C_t/4) plots.

[b]Melting temperature was calculated at a strand concentration of 100 μM.

The $\Delta G°_{25}$ values of ODN9 with higher A–T content, and of ODN10 with higher G–C content, in 4 M NaCl were −8.4 and −12.7 kcal mol^{-1}, respectively. In 4 M choline dhp, however, the stabilities were reversed: the $\Delta G°_{25}$ values of ODN9 and ODN10 were −10.1 and −8.4 kcal mol^{-1}, respectively. The higher A–T content duplex was more stable in the choline dhp-containing solution than in with NaCl-containing solution due to a favorable enthalpic contribution. Alkylammonium-derivative ions bind to single-stranded DNA, especially unpaired guanines, at high salt concentrations (Felitsky et al., 2004; Hong et al., 2004; Nordstrom et al., 2006; Vasudevamurthy et al., 2009). The destabilization of the higher G–C content duplex in the choline dhp-containing solution was caused by unfavorable enthalpy, implying that choline ions decrease duplex stability either by destabilizing hydrogen bond formation or by stabilizing the single-stranded conformation through preferential binding of choline ions to guanine bases in the single strand (Chaplin, 2006; Felitsky et al., 2004; Hong et al., 2004; Nakano et al., 1999; Nordstrom et al., 2006; Sugimoto et al., 1996). As the stabilization of ODN9 in the choline dhp-containing solution was enthalpically driven, we hypothesize that choline ions increase duplex stability due to an interaction between choline ions and the duplex conformation.

To confirm that preferential binding of choline ions to A–T base pairs affect DNA duplex stability, we carried out molecular dynamics calculations of ODN9 and ODN10 and used the CDOCKER program to model binding between one choline ion and ODN9 or ODN10. Several docking patterns between choline ion and OND9 were obtained, and the simulations show that choline ions bind to the major groove of the A–T-rich region. Figure 8.20 provides a specific example of choline ion binding to ODN9. The hydroxyl group of choline ion forms two hydrogen bonds with N7 and H6 of A_5, and the methyl group of choline ion is located close to methyl group of T_{15} on the opposite strand. Interaction of a choline ion with adenosine in a duplex with a different sequence was previously reported (Cardoso and Micaelo, 2011). Thus, choline ions bind tightly in the major groove of A–T regions but weakly to the G–C-rich regions. In contrast, choline ions bind preferentially to G–C-rich single-stranded DNA, relative to A- or T-rich strands. Cations bind preferentially to G (Holland et al., 2011), and choline ion binding to guanines effectively destabilizes G–C-rich DNA duplexes. The molecular dynamics calculations were carried out under the assumption that a single choline ion bound to the DNA. We do not rule out the possibility that there are other binding modes of the choline ion with DNA at high concentrations (4 M) of choline dhp. Our predicted structure is one of the stable binding styles.

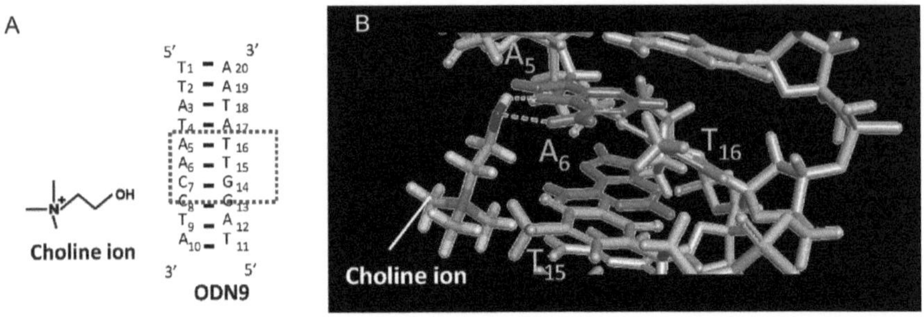

Figure 8.20 (A) Secondary structure of ODN9. The area surrounded by the dashed line was selected for molecular dynamics calculations. (B) The predicted interaction between ODN9 (gray) and choline ion (carbon, hydrogen, nitrogen, and oxygen atoms are indicated by yellow, white, blue, and red, respectively). Adenine and thymine bases close to choline ions are indicated by pink and light blue, respectively. Hydrogen bonds are rendered in dashed lines (light green). (For interpretation of the references to color in this figure legend, the reader is referred to the online version of this chapter.)

7. MOLECULAR CROWDING EFFECTS ON TRANSCRIPTION AND TRANSLATION

7.1. Transcription

Transcription is the highly regulated first step in gene expression. Transcription elongation in cells and *in vitro* can be interrupted by certain sequence motifs in the template DNA. This phenomenon is a ubiquitous aspect of transcription elongation and is known to play regulatory roles. Moreover, at some template positions, called "pause" sites, the polymerase may halt for as long as 1 min before continuing transcription. At other locations called "arrest" sites, a fraction of the transcribing polymerase is completely blocked and cannot continue RNA synthesis in the standard experimental time course (Cheung and Cramer, 2011; Datta et al., 2006; Kassavetis and Chamberlin, 1981; Kireeva and Kashlev, 2009; Santangelo and Artsimovitch, 2011; von Hippel, 2006). In this case, the transcript is cleaved and released (Cheung and Cramer, 2011). This cleaved RNA (arrested product) may misfold or may result in translation of mutant protein. Because short noncoding RNAs, such as siRNAs, ribozymes, and antisense RNAs (Endoh and Sugimoto, 2011; Nakano et al., 2009), play important roles in regulation of gene expression, it would be useful in diverse research fields including medical, pharmaceutical, and materials sciences fields, if induction of the arrested product could be controlled.

Previous reports have shown that transcription is arrested prior to stable structures in the DNA template such as a Z-form duplex

(Ditlevson et al., 2008), a triplex (Belotserkovskii et al., 2007; Ebbinghaus et al., 1993) or a G-quadruplex (Eddy et al., 2011; Tornaletti et al., 2008). Moreover, the transcription efficiency of a gene containing a G-rich sequence is drastically decreased by the addition of ligands known to bind and stabilize G-quadruplex structures (Cogoi and Xodo, 2006; Tsuji and Yoshikawa, 2010). We have shown recently that noncanonical structures are stabilized drastically under cell-like conditions induced by cosolutes (Miyoshi et al., 2006b, 2009), although a duplex is destabilized under the same conditions. Our results imply that noncanonical structures are likely present inside cells and play important role in gene regulation. However, there is no quantitative information regarding how stable noncanonical structures induce pausing or arrest of transcribing polymerases.

Transcriptional regulation is one of the central mechanisms for control of gene expression. For example, it is well known that on/off and termination switches play critical roles in biological processes (Farnham, 2009; Santangelo and Artsimovitch, 2011). Structural formation within DNA templates and within nascent RNAs (Balasubramanian et al., 2011; Brooks and Hurley, 2009; Du et al., 2008; Eddy et al., 2011; Wan et al., 2011; Waters and Storz, 2009), G-quadruplex formation by DNA templates in promoter regions (Balasubramanian et al., 2011; Brooks and Hurley, 2009), and formation of higher-order structures in nascent RNAs in the termination region regulate the amount and length of transcripts (Du et al., 2008; Eddy et al., 2011; Wan et al., 2011; Waters and Storz, 2009). Most characterized structures impact initiation or/and termination, but there are recent reports demonstrating that transcription can be regulated in the elongation phase (Larson et al., 2011; Min et al., 2011; Nechaev et al., 2010). The yields of transcripts that were capable of forming RNA hairpins with different stem lengths were determined in a T7 RNA polymerase (T7 RNAP) elongation assay (Nagatoishi et al., 2012). The amount of RNA transcript obtained decreased with stem length, demonstrating that transcriptional efficiency is regulated by secondary structure formation in the nascent RNA. However, there is limited information regarding the impact of structure in guanine-rich and trinucleotide-repeat sequences on transcriptional regulations (Belotserkovskii et al., 2010; Broxson et al., 2011; Salinas-Rios et al., 2011). The nascent RNA can form various types of secondary and tertiary structures during the transcription process (Pan and Sosnick, 2006; Zemora and Waldsich, 2010). A hairpin structure is one of the most simple and frequent secondary structures observed in RNAs. Formation of hairpins containing 5–7 base pair (bp) stems caused the elongation complex to pause (Datta and von Hippel, 2008; Toulokhonov et al., 2001). A model study of the elongation complex of RNA polymerase suggested that the RNA exit tunnel accommodates 4–6 nts of RNA in an extended conformation (Landick, 2006; Yin and Steitz, 2002). Therefore, it was proposed that hairpins with 5 bp stems would affect the elongation complex; however, an experimental analysis of how hairpin structures in the nascent RNA affect transcriptional efficiency had not been performed.

First, we designed ten different template DNAs (Fig. 8.21A, B) to know how noncanonical DNA structures and their stabilities affect transcription. Sequences H1–3 and Q1–6 were designed to form hairpins or G-quadruplexes, respectively, at 35 bases downstream from the T7 promoter region. The G-quadruplex-forming sequences are based on either the human telomeric sequence or the thrombin DNA aptamer sequence and have different numbers of G-quartets (Miyoshi et al., 2006b; Petraccone et al., 2012). To confirm the formation of the noncanonical

A

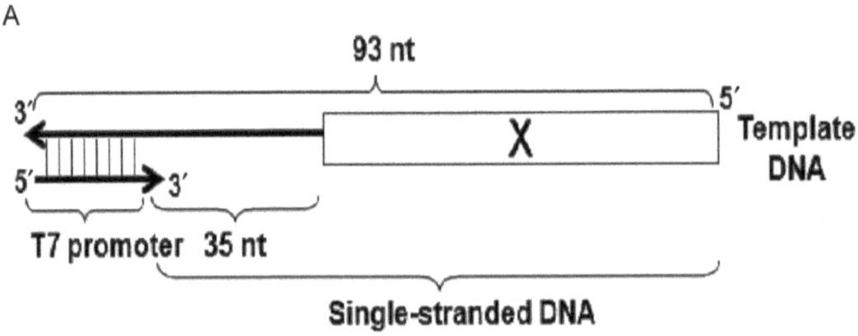

B

Sequence name	Sequence of X region (5′→3′)
Linear:	GCCGTTTCGTAGTATTTGGGTTGTAACTATCGAGG
H1:	GCCGTTTCGTAGTATTTCTATCC***GGACTTCGGTCC***
H2:	GCCGTTTCGTAGT***CTATCGGACTTCGGTCCGATAG***
H3:	GCCGT***CCAACTATCGGACTTCGGTCCGATAGTTGG***
Q1:	GCCGTTTCGTAGTATTTC***GGTTAGGTTAGGTTAGG***
Q2:	GCCGTTTCGTAGTA***GGGTTAGGGTTAGGGTTAGGG***
Q3:	GCCGTTTCGT***GGGGTTAGGGGTTAGGGGTTAGGGG***
Q4:	GCCGTTTCGTAGTATTTCTA***GGTTGGTGTGGTTGG***
Q5:	GCCGTTTCGTAGTATT***GGGTTGGGTGTGGGTTGGG***
Q6:	GCCGTTTCGTAG***GGGGTTGGGGTGTGGGGTTGGGG***

Figure 8.21 (A) Schematic of template DNA; the region denoted by the box containing an X forms a random coil or a noncanonical structure. (B) Sequence names and sequences of X regions. Sequences expected to form noncanonical structures are highlighted by italic and bold.

structures in the template DNAs, sequences that contain only the non-structured region were used as controls (Fig. 8.21B).

The thermodynamic parameters for the formation of G-quadruplexes and hairpins were obtained from their thermal melting curves obtained by monitoring UV absorbance at 295 or 260 nm (Fig. 8.22). The G-quadruplexes of q4, q5, and q6 were much more stable than those formed by q1, q2, and q3 due to differences in stacking interactions of loop regions;

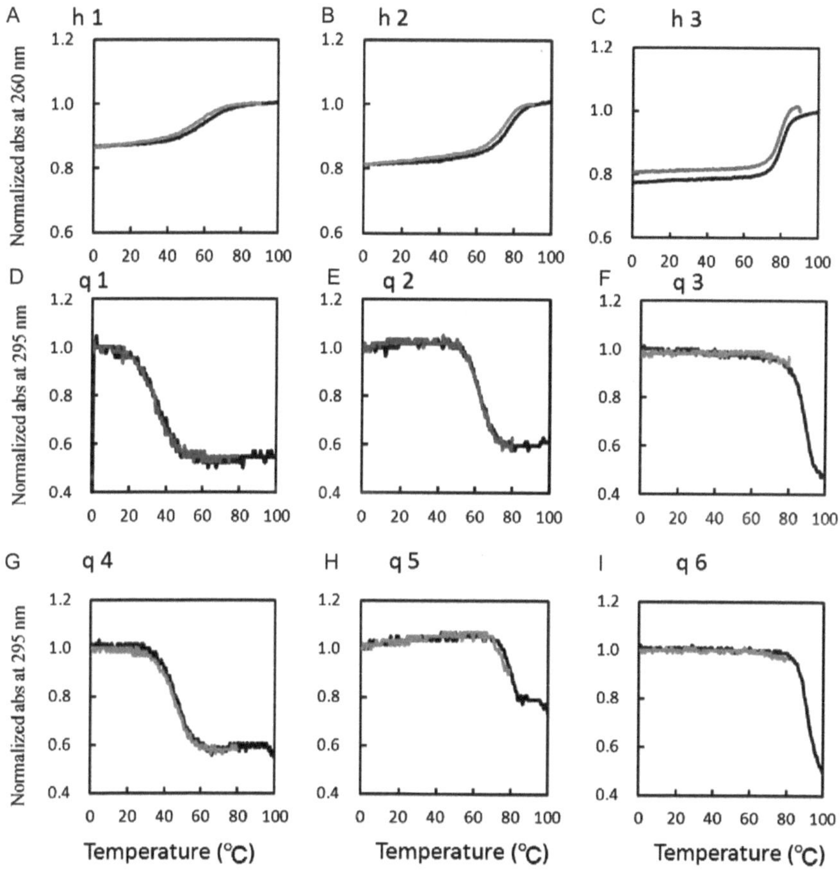

Figure 8.22 Normalized UV annealing (red) and melting (black) curves for 2 μM (A) h1, (B) h2, (C) h3, (D) q1, (E) q2, (F) q3, (G) q4, (H) q5, and (I) q6 in a buffer containing 30 mM KCl, 40 mM Tris–HCl (pH 8.0), 8 mM $MgCl_2$, and 2 mM spermidine. Melting and annealing were measured by UV absorbance at 260 nm (A–C) or 295 nm (D–I). The heating and cooling rates were 1 °C min^{-1}. (For color version of this figure, the reader is referred to the online version of this chapter.)

the hairpins were of intermediate stability. The thermal analyses indicated that all the noncanonical structures except Q1 should form in the template DNA at 37 °C.

Figure 8.23 shows the results of gel electrophoretic analysis of T7 RNAP-catalyzed multiple turnover transcription carried out for 90 min at 37 °C. The transcription of product RNA is saturated at 90 min; therefore, the inhibitory effects by noncanonical structures on transcription can be estimated from the amount of product formed at this time. RNA size was determined by running samples of each reaction in parallel with size markers and a 35-nt RNA (Fig. 8.23, lanes 1 and 2). When a template having no significant structure was used, transcription proceeded to the end of the DNA template resulting in the formation of an RNA of 70 nt (Fig. 8.23, lane 3). When templates able to form noncanonical structures were used, the amount of full-length RNA decreased dramatically (Fig. 8.23, lanes 4–12). The transcription efficiency decreased with increasing hairpin stability. Moreover, the transcription efficiencies of Q1, Q2, Q3, Q4, Q5, and Q6 also decreased with increasing G-quadruplex stability. The decrement of the transcription efficiency depended on the stability of the G-quadruplex but not the G-quadruplex topology. The inhibition of transcription efficiency due to G-quadruplex formation was larger than that due to hairpin formation. The decrease of transcription efficiency is likely due to a pause in transcription.

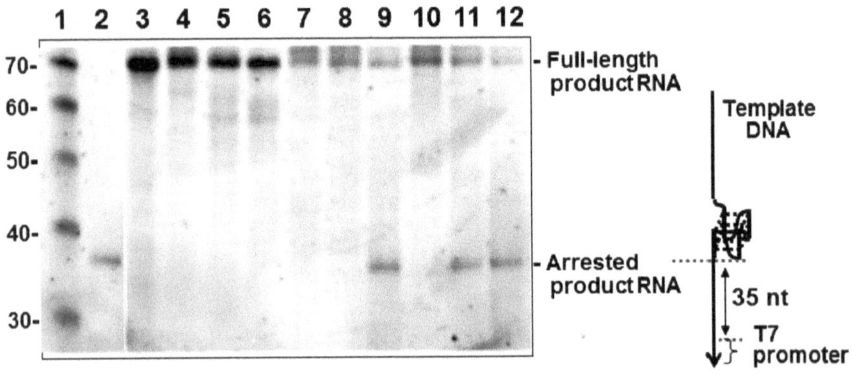

Figure 8.23 Denaturing gel electrophoresis of RNA products of a 90-min transcription reaction at 37 °C. Reactions contained 0.3 μM T7 polymerase and 1.5 μM DNA template in a buffer containing 30 mM KCl, 40 mM Tris–HCl (pH 8.0), 8 mM $MgCl_2$, and 2 mM spermidine. Lane 1, size markers; lane 2, 35-nt RNA, and lanes 3–12, transcription products for linear, H1, H2, H3, Q1, Q2, Q3, Q4, Q5, and Q6 templates.

Interestingly, template DNA able to form the stable G-quadruplexes induced production of an arrested product RNA, although the stable hairpin in H3 did not induce the arrest. The transcription efficiency of Q3, Q5, and Q6 decreased notably with a concomitant increase in the shorter RNA (Fig. 8.23, lanes 9, 11, and 12). The shorter RNAs correspond to 35-nt transcripts, and this length suggests that T7 transcription is arrested just prior to the G-quadruplex structure. Unlike dissociation of a hairpin structure, G-quadruplex unfolding is very slow. The G-quadruplex might therefore stall the polymerase long enough for the polymerase to dissociate. During transcription, the template DNA is bound inside the active site of the polymerase. It is possible that the polymerase bound to the template DNA may mimic the crowded conditions that we have shown to stabilize G-quadruplexes. These results indicate that stable noncanonical structures regulate transcription. Our results further our understanding of the transcription process involving noncanonical structures and may guide the design of transcription-regulating drugs.

Next, we chose T7 RNAP as a model RNA polymerase to know how the hairpins formed in the nascent RNA affect the transcription, because T7 RNAP is a single-subunit enzyme (Kochetkov et al., 1998; Sousa and Mukherjee, 2003). The DNA templates used in this study are shown in Fig. 8.24.

Their lengths were 69 bp, which includes a 6-bp extended sequence at the 5′ end, a 17-bp T7 promoter sequence and a 46-bp transcribed

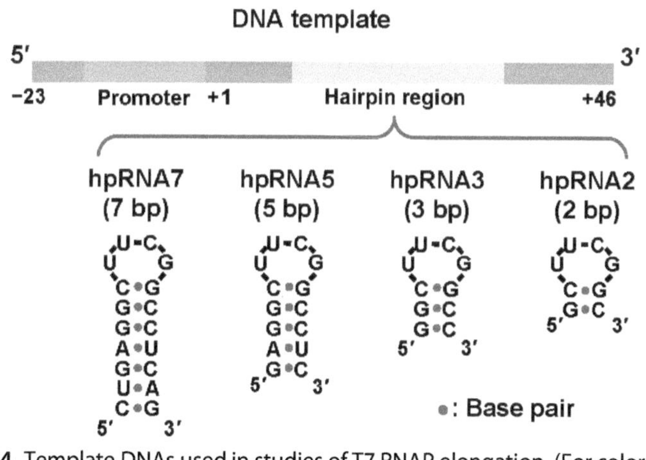

Figure 8.24 Template DNAs used in studies of T7 RNAP elongation. (For color version of this figure, the reader is referred to the online version of this chapter.)

sequence. The sequences in the transcribed region were designed to form hairpin structures once the elongation phase of the transcription reaction was begun. Templates were designed to yield transcripts that can form hairpin structures containing 7-, 5-, 3-, or 2-bp stems, each with a 4-nucleotide loop (Fig. 8.24). The hairpin with a 2-bp stem should be stable energetically under physiological conditions (Proctor et al., 2004). The sequences of DNA templates are shown in Table 8.3. The secondary structures of the transcribed RNA sequences were predicted using Mfold (Zuker, 2003). In the region from the 5′ end to the hairpin sequence, only the desired hairpin structures were predicted to form. As it has been reported that transcriptional efficiency depends on the sequence and the duplex stability of the DNA template (Palangat and Landick, 2001), we avoided sequence bias and designed sequences to contain roughly the same percentage of each base.

The structures of DNA templates are known to affect the transcriptional process (Broxson et al., 2011; Salinas-Rios et al., 2011). DNA templates have the potential to form cruciform structures in the region of the hairpin sequences in our system. To investigate the structure of the DNA templates, we performed CD experiments in a buffer containing 40 mM Tris–HCl (pH 8.0), 100 mM NaCl, and 10 mM $MgCl_2$ at 37 °C. All of the CD spectra showed the same peak positions and intensities. In addition, all of the UV melting curves monitored at 260 nm in this buffer were the same. In native PAGE, all of the DNA templates exhibited one band at a similar migration position. Furthermore, templates were treated with the restriction enzyme *Hae*III with the recognition site 5′-GGCC-3′, a sequence present in the loop regions of potential hairpins; cleavage should only occur if the DNA forms a duplex rather than a cruciform. Analysis by denaturing PAGE revealed a band of around 50 nt in DNA7, DNA5, and DNA3, indicating that these DNA templates form a duplex structure in the hairpin sequence regions. DNA2 does not have a *Hae*III cleavage site and so was not evaluated. Thus, DNA template structure did not impact the transcriptional efficiencies measured in this study.

To evaluate the structures of the RNA transcripts made from each template, we analyzed RNAs corresponding to expected transcripts (hpRNA7, hpRNA5, hpRNA3, and hpRNA2, Fig 8.24). CD spectra showed a positive peak at 260 nm and negative peaks at 240 and 210 nm in a buffer containing 40 mM Tris–HCl (pH 8.0) and 100 mM NaCl at 37 °C, indicating that the RNA strands form hairpin structures. The peak intensities decreased in the following order: hpRNA7>hpRNA5>

Table 8.3 Sequences of template DNAs

Template DNA	Sequence[a,b]
DNA7	5′-GATCACTAATACGACTCACTATA **GGGATGATATAGTGC** **CTGAGGCTTCGGCCTCAGTTTCAGACAGTAG**-3′
DNA5	5′-GATCACTAATACGACTCACTATA **GGGATGGTATACTGGCTGAGGCTTCGGCCTC** **TTTGAGACAACAATG**-3′
DNA3	5′-GATCACTAATACGACTCACTATA **GGGATGCTAATAGTGATACG** **GCTTCGGCCTTTAGACAGTGACGTGC**-3′
DNA2	5′-GATCACTAATACGACTCACTATA **GGGATGATACTAGTGCTTCGGC** **TTTCAGGACACGGTGACCTGATAG**-3′

[a]Bold fonts indicate a transcribed region.
[b]Underlines indicate a hairpin region.

hpRNA3>hpRNA2; this supports the hypothesis that the number of base pairs decreased, as expected from our design. In addition, UV melting experiments at 260 nm in a buffer containing 40 mM Tris–HCl (pH 8.0) and 0.1 mM NaCl indicated that the melting temperature of the hairpin structures decreased in the same order. The thermodynamic stabilities of hpRNA7, hpRNA5, hpRNA3, and hpRNA2 structures were estimated to be −9.1, −7.2, −4.6, and −1.6 kcal mol^{-1}, respectively. The results suggested that the RNA hairpins formed in transcription reactions have a range of stabilities.

The transcriptional efficiencies were inferred from PAGE analysis of samples of RNA transcription reactions taken at 240 min. All of the RNA transcripts were of the same length as indicated by similar migration on PAGE (Fig. 8.25A); thus no shorter RNA fragments were produced by the arrest of T7 RNAP. Interestingly, the amount of the RNA transcripts increased significantly (Fig. 8.25B) as the predicted stabilities of RNA hairpins that could be formed by the nascent transcript decreased (Fig. 8.26).

Our results suggest that nascent RNA structure affects elongation efficiency. Previous studies reported that a hairpin structure in an RNA transcript causes the RNA polymerase to pause (Datta and von Hippel, 2008; Landick, 2006; Nudler, 2009). The paused state can affect the total transcription rate, resulting in a decrement in yield of RNA products (Core and Lis, 2008). Datta and von Hippel, and Nudler proposed that

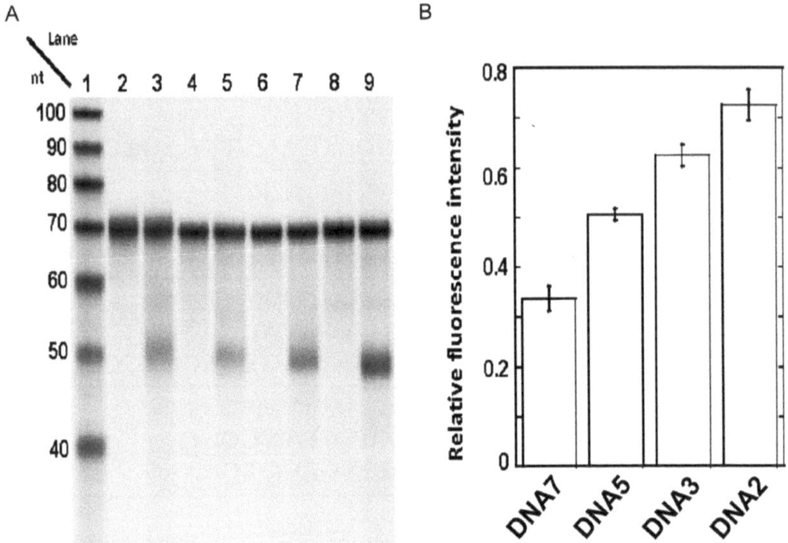

Figure 8.25 (A) Denaturing 12% PAGE containing 7 M urea for analysis of transcription reactions of DNA7 (lanes 2 and 3), DNA5 (lanes 4 and 5), DNA3 (lanes 6 and 7), and DNA2 (lanes 8 and 9). Samples in the absence of T7 RNAP were loaded in lanes 2, 4, 6, and 8 and samples in the presence of T7 RNAP in lanes 3, 5, 7, and 9. The 10-nt ladder was loaded in lane 1. The gel was stained with SYBR Gold. (B) The yields of RNA transcripts; error bars were obtained from at least four measurements. The relative fluorescence intensity is that of RNA divided by that of DNA.

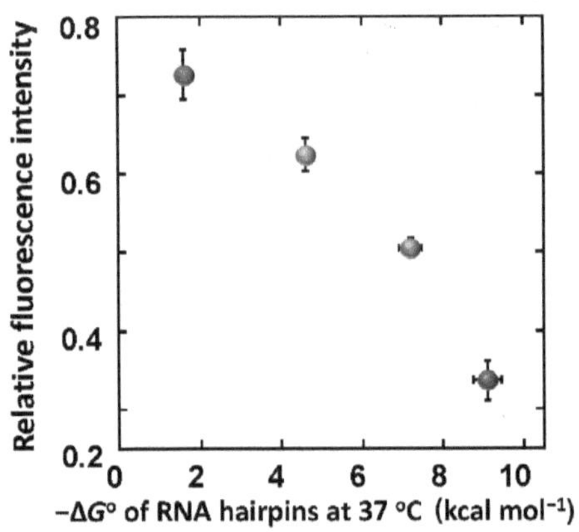

Figure 8.26 Plot of the transcriptional efficiency (relative fluorescence intensity of the product) versus the thermodynamic stability of RNA hairpins for DNA7 (red), DNA5 (orange), DNA3 (green), and DNA2 (blue). The error bars were obtained from four measurements. (For interpretation of the references to color in this figure legend, the reader is referred to the online version of this chapter.)

the aberrant paused state was induced by the disruption of the DNA scaffold triggered by the hairpin formation in the nascent RNA (Datta and von Hippel, 2008; Nudler, 2009). This may destabilize the elongation complex. The T7 RNAP binds to nascent RNA mainly through the N-terminal domain and a specificity loop (Tahirov et al., 2002). These domains also bind the promoter region of the DNA template (Cheetham and Steitz, 1999). Therefore, the destabilization induced by a disruption of the interaction between a nascent RNA and an RNA polymerase probably causes the inactivation of the transcription cycle. It has been reported that a conformational change in T7 RNAP can affect the reinitiation efficiency (Ferrari et al., 2004). Thus, hairpin structures formed by the nascent RNA likely affect the processivity of T7 RNAP.

A recent paper proposes that regions coding for RNA in the yeast genome form more stable secondary structures than do UTRs (David et al., 2011). Structures in the coding regions of an mRNA do affect gene expression levels (Kudla et al., 2009), and there are reports of regulation of transcriptional efficiency during the elongation phase (Larson et al., 2011; Min et al., 2011; Nechaev et al., 2010). Our analysis of the effect of hairpin structures in nascent transcripts provides evidence of a general correlation between transcriptional yield and stability of transcript structure, and suggests that transcription reaction efficiencies may be controlled by the design of transcript structures for *in vitro* applications (Franco et al., 2011; Sousa and Mukherjee, 2003).

7.2. Translation

During translation elongation, the ribosome decodes the mRNA sequence and converts this information into the amino acid sequence of a protein. Ribosome progression is arrhythmic rather than uniform due to features of mRNAs. Rare codons and certain stable secondary structures in open reading frames (ORFs) are known to slow or stall translation elongation (Buchan and Stansfield, 2007; Wen et al., 2008). Previous studies have revealed that stalling of translation elongation at a specific position regulates protein expression by affecting translation initiation of a second ORF or by inducing ribosomal frameshifts (Buchan and Stansfield, 2007; Farabaugh, 1996; Fernandez et al., 2005; Giedroc and Cornish, 2009). In addition, the positions of rare codons and stable secondary structures in mRNA correlate with flexible linker regions in protein structures (Komar, 2009; Watts et al., 2009; Zhang and Ignatova, 2009; Zhang et al., 2009). Since proteins partially fold during translation elongation (Kaiser et al., 2011; O'Brien et al., 2010, 2011), the rare codons and secondary structures of mRNA likely affect the structures of nascent proteins (Evans et al., 2008; Kimchi-Sarfaty et al., 2007; Komar, 2009; O'Brien et al., 2012; Thanaraj and Argos, 1996; Tsai et al., 2008; Zhang and Ignatova, 2011; Zhang et al., 2009). These observations suggest that mRNA sequence and structure might impact not only expression levels but also the processing of proteins.

It was recently suggested that G-quadruplex formation in the 5′ UTR of mRNAs potentially suppresses protein expression level by inhibiting scanning by the small ribosomal subunit, especially in the UTRs of oncogene mRNAs (Beaudoin and Perreault, 2010; Bugaut and Balasubramanian, 2012; Halder et al., 2009; Kumari et al., 2008). Human estrogen receptor (hER) is a ligand-inducible nuclear transcription factor that activates expression of genes responsible for growth, development, and maintenance of the reproductive, skeletal, neuronal, and immune systems in response to estrogens (Gibson and Saunders, 2012; Schultz-Norton et al., 2011). The *hER* mRNA contains a sequence with quadruplex-forming potential (QFP) in the region of the ORF that codes for domain D, a hinge region of the protein (Balkwill et al., 2009; Gibson and Saunders, 2012). The G-quadruplexes formed by RNAs are generally more stable than those formed by the equivalent DNA sequences and are also stabilized under molecular crowding conditions (Arora and Maiti, 2009; Joachimi et al., 2009; Zhang et al., 2010). Thus, if the QFP sequence in the *hER* mRNA adopts a G-quadruplex structure and slows or stalls translation elongation, it might affect the cotranslational folding of the preceding DNA binding domain (domain C) and the following ligand binding domain (domain E/F) (Balkwill et al., 2009; Gibson and Saunders, 2012).

We synthesized RNA oligonucleotides corresponding to the QFP sequence derived from the ORF of the *hER* mRNA and variants containing mutations that encode the same amino acid sequence (Fig. 8.27A), and their thermodynamic stabilities and structures were evaluated *in vitro*. Figure 8.27B and C shows normalized UV melting profiles of the QFP sequence variants analyzed from absorbance at 295 nm. All the QFP sequence variants except the A-mutant showed melting transitions typical of G-quadruplexes in 40 wt% PEG 200, whereas only the wild-type, C-mutant, and G-mutant RNAs showed this transition under dilute condition.

CD spectra of the QFP sequence variants under molecular crowding conditions at 37 °C were also obtained (Fig. 8.27D). The spectra of

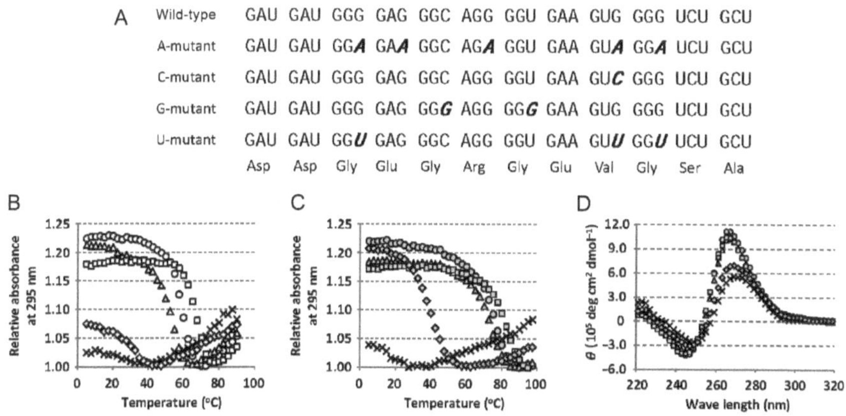

Figure 8.27 (A) RNA oligonucleotides corresponding to QFP sequences derived from the ORF of *hERα* mRNA and variants with a mutated G-rich region (indicated in bold and italic). Amino acid sequences encoded by the QFP sequence variants are indicated. (B, C) Relative absorbance at 295 nm of 5 μM QFP sequence variants, wild-type (circle), A-mutant (cross), C-mutant (triangle), G-mutant (square), and U-mutant (diamond), in a buffer containing 50 mM MES–LiOH, pH 7.0, and 3 mM KCl in the absence (B) or presence (C) of 40 wt% PEG 200. (D) CD spectra of 5 μM QFP sequence variants at 37 °C in a buffer containing 50 mM MES–LiOH, pH 7.0, 3 mM KCl, and 40 wt% PEG 200. Symbols are the same as in (B) and (C).

wild-type, C-mutant, G-mutant, and U-mutant RNAs were characterized by positive and negative peaks near 265 and 240 nm, respectively, which were larger than those of observed in the A-mutant spectrum. These results indicated that all QFP sequence variants with the exception of the A-mutant formed parallel G-quadruplex structures under molecular crowding conditions. The U-mutant was likely partially denatured at 37 °C as its T_m is 39.9 °C.

To evaluate the effect of the G-quadruplex formed by the QFP sequence derived from the *hER* mRNA on translation elongation, we analyzed the time course of translation *in vitro* by using a synchronized translation that we recently developed to evaluate a single turnover of the elongation reaction (Endoh et al., 2012). In the first step of the reaction, translation was artificially halted before a tyrosine codon in a reconstructed *in vitro* translation system (PUREfrex; GeneFrontier), in which release factor 1, release factor 2, and tyrosyl-tRNA synthetase were left out of the reaction mixture to synchronize all ribosomes. In the second step, translation elongation was restarted by addition of a reaction mixture containing all requisite proteins except the ribosome. Reporter mRNAs containing either the wild-type QFP sequence (wild-type mRNA) or the A-mutant sequence (A-mutant

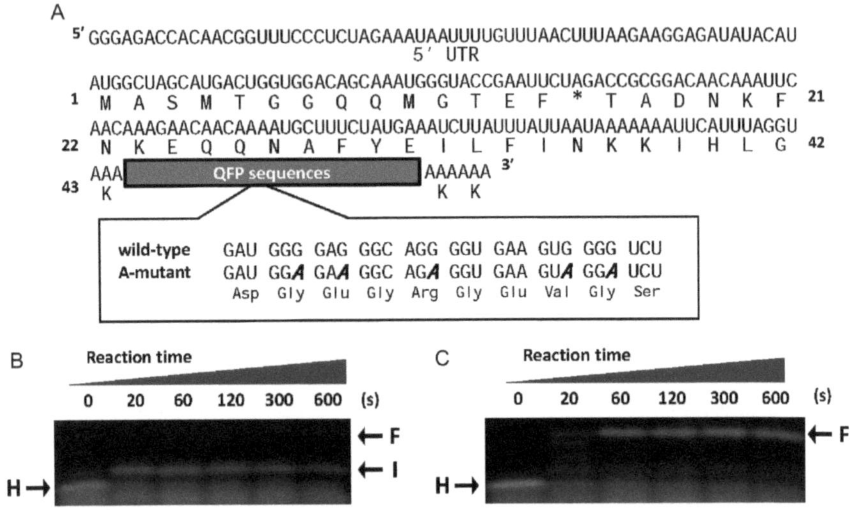

Figure 8.28 (A) Nucleotide and amino acid sequences of reporter mRNAs containing QFP sequence variants derived from *hERα*. The asterisk shows the amber codon for incorporation of a nonnatural fluorescent amino acid. Nucleotides mutated relative to the wild-type QFP sequence are indicated in bold and italic. (B, C) Time course of the synchronized translation of wild-type and A-mutant mRNAs. Samples taken from the synchronized translation at the indicated reaction times were resolved by 16% Tris–tricine SDS–PAGE. The fluorescence signal of the incorporated nonnatural amino acid (CR110-X-AF) was imaged using 473 nm excitation and 510 nm emission. Translated products due to ribosomal stalling (I) and complete elongation (F) are indicated. Strong signals near the bottom of the gel are due to unincorporated CR110-X-AF (Endoh et al., 2012). (For color version of this figure, the reader is referred to the online version of this chapter.)

mRNA) were constructed (Fig. 8.28A). The translated products from the reporter mRNAs were evaluated by Tris–tricine SDS–PAGE. Before the restart, the fluorescence signal due to the 29-residue peptide was similar in reactions with the wild-type and A-mutant mRNAs (Fig. 8.28B and C, indicated by H). The amino acid sequences encoded by both the reporter mRNAs are the same, so positions of products in the gel can be directly compared. The products after restart of the reaction from the two templates differed; the main translation product of the wild-type mRNA migrated at a lower position in the gel (Fig. 8.28B, indicated by I) than that of the A-mutant mRNA (Fig. 8.28C, indicated by F). In addition, in the case of the wild-type mRNA, fluorescence signals at positions I and F gradually decreased and increased, respectively, with increasing reaction time. The fluorescence signal indicated by I corresponds to translation stalled at an intermediate position, and that indicated by F corresponds to a product in which the translation elongation progressed through the 3′ end of the mRNA. These data indicate that translation elongation was stalled in the wild-type sequence, presumably because of the presence of a G-quadruplex structure. The parallel G-quadruplex formed by the QFP sequence likely inhibits the helicase activity of the ribosome (Takyar et al., 2005) and blocks mRNA entry into the ribosome. The translation suppression was temporary rather than permanent, because a full-length product was observed with increasing reaction time (Fig. 8.28B).

We then cloned the coding sequence of *hER* into a vector, and variants containing synonymous mutations in the QFP sequence were constructed. The plasmid vectors were transfected into HepG2 human liver carcinoma cells. Figure 8.29A shows results of western blotting using anti-hER antibody specific for the N-terminal domain of the protein (Thermo Scientific). In cells transfected with QFP sequence variants, a signal corresponding to full-length hER was observed located between the 50 kDa and the 75 kDa marker proteins (Fig. 8.29A, indicated by F). A signal migrated slightly slower than the 25 kDa marker protein, and the amount of this product depended on the template sequence (Fig. 8.29A, indicated by N).

To quantitatively evaluate the effect of the G-quadruplex stability on the production of the short hER, the coding sequence of the fluorescent protein (AcGFP) was inserted either into the 5′ or the 3′ region of hER to express AcGFP fused to hER at the N-terminus (AcGFP-hER) or the C-terminus (hER-AcGFP), respectively. These vectors were transfected into MCF7 human breast carcinoma cells that constitutively express hER. Figures 8.29B and 8.29C show fluorescence signals of AcGFP-hER and hER-AcGFP, respectively, in 8% SDS–PAGE. Fluorescence signals due to the full-length protein were observed near the 75 kDa marker protein in all the QFP sequence variants, and for certain QFP sequence variants a signal was observed between the 37 and 50 kDa maker proteins (Fig. 8.29B, indicated by N). In the case of hER-AcGFP, although several signals appeared uniformly in all the QFP sequence variants, a fluorescence signal near the 50 kDa marker was observed for certain QFP sequence variants (Fig. 8.29C, indicated by C). These results suggested that the short hER results from proteolysis.

Ratios of the cleaved AcGFP-hER (Fig. 8.29B, indicated by N) and hER-AcGFP (Fig. 8.29C, indicated by C) relative to full-length protein (Fig. 8.29B and C, indicated by F) were calculated from the fluorescence intensities. The ratios calculated for the A-mutant were used as standards, because the A-mutant did not form a G-quadruplex even under the

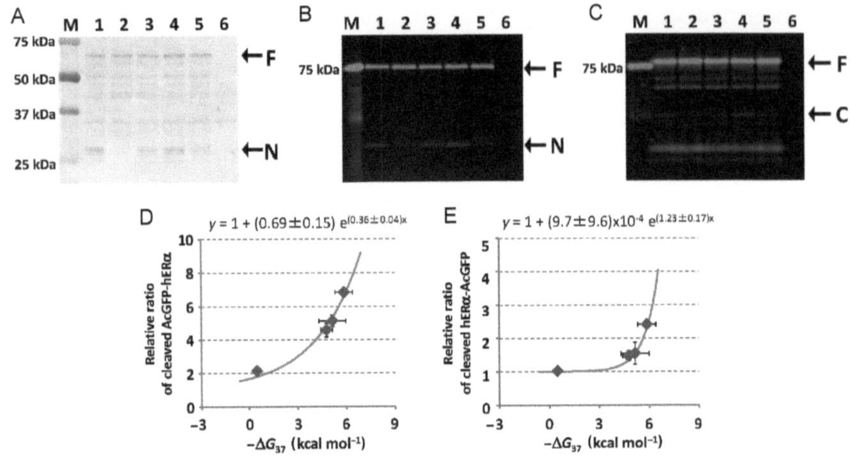

Figure 8.29 (A) Expression of hERα in HepG2 cells. Vectors for expression of hERα containing the QFP sequence variants wild-type (lane 1), A-mutant (lane 2), C-mutant (lane 3), G-mutant (lane 4), and U-mutant (lane 5) or the control plasmid (pcDNA5/FRT, Invitrogen) (lane 6) were transfected into HepG2 cells. Expression of hERα was analyzed by western blotting using an anti-hERα antibody specific for the N-terminal domain. Lane M shows marker proteins. Signals of hERα corresponding to full-length protein and shorter forms are indicated by F and N, respectively. (B, C) Protein expression patterns of (B) AcGFP-hERα and (C) hERα-AcGFP in MCF7 cells. Cells were transfected with vectors for expression of AcGFP-hERα or hERα-AcGFP containing the QFP sequence variants wild-type (lane 1), A-mutant (lane 2), C-mutant (lane 3), G-mutant (lane 4), and U-mutant (lane 5), or were not transfected (lane 6). Cell lysates were resolved on 8% SDS–PAGE without denaturing, and fluorescence signals were imaged using 473 nm excitation and 510 nm emission. (D, E) Plots of the relative ratios of (D) AcGFP-hERα and (E) hERα-AcGFP normalized to the ratio of A-mutant versus $-\Delta G°_{37}$ of the QFP sequence variants calculated from the UV melting curve. Fits to a single exponential equation are indicated. (For color version of this figure, the reader is referred to the online version of this chapter.)

molecular crowding conditions. Figures 8.29D and E show plots of the relative ratio values of QFP sequence variants normalized to the value of the A-mutant versus the thermodynamic stability of the QFP sequence variants. The data from both the cleaved AcGFP-hER and hER-AcGFP fit a single exponential function, as was observed during *in vitro* analysis of the effect of mechanical strength of RNA pseudoknots on frameshifting (Chen et al., 2009). If the activation energy necessary for the ribosome to unwind the G-quadruplex is proportional to the thermodynamic stability of the structure, the rate to overcome the stalling should depend on the G-quadruplex stability in a single exponential manner, and this is indeed what we observe. Thus, temporary stalling of translation elongation before the G-quadruplex is likely a main determinant of the proteolysis sensitivity of hER. Stalling may result in misfolding of the nascent hER. Previous studies have shown that proteolysis of hER by chymotrypsin differs in rabbit and wheat germ *in vitro* translation systems (Horjales et al., 2007). This might be due to differences in G-quadruplex stability in the two systems.

Although the proteolysis position has not been accurately determined, the short hER had a MW between 25 and 30 kDa (Fig. 8.29A, indicated by N). We assume that hER is cleaved after the DNA binding domain, because the MW of the N-terminal 251 amino acids containing the activation function-1 domain (domain A/B) and the DNA binding domain should be 27.6 kDa (Gibson and Saunders, 2012). The *hER* pre-mRNA is known to be alternatively spliced resulting in hER variants that confer dominant-positive and dominant-negative effects on the protein functions (Barone et al., 2010), and these variants are hypothesized to contribute to progression of certain cancers (Bollig and Miksicek, 2000; Chaidarun and Alexander, 1998; Flouriot et al., 2000; Garcia Pedrero et al., 2003). Thus, our data suggest that G-quadruplex formation in the mRNA contributes to protein polymorphism.

8. PERSPECTIVES

A number of cosolute studies have examined the interactions, structures, and functions of nucleic acids. Studies using DNA and RNA oligonucleotides have shown that Watson–Crick base pairs are destabilized in the presence of small cosolutes. In contrast, small cosolutes stabilize certain protein secondary and tertiary structures (Davis-Searles et al., 2001; Roque et al., 2007; Stanley and Strey, 2008; Tokuriki et al., 2004). The difference in the cosolute effects between nucleic acids and proteins is of great interest and could be significant, given their different roles in living cells. Nucleic acid interactions appear to be more sensitive than protein interactions to the solution components, and therefore the polymorphic properties of nucleic acid structure and stability may be regulated by alterations in the molecular environment in the cell. This property makes it difficult to make predictions regarding the structures adopted by nucleic acids based on data obtained *in vitro*. Molecular crowding experiments using cosolutes are useful for constructing realistic models of nucleic acid structures and interactions in cells. There is considerable heterogeneity in the molecular environment within cells and intracellular organelles, and thus studies using *in vitro* model systems allow us to examine the determinant factors for molecular interactions and reactions under unusual conditions such as those found in living cells or on solid surfaces.

Since reality often conflicts with theory, the concept of molecular crowding can help us reconcile the two. Molecularly crowded systems also show how we can adjust experimental results to reproduce *in cellulo* biomolecular reactions. An *in vitro* experimental system mimicking intracellular crowding conditions can be a translational system between *in cellulo* and *in vitro*, but experiments using synthetic cosolutes are still not optimal. More sophisticated translational experiments are obviously needed, not only to predict biomolecular behaviors *in vivo* but also to design devices and materials to be used for bio- and nanobiotechnology. For example, the influence of the nature of the cosolute (e.g., the chemical structure, size, shape, charge, and conformational rigidity) should be addressed to more precisely simulate intracellular conditions. Quantitative data for nucleic acid interactions under molecular crowding conditions are useful not only for understanding nucleic acid structures and functions in living cells, but also for designing oligonucleotides that exhibit desired functions in cells or that will serve as probes immobilized onto material surfaces and cell surfaces. The use of neutral cosolutes for nucleic acid and protein reactions to enhance their reaction efficiency is also promising for therapeutic and diagnostic purposes.

ACKNOWLEDGMENTS

The author is grateful to the colleagues named in the cited papers from my laboratory and institute (FIBER), especially Drs Nakano, Miyoshi, Endo, Tateishi-Karimata, Nagatoishi, Pramanik, my technical assistant Ms. Yamaguchi, and my secretary Ms. Yoshioka. This work was supported in part by Grants-in-Aid for Scientific Research, the "Academic Frontier" Project (2004–2009), and the "Strategic Research Foundation at Private Universities" (2009–2014) from the Ministry of Education, Culture, Sports, Science and Technology (MEXT), Japan, the Hirao Taro Foundation of the Konan University Association for Academic Research, and Nagase Science and Technology Foundation.

REFERENCES

Abel, S., Waks, M., Urbach, W., Marchi, M., 2006. Structure, stability, and hydration of a polypeptide in AOT reverse micelles. J. Am. Chem. Soc. 128, 382–383.

Al-Habori, M., 2001. Macromolecular crowding and its role as intracellular signalling of cell volume regulation. Int. J. Biochem. Cell Biol. 33, 844–864.

Arora, A., Maiti, S., 2009. Differential biophysical behavior of human telomeric RNA and DNA quadruplex. J. Phys. Chem. 113, 10515–10520.

Arthanari, H., Bolton, P.H., 2001. Functional and dysfunctional roles of quadruplex DNA in cells. Chem. Biol. 8, 221–230.

Asami, K., Hanai, T., Koizumi, N., 1976. Dielectric properties of yeast cells. J. Membr. Biol. 28, 169–180.

Attri, P., Venkatesu, P., 2011. Thermodynamic characterization of the biocompatible ionic liquid effects on protein model compounds and their functional groups. Phys. Chem. Chem. Phys. 13, 6566–6575.

Babu, C.R., Flynn, P.F., Wand, A.J., 2001. Validation of protein structure from preparations of encapsulated proteins dissolved in low viscosity fluids. J. Am. Chem. Soc. 123, 2691–2692.

Bacolla, A., Larson, J.E., Collins, J.R., Li, J., Milosavljevic, A., Stenson, P.D., Cooper, D.N., Wells, R.D., 2008. Abundance and length of simple repeats in vertebrate genomes are determined by their structural properties. Genome Res. 18, 1545–1553.

Balasubramanian, S., Hurley, L.H., Neidle, S., 2011. Targeting G-quadruplexes in gene promoters: a novel anticancer strategy? Nat. Rev. Drug Discov. 10, 261–275.

Balkwill, G.D., Derecka, K., Garner, T.P., Hodgman, C., Flint, A.P., Searle, M.S., 2009. Repression of translation of human estrogen receptor alpha by G-quadruplex formation. Biochemistry 48, 11487–11495.

Ball, P., 2008. Water as an active constituent in cell biology. Chem. Rev. 108, 74–108.

Barone, I., Brusco, L., Fuqua, S.A., 2010. Estrogen receptor mutations and changes in downstream gene expression and signaling. Clin. Cancer Res. 16, 2702–2708.

Barrick, J.E., Breaker, R.R., 2007. The distributions, mechanisms, and structures of metabolite-binding riboswitches. Genome Biol. 8, R239.

Beaudoin, J.D., Perreault, J.P., 2010. 5′-UTR G-quadruplex structures acting as translational repressors. Nucleic Acids Res. 38, 7022–7036.

Belotserkovskii, B.P., De Silva, E., Tornaletti, S., Wang, G., Vasquez, K.M., Hanawalt, P.C., 2007. A triplex-forming sequence from the human c-MYC promoter interferes with DNA transcription. J. Biol. Chem. 282, 32433–32441.

Belotserkovskii, B.P., Liu, R., Tornaletti, S., Krasilnikova, M.M., Mirkin, S.M., Hanawalt, P.C., 2010. Mechanisms and implications of transcription blockage by guanine-rich DNA sequences. Proc. Natl. Acad. Sci. U. S. A. 107, 12816–12821.

Bevilacqua, P.C., Brown, T.S., Nakano, S., Yajima, R., 2004. Catalytic roles for proton transfer and protonation in ribozymes. Biopolymers 73, 90–109.

Blackburn, E.H., 1991. Structure and function of telomeres. Nature 350, 569–573.

Bloomfield, V.A., 1996. DNA condensation. Curr. Opin. Struct. Biol. 6, 334–341.

Bloomfield, V.A., Crothers, D.M., Tinoco Jr., I., 2000. Nucleic Acids: Structures, Properties, and Functions. University Science Books, Mill Valley, CA, USA.

Blount, K.F., Breaker, R.R., 2006. Riboswitches as antibacterial drug targets. Nat. Biotechnol. 24, 1558–1564.

Bollig, A., Miksicek, R.J., 2000. An estrogen receptor-alpha splicing variant mediates both positive and negative effects on gene transcription. Mol. Endocrinol. 14, 634–649.

Brenner, M.D., Scanlan, M.S., Nahas, M.K., Ha, T., Silverman, S.K., 2010. Multivector fluorescence analysis of the xpt guanine riboswitch aptamer domain and the conformational role of guanine. Biochemistry 49, 1596–1605.

Bridge, N.J., Fletcher, P.D.I., 1983. Time resolved studies of fluorescence quenching in a water-in-oil microemulsion. J. Chem. Soc. Faraday Trans. 79, 2161–2169.

Brooks, T.A., Hurley, L.H., 2009. The role of supercoiling in transcriptional control of MYC and its importance in molecular therapeutics. Nat. Rev. Cancer 9, 849–861.

Broxson, C., Beckett, J., Tornaletti, S., 2011. Transcription arrest by a G quadruplex forming-trinucleotide repeat sequence from the human c-myb gene. Biochemistry 50, 4162–4172.

Buchan, J.R., Stansfield, I., 2007. Halting a cellular production line: responses to ribosomal pausing during translation. Biol. Cell 99, 475–487.

Bugaut, A., Balasubramanian, S., 2012. 5′-UTR RNA G-quadruplexes: translation regulation and targeting. Nucleic Acids Res. 40, 4727–4741.

Butcher, S.E., Allain, F.H., Feigon, J., 1999. Solution structure of the loop B domain from the hairpin ribozyme. Nat. Struct. Biol. 6, 212–216.

Byrne, N., Angell, C.A., 2008. Protein unfolding, and the "tuning in" of reversible intermediate states in protic ionic liquid media. J. Mol. Biol. 378, 707–714.
Cardoso, L., Micaelo, N.M., 2011. DNA molecular solvation in neat ionic liquids. ChemPhysChem 12, 275–277.
Cayley, S., Record Jr., M.T., 2003. Roles of cytoplasmic osmolytes, water, and crowding in the response of Escherichia coli to osmotic stress: biophysical basis of osmoprotection by glycine betaine. Biochemistry 42, 12596–12609.
Chaidarun, S.S., Alexander, J.M., 1998. A tumor-specific truncated estrogen receptor splice variant enhances estrogen-stimulated gene expression. Mol. Endocrinol. 12, 1355–1366.
Chang, G.G., Huang, T.M., Hung, H.C., 2000. Reverse micelles as life-mimicking systems. Proc. Natl. Sci. Counc. Repub. China B 24, 89–100.
Chaplin, M., 2006. Do we underestimate the importance of water in cell biology? Nat. Rev. Mol. Cell Biol. 7, 861–866.
Cheetham, G.M., Steitz, T.A., 1999. Structure of a transcribing T7 RNA polymerase initiation complex. Science 286, 2305–2309.
Chen, G., Chang, K.Y., Chou, M.Y., Bustamante, C., Tinoco Jr., I., 2009. Triplex structures in an RNA pseudoknot enhance mechanical stability and increase efficiency of -1 ribosomal frameshifting. Proc. Natl. Acad. Sci. U. S. A. 106, 12706–12711.
Cheung, A.C., Cramer, P., 2011. Structural basis of RNA polymerase II backtracking, arrest and reactivation. Nature 471, 249–253.
Cheung, M.S., Klimov, D., Thirumalai, D., 2005. Molecular crowding enhances native state stability and refolding rates of globular proteins. Proc. Natl. Acad. Sci. U. S. A. 102, 4753–4758.
Citti, L., Rainaldi, G., 2005. Synthetic hammerhead ribozymes as therapeutic tools to control disease genes. Curr. Gene Ther. 5, 11–24.
Cogoi, S., Xodo, L.E., 2006. G-quadruplex formation within the promoter of the KRAS proto-oncogene and its effect on transcription. Nucleic Acids Res. 34, 2536–2549.
Core, L.J., Lis, J.T., 2008. Transcription regulation through promoter-proximal pausing of RNA polymerase II. Science 319, 1791–1792.
Damadian, R., 1971. Tumor detection by nuclear magnetic resonance. Science 171, 1151–1153.
Datta, K., von Hippel, P.H., 2008. Direct spectroscopic study of reconstituted transcription complexes reveals that intrinsic termination is driven primarily by thermodynamic destabilization of the nucleic acid framework. J. Biol. Chem. 283, 3537–3549.
Datta, K., Johnson, N.P., von Hippel, P.H., 2006. Mapping the conformation of the nucleic acid framework of the T7 RNA polymerase elongation complex in solution using low-energy CD and fluorescence spectroscopy. J. Mol. Biol. 360, 800–813.
David, L., Clauder-Munster, S., Steinmetz, L.M., 2011. Genome-wide transcriptome analysis in yeast using high-density tiling arrays. Methods Mol. Biol. 759, 107–123.
Davis, J.T., 2004. G-quartets 40 years later: from 5′-GMP to molecular biology and supramolecular chemistry. Angew. Chem. Int. Ed. Engl. 43, 668–698.
Davis-Searles, P.R., Saunders, A.J., Erie, D.A., Winzor, D.J., Pielak, G.J., 2001. Interpreting the effects of small uncharged solutes on protein-folding equilibria. Annu. Rev. Biophys. Biomol. Struct. 30, 271–306.
Del Vecchio, P., Esposito, D., Ricchi, L., Barone, G., 1999. The effects of polyols on the thermal stability of calf thymus DNA. Int. J. Biol. Macromol. 24, 361–369.
Denesyuk, N.A., Thirumalai, D., 2011. Crowding promotes the switch from hairpin to pseudoknot conformation in human telomerase RNA. J. Am. Chem. Soc. 133, 11858–11861.
Ding, Y., Zhang, L., Xie, J., Guo, R., 2010. Binding characteristics and molecular mechanism of interaction between ionic liquid and DNA. J. Phys. Chem. B 114, 2033–2043.

Ditlevson, J.V., Tornaletti, S., Belotserkovskii, B.P., Teijeiro, V., Wang, G., Vasquez, K.M., Hanawalt, P.C., 2008. Inhibitory effect of a short Z-DNA forming sequence on transcription elongation by T7 RNA polymerase. Nucleic Acids Res. 36, 3163–3170.

Dix, J.A., Verkman, A.S., 2008. Crowding effects on diffusion in solutions and cells. Annu. Rev. Biophys. 37, 247–263.

Doktycz, M.J., Larimer, F.W., Pastrnak, M., Stevens, A., 1998. Comparative analyses of the secondary structures of synthetic and intracellular yeast MFA2 mRNAs. Proc. Natl. Acad. Sci. U. S. A. 95, 14614–14621.

Dong, Q., Stellwagen, E., Stellwagen, N.C., 2009. Monovalent cation binding in the minor groove of DNA A-tracts. Biochemistry 48, 1047–1055.

Doudna, J.A., Lorsch, J.R., 2005. Ribozyme catalysis: not different, just worse. Nat. Struct. Mol. Biol. 12, 395–402.

Downey, C.D., Crisman, R.L., Randolph, T.W., Pardi, A., 2007. Influence of hydrostatic pressure and cosolutes on RNA tertiary structure. J. Am. Chem. Soc. 129, 9290–9291.

Du, Z., Zhao, Y., Li, N., 2008. Genome-wide analysis reveals regulatory role of G4 DNA in gene transcription. Genome Res. 18, 233–241.

DuPre, A.M., Hempling, H.G., 1978. Osmotic properties of Ehrlich ascites tumor cells during the cell cycle. J. Cell. Physiol. 97, 381–396.

Earle, M.J., Seddon, K.R., 2000. Ionic liquids. Green solvents for the future. Pure Appl. Chem. 72, 1391–1398.

Ebbinghaus, S.W., Gee, J.E., Rodu, B., Mayfield, C.A., Sanders, G., Miller, D.M., 1993. Triplex formation inhibits HER-2/neu transcription in vitro. J. Clin. Invest. 92, 2433–2439.

Eddy, J., Vallur, A.C., Varma, S., Liu, H., Reinhold, W.C., Pommier, Y., Maizels, N., 2011. G4 motifs correlate with promoter-proximal transcriptional pausing in human genes. Nucleic Acids Res. 39, 4975–4983.

Ellis, R.J., 2001. Macromolecular crowding: an important but neglected aspect of the intracellular environment. Curr. Opin. Struct. Biol. 11, 114–119.

Endoh, T., Sugimoto, N., 2011. Gene regulation system with an artificial RNA switch operating in human cells. ChemBioChem 12, 1174–1178.

Endoh, T., Kawasaki, Y., Sugimoto, N., 2012. Synchronized translation for detection of temporal stalling of ribosome during single-turnover translation. Anal. Chem. 84, 857–861.

Evans, M.S., Sander, I.M., Clark, P.L., 2008. Cotranslational folding promotes beta-helix formation and avoids aggregation in vivo. J. Mol. Biol. 383, 683–692.

Farabaugh, P.J., 1996. Programmed translational frameshifting. Microbiol. Rev. 60, 103–134.

Farnham, P.J., 2009. Insights from genomic profiling of transcription factors. Nat. Rev. Genet. 10, 605–616.

Felitsky, D.J., Cannon, J.G., Capp, M.W., Hong, J., Van Wynsberghe, A.W., Anderson, C.F., Record Jr., M.T., 2004. The exclusion of glycine betaine from anionic biopolymer surface: why glycine betaine is an effective osmoprotectant but also a compatible solute. Biochemistry 43, 14732–14743.

Fernandez, J., Yaman, I., Huang, C., Liu, H., Lopez, A.B., Komar, A.A., Caprara, M.G., Merrick, W.C., Snider, M.D., Kaufman, R.J., Lamers, W.H., Hatzoglou, M., 2005. Ribosome stalling regulates IRES-mediated translation in eukaryotes, a parallel to prokaryotic attenuation. Mol. Cell 17, 405–416.

Ferrari, R., Rivetti, C., Dieci, G., 2004. Transcription reinitiation properties of bacteriophage T7 RNA polymerase. Biochem. Biophys. Res. Commun. 315, 376–380.

Flouriot, G., Brand, H., Denger, S., Metivier, R., Kos, M., Reid, G., Sonntag-Buck, V., Gannon, F., 2000. Identification of a new isoform of the human estrogen receptor-alpha

(hER-alpha) that is encoded by distinct transcripts and that is able to repress hER-alpha activation function 1. EMBO J. 19, 4688–4700.
Forster, A.C., Symons, R.H., 1987. Self-cleavage of virusoid RNA is performed by the proposed 55-nucleotide active site. Cell 50, 9–16.
Franco, E., Friedrichs, E., Kim, J., Jungmann, R., Murray, R., Winfree, E., Simmel, F.C., 2011. Timing molecular motion and production with a synthetic transcriptional clock. Proc. Natl. Acad. Sci. U. S. A. 108, E784–E793.
Fujita, K., MacFarlane, D.R., Forsyth, M., 2005. Protein solubilising and stabilising ionic liquids. Chem. Commun. 4804–4806.
Fulton, A.B., 1982. How crowded is the cytoplasm? Cell 30, 345–347.
Garcia Pedrero, J.M., Zuazua, P., Martinez-Campa, C., Lazo, P.S., Ramos, S., 2003. The naturally occurring variant of estrogen receptor (ER) ERDeltaE7 suppresses estrogen-dependent transcriptional activation by both wild-type ERalpha and ERbeta. Endocrinology 144, 2967–2976.
Garlid, K.D., 2000. The state of water in biological systems. Int. Rev. Cytol. 192, 281–302.
Gehring, K., Leroy, J.L., Gueron, M., 1993. A tetrameric DNA structure with protonated cytosine-cytosine base pairs. Nature 363, 561–565.
Gellert, M., Lipsett, M.N., Davies, D.R., 1962. Helix formation by guanylic acid. Proc. Natl. Acad. Sci. U. S. A. 48, 2013–2018.
Gibson, D.A., Saunders, P.T., 2012. Estrogen dependent signaling in reproductive tissues—a role for estrogen receptors and estrogen related receptors. Mol. Cell. Endocrinol. 348, 361–372.
Giedroc, D.P., Cornish, P.V., 2009. Frameshifting RNA pseudoknots: structure and mechanism. Virus Res. 139, 193–208.
Gluick, T.C., Yadav, S., 2003. Trimethylamine N-oxide stabilizes RNA tertiary structure and attenuates the denaturating effects of urea. J. Am. Chem. Soc. 125, 4418–4419.
Goobes, R., Minsky, A., 2001. Thermodynamic aspects of triplex DNA formation in crowded environments. J. Am. Chem. Soc. 123, 12692–12693.
Goobes, R., Cohen, O., Minsky, A., 2002. Unique condensation patterns of triplex DNA: physical aspects and physiological implications. Nucleic Acids Res. 30, 2154–2161.
Goobes, R., Kahana, N., Cohen, O., Minsky, A., 2003. Metabolic buffering exerted by macromolecular crowding on DNA-DNA interactions: origin and physiological significance. Biochemistry 42, 2431–2440.
Goodsell, D.S., 1991. Inside a living cell. Trends Biochem. Sci. 16, 203–206.
Gu, X.B., Nakano, S., Sugimoto, N., 2007. Consecutive GC base pairs determine the energy barrier of DNA duplex formation under molecularly crowded conditions. Chem. Commun. 2750–2752.
Guigas, G., Weiss, M., 2008. Sampling the cell with anomalous diffusion—the discovery of slowness. Biophys. J. 94, 90–94.
Guthrie, C., Patterson, B., 1988. Spliceosomal snRNAs. Annu. Rev. Genet. 22, 387–419.
Halder, K., Wieland, M., Hartig, J.S., 2009. Predictable suppression of gene expression by 5′-UTR-based RNA quadruplexes. Nucleic Acids Res. 37, 6811–6817.
Hancock, R., 2004. A role for macromolecular crowding effects in the assembly and function of compartments in the nucleus. J. Struct. Biol. 146, 281–290.
Hancock, R., 2007. Packing of the polynucleosome chain in interphase chromosomes: evidence for a contribution of crowding and entropic forces. Semin. Cell Dev. Biol. 18, 668–675.
Handwerger, K.E., Cordero, J.A., Gall, J.G., 2005. Cajal bodies, nucleoli, and speckles in the Xenopus oocyte nucleus have a low-density, sponge-like structure. Mol. Biol. Cell 16, 202–211.
Haseloff, J., Gerlach, W.L., 1988. Simple RNA enzymes with new and highly specific endoribonuclease activities. Nature 334, 585–591.
Herve, G., Tobe, S., Heams, T., Vergne, J., Maurel, M.C., 2006. Hydrostatic and osmotic pressure study of the hairpin ribozyme. Biochim. Biophys. Acta 1764, 573–577.

Holland, J.G., Malin, J.N., Jordan, D.S., Morales, E., Geiger, F.M., 2011. Specific and nonspecific metal ion-nucleotide interactions at aqueous/solid interfaces functionalized with adenine, thymine, guanine, and cytosine oligomers. J. Am. Chem. Soc. 133, 2567–2570.

Hong, J., Capp, M.W., Anderson, C.F., Saecker, R.M., Felitsky, D.J., Anderson, M.W., Record Jr., M.T., 2004. Preferential interactions of glycine betaine and of urea with DNA: implications for DNA hydration and for effects of these solutes on DNA stability. Biochemistry 43, 14744–14758.

Horjales, S., Cota, G., Senorale-Pose, M., Rovira, C., Roman, E., Artagaveytia, N., Ehrlich, R., Marin, M., 2007. Translational machinery and protein folding: evidence of conformational variants of the estrogen receptor alpha. Arch. Biochem. Biophys. 467, 139–143.

Hud, N.V., Sklenar, V., Feigon, J., 1999. Localization of ammonium ions in the minor groove of DNA duplexes in solution and the origin of DNA A-tract bending. J. Mol. Biol. 286, 651–660.

Hutchins, C.J., Rathjen, P.D., Forster, A.C., Symons, R.H., 1986. Self-cleavage of plus and minus RNA transcripts of avocado sunblotch viroid. Nucleic Acids Res. 14, 3627–3640.

Jensch, F., Kemper, B., 1986. Endonuclease VII resolves Y-junctions in branched DNA in vitro. EMBO J. 5, 181–189.

Joachimi, A., Benz, A., Hartig, J.S., 2009. A comparison of DNA and RNA quadruplex structures and stabilities. Bioorg. Med. Chem. 17, 6811–6815.

Johansson, G., Walter, H., 2000. Partitioning and concentrating biomaterials in aqueous phase systems. Int. Rev. Cytol. 192, 33–60.

Kaiser, C.M., Goldman, D.H., Chodera, J.D., Tinoco Jr., I., Bustamante, C., 2011. The ribosome modulates nascent protein folding. Science 334, 1723–1727.

Kan, Z.Y., Yao, Y., Wang, P., Li, X.H., Hao, Y.H., Tan, Z., 2006. Molecular crowding induces telomere G-quadruplex formation under salt-deficient conditions and enhances its competition with duplex formation. Angew. Chem. Int. Ed. Engl. 45, 1629–1632.

Kan, Z.Y., Lin, Y., Wang, F., Zhuang, X.Y., Zhao, Y., Pang, D.W., Hao, Y.H., Tan, Z., 2007. G-quadruplex formation in human telomeric (TTAGGG)4 sequence with complementary strand in close vicinity under molecularly crowded condition. Nucleic Acids Res. 35, 3646–3653.

Kassavetis, G.A., Chamberlin, M.J., 1981. Pausing and termination of transcription within the early region of bacteriophage T7 DNA in vitro. J. Biol. Chem. 256, 2777–2786.

Khvorova, A., Lescoute, A., Westhof, E., Jayasena, S.D., 2003. Sequence elements outside the hammerhead ribozyme catalytic core enable intracellular activity. Nat. Struct. Mol. Biol. 10, 708–712.

Kilburn, D., Roh, J.H., Guo, L., Briber, R.M., Woodson, S.A., 2010. Molecular crowding stabilizes folded RNA structure by the excluded volume effect. J. Am. Chem. Soc. 132, 8690–8696.

Kim, J.N., Breaker, R.R., 2008. Purine sensing by riboswitches. Biol. Cell 100, 1–11.

Kim, D., Lee, Y.H., Hwang, H.Y., Kim, K.K., Park, H.J., 2010. Z-DNA binding proteins as targets for structure-based virtual screening. Curr. Drug Targets 11, 335–344.

Kimchi-Sarfaty, C., Oh, J.M., Kim, I.W., Sauna, Z.E., Calcagno, A.M., Ambudkar, S.V., Gottesman, M.M., 2007. A "silent" polymorphism in the MDR1 gene changes substrate specificity. Science 315, 525–528.

Kireeva, M.L., Kashlev, M., 2009. Mechanism of sequence-specific pausing of bacterial RNA polymerase. Proc. Natl. Acad. Sci. U. S. A. 106, 8900–8905.

Kitts, P.A., Nash, H.A., 1987. Homology-dependent interactions in phage lambda site-specific recombination. Nature 329, 346–348.

Knowles, D.B., LaCroix, A.S., Deines, N.F., Shkel, I., Record Jr., M.T., 2011. Separation of preferential interaction and excluded volume effects on DNA duplex and hairpin stability. Proc. Natl. Acad. Sci. U. S. A. 108, 12699–12704.

Kochetkov, S.N., Rusakova, E.E., Tunitskaya, V.L., 1998. Recent studies of T7 RNA polymerase mechanism. FEBS Lett. 440, 264–267.

Komar, A.A., 2009. A pause for thought along the co-translational folding pathway. Trends Biochem. Sci. 34, 16–24.

Kotlarchyk, M., Huang, J.S., Chen, S.H., 1985. Structure of AOT reversed micelles determined by small-angle neutron scattering. J. Phys. Chem. 89, 4382–4386.

Kraeva, R.I., Krastev, D.B., Roguev, A., Ivanova, A., Nedelcheva-Veleva, M.N., Stoynov, S.S., 2007. Stability of mRNA/DNA and DNA/DNA duplexes affects mRNA transcription. PLoS One 2, e290.

Krzyzaniak, A., Salanski, P., Jurczak, J., Barciszewski, J., 1991. B-Z DNA reversible conformation changes effected by high pressure. FEBS Lett. 279, 1–4.

Krzyzaniak, A., Barciszewski, J., Furste, J.P., Bald, R., Erdmann, V.A., Salanski, P., Jurczak, J., 1994. A-Z-RNA conformational changes effected by high pressure. Int. J. Biol. Macromol. 16, 159–162.

Kudla, G., Murray, A.W., Tollervey, D., Plotkin, J.B., 2009. Coding-sequence determinants of gene expression in Escherichia coli. Science 324, 255–258.

Kumar, N., Maiti, S., 2004. Quadruplex to Watson-Crick duplex transition of the thrombin binding aptamer: a fluorescence resonance energy transfer study. Biochem. Biophys. Res. Commun. 319, 759–767.

Kumar, N., Maiti, S., 2005. The effect of osmolytes and small molecule on Quadruplex-WC duplex equilibrium: a fluorescence resonance energy transfer study. Nucleic Acids Res. 33, 6723–6732.

Kumar, V., Endoh, T., Murakami, K., Sugimoto, N., 2012. Dehydration from conserved stem regions is fundamental for ligand-dependent conformational transition of the adenine-specific riboswitch. Chem. Commun. 48, 9693–9695.

Kumari, S., Bugaut, A., Balasubramanian, S., 2008. Position and stability are determining factors for translation repression by an RNA G-quadruplex-forming sequence within the 5′ UTR of the NRAS proto-oncogene. Biochemistry 47, 12664–12669.

Lambert, D., Leipply, D., Draper, D.E., 2010. The osmolyte TMAO stabilizes native RNA tertiary structures in the absence of Mg2+: evidence for a large barrier to folding from phosphate dehydration. J. Mol. Biol. 404, 138–157.

Landick, R., 2006. The regulatory roles and mechanism of transcriptional pausing. Biochem. Soc. Trans. 34, 1062–1066.

Lang, F., Busch, G.L., Ritter, M., Volkl, H., Waldegger, S., Gulbins, E., Haussinger, D., 1998. Functional significance of cell volume regulatory mechanisms. Physiol. Rev. 78, 247–306.

Larson, D.R., Zenklusen, D., Wu, B., Chao, J.A., Singer, R.H., 2011. Real-time observation of transcription initiation and elongation on an endogenous yeast gene. Science 332, 475–478.

Lee, T.S., Lopez, C.S., Giambasu, G.M., Martick, M., Scott, W.G., York, D.M., 2008. Role of Mg(2+) in hammerhead ribozyme catalysis from molecular simulation. J. Am. Chem. Soc. 130, 3053–3064.

Lemay, J.F., Lafontaine, D.A., 2007. Core requirements of the adenine riboswitch aptamer for ligand binding. RNA 13, 339–350.

Lemay, J.F., Penedo, J.C., Tremblay, R., Lilley, D.M., Lafontaine, D.A., 2006. Folding of the adenine riboswitch. Chem. Biol. 13, 857–868.

Levinger, N.E., 2002. Chemistry. Water in confinement. Science 298, 1722–1723.

Lewis, J.D., Tollervey, D., 2000. Like attracts like: getting RNA processing together in the nucleus. Science 288, 1385–1389.

Li, W., Miyoshi, D., Nakano, S., Sugimoto, N., 2003. Structural competition involving G-quadruplex DNA and its complement. Biochemistry 42, 11736–11744.

Li, J., Correia, J.J., Wang, L., Trent, J.O., Chaires, J.B., 2005. Not so crystal clear: the structure of the human telomere G-quadruplex in solution differs from that present in a crystal. Nucleic Acids Res. 33, 4649–4659.

Li, S., Nosrati, M., Kashani-Sabet, M., 2008. Knockdown of telomerase RNA using hammerhead ribozymes and RNA interference. Methods Mol. Biol. 405, 113–131.

Lilley, D.M., 2005. Structure, folding and mechanisms of ribozymes. Curr. Opin. Struct. Biol. 15, 313–323.

Liu, J., Lu, Y., 2007. Rational design of "turn-on" allosteric DNAzyme catalytic beacons for aqueous mercury ions with ultrahigh sensitivity and selectivity. Angew. Chem. Int. Ed. Engl. 46, 7587–7590.

Low, J.T., Weeks, K.M., 2010. SHAPE-directed RNA secondary structure prediction. Methods 52, 150–158.

Lu, M., Guo, Q., Marky, L.A., Seeman, N.C., Kallenbach, N.R., 1992. Thermodynamics of DNA branching. J. Mol. Biol. 223, 781–789.

Luby-Phelps, K., 2000. Cytoarchitecture and physical properties of cytoplasm: volume, viscosity, diffusion, intracellular surface area. Int. Rev. Cytol. 192, 189–221.

Luger, K., Hansen, J.C., 2005. Nucleosome and chromatin fiber dynamics. Curr. Opin. Struct. Biol. 15, 188–196.

Luger, K., Mader, A.W., Richmond, R.K., Sargent, D.F., Richmond, T.J., 1997. Crystal structure of the nucleosome core particle at 2.8 A resolution. Nature 389, 251–260.

Luisi, P.L., Giomini, M., Pileni, M.P., Robinson, B.H., 1988. Reverse micelles as hosts for proteins and small molecules. Biochim. Biophys. Acta 947, 209–246.

Lukacs, G.L., Haggie, P., Seksek, O., Lechardeur, D., Freedman, N., Verkman, A.S., 2000. Size-dependent DNA mobility in cytoplasm and nucleus. J. Biol. Chem. 275, 1625–1629.

Luo, D., Condon, C., Grunberg-Manago, M., Putzer, H., 1998. In vitro and in vivo secondary structure probing of the thrS leader in Bacillus subtilis. Nucleic Acids Res. 26, 5379–5387.

MacMillen, R.E., Lee, A.K., 1967. Australian desert mice: independence of exogenous water. Science 158, 383–385.

Mandal, M., Breaker, R.R., 2004. Adenine riboswitches and gene activation by disruption of a transcription terminator. Nat. Struct. Mol. Biol. 11, 29–35.

Mandal, M., Boese, B., Barrick, J.E., Winkler, W.C., Breaker, R.R., 2003. Riboswitches control fundamental biochemical pathways in Bacillus subtilis and other bacteria. Cell 113, 577–586.

Mao, C., Sun, W., Shen, Z., Seeman, N.C., 1999. A nanomechanical device based on the B-Z transition of DNA. Nature 397, 144–146.

Marky, L.A., Kupke, D.W., 2000. Enthalpy-entropy compensations in nucleic acids: contribution of electrostriction and structural hydration. Methods Enzymol. 323, 419–441.

Martick, M., Scott, W.G., 2006. Tertiary contacts distant from the active site prime a ribozyme for catalysis. Cell 126, 309–320.

Martick, M., Horan, L.H., Noller, H.F., Scott, W.G., 2008. A discontinuous hammerhead ribozyme embedded in a mammalian messenger RNA. Nature 454, 899–902.

Mathews, D.H., 2004. Using an RNA secondary structure partition function to determine confidence in base pairs predicted by free energy minimization. RNA 10, 1178–1190.

Mathews, D.H., Sabina, J., Zuker, M., Turner, D.H., 1999. Expanded sequence dependence of thermodynamic parameters improves prediction of RNA secondary structure. J. Mol. Biol. 288, 911–940.

Matveeva, O.V., Kang, Y., Spiridonov, A.N., Saetrom, P., Nemtsov, V.A., Ogurtsov, A.Y., Nechipurenko, Y.D., Shabalina, S.A., 2010. Optimization of duplex stability and terminal asymmetry for shRNA design. PLoS One 5, e10180.

McIntosh, R., Nicastro, D., Mastronarde, D., 2005. New views of cells in 3D: an introduction to electron tomography. Trends Cell Biol. 15, 43–51.
Melchior Jr., W.B., Von Hippel, P.H., 1973. Alteration of the relative stability of dA-dT and dG-dC base pairs in DNA. Proc. Natl. Acad. Sci. U. S. A. 70, 298–302.
Mereau, A., Fournier, R., Gregoire, A., Mougin, A., Fabrizio, P., Luhrmann, R., Branlant, C., 1997. An in vivo and in vitro structure-function analysis of the Saccharomyces cerevisiae U3A snoRNP: protein-RNA contacts and base-pair interaction with the pre-ribosomal RNA. J. Mol. Biol. 273, 552–571.
Meyer, B., Benichou, O., Kafri, Y., Voituriez, R., 2012. Geometry-induced bursting dynamics in gene expression. Biophys. J. 102, 2186–2191.
Miller, M.C., Buscaglia, R., Chaires, J.B., Lane, A.N., Trent, J.O., 2010. Hydration is a major determinant of the G-quadruplex stability and conformation of the human telomere 3′ sequence of d(AG3(TTAG3)3). J. Am. Chem. Soc. 132, 17105–17107.
Min, I.M., Waterfall, J.J., Core, L.J., Munroe, R.J., Schimenti, J., Lis, J.T., 2011. Regulating RNA polymerase pausing and transcription elongation in embryonic stem cells. Genes Dev. 25, 742–754.
Minsky, A., Ghirlando, R., Reich, Z., 1997. Nucleosomes: a solution to a crowded intracellular environment? J. Theor. Biol. 188, 379–385.
Minton, A.P., 1998. Molecular crowding: analysis of effects of high concentrations of inert cosolutes on biochemical equilibria and rates in terms of volume exclusion. Methods Enzymol. 295, 127–149.
Misteli, T., 2008. Physiological importance of RNA and protein mobility in the cell nucleus. Histochem. Cell Biol. 129, 5–11.
Miyoshi, D., Nakao, A., Sugimoto, N., 2002. Molecular crowding regulates the structural switch of the DNA G-quadruplex. Biochemistry 41, 15017–15024.
Miyoshi, D., Matsumura, S., Li, W., Sugimoto, N., 2003. Structural polymorphism of telomeric DNA regulated by pH and divalent cation. Nucleosides Nucleotides Nucleic Acids 22, 203–221.
Miyoshi, D., Matsumura, S., Nakano, S., Sugimoto, N., 2004. Duplex dissociation of telomere DNAs induced by molecular crowding. J. Am. Chem. Soc. 126, 165–169.
Miyoshi, D., Karimata, H., Sugimoto, N., 2005. Drastic effect of a single base difference between human and tetrahymena telomere sequences on their structures under molecular crowding conditions. Angew. Chem. Int. Ed. Engl. 44, 3740–3744.
Miyoshi, D., Inoue, M., Sugimoto, N., 2006a. DNA logic gates based on structural polymorphism of telomere DNA molecules responding to chemical input signals. Angew. Chem. Int. Ed. Engl. 45, 7716–7719.
Miyoshi, D., Karimata, H., Sugimoto, N., 2006b. Hydration regulates thermodynamics of G-quadruplex formation under molecular crowding conditions. J. Am. Chem. Soc. 128, 7957–7963.
Miyoshi, D., Karimata, H., Wang, Z.M., Koumoto, K., Sugimoto, N., 2007. Artificial G-wire switch with 2,2′-bipyridine units responsive to divalent metal ions. J. Am. Chem. Soc. 129, 5919–5925.
Miyoshi, D., Nakamura, K., Tateishi-Karimata, H., Ohmichi, T., Sugimoto, N., 2009. Hydration of Watson-Crick base pairs and dehydration of Hoogsteen base pairs inducing structural polymorphism under molecular crowding conditions. J. Am. Chem. Soc. 131, 3522–3531.
Montange, R.K., Batey, R.T., 2008. Riboswitches: emerging themes in RNA structure and function. Annu. Rev. Biophys. 37, 117–133.
Moyzis, R.K., Buckingham, J.M., Cram, L.S., Dani, M., Deaven, L.L., Jones, M.D., Meyne, J., Ratliff, R.L., Wu, J.R., 1988. A highly conserved repetitive DNA sequence, (TTAGGG)n, present at the telomeres of human chromosomes. Proc. Natl. Acad. Sci. U. S. A. 85, 6622–6626.

Muhuri, S., Mimura, K., Miyoshi, D., Sugimoto, N., 2009. Stabilization of three-way junctions of DNA under molecular crowding conditions. J. Am. Chem. Soc. 131, 9268–9280.
Nagatoishi, S., Ono, R., Sugimoto, N., 2012. The yields of transcripts for a RNA polymerase regulated by hairpin structures in nascent RNAs. Chem. Commun. 48, 5121–5123.
Nakano, S., Fujimoto, M., Hara, H., Sugimoto, N., 1999. Nucleic acid duplex stability: influence of base composition on cation effects. Nucleic Acids Res. 27, 2957–2965.
Nakano, S., Karimata, H., Ohmichi, T., Kawakami, J., Sugimoto, N., 2004. The effect of molecular crowding with nucleotide length and cosolute structure on DNA duplex stability. J. Am. Chem. Soc. 126, 14330–14331.
Nakano, S., Kirihata, T., Sugimoto, N., 2008b. Capture of cationic ligands bound diffusely to base pairs during the DNA refolding. Chem. Commun. 14, 700–702.
Nakano, S., Wu, L., Oka, H., Karimata, H.T., Kirihata, T., Sato, Y., Fujii, S., Sakai, H., Kuwahara, M., Sawai, H., Sugimoto, N., 2008a. Conformation and the sodium ion condensation on DNA and RNA structures in the presence of a neutral cosolute as a mimic of the intracellular media. Mol. BioSyst. 4, 579–588.
Nakano, S., Karimata, H.T., Kitagawa, Y., Sugimoto, N., 2009. Facilitation of RNA enzyme activity in the molecular crowding media of cosolutes. J. Am. Chem. Soc. 131, 16881–16888.
Nakano, S., Yamaguchi, D., Tateishi-Karimata, H., Miyoshi, D., Sugimoto, N., 2012. Hydration changes upon DNA folding studied by osmotic stress experiments. Biophys. J. 102, 2808–2817.
Nakano, S., Miyoshi, D., Sugimoto, N., 2013. Effects of molecular crowding on the structures, interactions, and functions of nucleic acids. Chem. Rev. (in press).
Nechaev, S., Fargo, D.C., dos Santos, G., Liu, L., Gao, Y., Adelman, K., 2010. Global analysis of short RNAs reveals widespread promoter-proximal stalling and arrest of Pol II in Drosophila. Science 327, 335–338.
Nishimura, N., Ohno, H., 2002. Design of successive ion conduction paths in DNA films with ionic liquids. J. Mater. Chem. 12, 2299–2304.
Noeske, J., Schwalbe, H., Wohnert, J., 2007. Metal-ion binding and metal-ion induced folding of the adenine-sensing riboswitch aptamer domain. Nucleic Acids Res. 35, 5262–5273.
Nordstrom, L.J., Clark, C.A., Andersen, B., Champlin, S.M., Schwinefus, J.J., 2006. Effect of ethylene glycol, urea, and N-methylated glycines on DNA thermal stability: the role of DNA base pair composition and hydration. Biochemistry 45, 9604–9614.
Nucci, N.V., Pometun, M.S., Wand, A.J., 2011. Site-resolved measurement of water-protein interactions by solution NMR. Nat. Struct. Mol. Biol. 18, 245–249.
Nudler, E., 2009. RNA polymerase active center: the molecular engine of transcription. Annu. Rev. Biochem. 78, 335–361.
O'Brien, T.P., Bult, C.J., Cremer, C., Grunze, M., Knowles, B.B., Langowski, J., McNally, J., Pederson, T., Politz, J.C., Pombo, A., Schmahl, G., Spatz, J.P., van Driel, R., 2003. Genome function and nuclear architecture: from gene expression to nanoscience. Genome Res. 13, 1029–1041.
O'Brien, E.P., Hsu, S.T., Christodoulou, J., Vendruscolo, M., Dobson, C.M., 2010. Transient tertiary structure formation within the ribosome exit port. J. Am. Chem. Soc. 132, 16928–16937.
O'Brien, E.P., Christodoulou, J., Vendruscolo, M., Dobson, C.M., 2011. New scenarios of protein folding can occur on the ribosome. J. Am. Chem. Soc. 133, 513–526.
O'Brien, E.P., Vendruscolo, M., Dobson, C.M., 2012. Prediction of variable translation rate effects on cotranslational protein folding. Nat. Commun. 3, 868.
Oganesian, L., Bryan, T.M., 2007. Physiological relevance of telomeric G-quadruplex formation: a potential drug target. Bioessays 29, 155–165.

Ohmichi, T., Kawamoto, Y., Wu, P., Miyoshi, D., Karimata, H., Sugimoto, N., 2005. DNA-based biosensor for monitoring pH in vitro and in living cells. Biochemistry 44, 7125–7130.

Ovádi, J., Srere, P.A., 2000. Macromolecular compartmentation and channeling. Int. Rev. Cytol. 192, 255–280.

Palangat, M., Landick, R., 2001. Roles of RNA:DNA hybrid stability, RNA structure, and active site conformation in pausing by human RNA polymerase II. J. Mol. Biol. 311, 265–282.

Pan, T., Sosnick, T., 2006. RNA folding during transcription. Annu. Rev. Biophys. Biomol. Struct. 35, 161–175.

Park, L.C., Maruyama, T., Goto, M., 2003. DNA hybridization in reverse micelles and its application to mutation detection. Analyst 128, 161–165.

Parsegian, V.A., Rau, D.C., 1984. Water near intracellular surfaces. J. Cell Biol. 99, 196–200.

Partikian, A., Olveczky, B., Swaminathan, R., Li, Y., Verkman, A.S., 1998. Rapid diffusion of green fluorescent protein in the mitochondrial matrix. J. Cell Biol. 140, 821–829.

Petraccone, L., Spink, C., Trent, J.O., Garbett, N.C., Mekmaysy, C.S., Giancola, C., Chaires, J.B., 2012. Structure and stability of higher-order human telomeric quadruplexes. J. Am. Chem. Soc. 133, 20951–20961.

Phan, A.T., Mergny, J.L., 2002. Human telomeric DNA: G-quadruplex, i-motif and Watson-Crick double helix. Nucleic Acids Res. 30, 4618–4625.

Pincus, D.L., Hyeon, C., Thirumalai, D., 2008. Effects of trimethylamine N-oxide (TMAO) and crowding agents on the stability of RNA hairpins. J. Am. Chem. Soc. 130, 7364–7372.

Politi, M.J., Chaimovich, H., 1986. Water activity in reversed sodium bis(2-ethylhexyl) sulfosuccinate micelles. J. Phys. Chem. 90, 282–287.

Pramanik, S., Nagatoishi, S., Sugimoto, N., 2012. DNA tetraplex structure formation from human telomeric repeat motif (TTAGGG):(CCCTAA) in nanocavity water pools of reverse micelles. Chem. Commun. 48, 4815–4817.

Praseuth, D., Guieysse, A.L., Helene, C., 1999. Triple helix formation and the antigene strategy for sequence-specific control of gene expression. Biochim. Biophys. Acta 1489, 181–206.

Preisler, R.S., Chen, H.H., Colombo, M.F., Choe, Y., Short Jr., B.J., Rau, D.C., 1995. The B form to Z form transition of poly(dG-m5dC) is sensitive to neutral solutes through an osmotic stress. Biochemistry 34, 14400–14407.

Proctor, D.J., Ma, H., Kierzek, E., Kierzek, R., Gruebele, M., Bevilacqua, P.C., 2004. Folding thermodynamics and kinetics of YNMG RNA hairpins: specific incorporation of 8-bromoguanosine leads to stabilization by enhancement of the folding rate. Biochemistry 43, 14004–14014.

Puglisi, J.D., Tinoco Jr., I., 1989. Absorbance melting curves of RNA. Methods Enzymol. 180, 304–325.

Qin, Y., Hurley, L.H., 2008. Structures, folding patterns, and functions of intramolecular DNA G-quadruplexes found in eukaryotic promoter regions. Biochimie 90, 1149–1171.

Rajendran, A., Nakano, S., Sugimoto, N., 2010. Molecular crowding of the cosolutes induces an intramolecular i-motif structure of triplet repeat DNA oligomers at neutral pH. Chem. Commun. 46, 1299–1301.

Rangan, P., Woodson, S.A., 2003. Structural requirement for Mg2+ binding in the group I intron core. J. Mol. Biol. 329, 229–238.

Record Jr., M.T., Courtenay, E.S., Cayley, S., Guttman, H.J., 1998a. Biophysical compensation mechanisms buffering E. coli protein-nucleic acid interactions against changing environments. Trends Biochem. Sci. 23, 190–194.

Record Jr., M.T., Zhang, W., Anderson, C.F., 1998b. Analysis of effects of salts and uncharged solutes on protein and nucleic acid equilibria and processes: a practical guide to recognizing and interpreting polyelectrolyte effects, Hofmeister effects, and osmotic effects of salts. Adv. Protein Chem. 51, 281–353.

Richmond, T.J., Davey, C.A., 2003. The structure of DNA in the nucleosome core. Nature 423, 145–150.

Richter, K., Nessling, M., Lichter, P., 2007. Experimental evidence for the influence of molecular crowding on nuclear architecture. J. Cell Sci. 120, 1673–1680.

Richter, K., Nessling, M., Lichter, P., 2008. Macromolecular crowding and its potential impact on nuclear function. Biochim. Biophys. Acta 1783, 2100–2107.

Rieder, R., Lang, K., Graber, D., Micura, R., 2007. Ligand-induced folding of the adenosine deaminase A-riboswitch and implications on riboswitch translational control. ChemBioChem 8, 896–902.

Ringrose, L., Chabanis, S., Angrand, P.O., Woodroofe, C., Stewart, A.F., 1999. Quantitative comparison of DNA looping in vitro and in vivo: chromatin increases effective DNA flexibility at short distances. EMBO J. 18, 6630–6641.

Roque, A., Ponte, I., Suau, P., 2007. Macromolecular crowding induces a molten globule state in the C-terminal domain of histone H1. Biophys. J. 93, 2170–2177.

Rosen, M.A., Patel, D.J., 1993. Conformational differences between bulged pyrimidines (C-C) and purines (A-A, I-I) at the branch point of three-stranded DNA junctions. Biochemistry 32, 6563–6575.

Rösgen, J., Pettitt, B.M., Bolen, D.W., 2005. Protein folding, stability, and solvation structure in osmolyte solutions. Biophys. J. 89, 2988–2997.

Rosi, N.L., Giljohann, D.A., Thaxton, C.S., Lytton-Jean, A.K., Han, M.S., Mirkin, C.A., 2006. Oligonucleotide-modified gold nanoparticles for intracellular gene regulation. Science 312, 1027–1030.

Saenger, K., 1984. Principles of Nucleic Acid Structure. Springer Advanced Texts in Chemistry. Springer, New York.

Salehi-Ashtiani, K., Luptak, A., Litovchick, A., Szostak, J.W., 2006. A genomewide search for ribozymes reveals an HDV-like sequence in the human CPEB3 gene. Science 313, 1788–1792.

Salinas-Rios, V., Belotserkovskii, B.P., Hanawalt, P.C., 2011. DNA slip-outs cause RNA polymerase II arrest in vitro: potential implications for genetic instability. Nucleic Acids Res. 39, 7444–7454.

Santangelo, T.J., Artsimovitch, I., 2011. Termination and antitermination: RNA polymerase runs a stop sign. Nat. Rev. Microbiol. 9, 319–329.

Schoeftner, S., Blasco, M.A., 2009. A 'higher order' of telomere regulation: telomere heterochromatin and telomeric RNAs. EMBO J. 28, 2323–2336.

Schoen, I., Krammer, H., Braun, D., 2009. Hybridization kinetics is different inside cells. Proc. Natl. Acad. Sci. U. S. A. 106, 21649–21654.

Schultz-Norton, J.R., Ziegler, Y.S., Nardulli, A.M., 2011. ERα-associated protein networks. Trends Endocrinol. Metab. 22, 124–129.

Seddon, K.R., 1997. Ionic liquids for clean technology. J. Chem. Technol. Biotechnol. 68, 351–356.

Seeman, N.C., 2003a. At the crossroads of chemistry, biology, and materials: structural DNA nanotechnology. Chem. Biol. 10, 1151–1159.

Seeman, N.C., 2003b. DNA in a material world. Nature 421, 427–431.

Semrad, K., Green, R., 2002. Osmolytes stimulate the reconstitution of functional 50S ribosomes from in vitro transcripts of Escherichia coli 23S rRNA. RNA 8, 401–411.

Senecoff, J.F., Meagher, R.B., 1992. In vivo analysis of plant RNA structure: soybean 18S ribosomal and ribulose-1,5-bisphosphate carboxylase small subunit RNAs. Plant Mol. Biol. 18, 219–234.

Serganov, A., Yuan, Y.R., Pikovskaya, O., Polonskaia, A., Malinina, L., Phan, A.T., Hobartner, C., Micura, R., Breaker, R.R., Patel, D.J., 2004. Structural basis for

discriminative regulation of gene expression by adenine- and guanine-sensing mRNAs. Chem. Biol. 11, 1729–1741.
Serganov, A., Huang, L., Patel, D.J., 2009. Coenzyme recognition and gene regulation by a flavin mononucleotide riboswitch. Nature 458, 233–237.
Shiman, R., Draper, D.E., 2000. Stabilization of RNA tertiary structure by monovalent cations. J. Mol. Biol. 302, 79–91.
Simonsson, T., 2001. G-quadruplex DNA structures—variations on a theme. Biol. Chem. 382, 621–628.
Singleton, M.R., Scaife, S., Wigley, D.B., 2001. Structural analysis of DNA replication fork reversal by RecG. Cell 107, 79–89.
Sousa, R., Mukherjee, S., 2003. T7 RNA polymerase. Prog. Nucleic Acid Res. Mol. Biol. 73, 1–41.
Spink, C.H., Chaires, J.B., 1995. Selective stabilization of triplex DNA by poly(ethylene glycols). J. Am. Chem. Soc. 117, 12887–12888.
Spink, C.H., Chaires, J.B., 1999. Effects of hydration, ion release, and excluded volume on the melting of triplex and duplex DNA. Biochemistry 38, 496–508.
Spink, C.H., Garbett, N., Chaires, J.B., 2007. Enthalpies of DNA melting in the presence of osmolytes. Biophys. Chem. 126, 176–185.
Srere, P.A., 1980. The infrastructure of the mitochondrial matrix. Trends Biochem. Sci. 5, 120–121.
Srere, P.A., 1981. Protein crystals as a model for mitochondrial matrix proteins. Trends Biochem. Sci. 6, 4–7.
Stanley, C.B., Strey, H.H., 2008. Osmotically induced helix-coil transition in poly(glutamic acid). Biophys. J. 94, 4427–4434.
Stellwagen, E., Dong, Q., Stellwagen, N.C., 2007. Quantitative analysis of monovalent counterion binding to random-sequence, double-stranded DNA using the replacement ion method. Biochemistry 46, 2050–2058.
Stoddard, C.D., Montange, R.K., Hennelly, S.P., Rambo, R.P., Sanbonmatsu, K.Y., Batey, R.T., 2010. Free state conformational sampling of the SAM-I riboswitch aptamer domain. Structure 18, 787–797.
Sugimoto, N., Kierzek, R., Turner, D.H., 1988. Kinetics for reaction of a circularized intervening sequence with CU, UCU, CUCU, and CUCUCU: mechanistic implications from the dependence on temperature and on oligomer and Mg2+ concentrations. Biochemistry 27, 6384–6392.
Sugimoto, N., Tomka, M., Kierzek, R., Bevilacqua, P.C., Turner, D.H., 1989. Effects of substrate structure on the kinetics of circle opening reactions of the self-splicing intervening sequence from Tetrahymena thermophila: evidence for substrate and Mg2+ binding interactions. Nucleic Acids Res. 17, 355–371.
Sugimoto, N., Nakano, S., Katoh, M., Matsumura, A., Nakamuta, H., Ohmichi, T., Yoneyama, M., Sasaki, M., 1995. Thermodynamic parameters to predict stability of RNA/DNA hybrid duplexes. Biochemistry 34, 11211–11216.
Sugimoto, N., Nakano, S., Yoneyama, M., Honda, K., 1996. Improved thermodynamic parameters and helix initiation factor to predict stability of DNA duplexes. Nucleic Acids Res. 24, 4501–4505.
Sugimoto, N., Nakano, M., Nakano, S., 2000. Thermodynamics-structure relationship of single mismatches in RNA/DNA duplexes. Biochemistry 39, 11270–11281.
Tahirov, T.H., Temiakov, D., Anikin, M., Patlan, V., McAllister, W.T., Vassylyev, D.G., Yokoyama, S., 2002. Structure of a T7 RNA polymerase elongation complex at 2.9 A resolution. Nature 420, 43–50.
Takyar, S., Hickerson, R.P., Noller, H.F., 2005. mRNA helicase activity of the ribosome. Cell 120, 49–58.
Tanizaki, S., Clifford, J., Connelly, B.D., Feig, M., 2008. Conformational sampling of peptides in cellular environments. Biophys. J. 94, 747–759.

Tateishi-Karimata, H., Sugimoto, N., 2012. A-T base pairs are more stable than G-C base pairs in a hydrated ionic liquid. Angew. Chem. Int. Ed. Engl. 51, 1416–1419. http://dx.doi.org/10.1002/anie. 201108253.

Taton, T.A., Mirkin, C.A., Letsinger, R.L., 2000. Scanometric DNA array detection with nanoparticle probes. Science 289, 1757–1760.

Teixeira, A., Tahiri-Alaoui, A., West, S., Thomas, B., Ramadass, A., Martianov, I., Dye, M., James, W., Proudfoot, N.J., Akoulitchev, A., 2004. Autocatalytic RNA cleavage in the human beta-globin pre-mRNA promotes transcription termination. Nature 432, 526–530.

Thanaraj, T.A., Argos, P., 1996. Ribosome-mediated translational pause and protein domain organization. Protein Sci. 5, 1594–1612.

Tobe, S., Heams, T., Vergne, J., Herve, G., Maurel, M.C., 2005. The catalytic mechanism of hairpin ribozyme studied by hydrostatic pressure. Nucleic Acids Res. 33, 2557–2564.

Tokuriki, N., Kinjo, M., Negi, S., Hoshino, M., Goto, Y., Urabe, I., Yomo, T., 2004. Protein folding by the effects of macromolecular crowding. Protein Sci. 13, 125–133.

Toor, N., Keating, K.S., Taylor, S.D., Pyle, A.M., 2008. Crystal structure of a self-spliced group II intron. Science 320, 77–82.

Tornaletti, S., Park-Snyder, S., Hanawalt, P.C., 2008. G4-forming sequences in the non-transcribed DNA strand pose blocks to T7 RNA polymerase and mammalian RNA polymerase II. J. Biol. Chem. 283, 12756–12762.

Toulokhonov, I., Artsimovitch, I., Landick, R., 2001. Allosteric control of RNA polymerase by a site that contacts nascent RNA hairpins. Science 292, 730–733.

Tsai, C.J., Sauna, Z.E., Kimchi-Sarfaty, C., Ambudkar, S.V., Gottesman, M.M., Nussinov, R., 2008. Synonymous mutations and ribosome stalling can lead to altered folding pathways and distinct minima. J. Mol. Biol. 383, 281–291.

Tsuji, A., Yoshikawa, K., 2010. ON-OFF switching of transcriptional activity of large DNA through a conformational transition in cooperation with phospholipid membrane. J. Am. Chem. Soc. 132, 12464–12471.

Turner, D.H., Sugimoto, N., Freier, S.M., 1988. RNA structure prediction. Annu. Rev. Biophys. Biophys. Chem. 17, 167–192.

Uhlenbeck, O.C., 1987. A small catalytic oligoribonucleotide. Nature 328, 596–600.

Valentine, K.G., Peterson, R.W., Saad, J.S., Summers, M.F., Xu, X., Ames, J.B., Wand, A.J., 2010. Reverse micelle encapsulation of membrane-anchored proteins for solution NMR studies. Structure 18, 9–16.

Van Horn, W.D., Ogilvie, M.E., Flynn, P.F., 2009. Reverse micelle encapsulation as a model for intracellular crowding. J. Am. Chem. Soc. 131, 8030–8039.

Vasudevamurthy, M.K., Lever, M., George, P.M., Morison, K.R., 2009. Betaine structure and the presence of hydroxyl groups alters the effects on DNA melting temperatures. Biopolymers 91, 85–94.

Vazquez, A., Oltvai, Z.N., 2011. Molecular crowding defines a common origin for the Warburg effect in proliferating cells and the lactate threshold in muscle physiology. PLoS One 6, e19538.

Venkatesan, N., Seo, Y.J., Kim, B.H., 2008. Quencher-free molecular beacons: a new strategy in fluorescence based nucleic acid analysis. Chem. Soc. Rev. 37, 648–663.

Vijayaraghavan, R., Izgorodin, A., Ganesh, V., Surianarayanan, M., MacFarlane, D.R., 2010. Long-term structural and chemical stability of DNA in hydrated ionic liquids. Angew. Chem. Int. Ed. Engl. 49, 1631–1633.

von Hippel, P.H., 2006. Transcriptional pausing caught in the act. Cell 125, 1027–1028.

Wan, Y., Kertesz, M., Spitale, R.C., Segal, E., Chang, H.Y., 2011. Understanding the transcriptome through RNA structure. Nat. Rev. Genet. 12, 641–655.

Wand, A.J., Ehrhardt, M.R., Flynn, P.F., 1998. High-resolution NMR of encapsulated proteins dissolved in low-viscosity fluids. Proc. Natl. Acad. Sci. U. S. A. 95, 15299–15302.

Wang, G., Vasquez, K.M., 2006. Non-B DNA structure-induced genetic instability. Mutat. Res. 598, 103–119.
Wang, G., Vasquez, K.M., 2007. Z-DNA, an active element in the genome. Front. Biosci. 12, 4424–4438.
Waters, L.S., Storz, G., 2009. Regulatory RNAs in bacteria. Cell 136, 615–628.
Watkins Jr., N.E., Kennelly, W.J., Tsay, M.J., Tuin, A., Swenson, L., Lee, H.R., Morosyuk, S., Hicks, D.A., Santalucia Jr., J., 2011. Thermodynamic contributions of single internal rA.dA, rC.dC, rG.dG and rU.dT mismatches in RNA/DNA duplexes. Nucleic Acids Res. 39, 1894–1902.
Watts, J.M., Dang, K.K., Gorelick, R.J., Leonard, C.W., Bess Jr., J.W., Swanstrom, R., Burch, C.L., Weeks, K.M., 2009. Architecture and secondary structure of an entire HIV-1 RNA genome. Nature 460, 711–716.
Wells, R.D., 2007. Non-B DNA conformations, mutagenesis and disease. Trends Biochem. Sci. 32, 271–278.
Welton, T., 1999. Room-temperature ionic liquids. Solvents for synthesis and catalysis. Chem. Rev. 99, 2071–2084.
Wen, J.D., Lancaster, L., Hodges, C., Zeri, A.C., Yoshimura, S.H., Noller, H.F., Bustamante, C., Tinoco, I., 2008. Following translation by single ribosomes one codon at a time. Nature 452, 598–603.
Westhof, E., 1988. Water: an integral part of nucleic acid structure. Annu. Rev. Biophys. Biophys. Chem. 17, 125–144.
Wilson, W.D., Li, K., 2000. Targeting RNA with small molecules. Curr. Med. Chem. 7, 73–98.
Wimberly, B.T., Brodersen, D.E., Clemons Jr., W.M., Morgan-Warren, R.J., Carter, A.P., Vonrhein, C., Hartsch, T., Ramakrishnan, V., 2000. Structure of the 30S ribosomal subunit. Nature 407, 327–339.
Winkler, W.C., Nahvi, A., Roth, A., Collins, J.A., Breaker, R.R., 2004. Control of gene expression by a natural metabolite-responsive ribozyme. Nature 428, 281–286.
Wong, B., Chen, S., Kwon, J.A., Rich, A., 2007. Characterization of Z-DNA as a nucleosome-boundary element in yeast Saccharomyces cerevisiae. Proc. Natl. Acad. Sci. U. S. A. 104, 2229–2234.
Woolley, P., Wills, P.R., 1985. Excluded-volume effect of inert macromolecules on the melting of nucleic acids. Biophys. Chem. 22, 89–94.
Workman, H., Flynn, P.F., 2009. Stabilization of RNA oligomers through reverse micelle encapsulation. J. Am. Chem. Soc. 131, 3806–3807.
Xia, T., SantaLucia Jr., J., Burkard, M.E., Kierzek, R., Schroeder, S.J., Jiao, X., Cox, C., Turner, D.H., 1998. Thermodynamic parameters for an expanded nearest-neighbor model for formation of RNA duplexes with Watson-Crick base pairs. Biochemistry 37, 14719–14735.
Xue, Y., Kan, Z.Y., Wang, Q., Yao, Y., Liu, J., Hao, Y.H., Tan, Z., 2007. Human telomeric DNA forms parallel-stranded intramolecular G-quadruplex in K+ solution under molecular crowding condition. J. Am. Chem. Soc. 129, 11185–11191.
Yancey, P.H., Clark, M.E., Hand, S.C., Bowlus, R.D., Somero, G.N., 1982. Living with water stress: evolution of osmolyte systems. Science 217, 1214–1222.
Yen, L., Svendsen, J., Lee, J.S., Gray, J.T., Magnier, M., Baba, T., D'Amato, R.J., Mulligan, R.C., 2004. Exogenous control of mammalian gene expression through modulation of RNA self-cleavage. Nature 431, 471–476.
Yin, Y.W., Steitz, T.A., 2002. Structural basis for the transition from initiation to elongation transcription in T7 RNA polymerase. Curr. Opin. Struct. Biol. 14, 4–9.
Zaug, A.J., Cech, T.R., 1995. Analysis of the structure of Tetrahymena nuclear RNAs in vivo: telomerase RNA, the self-splicing rRNA intron, and U2 snRNA. RNA 1, 363–374.
Zemora, G., Waldsich, C., 2010. RNA folding in living cells. RNA Biol. 7, 634–641.

Zhang, G., Ignatova, Z., 2009. Generic algorithm to predict the speed of translational elongation: implications for protein biogenesis. PLoS One 4, e5036.
Zhang, G., Ignatova, Z., 2011. Folding at the birth of the nascent chain: coordinating translation with co-translational folding. Curr. Opin. Struct. Biol. 21, 25–31.
Zhang, G., Hubalewska, M., Ignatova, Z., 2009. Transient ribosomal attenuation coordinates protein synthesis and co-translational folding. Nat. Struct. Mol. Biol. 16, 274–280.
Zhang, D.H., Fujimoto, T., Saxena, S., Yu, H.Q., Miyoshi, D., Sugimoto, N., 2010. Monomorphic RNA G-quadruplex and polymorphic DNA G-quadruplex structures responding to cellular environmental factors. Biochemistry 49, 4554–4563.
Zheng, K.W., Chen, Z., Hao, Y.H., Tan, Z., 2010. Molecular crowding creates an essential environment for the formation of stable G-quadruplexes in long double-stranded DNA. Nucleic Acids Res. 38, 327–338.
Zhong, M., Marky, L.A., Kallenbach, N.R., Kupke, D.W., 1997. Thermodynamics of dT–dT base pair mismatching in linear DNA duplexes and three-arm DNA junctions. Biochemistry 36, 2485–2491.
Zhou, H.X., Rivas, G., Minton, A.P., 2008. Macromolecular crowding and confinement: biochemical, biophysical, and potential physiological consequences. Annu. Rev. Biophys. 37, 375–397.
Zhou, J., Wei, C., Jia, G., Wang, X., Feng, Z., Li, C., 2009. Human telomeric G-quadruplex formed from duplex under near physiological conditions: spectroscopic evidence and kinetics. Biochimie 91, 1104–1111.
Zhou, J., Wei, C., Jia, G., Wang, X., Feng, Z., Li, C., 2010. Formation and stabilization of G-quadruplex in nanosized water pools. Chem. Commun. 46, 1700–1702.
Zimmerman, S.B., Murphy, L.D., 1996. Macromolecular crowding and the mandatory condensation of DNA in bacteria. FEBS Lett. 390, 245–248.
Zimmerman, S.B., Trach, S.O., 1991. Estimation of macromolecule concentrations and excluded volume effects for the cytoplasm of eschericha coli. J. Mol. Biol. 222, 599–620.
Zuker, M., 2003. Mfold web server for nucleic acid folding and hybridization prediction. Nucleic Acid Res. 31, 3406–3415.

CHAPTER NINE

Computational Models of Large-Scale Genome Architecture

Angelo Rosa[*,1], **Christophe Zimmer**[†,1]
[*]Scuola Internazionale Superiore di Studi Avanzati, Trieste, Italy
[†]Institut Pasteur, Unité Imagerie et Modélisation; CNRS URA 2582, Paris, France
[1]Corresponding authors: e-mail address: anrosa@sissa.it; czimmer@pasteur.fr

Contents

Abstract

The spatial architecture and dynamics of the genomic material in the limited volume of the nucleus plays an important role in biological processes ranging from gene expression to DNA repair. Yet, detailed descriptions of dynamic genome architecture are still lacking and its governing principles and functional implications remain largely unknown. Powerful experimental methods have been developed to address this gap, including single-cell imaging and chromosome conformation capture methods, leading to rapidly growing quantitative data sets. Despite their importance, however, these data are insufficient to provide a full understanding of genome architecture and function. Computational models are becoming an increasingly indispensable complement in order to make sense of the experimental data and to allow a quantitative understanding of how chromosomes fold, move and interact. Here, we review efforts, developed over the last ~25 years, to model the large-scale 3D organization and dynamics of chromosomes or genomes quantitatively. We discuss models based on theories and simulations of polymer physics or computational reconstruction methods, highlighting similarities and differences between models, as well as limitations and possible improvements.

International Review of Cell and Molecular Biology, Volume 307
ISSN 1937-6448
http://dx.doi.org/10.1016/B978-0-12-800046-5.00009-6

1. INTRODUCTION

Eukaryotic genomes are organized in discrete functional units called chromosomes, which consist of a single filament of the DNA double helix wound around complexes of histone proteins called nucleosomes (Alberts et al., 2008). During mitosis, chromosomes assume a characteristic rod-like shape, which make them suitable for transport inside the cytoplasm. At the onset of interphase, chromosomes are sequestered and swell inside the nuclear compartment, whose size is much smaller than the length of the DNA if it was stretched out: in human cells, roughly 2 m of linear DNA are folded in a nucleus of ~10 μm diameter. Much current research is directed at describing and understanding the folding and the spatial arrangements of chromosomes in the nucleus. Interest in this field is driven by a host of evidence that spatial genome architecture matters for how cells work or go awry. Indeed, chromosome organization and its alterations have been implicated in various aspects of gene expression, DNA repair, recombination, and replication in organisms ranging from yeast to human and in biomedically important processes such as translocations in cancer cells or accelerated aging caused by defects in the nuclear lamina (Misteli, 2010). For recent reviews on the functional importance of genome architecture see for example, Bickmore and van Steensel (2013), Cavalli and Misteli (2013), and Edelman and Fraser (2012).

Current investigations of genome architectures leverage powerful experimental methods, which can be divided into two broad categories: imaging methods and biochemical techniques. Imaging by electron microscopy allows uniquely detailed visualizations of regions of condensed versus decondensed chromatin in individual cell nuclei and even to analyze chromatin fiber structure *in situ* (Eltsov et al., 2008), but mostly lacks the ability to highlight specific DNA sequences. Light microscopy, by contrast, in combination with fluorescent tags, can visualize specific chromatin loci or chromosomes, although at lower resolution, and enables live cell imaging of chromatin dynamics (Belmont, 2001; Cremer and Cremer, 2010; Gasser, 2002). Super-resolution light microscopy, correlative methods combining light and electron microscopy on the same sample, and combinatorial labeling methods to distinguish many different loci simultaneously offer important new perspectives for imaging genome architecture. Despite these promises, the throughput of imaging methods currently allows to analyze only a limited number of loci. This limitation is addressed by the emergence of very different experimental approaches based on chromosome conformation capture (3C), a biochemical method that allows to detect and quantify physical contacts between DNA segments (Dekker et al., 2002). This is achieved by analyzing chimeric DNA sequences obtained after an experimental protocol that involves fixation, cross-linking, digestion by restriction enzymes, and dilution. In recent years, this approach has been extended to the analysis of contacts across the entire genome, yielding massive quantitative data about genome architecture at the level of large cell populations (De Laat and Dekker, 2012; Lieberman-Aiden et al., 2009). For recent reviews on imaging nuclear architecture see for example, Flors (2011), Markaki et al. (2012), and Rouquette et al. (2010); 3C-based techniques and insights derived from them have been reviewed recently in De Laat and Dekker (2012), Dostie and Bickmore (2012), and De Wit and De Laat (2012).

Although advances in experimental studies have been impressive, it is increasingly clear that computational models are an indispensable complement for a better understanding of genome architecture (Dekker et al., 2013). Indeed, none of the available experimental methods directly provides a detailed quantitative description of how chromosomes fold, move, and interact within the nucleus. Moreover, mere descriptions, however accurate, do not directly reveal the mechanisms that govern genome architecture or predict how this architecture changes in different conditions or organisms. Computational approaches have the potential to address these gaps.

The study of chromosome organization is a particularly attractive field for physicists and computational researchers and has motivated the development of two broad types of models (Marti-Renom and Mirny, 2011). One reason for this attraction is that chromosomes are long polymers. As such, the study of chromosomes can capitalize on a large body of preexisting theoretical and computational work in statistical physics of polymers. Models based on these theories have the unique potential to offer predictive mechanistic insights into the architecture of chromosomes at a quantitative level (Fig. 9.1, top). The availability of large quantitative data sets from chromosome conformation capture or imaging techniques offers the possibility to test theoretical models in detail, allowing to discriminate among competing models and to improve them. Another approach, leading to a second broad group of models, takes inspiration from structural biology methods more traditionally used for protein structure determination. In these models, the experimental data are used to reconstruct detailed 3D chromosome configurations *in silico* that are as consistent as possible with these data (Fig. 9.1, bottom). These models can then be used to learn about other, not directly observable, features of genomic architecture.

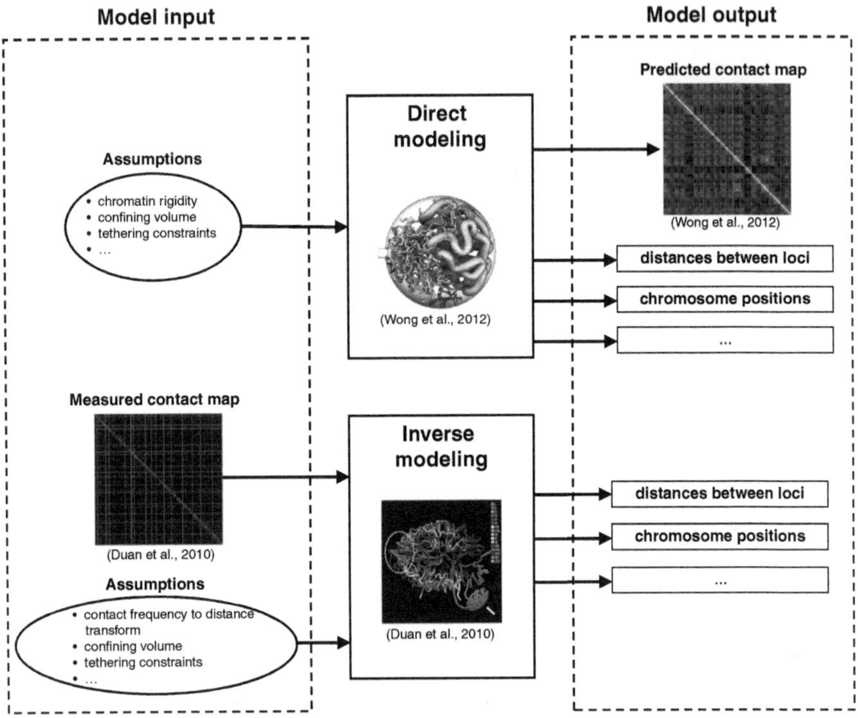

Figure 9.1 Two broad approaches to computational modeling of genome architecture. Direct modeling approaches (top) use a relatively small number of assumptions and quantitative parameters. The behavior of chromosomes is then typically computed using polymer physics theories and/or numerical simulations. Such models can be used to predict a variety of observable quantities such as contact frequency maps or distances between loci, which can then be compared to experimental measurements. Models of this type are reviewed in Section **2**. Inverse modeling approaches (bottom) use rich experimental data sets (such as genome-wide contact matrices measured by chromosome conformation capture) to reconstruct chromosome structures. Optimization methods are typically used to determine structures that are as much as possible consistent with these experimental data and with a small number of data-independent assumptions (some of which may be similar to those used to build the direct models). These reconstructed models can then be used to determine quantities not directly accessible in the experimental data, such as positions of loci or chromosomes. Models of this type are reviewed in Section **3**. *Figures are reprinted with permission from Duan et al. (2010) © (2010) Macmillan Publishers Ltd and from Wong et al. (2012) © (2012) with permission from Elsevier.* (For color version of this figure, the reader is referred to the online version of this chapter.)

In this chapter, we attempt to provide an overview of a representative selection of computational models of genome architectures that have been developed over the last ~25 years. Some of these models may seem similar, but in some cases in fact differ widely in their internal make-up, their underlying assumptions, the numerical or mathematical tools used for computations, and the experimental data they attempt to explain or describe, or the data that are used to construct the model. We therefore tried to emphasize what each model assumes, how it works, and what it can (or cannot) explain.

A discussion of quantitative chromosome models requires reference to concepts and terms from statistical and computational physics and introduction of quantitative relationships. However, in order to keep the review readable for a larger audience, we provide a brief introduction to polymer physics in Box 9.1, define and discuss these more technical aspects separately in Box 9.2, and refer to these in the main text using asterisks (*).

2. DIRECT MODELS OF GENOME ARCHITECTURE

In this section, we review theoretical models that rely on minimal physical assumptions and a very limited number of parameters. In these "direct" models, the behavior of chromosomes is essentially deduced from the laws of polymer physics. This is in contrast with "inverse" models (see Section 3), where simulated chromosome conformations are instead deduced from experimental data, via the satisfaction of data-derived constraints imposed on a 3D model of chromatin fibers. Polymer physics models aim to explain essential properties of the large-scale organization and dynamics of chromosomes from the mere fact that they consist of extremely long semi-flexible filaments. At first, this might seem like a hopeless endeavor. Indeed, given the complexity of chromatin on the molecular scale (Kouzarides, 2007), it is not *a priori* obvious that polymer physics—which entirely neglects this complexity—has relevant quantitative or even qualitative insights to offer.

2.1. Random polymers and chromosome size

Before describing any specific model in more detail, we begin by illustrating simple concepts from polymer physics and their relevance to genome architecture. Specifically, we discuss how basic quantitative results from polymer theory bear on a simple but important biological property: the spatial extent of chromosomes. In polymer physics, this size is often measured by the square root of the average (squared) end-to-end* distance $\langle R^2(L)\rangle^{1/2}$ as a function of the chain contour* length L.

In general, as mentioned in the introduction, eukaryotic chromosomes are linear chromatin fibers which, if stretched out, would be much longer than the nucleus. Hence, a suitable starting point for describing their physical properties is a dense solution (a melt*) of linear polymer chains (Kreth et al., 2004).

The spatial arrangements (conformations) of such polymers are known to be described by the freely* jointed chain (FJC) (Fig. 9.22), which models the chain as a random walk of constant step length, or the closely related worm-like* chain (WLC) model (Doi and Edwards, 1988). The average size of a FJC or WLC is given by $\langle R^2(L)\rangle^{1/2} = (L \times L_K)^{1/2}$, where L_K is the Kuhn* length, a measure of the chain bending rigidity. In order to apply this relationship to estimate chromosome sizes, we need to specify these two parameters, both of which depend on the still debated structure of the chromatin fiber (Maeshima et al., 2010). We will therefore consider three cases by assuming that each chromosome is described, either as a homogeneous long filament of naked DNA, a 10 nm chromatin fiber or a 30 nm fiber.

The genomic* length s associated with an average human chromosome is of the order of $s \sim 10^8$ basepairs (bp). Because 3 bp of naked DNA span approximately 1 nm in length (Alberts et al., 2008), this implies a contour* length $L \sim 30$ mm. Since the nominal bending rigidity of DNA is approximately $L_K \sim 100$ nm (Marko and Siggia, 1995), a single, chromosome-sized molecule of DNA would span a region of linear size $\langle R^2(L)\rangle^{1/2} \sim (0.1\ \mu\text{m} \times 30{,}000\ \mu\text{m})^{1/2} \sim 55\ \mu\text{m}$, which exceeds the typical nuclear size of $\sim 10\ \mu\text{m}$.

In eukaryotic nuclei, however, chromosomal DNA exists in a complex with histone proteins which results in a tighter compaction of DNA in the chromatin fiber (Alberts et al., 2008). The main structural unit of the chromatin fiber, the nucleosome, is formed by ~150 bp of DNA wrapped around eight histone proteins in a spool-like configuration of ~10 nm of diameter. Consecutive spools are linked to each other by ~50 bp of DNA (linker DNA). The overall spatial extent of nucleosome plus linker DNA is ~25 nm, thus implying a ~2–3 fold compaction compared to naked DNA. In textbooks (Alberts et al., 2008), this chain of nucleosomes is commonly known as the "10 nm" chromatin fiber or "beads on a string" structure (Kornberg, 1974). Assuming that the 10 nm fiber can be described as a flexible chain of nucleosomes ($L_K = 25$ nm), we find that a 10 nm fiber chromosome would have a size $\langle R^2(L)\rangle^{1/2} \sim (0.025\ \mu\text{m} \times 10{,}000\ \mu\text{m})^{1/2} \sim 16\ \mu\text{m}$.

In vitro, the 10 nm fiber appears to fold into a thicker fiber with a diameter of ~30 nm (Alberts et al., 2008). In this state, DNA acquires ~30 times more compaction compared to the simple double helix (Maeshima et al., 2010). Thus, the typical contour length of a human chromosome made up of 30 nm chromatin fiber is $L \sim 1$ mm. Assuming that the bending rigidity of the 30 nm fiber is such that $L_K \sim 340$–440 nm (Bystricky et al., 2004), human chromosomes would span a region of size $\langle R^2(L) \rangle^{1/2} \sim (0.34\text{–}0.44\ \mu\text{m} \times 1000\ \mu\text{m})^{1/2} \sim 18\text{–}21\ \mu\text{m}$.

Provided that the rigidities assumed above are roughly correct, this simple exercise suggests that the expected size of the chromosome only weakly depends on its exact structure, and predicts a size in the range of a few tens of μms. Thus, it appears that, intriguingly, the small-scale organization of chromatin (at the scale of a few nucleosomes) might have little influence on chromosome behavior at the micrometer scale. Another prediction is that each chromosome is expected to occupy the entire nuclear volume. This, however, highlights a first shortcoming of this simple model: in fact, experimental studies have shown that different chromosomes tend to occupy distinct nuclear territories, approximately ~1 μm in size, which represents only a small fraction of the typical nuclear diameter (Cremer and Cremer, 2001). Thus, this exercise fails to reproduce the known behavior of chromosomes during interphase.

Despite this failure of the simplest random polymer model, is it still possible to understand chromosome behavior, at least in part, using the language and theories of polymer physics? Many studies have attempted to provide answers to this question, several of which are reviewed in the following subsections. We have classified these polymer models in three categories: those that invoke (protein-mediated) intrachromosomal interactions (*cis* interactions) to account for chromosome folding (Section 2.2), those that model chromosome folding as the result of topological interactions (Section 2.3), and those that introduce some DNA sequence specificity (Section 2.4).

2.2. Chromosome folding by *cis* interactions

As discussed above, the size of chromosomes as predicted by the simplest random polymer models is too large to account for observations. Therefore, investigators looked for ways to induce a more compact folding of chromosomes modeled as polymers. One possibility is to introduce physical links between distant monomers of the same polymer chain (*cis* interactions), leading to the formation of chromatin loops. The compaction of the chromosome then depends on the number and size of the loops. A sufficiently large number of loops may yield considerable compaction, and loops have thus been advocated as a possibility to explain the formation of chromosome territories. We discuss individual chromosome models based on looped polymers in the following subsections.

2.2.1 Random-walk/giant-loop model (RW/GL)

In 1995, two studies reported important results from fluorescence *in situ* hybridization (FISH) experiments that measured the average squared spatial distances, $\langle R^2(s) \rangle$, between tens of fluorescently labeled loci positioned at various genomic intervals s on human chromosome 4 (Sachs et al., 1995; Yokota et al., 1995). Since this quantity is also a standard observable in polymer physics, by comparing experimental data to known polymer models, one should be able to deduce important features of chromosome folding. Interestingly, it was found that the increase of $\langle R^2(s) \rangle$ with s showed two very distinct linear slopes for genomic separations smaller or larger than ~3 Mbp (Fig. 9.2B, C) (Sachs et al., 1995).

This led the authors to formulate a new phenomenological polymer model, which they baptized "random-walk/giant-loop" (RW/GL) model. The model is a generalization of the polymer description as a random walk provided, for example, by the FJC. In the RW/GL model, flexible chromatin loops are attached one after the other along a hypothetical random-walk-like protein scaffold (Fig. 9.2A). The model, which was formulated analytically as an equation relating $\langle R^2(s) \rangle$ to s, contained essentially three free parameters: (i) the flexibility of the loops, measured by the Kuhn* length, $L_{K,loop}$; (ii) the flexibility of the backbone, given by its own Kuhn length, $L_{K,backbone}$; and (iii) the genomic length of the loops, d_0. According to this model, spatial distances measured between FISH-labeled chromatin loci at small genomic separations s arise from the loop structures (Fig. 9.2B), while distances between loci at large genomic separations reflect the flexibility of the giant backbone (Fig. 9.2C). The model parameters were determined by fitting the analytical expression to the FISH data. The fitted parameters (originally expressed by different quantities) correspond to a loop flexibility $L_{K,loop} \sim 300$ nm, a backbone flexibility of $L_{K,backbone} \sim 500$–600 nm, and a loop size of $d_0 \sim 3$ Mbp.

An essential feature of the RW/GL model is the presence of giant loops of chromatin protruding from an underlying backbone, the nature of which remains to be clarified.

2.2.2 Multi-loop/subcompartment model (MLS)

A straightforward consequence of the RW/GL model was that $\langle R^2(s) \rangle$ should be proportional to genomic separation s for either short or large s (with different proportionality constants), reflecting the random behavior of the loops and the assumed backbone, respectively. However, a reanalysis of the above-mentioned FISH data (Sachs et al., 1995) concluded that for

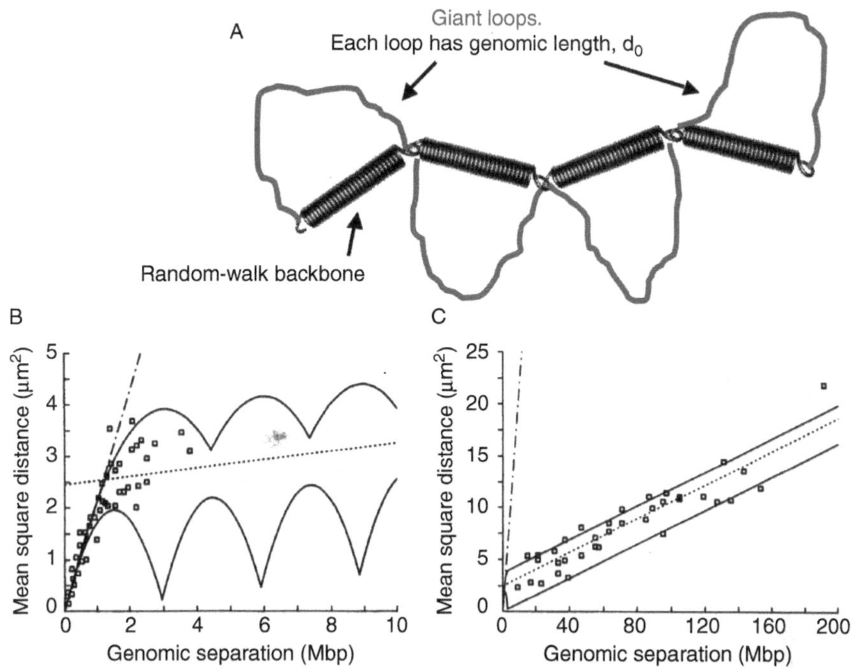

Figure 9.2 Random walk giant-loop model. (A) Schematic illustration of the random-walk/giant-loop model. Four chromatin loops (red), each of genomic length $d_0=3$ Mbp, protrude from a random-walk backbone supposed to be formed by protein links (schematically drawn as a sequence of springs). (B, C) Physical distances and genomic separations. Both plots show average squared spatial distances measured between FISH-labeled loci (symbols) on human chromosome 4. Panel B shows data for loci located within $a \sim 4$ Mbp-wide region at one chromosome end (chromosome region 4p16.3). Panel (C) shows data for loci scattered over the entire chromosome length (192 Mbp). Dot-dashed and dotted lines show the apparent slopes at small and large genomic separations, respectively. Solid curves in (B) show the upper and lower bounds predicted by the RW/GL model. These are distinct because spatial distances between loci depend sensitively on their specific intra-loop locations. Solid lines in (C) are the envelopes of these bounds. See Sachs et al. (1995) for details. *Panels (B) and (C) are reprinted with permission from Sachs et al. (1995) © (1995) National Academy of Sciences, USA.* (For interpretation of the references to color in this figure legend, the reader is referred to the online version of this chapter.)

large genomic separations s, the increase in spatial distances is better described by a more compact power law, with $\langle R^2(s)\rangle$ proportional to $s^{2/3}$ (Fig. 9.3, left) (Münkel and Langowski, 1998). This behavior cannot be explained by the RW/GL model, nor by any other random polymer model.

To address this, the authors proposed an alternative coarse-grained polymer model, the so-called "multi-loop/subcompartment" (MLS) model

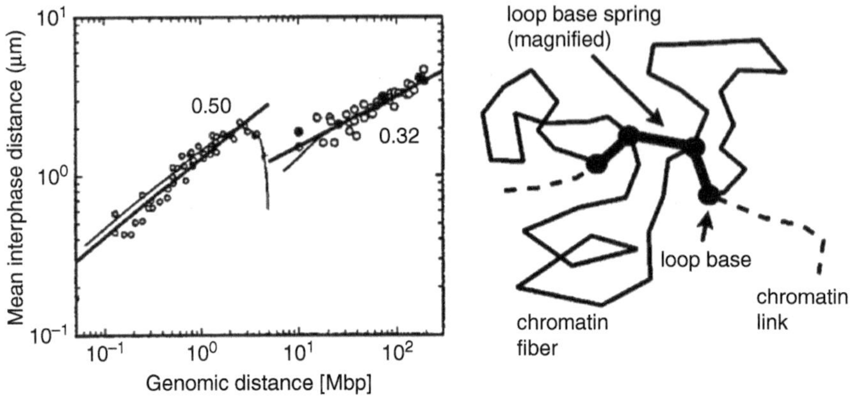

Figure 9.3 Multi-loop subcompartment (MLS) model. Left: Log–log plot of spatial distances between FISH-labeled chromosome loci as function of genomic separation s (symbols). The data are from Sachs et al. (1995) and were plotted differently (squared and using a linear scale) in Fig. 9.2B,C. The plot shown here highlights the random-walk-like small-scale behavior, characterized by an exponent 1/2, and a compact large-scale behavior, characterized by the nontrivial exponent 1/3. Thick lines show the corresponding power law behavior, and thin lines show the predictions of the MLS model. The data (and corresponding model curves) at short and large genomic distances correspond to different regions of the human chromosome 4. Empty and solid circles come from different experimental preparations (see Münkel et al., 1999, for details). Reprinted from Münkel et al. (1999) © (1999) with permission from Elsevier. Right: Schematic illustration of the MLS model. Compartments consisting of ~10 chromatin loops are connected to each other by chromatin linkers (dotted). Within each compartment, the MLS model is similar to the RW/GL model, except that loops are much smaller (~120 kbp). Different loops are connected by harmonic springs. In addition, chromosome swelling is hindered by an external potential that maintains the size of chromosomes to the single micrometer range, that is, the typical size of a chromosome territory. *Reprinted with permission from Münkel and Langowski (1998)) © (1998) by the American Physical Society.*

(Fig. 9.3, right). In this model, chromosomes are still organized as linear arrays of loops, but the genomic length associated with each loop is now smaller, namely 120 kbp. Approximately 10 chromatin loops form an Mbp-sized "subcompartment" and different subcompartments are connected by chromatin linkers (Münkel and Langowski, 1998). Much as in the RW/GL model, the molecular mechanisms responsible for creating and maintaining the loops were left open.

Another feature of the model by Münkel and Langowski (1998) is a soft repulsive barrier between spatially close chromatin monomers. Here, "soft" means that two monomers in close spatial proximity have a nonzero probability to pass through each other (as in a phantom* chain), allowing faster chain relaxation*. This aspect of the model was justified biologically by the presence of topoisomerase II, an enzyme capable of cutting the two strands of DNA, thereby enabling double-strand crossings. A technical advantage of assuming a soft interaction barrier is that chromosomes relax to an equilibrium much more rapidly than with a hard barrier (which would hinder *de facto* chain crossing) (De Gennes, 1979; De Gennes, 1971).

Without any further constraint, the MLS model would fail to predict chromosome territories, since the relaxed chain would not be ultimately very different from an ordinary (unlooped) random polymer, with a renormalized flexibility. In order to remedy this, the authors introduced an additional energy barrier that effectively confined each segment of a chromosome chain inside a micrometer-sized spherical territory. This final ingredient was needed to reproduce the $s^{2/3}$ power law mentioned above. The model was then studied numerically using Monte Carlo* simulations, but the authors mentioned that the model could also be efficiently studied with Brownian dynamics simulations (Münkel and Langowski, 1998).

The MLS model was found to well reproduce the experimental FISH data already mentioned (Fig. 9.3, left) (Münkel and Langowski, 1998). In a follow-up study, the model was also employed to describe compartmentalization of interphase chromosomes in early and late replicating chromatin domains (Münkel et al., 1999).

2.2.3 Micelles model

In the same year, Ostashevky also noted a contradiction between the experimentally observed nonoverlapping chromosome territories and the expected behavior of random-walk-like polymers, which are known to be highly overlapping (De Gennes, 1979) (Ostashevsky, 1998). Nevertheless, as discussed above, for genomic separations up to a few Mbp, chromosomes seem indeed to obey a simple random-chain behavior (Fig. 9.3, left). In order to resolve this apparent discrepancy, Ostashevsky proposed an alternative polymer model, where chromatin fibers are organized as linear strings of "micelles" (Fig. 9.4) (Ostashevsky, 1998). This model is based on the observation that mammalian chromosomes can be schematically represented as copolymers* containing alternating ~1 Mbp long blocks of different GC content: GC-rich "R-blocks" and GC-poor "G-blocks." Ostashevsky argued that these blocks corresponded to interphase R and G minibands detected after Giemsa staining (Alberts et al., 2008).

The formation of micelles in multi-block copolymers is a very well-known energy-favored process that occurs, for example, if monomers of

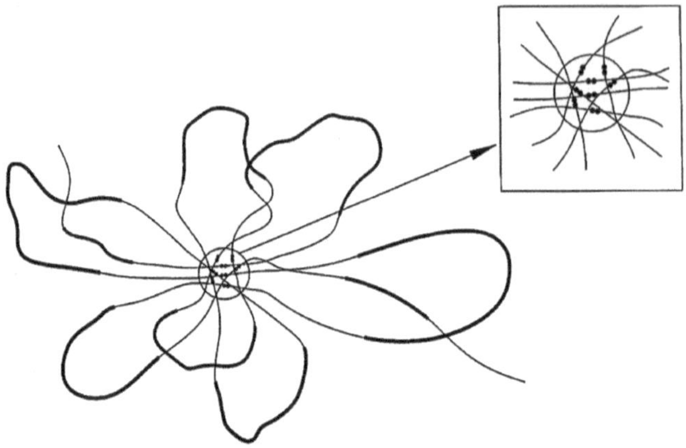

Figure 9.4 Chromatin micelles. Schematic illustration of the "micelles" model, in which chromosomes are modeled as block copolymers. Chemical differences due to the heterogeneous GC content of chromatin fibers leads to phase separation with the formation of nonoverlapping micelles. Thick and thin lines indicate G and R minibands, respectively. The circle represents the micelle core, where loop termini are located. Dots represent multi-protein complexes whose presence could further stabilize the micelle structures. *Reprinted with permission from Ostashevsky (1998) © (1998) by the American Society for Cell Biology.*

the same species attract each other, while monomers of different species repel each other (or attract each other less) (Halperin, 1991; Israelachvili, 2011). Monomers of the same species will then tend to cluster together and phase-separate from monomers of the other species. This process produces rosette-like structures known as "micelles," in which cores and loops are thus mainly populated by a different species of monomers. The cores are populated by the species of monomers for which the mutual attraction is strongest. For example, in aqueous solutions of copolymers made of hydrophobic and hydrophilic blocks, the hydrophobic groups, which tend to cluster to minimize interactions with the solvent, will tend to form the micelle cores, while hydrophilic groups will loop out from the cores. In a similar fashion, Ostashevsky suggested that incompatibility between GC-rich and AT-rich blocks might lead to "chromatin micelles." Since the size of each block is in the Mbp range, even a small difference in interaction energies between chromatin fibers from two different kinds of blocks can lead to phase separation: blocks of one kind will merge and form loop termini in the micelle core, while blocks of the other kind would move to loop apices (Fig. 9.4). Such differences among interaction energies can emerge, for instance, by chemical modifications as the consequence of, for example, histone acetylation (Ostashevsky, 1998).

Ostashevsky remarked that chromatin R-blocks are less condensed than G-blocks: hence, he naturally associated R-blocks to micelle cores and G-blocks to micelle loops. The author suggested that the micelle structures could be further stabilized by the presence of multi-protein complexes (see Fig. 9.4). For instance, because R-blocks have higher gene densities than G-blocks (Alberts et al., 2008), they could arguably display a higher concentration of transcription factors. This would enhance structure inhomogeneity further, and produce more stable micelles. Other protein factors acting during specific nuclear processes (transcription and replication, for instance) would act similarly.

Interestingly, the genomic size of the G and R-bands is ~1 Mbp, the scale up to which the FISH data are consistent with a random-chain behavior. According to Ostashevsky, this would also be the typical genomic length associated with one micellar loop.

Are there experimental facts supporting the "micellar" hypothesis? Indeed, micelles display many of the known features of interphase chromosomes. Notably, different micelles do not mix, as they tend to topologically repel each other, and the distances between pairs of loci at short genomic scales (below the ~1 Mbp size of a single loop) are predicted to exhibit random-chain behavior. Thus the model, which was formulated analytically, reconciles the observations of chromosome territories and the random-chain like trends in distances between FISH-labeled loci (Cremer and Cremer, 2010; Sachs et al., 1995; Yokota et al., 1995). Moreover, assuming that cores replicate earlier, the model implies that R-blocks replicate earlier than G-blocks, a prediction in agreement with experiments (Ostashevsky, 1998). Finally, the model predicts a number of micelle cores in the order of a few thousands, in line with the number of experimentally estimated replication foci in mammalian nuclei (Jackson, 1998). Thus, despite its simplicity, the micelle model captured a remarkable number of experimental features of chromosome organization.

2.2.4 A possible mechanism of loop formation: Depletion effects in genome organization

In the models discussed above, the formation of loops depended on specific energetically favored interactions between the loop termini. However, a different and generic mechanism for loop formation has been advocated based on an entropic effect caused by macromolecular crowding. Indeed, it can be estimated that nanometer-sized molecules occupy ~20% of the nuclear volume. Based on this simple observation, Marenduzzo et al. (2006) proposed that genome organization and looping of chromatin fibers might be driven by a well-known physical phenomenon known as "depletion attraction."

This effect can be easily explained with the example of a solution consisting of a mixture of small and big spheres (Fig. 9.5A). These spheres frequently collide each other because of Brownian motion. Consider an isolated large sphere (Fig. 9.5A, top left): this sphere is subject to collisions from small spheres coming from any direction of space; therefore, the net force exerted on it by these collisions is zero. However, if two big spheres are in close proximity, there is a spatial region between the two that is effectively "depleted" of small spheres. As a consequence, the net force exerted

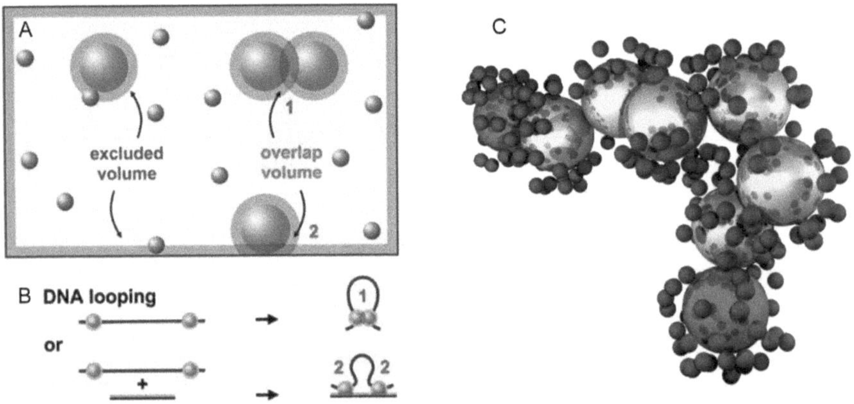

Figure 9.5 Depletion attraction and chromosome organization. (A, B) Schematic illustration of the depletion effect in a crowded environment. (A) Small and large spheres undergoing thermal Brownian motion coexist in a confined volume (rectangle). Because of excluded volume effects, the centers of small spheres cannot penetrate the gray corona surrounding the large spheres (or the gray stripe along the confining boundary). Isolated large spheres (top left) are hit by small spheres coming from all possible directions, thus collision forces cancel each other out. By contrast, when two large spheres touch each other (top right), small spheres cannot access the region shown in green, creating an asymmetry of collisions that results in a net force exerted on each large sphere toward the other. This force is called depletion attraction. A similar force pushes large spheres toward the boundary (bottom). (B) If the two large spheres are connected by a polymer chain, depletion might overcome the tendency of the polymer chain to swell and enhance looping. (C) The effects of depletion attraction were studied in a numerical model for a polymer chain made of large monomers (yellow and red spheres), and surrounded by a "sea" of small crowding agents (blue spheres). *Panels (A) and (B) are reprinted from Marenduzzo et al. (2006) © (2006) with permission from Elsevier. Panel (C) is reprinted with permission from Toan et al. (2006) © (2006) by the American Physical Society.* (For interpretation of the references to color in this figure legend, the reader is referred to the online version of this chapter.)

on one of the large spheres by the colliding small spheres is directed toward the other big sphere. This has the effect to keep the two big spheres glued together, as if they experienced an attractive interaction. (The same effect tends to glue the large spheres to the confining boundary.)

Moving to polymers, Marenduzzo et al. (2006) showed that when the two large spheres are part of the same chain, depletion attraction can counteract the natural tendency of the polymer chain to swell (Fig. 9.5B). A biological example of such a situation is provided by RNA polymerase II actively bound to chromosomal DNA during transcription and could potentially explain the presence of loops (even up to sizes of 1 Mbp) in eukaryotic chromosomes.

In a later study by the same group, the role of depletion attraction on the looping of a polymer chain was explored by a numerical model (Fig. 9.5C) (Toan et al., 2006). Specifically, the authors employed Brownian dynamics simulations of a self-avoiding* chain consisting entirely of large spheres, in absence and in presence of small spheres acting as crowding agents. This work demonstrated that crowding makes the chain more compact, although looping was affected differently for short and long chains. While crowding favors looping of long chains, the looping dynamics of short chains is mostly dominated by the large friction caused by the crowding agents, resulting in longer times for chain ends to meet. According to the authors, crowding might play a role in the formation of transcription or replication factories: when polymerases and other proteins bind to specific DNA sequences at multiple loci, the resulting increase in the local diameter of the fiber, in combination with nuclear crowding, could help concentrate these loci and the DNA processing machineries in nuclear foci (Cook, 1999).

2.2.5 Random-loop model (RL)

The RW/GL and MLS models summarized above (Sections 2.2.1 and 2.2.2) assumed that loops originate at specific loci along chromosomes and form at characteristic genomic scales (~1 Mbp for RW/GL, ~120 kbp for MLS). In both models, the loops were effectively "frozen" along the sequence, that is, the positions of their termini along the sequence did not change in the course of the simulations. The rearrangements of chromosomes thus proceeded only from the spatial displacements of chain monomers. By contrast, in 2007, it was suggested that chromatin loops forming at random loci and at all genomic scales are necessary to explain the observed folding of the chromatin fiber inside the nucleus (Bohn et al., 2007). In the "random loop" (RL) model proposed by the authors, different model chromosome configurations differ not only in the spatial positions of the monomers forming the chain, but also in the specific realization of the loops. In this model, loops are said to be "annealed" along the chain, that is, they form and disappear randomly (see Fig. 9.6A and B). Thus, observables need to be

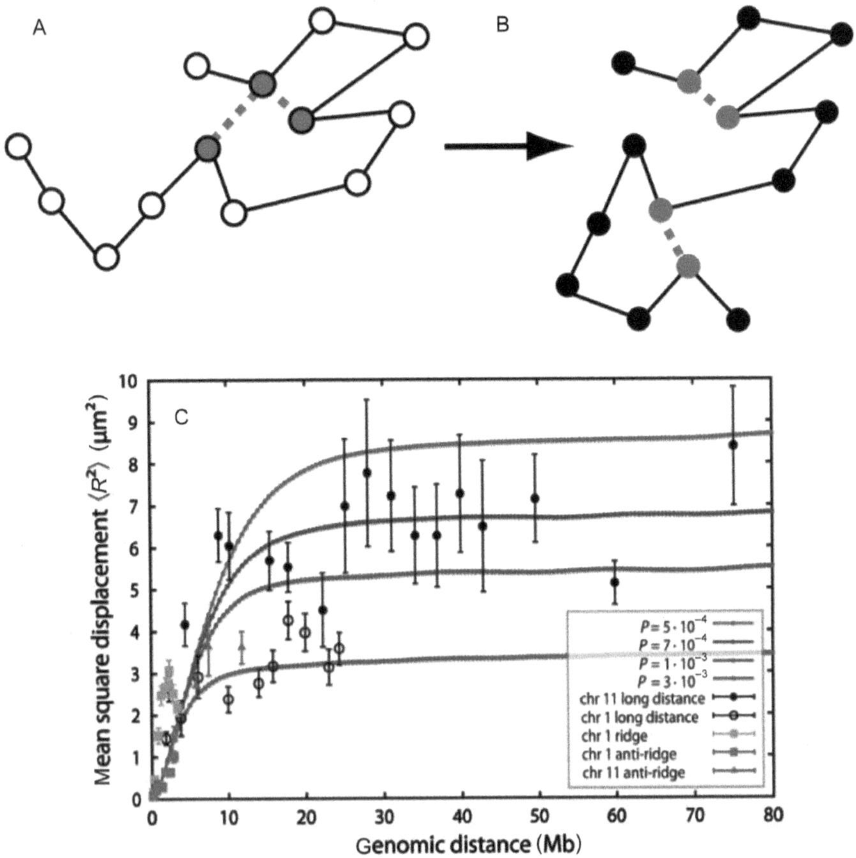

Figure 9.6 Random loop model. (A, B) In the RL model, and unlike in the RW/GL (Section 2.2.1) and MLS (Section 2.2.2) models, the position of loops (dashed red lines) along the chain can change over time. *Reprinted with permission from Bohn et al. (2007) © (2007) by the American Physical Society.* (C) Mean-square spatial distance between chromatin loci as function of the genomic distance. The plot shows theoretical predictions from the RL model for different values of the looping probability P (colored curves) and experimental FISH data (symbols). The simulated chain is made of 300 monomer beads. Given the total size of ~135 Mbp for human chromosome 11, each bead maps to ~450 kbp. Because predicted distances were initially in arbitrary units, they were multiplied by a constant to fit the experimental data. *Reprinted with permission from Mateos-Langerak et al. (2009) © (2009) National Academy of Sciences, USA.* (For interpretation of the references to color in this figure legend, the reader is referred to the online version of this chapter.)

averaged not only over all possible configurations of the chain at fixed loops positions, but also over all possible loop positions. The model features two new parameters: (1) the probability P that two monomers interact to form a loop; and (2) an upper bound for the contour* length of the loops. The RL model was first proposed in a version which did not account for excluded chromatin volumes, thus allowing for an analytically tractable solution (i.e., fully described by mathematical equations) (Bohn et al., 2007).

What is the experimental evidence for the RL model? Random loops on all scales were introduced essentially in order to reproduce the plateau in the dependence of internal distances, $\langle R^2(s) \rangle$ as a function of genomic* distance s at intermediate scales (25–75 Mbp), which had been observed in human primary female fibroblasts (Fig. 9.6C, full black circles) (Mateos-Langerak et al., 2009). Solid lines in Fig. 9.6C show the results of Brownian dynamics simulations for a RL polymer model made up of 300 beads, which explicitly accounts for excluded volume interactions (Mateos-Langerak et al., 2009). The relative agreement between the RL model and the experimental data is apparent from Fig. 9.6C.

The model was later generalized into the so-called Dynamic Loop (DL) model with the goal of providing a theoretical framework for *mitotic* (rather than interphase) chromosomes (Zhang and Heermann, 2011). In the DL model, the protein–chromatin interactions that underly looping were included by a probabilistic and dynamic mechanism. Random polymer chains and their internal rearrangements were simulated by Monte Carlo* simulations. When in the course of the simulations, two chain monomers came close to each other, a physical bond was placed between them with a certain probability, P, which is a tunable parameter of the model. This bond was not permanent, but was given a finite lifetime characterized by an additional free parameter. As in the RL model, loops were not allowed to form beyond a maximum contour* length.

The results from this study supported the idea that chromatin loops might explain the tight compaction and bending rigidity of mitotic chromosomes, and suggested that the internal structure of mitotic chromosomes is based on self-organization of the chromatin fiber rather than attachment of chromatin to a protein scaffold. In this way, the authors proposed that changes in the mechanical characteristics of chromosomes during different stages of the cell cycle could result simply from alterations in the internal loop structure.

We conclude this section by noting that the RL model, as well as the RW/GL and MLS models, assume that binding interactions occur only in *cis* (within chromosomes) and not in *trans* (between different chromosomes).

2.2.6 Strings-and-binders-switch model (SBS)

In the RW/GL, MLS, and RL models discussed in Sections 2.2.1, 2.2.2, and 2.2.5, looping interactions are taken into account implicitly, that is, they are built inside the polymer model. In reality, however, these interactions are mediated by chromatin-binding proteins freely diffusing in the nucleoplasm. The concentration of these binding proteins, ignored in these models, can be expected to significantly affect chromosome looping and hence folding.

Barbieri et al. (2012) introduced the "strings-and-binders-switch" (SBS) model, which explicitly models the action of binding proteins. A further motivation for this new model was the experimental observation that chromatin can adopt many different states. In particular, the so-called crumpled globule model (see Sections 2.3.1 and 2.3.4) turns out to be just one out of many possible states captured by the SBS model (see below).

In the SBS model, a single chromosome is described by a coarse self-avoiding* polymer chain composed of ~500 beads of two species, which are evenly distributed along the chain. Monomers of the two species have different binding affinities for protein complexes, which mediate attractive interactions between genomically distant sites along the chromatin fiber (Fig. 9.7A). These proteins are described as particles undergoing Brownian motion. Thus, besides the standard parameters of the polymer chain, the model includes two new parameters: (1) the binding affinity between model protein complexes (E_x, expressed in units of room temperature energy, $k_B T$), and (2) the molecular concentration of binders (c_m, expressed in units of nmol/l). The values of these parameters strongly affect the behavior of the polymer, as determined using Monte Carlo* simulations.

The results are best summarized by the "phase diagram" shown in Fig. 9.7B. At a fixed binding affinity E_x, and below a critical concentration of binders $c_{tr}(E_x)$, the fiber remains swollen (as in the absence of binders), and reproduces the known features of a generic self-avoiding* chain. At the critical concentration $c_{tr}(E_x)$, when repulsive interactions between beads are exactly compensated by the attractive action of binder proteins, the chain behaves as a generic random walk like polymer (De Gennes, 1979). At binder concentrations exceeding $c_{tr}(E_x)$, the polymer collapses to a globular state. This change in polymer behavior changes sharply around the critical concentration, hence binder concentration acts as a "switch." The critical concentration c_{tr} depends on the binding affinity E_x and is higher for lower E_x (Fig. 9.7B). The behavior of the polymer for different regimes of binder concentration can be further characterized by the scaling laws

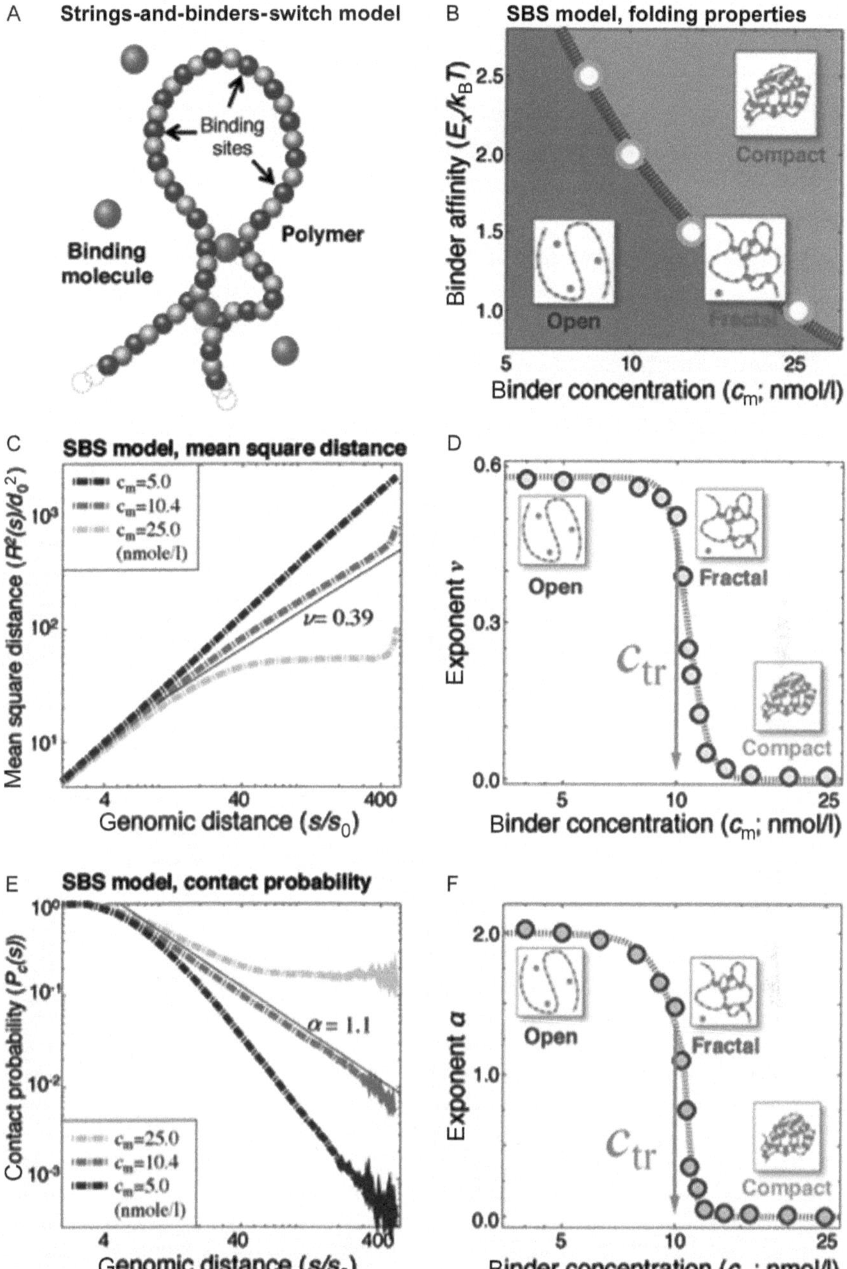

Figure 9.7 Strings and binders switch model (SBS). (A) In the SBS model, chromatin fibers are described as copolymer* chains made up of two species of monomers, with

(Continued)

(with exponents ν and α) that quantify how internal distances $\langle R(s)^2 \rangle$ and contact frequencies $P_c(s)$ between polymer loci vary with genomic distance s (Fig. 9.7C–F). Note in particular, how internal distances and contact frequencies reach a plateau ($\nu \sim 0$, $\alpha \sim 0$) for binder concentrations above c_{tr}, where the polymer collapses.

The authors point to experimental data from different organisms and cell types characterized by different degrees of compactness and different scaling laws. They argue against the existence of a unique chromatin state and in favor of a multitude of chromatin states. In particular, the authors point out that the $P_c(s) \sim 1/s$ decay for genomic distances s between 0.5 and 7 Mbp (more exactly, $\alpha = 1.08$) observed in human lymphoblasts (see Section 2.3.4) (Lieberman-Aiden et al., 2009) is an ensemble average over many chromosomes and cells, and could be explained by a mixture of compact ($\alpha = 0$) and open ($\alpha = 2.1$) chromatin states. Reanalyzing the experimental data, they also note that for some chromosomes, the contact frequency decay significantly deviates from the genome-wide average of $\alpha = 1.08$ (with α ranging from 0.93 to 1.3) and that very different average behavior ($\alpha \sim 1.6$) has been determined for human embryonic cells (Dixon et al., 2012). Barbieri et al. (2012) argue that this multitude of chromatin states might be explained by the SBS model, owing to its ability to generate a range of different chromosome structures (in particular ν varying between 0 and 0.6, α between 0 and 2.1) depending only on the concentration of the binding proteins.

Figure 9.7—Cont'd different binding affinities for diffusing molecules (red particles), which act as "condensing agents." The system is let to evolve and achieve equilibration by Monte Carlo* simulations. (B) Below a critical concentration of the condensing agents (compatible with assumed *in vivo* conditions), polymers obey the standard self-avoiding* chain behavior (blue region). Above this threshold, the polymer condenses to a compact state (grey region). At the critical threshold, the polymer has a fractal structure (red line). (C) Average square internal distance between polymer loci, $\langle R(s)^2 \rangle$ (in units of the bead square diameter, d_0^2), as a function of genomic* distance, s (in units of the bead genomic content, s_0). (D) Polymer metric exponent, ν as function of binder concentration. The exponent ν is defined by $\langle R(s)^2 \rangle \sim s^{2\nu}$ (where $\sim$means proportional to). (E) Average contact probability, $P_c(s)$ between polymer loci, as a function of genomic* distance, s. (F) Contact probability exponent, α, as function of binder concentration. The exponent α is defined by $P_c(s) \sim s^{-\alpha}$. In panels (C) and (E), the three curves correspond to binder concentrations c_m below, at, and above the concentration threshold c_{tr} (here, the binding affinity was assumed to be: $E_x = 2\, k_B T$). *Reprinted with permission from Barbieri et al. (2012) © (2012) National Academy of Sciences, USA.* (For interpretation of the references to color in this figure legend, the reader is referred to the online version of this chapter.)

2.2.7 *Lattice animal model (LA)*

In an effort to simultaneously describe looping of chromatin fibers and the experimentally observed fractal-like behavior of contact frequencies (Section 2.3.4), Iyer and Arya proposed a so-called lattice animal (LA) model for chromosome organization (Iyer and Arya, 2012). An LA is a generic, branched macromolecular-like structure built on the cubic lattice. It consists of nodes, located at the sites of the lattice, and links that connect different nodes together (Fig. 9.8A). On the standard cubic lattice, each node can be connected to up to six neighboring nodes. Each link of the structure represents two parallel strands of chromatin running in opposite directions, while each node represents the extremes of a chromatin loop (Fig. 9.8B).

As acknowledged by the authors, this model is very abstract. In their own words, it "is not amenable to studying detailed features of chromosomal organization that require a mapping of the lattice animal structure to specific features of DNA and higher-order structures of chromatin" (Iyer and Arya, 2012). Nevertheless, the LA model captures in a simple and transparent way properties of the chromatin fiber that were also assumed by other models discussed above. In particular, the model naturally incorporates loops at all genomic length scales, without further arbitrary assumptions.

Even in the simplest case of ideal lattice animals, the model is analytically untractable. Thus, the authors resorted to Monte Carlo computer simulations of single LA polymer configurations. They adopted an algorithm that implemented stochastic moves in order to continually rearrange LA sites (Van Rensburg and Madras, 1992). This algorithm picks a site at random, and—provided the connectivity and other constraints detailed below are satisfied—moves it next to another random site of the LA. In this way, the algorithm simulates the rearrangements of the chromatin fiber due to, for example, regulatory mechanisms inside the cell. In order to make quantitative predictions that can be compared to experiments, the authors considered four different versions of the LA model, with increasing levels of sophistication (see Fig. 9.8C,D):

a. Ideal LA: different sites of the walk can occupy the same site of the underlying lattice (as for a phantom* chain).

b. Self-avoiding LA (LA with excluded volume effects): now, a lattice site can be occupied by only one walk site. This mimics the effect that chromatin fibers cannot occupy the same spatial region.

c. LA with excluded volume effects and spherical confinement: the added confinement takes into account the fact that chromosome are confined to roughly spherical territories (Cremer and Cremer, 2001).

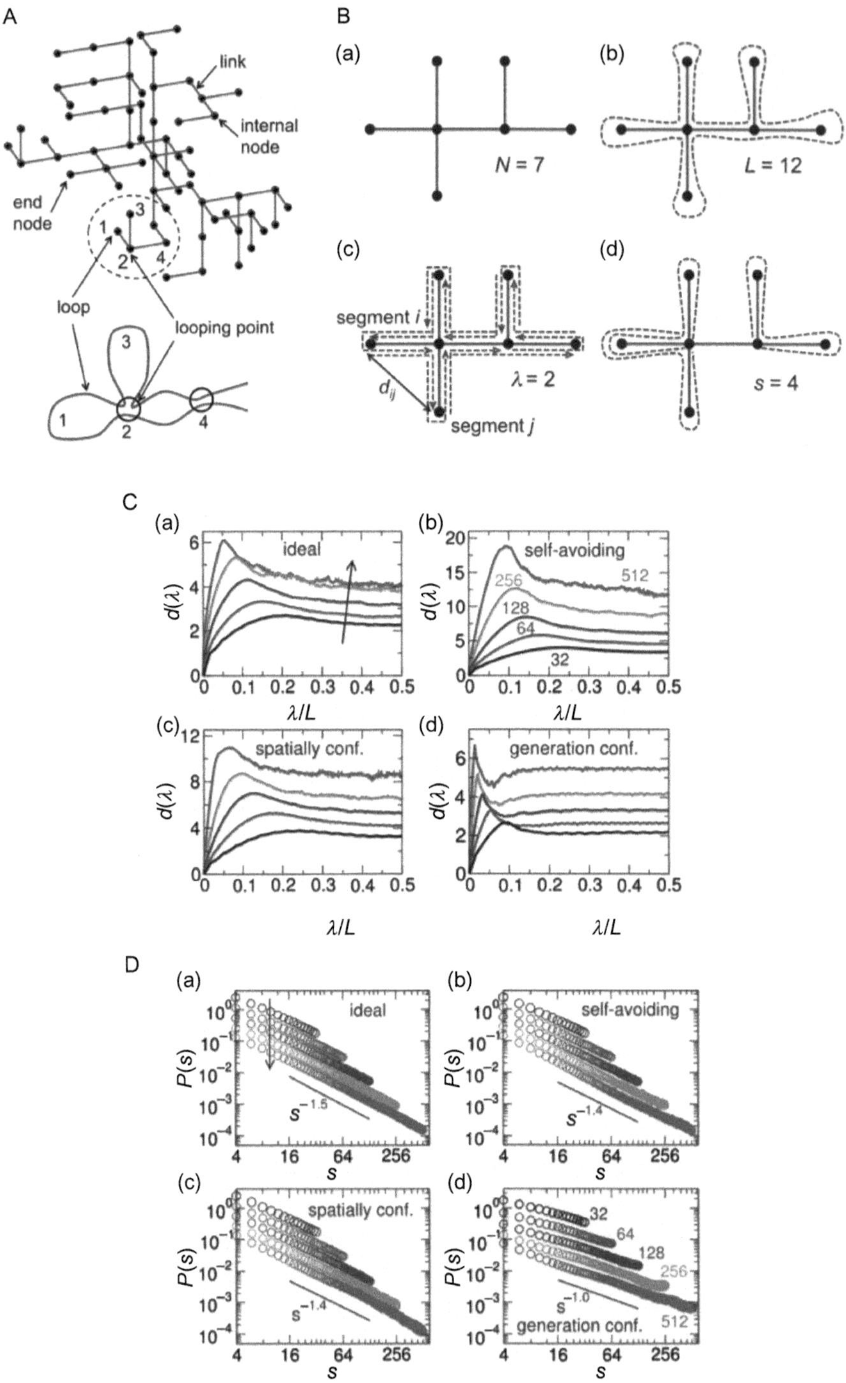

Figure 9.8 See figure caption on next page bottom.

d. LA with excluded volume effects and a constraint limiting the size of loops: now, only loops up to a certain maximal size are allowed. The authors argue that this limit might arise from difficulty in bending higher-order structures of chromatin. Note that a similar assumption was made in the DL model discussed above (Zhang and Heermann, 2011) (Section 2.2.5).

Iyer and Arya found that all models described in a correct, albeit qualitative, manner the experimental behavior of average spatial distances between chromatin loci, suggesting that loops should be sufficient *per se* (Fig. 9.8C). However, only model (d) above was capable of describing also the experimental power law decay $\sim s^{-1}$ of contact frequencies between chromatin loci (Fig. 9.8D) (Lieberman-Aiden et al., 2009). In fact, excluded volume effects alone led to excessive polymer swelling, which therefore had to be compensated by constraining the maximal loop size.

2.3. Chromosome folding by topological mechanisms

Although looping of chromatin fibers due to *cis* interactions has been repeatedly advocated as a key reason for chromosome compaction, alternative

Figure 9.8 Lattice animal (LA) model of chromosomes, reproduced from Iyer and Arya (2012). (A) Schematic illustration of the LA. In the model, each segment of the walk (red lines) represents two strands of chromatin running in opposite directions, while nodes (black dots) represent either the middle portion or the termini of chromatin loops. Thus, for example, nodes 1 and 3 represent the middle portions of two different loops, both protruding from node 2, which is therefore a looping point. Similarly, node 4 is the basis of a larger chromatin loop. (B) Parameters describing LA structure. Schematic illustration of total "chromatin" length, L, path length, λ, and loop length, s. (a) The total number of nodes is noted N. Here, $N=7$. (b) Total length of the chromatin fiber (dashed blue curve) L is $L=2(N-1)$ in lattice units. Here, $L=12$. (c) The path length, λ, is the minimal number of bonds connecting two nodes. Dashed blue arrows show all possible pairs of nodes with $\lambda=2$. (d) Similarly, dashed blue curves show all possible pairs of nodes with loop length $s=4$. (C,D) Quantitative predictions for four versions of the model: (a) an ideal LA (i.e., without excluded volume interactions), (b) a self-avoiding LA, (c) a spatially confined LA, (d) a spatially confined LA with a maximum loop size. Each colored curve corresponds to a distinct number of nodes N, as indicated. (C) Internal distances d between polymer loci, as a function of the scaled path length (or genomic distance), λ/L. All four models predict that internal distances between chromatin loci flatten at large genomic distances, in qualitative agreement with recent FISH data (Mateos-Langerak et al., 2009) (see Section 2.2.5). (D) Contact probability P as function of the loop size, s. Only model (d) faithfully reproduces the "1/s" scaling observed in HiC data (Lieberman-Aiden et al., 2009). *Reprinted with permission from Iyer and Arya (2012) © (2012) by the American Physical Society.* (For interpretation of the references to color in this figure legend, the reader is referred to the online version of this chapter.)

polymer pictures have also been proposed. In particular, models based on *topological* constraints seem to provide a robust framework, capable of explaining chromosome territories and other features that emerged from recent conformation capture experiments.

A notion of the importance of topology in chromatin fibers is apparent from the following. Chromatin is highly dynamic, as reflected most spectacularly by the cell cycle, when the genome is periodically replicated (every ~24 h in most animal cells) in order to transmit the same amount of genomic content to two daughter cells (Alberts et al., 2008). This important task is accomplished through the regular alternation of the phase of chromosome condensation (mitosis) and the phase of normal cell activity (interphase), where chromosomes swell inside the nucleus and form territories. One may wonder how the cell manages to untangle chromosomes for mitosis from the ball of wool of the interphase genome. Indeed, it is widely known that the effects of topological entanglement* between chains dominate the visco-elastic properties of polymer solutions and melts* (De Gennes, 1979; Doi and Edwards, 1988; Khokhlov and Grosberg, 1994; Rubinstein and Colby, 2003). What then is the effect of topology on chromosome folding and genome architecture? In the following, we discuss several models that address this question.

2.3.1 Crumpled-globule model

Seminal publications in 1988 and 1993 argued for the first time that because of topological constraints, chromosomal DNA could potentially exist in an out-of-equilibrium, long-lived, and unknotted state, which the authors called a "crumpled globule" (Fig. 9.9, left) (Grosberg et al., 1988; Grosberg et al., 1993).

According to the authors, a crumpled globule can be "constructed" from the fast collapse (Fig. 9.9, right) of a knot-free self-avoiding polymer chain where interactions between single chain monomers are rapidly switched from repulsive to attractive (Grosberg et al., 1988). They argued that chain relaxation to equilibrium is a process occurring in two stages:

1. Rapid stage of chain collapse: During this stage, the chain is out-of-equilibrium*, collapses due to self-attractions and, because of topological constraints, polymer crumples form at all length scales (Fig. 9.9, right). The chain crumples interact only at their surfaces, while monomers from different crumples barely interact with each other. This stage proceeds up to a Rouse*-like timescale of order $\tau_R \sim L^2/D$, where L is the chromosome contour* length and D is the diffusion coefficient of a single monomer. A crude estimate of this timescale for naked DNA can be

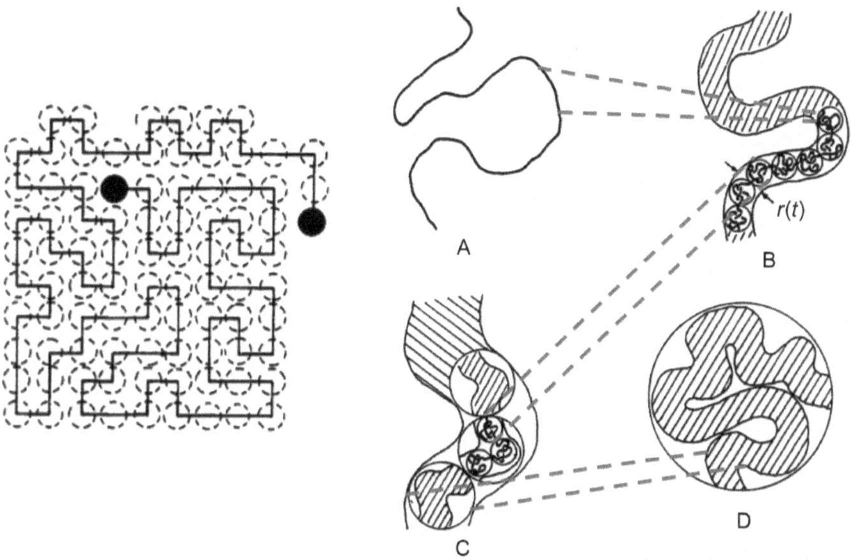

Figure 9.9 The crumpled globule model. (Left) Schematic illustration of a "crumpled globule" on a two-dimensional lattice. Chain monomers are represented by dashed circles, connected by the solid line. End monomers are represented by filled circles. The curve is a space-filling fractal (Vanderzande, 1998). The three-dimensional analog of this curve contains no knots, and it is thus very different from a random, space-filling globule which is expected to be highly knotted (Sumners and Whittington, 1988). *Reprinted with permission from Grosberg et al. (1993) © (1993) by EDP Sciences.* (Right) Physical mechanism leading to the crumpled globule. A swollen, unknotted chain collapses rapidly after switching from repulsive to attractive interactions between chain monomers. This switch can be triggered by changing the "quality" of the solvent surrounding the chain, typically by making it repelling for the polymer, which can thus fold onto itself (De Gennes, 1979). The self-collapse of the chain is accompanied by the formation of crumples *at all length scales*. For example, panel (A) shows the original chain collapsed into small-scale crumples forming configuration (B). Each crumple in (B) fills a larger crumple in (C), and so on—in a self-similar mechanism—up to configuration (D). Reversing the picture, the crumpled globule regime can be viewed as a chain of blobs made of smaller blobs made of even smaller blobs, and so on down to single monomers. *Reprinted with permission from Grosberg et al. (1988) © (1988) by EDP Sciences.* (For color version of this figure, the reader is referred to the online version of this chapter.)

obtained assuming a chromosome contour length of $L=30$ mm (see Section 2.1), and assuming that 1 nm of DNA (3 bp) has a diffusion coefficient $D \sim 10^8$ nm^2/s (obtained using the Stokes–Einstein relation: $D = k_B T/(6\pi r \eta)$, where $T \sim 293$ K is the room temperature, $r \sim 1$ nm is the monomer radius, and $\eta \sim 10^{-3}$ Pa s is the viscosity of water). Thus the stage of chain collapse is expected to last $\sim L^2/D \sim 100$ days.

Interestingly, this crude estimate tells us that even this first stage could already take much longer than the entire cell life.

2. Slow stage of chain self-interpenetration: This second stage occurs by a "reptation*-like" mechanism on a timescale of order $\sim\tau_R\ (L/L_e)$, where $L_e \ll L$ is the so-called entanglement* length. Hence, this stage is typically much longer than the first stage (De Gennes, 1979; De Gennes, 1971). Thus, especially for large chromosomes such as human ones, the process of self-interpenetration is expected to be much longer than any observable event during cell life.

A "crumpled globule" of chromatin fibers effectively has no time to develop knots. This property makes this model an appealing solution to the "chromosome folding" problem.

Interestingly, in 1994, Sikorav and Jannink employed similar considerations, but in fact reversed the argument (Sikorav and Jannink, 1994): they showed that chromosome condensation at the end of interphase would proceed on a timescale of hundreds of years that is incompatible with the natural duration of cell life. For this reason, they claimed that chromosome condensation requires the assistance of topoisomerase II, an enzyme capable of cutting both strands of the DNA. From the physical point of view, this would be equivalent to the relaxation of a phantom* chain, instead of an entangled* chain (De Gennes, 1971; Doi and Edwards, 1988). Thus, breaking of topological constraints could boost condensation by several orders of magnitude (Sikorav and Jannink, 1994).

In spite of its relevance for understanding chromosome organization, the "crumpled globule" model appears to have received little attention from the biology community until almost 20 years later, when solid computational and experimental evidence brought strong support to the idea that kinetic and topological effects likely play a fundamental role in chromosome organization (see Sections 2.3.3 and 2.3.4) (Rosa and Everaers, 2008; Lieberman-Aiden et al., 2009).

2.3.2 Entropy-driven chromosome organization in bacteria

Although this review focuses on chromosome organization in eukaryotes, it is worth mentioning that there are interesting attempts to model chromosome behavior in bacteria, as well. Due to the simplicity of their genomes, bacteria represent interesting targets for modeling: *E. coli*, for example, one of the most important model organisms, has only one chromosome made up of a circular DNA filament of ~4 Mbp, roughly 1000 times smaller than the human genome.

During DNA replication in *E. coli*, the two chromosomes segregate to opposite poles of the nucleoid—the approximately cylindrical, membrane-less cellular compartment that contains the genetic material and grows in size simultaneously with its duplication. Jun and Mulder (2006) suggested an interesting physical mechanism for chromosome segregation in bacteria. During replication, the two new chromosomes are necessarily highly entangled with each other. In addition, the chains stay confined inside the nucleoid, whose transversal size is much smaller than the typical size of free, unconfined DNA (see Section 2.1). The combination of entanglement and confinement leads to a strong entropic repulsion between the two chromosomes, finally resulting in DNA segregation. This effect can be explained as follows: on the scale of the nucleoid diameter, chain repulsion is weak (of the order of $\sim k_B T$). However, these small contributions add up along the significantly longer axis of the nucleoid, leading to strong effective repulsion much larger than the room temperature.

To quantify this effect, Jun and Mulder performed Monte Carlo computer simulations of a coarse-grained polymer model for DNA replication, in a confining cylinder mimicking the nucleoid geometry (Fig. 9.10). Remarkably, they were able to reproduce the experimentally observed dynamical reorganization of DNA during replication, without any need for sophisticated molecular machineries (Jun and Mulder, 2006).

Physically, this process is very similar to the segregation of circular, unlinked polymers in concentrated solutions, a mechanism advocated as a possible explanation for chromosome territories, as discussed in the next section (Rosa and Everaers, 2008).

Despite the attractiveness of the entropic model, we note that several experimental features have been cited to challenge the random polymer model and support alternative views, such as elastic filaments formed by extensive protein cross-links (Hadizadeh Yazdi et al., 2012; Wiggins et al., 2010). As a result, the mechanisms of bacterial chromosome segregation currently remain debated (Pelletier et al., 2012; Possoz et al., 2012; Wang et al., 2013).

2.3.3 Entagled polymer model of chromosome territories

Rosa and Everaers (2008) provided computational evidence supporting a "topological origin" for the experimentally observed structure and dynamics of eukaryotic chromosomes. In particular, they concluded that territories should form even in the absence of looping interactions between chromatin fibers—such interactions could still stabilize chromosome territories further, but without being at their origin. The authors reversed the earlier

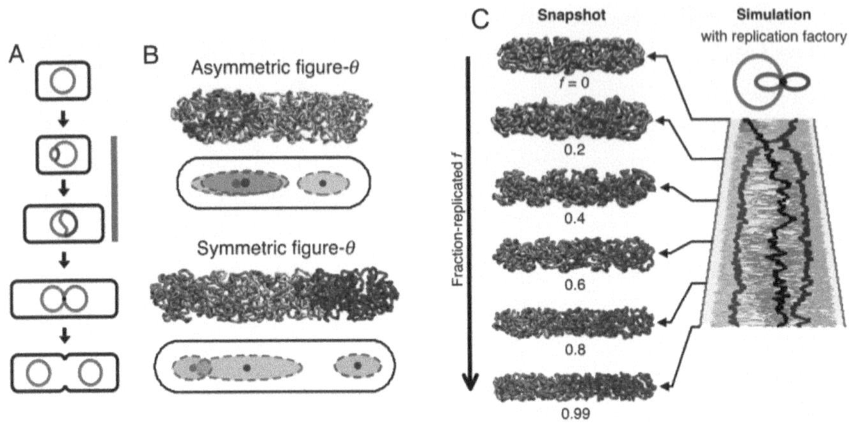

Figure 9.10 Entropic segregation of bacterial chromosomes. (A) Schematic representation of DNA replication in *E. coli*. Replication starts at one replication origin, and proceeds until the entire chromosome is replicated. Finally, the two daughter chromosomes segregate to the two daughter cells. (B, C) Model of the duplicating chromosome as a coarse-grained polymer chain. The division dynamics was followed by Monte Carlo* computer simulations. The mother chromosome is shown in gray, the two daughter chromosomes are shown in red and blue. (B) Top and bottom configurations correspond to the two stages indicated by a grey bar in panel (A). For each, a snapshot of the simulation is shown on top, and the corresponding average spatial positions of the midpoints of the colored segments are shown below. (C) Simulated replication and resulting segregation of bacterial DNA, with one replication origin (black dot). The internal "clock" is indicated by the replicated genome fraction, *f*. On the right are shown the instantaneous (thin light lines) and average trajectories (thick dark lines) of the origin and terminus (ori-ter) of replication. *Reprinted with permission from Jun and Mulder (2006) © (2006) National Academy of Sciences, USA.* (For interpretation of the references to color in this figure legend, the reader is referred to the online version of this chapter.)

observation that topological constraints would not allow chromosomes to condense from interphase to mitosis within the observed timescales (Sikorav and Jannink, 1994). The model of Rosa and Everaers started at the beginning of interphase, when chromosomes are tightly compacted and well separated from each other. Upon swelling, chromatin fibers are expected to come closer to each other and start mixing. How long does it take before achieving "perfect mixing" of chromosomes? Because of topological constraints, Rosa and Everaers (2008) estimated that this process would take hundreds of years for long chromosomes such as human chromosomes.

In order to test this prediction, and to quantify the role of topological constraints in interphase chromosomes, Rosa and Everaers performed molecular* dynamics simulations of a minimal polymer model of decondensing chromosomes, which accounted for the linear connectivity and self-avoidance of the chromatin fiber. The model assumed a Kuhn* length of 300 nm and a DNA compaction of 100 bp/nm, but was otherwise parameter-free. The simulation results are summarized in Fig. 9.11 (Rosa and Everaers, 2008).

Qualitatively, the model reproduces known features of eukaryotic chromosomes for three different organisms: yeast, *Drosophila melanogaster*

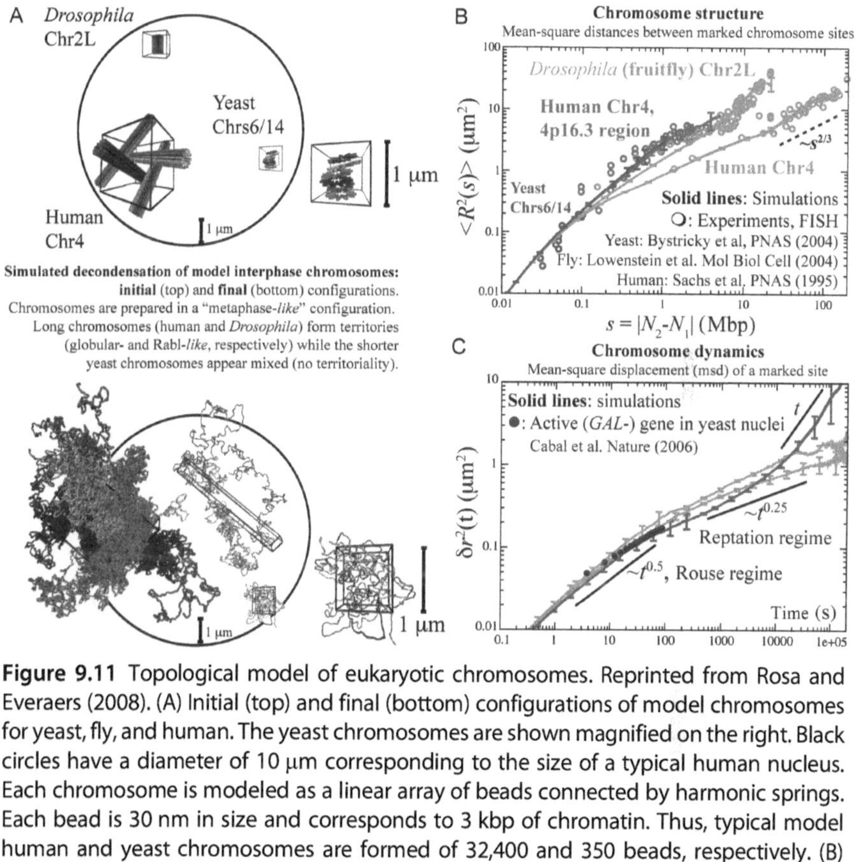

Figure 9.11 Topological model of eukaryotic chromosomes. Reprinted from Rosa and Everaers (2008). (A) Initial (top) and final (bottom) configurations of model chromosomes for yeast, fly, and human. The yeast chromosomes are shown magnified on the right. Black circles have a diameter of 10 μm corresponding to the size of a typical human nucleus. Each chromosome is modeled as a linear array of beads connected by harmonic springs. Each bead is 30 nm in size and corresponds to 3 kbp of chromatin. Thus, typical model human and yeast chromosomes are formed of 32,400 and 350 beads, respectively. (B) Average square spatial distances, $\langle R^2(s) \rangle$, between chromosome sites located N_1 and N_2 Mbp away from one end of the fiber (their genomic distance is thus given by $s = |N_2 - N_1|$). Experimental data are shown by symbols and simulation predictions by solid lines. (C) Temporal, mean square displacements (MSD) $\delta r^2(t)$, as function of time interval, as measured experimentally (symbols) and as predicted by simulations (solid lines), for individual chromosome sites. The two indicated scaling regimes (Rouse* and reptation*) are characteristic features of polymer relaxation in entangled solutions (De Gennes, 1979; Doi and Edwards, 1988; Khokhlov and Grosberg, 1994; Rubinstein and Colby, 2003). (For color version of this figure, the reader is referred to the online version of this chapter.)

(fruit-fly), and human (Fig. 9.11A). In particular: (1) large human chromosomes appear compartmentalized in territories; (2) fruit-fly chromosomes assume the elongated shape typical of a Rabl-like configuration; and finally (3) small yeast chromosomes do not form clear territories. *Quantitatively*, the model is able to reproduce experimental data on chromosome three-dimensional structure (Fig. 9.11B) and dynamics (Fig. 9.11C). In addition, a later study showed that the model also predicted the experimental $\sim 1/s$ decay for intra-chromosome contacts between genomic loci and the authors provided a theoretical justification for this behavior, based on scaling arguments (Rosa et al., 2010; Lieberman-Aiden et al., 2009).

What is the physics behind the territory organization of long chromosomes during interphase? Rosa and Everaers argued that it could be related to "the segregation of unentangled ring polymers in concentrated solutions due to topological barriers" (Cates and Deutsch, 1986; Müller et al., 1996; Vettorel et al., 2009). By construction, it is topologically impossible for unlinked ring polymers (i.e., circular polymers) to interpenetrate each other. Linear polymers in principle do not face this impossibility, but for giant polymers like human chromosomes, which are virtually impenetrable and well segregated in mitosis, full relaxation and mixing would require time-scales not compatible with ordinary life and are consequently never observed. Thus, for long linear chromosomes, topological constraints play a similar role as for ring polymers.

The model was adapted later by another group, who used it in order to describe the different stages of chromosome condensation during interphase (Kim et al., 2011). In particular, they connected chromatin density fluctuations to the altered nuclear architecture described in some medical studies on early carcinogenesis.

Very recently, the model was employed in order to explore the hypothesis formulated elsewhere (e.g., Cavalli, 2007) that coexpressed genes are in close spatial proximity, potentially allowing them to share transcription factories (Di Stefano et al., 2013). The authors performed steered molecular* dynamics simulations on a coarse-grained polymer model for human chromosome 19 in order to enforce the colocalization of thousands of coexpressed gene sequences. The gene pairs were identified through an extensive statistical study (based on mutual information) aiming at detecting coordinated expression patterns from microarray data (Di Stefano et al., 2013). In the model, the tendency of these sequence pairs to colocalize was enforced by the addition of harmonic springs connecting the pairs (with zero equilibrium length). Remarkably, the authors found that nearly 80% of imposed constraints could be satisfied without violating topological constraints. Moreover, they also demonstrated that constrained model chromosome conformations showed a remarkable organization into macrodomains very similar to that observed in recent genome-wide chromosome conformation capture experiments (Dixon et al., 2012).

2.3.4 Incomplete relaxation of unknotted linear polymers after fast spherical compaction

In 2009, a high-throughput technique to detect chromatin–chromatin contacts genome-wide in large populations of cells was presented under the name Hi-C and applied to human lymphoblastoid cells (Lieberman-Aiden et al., 2009). A salient result of this study was that, in the genomic range between 0.5 and 7 Mbp, intrachromosomal (*cis*) contacts seem to decay as a power law $\sim s^{-1}$ of the genomic distance s (Fig. 9.12A). This exponent is remarkable, because confined polymers at equilibrium are expected instead to display a power law $\sim s^{-3/2}$ (Fig. 9.12B) (Rosa et al., 2010). As a possible explanation, the authors suggested that—at least on the above-mentioned genomic scales—chromosomes are better described by the "crumpled globule" model (which the authors referred to as "fractal globule"), already reviewed in Section 2.3.1 (Grosberg et al., 1993).

To verify this, the authors performed Monte Carlo* simulations of a long, swollen, and initially unknotted polymer chain, which was forced to collapse by confining it to a sphere. Confinement was performed very fast, in order to avoid chain relaxation and self-entanglements. A close look at the typical polymer configuration obtained by this simulation is shown in Fig. 9.12D. For comparison, Fig. 9.12C shows a configuration obtained when applying a slow confinement, which allows the chain to relax to an equilibrium polymer. The polymer that underwent fast confinement was shown to exhibit many properties expected of a "crumpled globule." In particular, contact frequencies displayed the predicted $\sim s^{-1}$ behavior, at odds with the $\sim s^{-3/2}$ behavior observed for the equilibrium globule (Fig. 9.12B). The former polymer was unknotted, while the latter was highly entangled (Fig. 9.12C and D). As a consequence, once the external confinement imposed on the chain was removed from the simulation, the polymer that underwent fast confinement also expanded very fast. This is in contrast with the equilibrium globule, which expanded briefly, but then remained trapped due to the presence of many internal knots formed during chain collapse (see fig. S28 in Lieberman-Aiden et al., 2009). As pointed out by the authors, the ability of the unknotted polymer to unfold without being hindered by entanglements is an attractive feature that might

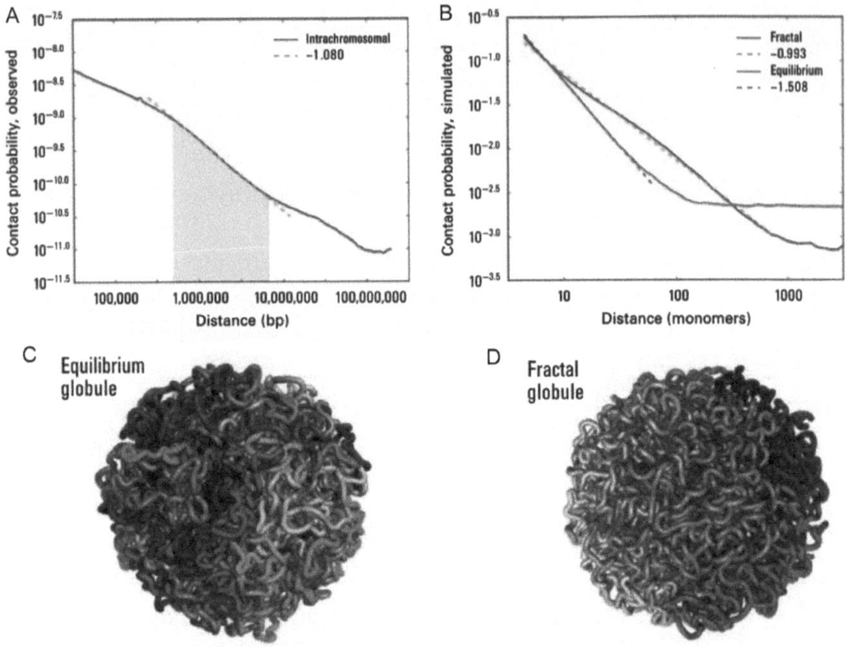

Figure 9.12 Hi-C and the crumpled globule. *From Lieberman-Aiden et al. (2009). Reprinted with permission from AAAS.* (A, B) Average intrachromosomal contact frequency $P_c(s)$ versus genomic separation s (log–log plot), as determined experimentally from Hi-C data (A) and as predicted by simulations (B). (A) In the range 0.5–7 Mb (shaded region), $P_c(s)$ decays approximately as $\sim s^{-1}$ (dashed line). (B) In a simulated equilibrium polymer, $P_c(s)$ decays as $\sim s^{-3/2}$ before reaching a plateau (red curve). In the simulated crumpled globule, $P_c(s)$ decays approximately as $\sim s^{-1}$. (C, D) Folding of a polymer chain in the equilibrium globule (C) and the crumpled globule (D). The chain configurations are obtained by Monte Carlo* simulations of an initially unknotted linear polymer subjected to slow relaxation (C) or to fast spherical confinement (D). The chain is colored along its contour* length following the colors of the rainbow. In the equilibrium globule (C) the polymer is highly entangled so that genomically close loci can be found at large spatial distances, as apparent from the presence of similar colors in different regions of the globule. By contrast, in the crumpled globule (D), the polymer has no time to relax and remains in the same (unknotted) topological state as the initial configuration. Genomically close loci also tend to remain close in space, as apparent from the large homogenous color patches. (For interpretation of the references to color in this figure legend, the reader is referred to the online version of this chapter.)

facilitate local decondensation of chromatin for example, during gene activation.

How robust are these results? A recent study addressed the stability of the crumpled globule structure (Schram et al., 2013). In this work, the authors performed Monte Carlo* simulations of polymer chains (with roughly similar contour* lengths) that rapidly collapse under the action of attractive monomer–monomer interactions. Quite surprisingly, on timescales comparable to the Rouse* time, for which the "crumpled globule" regime is expected (see Section 2.3.1), the system achieved a metastable (long-lived out-of-equilibrium) state that is almost undistinguishable from the "equilibrium" globule (see Fig. 13 of Schram et al., 2013). In particular, the *cis* contact frequencies obeyed a $P(s) \sim s^{-\alpha}$ dependence with α very close to 3/2 (the theoretically predicted exponent for an equilibrium globule) instead of $\alpha = 1$. The same conclusion was obtained when simulating the relaxation of a fractal, space-filling polymer configuration, underscoring the robustness of this result. According to the authors, the agreement between simulations and experiments in Lieberman-Aiden et al. (2009) might result from the use of short simulation times (Schram et al., 2013).

2.4. Introducing sequence specificity

As apparent from the previous sections, models based on polymer physics can explain many important features of genome architecture and dynamics, despite the fact that they ignore the complex information contained in the DNA sequence, post-translational modifications of histones, chromatin-binding proteins, etc. We have discussed the success of these models in reproducing quantitative properties such as scaling laws with genomic distances of average spatial distances between loci measured by FISH or average contact frequencies measured by cross-linking experiments. However, beyond these *statistical* quantities, the discussed models typically cannot make detailed predictions concerning the nuclear positions, relative distances, or contact frequencies involving *specific* loci. Such information is important, however, to fully relate genome architecture and dynamics to biological functions, for example to determine how frequently coexpressed loci from different chromosomes can come into physical proximity. Below, we discuss an example of a model that incorporates a limited amount of DNA sequence specificity into a generic polymer model and is able to provide detailed predictions for any sets of specific loci in the genome.

2.4.1 A Minimal polymer model of yeast genome architecture

Budding yeast is an attractive organism for computational models of chromosomes, owing to its well studied and comparatively simple nuclear architecture, and to the availability of large quantitative data sets acquired in several experimental studies (Duan et al., 2010; Taddei et al., 2010;

Zimmer and Fabre, 2011). Accordingly, several groups have recently developed independent computational models of yeast genome architecture. Here, we summarize one of these studies, which investigated whether a polymer model with a minimal set of assumptions could accurately predict the architecture of the ~12 Mbp long yeast genome (Wong et al., 2012). Other yeast models will be discussed in Sections 3.1.3, 3.2.4, and 3.2.7 (Duan et al., 2010; Gehlen et al., 2012; Tjong et al., 2012; Tokuda et al., 2012).

In the model of Wong et al. (2012), the 16 chromosomes of haploid yeast were represented by self-avoiding* chains consisting of jointed rigid segments with an assumed Kuhn* length of $L_K = 60$ nm, a diameter 20 nm, and a DNA content of 5 kbp (implying a compaction of 83 bp/nm, i.e., 12 kb/nm) (Fig. 9.13A). The 16 chains were enclosed by a 1 μm radius sphere representing the yeast nuclear envelope, and their motions were simulated using Brownian* dynamics, while respecting topological constraints within and between chains. On top of this generic polymer model, the model added constraints for three specific DNA sequences: centromeres, telomeres, and the rDNA. First, because each budding yeast centromere is known to be tethered by a single microtubule to the spindle pole body (SPB), a protein complex embedded in the nuclear envelope, the model included a rigid segment linking the corresponding chromosome segment to a model SPB (Fig. 9.13A). Second, because the 32 telomeric extremities of yeast chromosomes are known to be anchored to the nuclear envelope, a radial outward force was applied to maintain the corresponding chain segments in the vicinity of the nuclear envelope (without imposing any specific location on the surface) (Fig. 9.13A). Third, the model accounted for the specificity of the rDNA locus, a ~1.5 Mbp long region on chromosome 12 consisting of ~150 tandem repeats that encode ribosomal RNA. The nucleolus is the site of rDNA transcription and, in yeast, consists of a large membrane-less compartment with a characteristic crescent shape located opposite the SPB. In the model, chain segments corresponding to the rDNA locus were given a larger diameter than the 20 nm diameter assumed elsewhere (Fig. 9.13A). This modification was introduced to account for the intense transcriptional activity of the rDNA, which leads to a strong accumulation of RNA and proteins at this locus. The diameter of the rDNA segments was adjusted to 200 nm, such that the effective volume occupied by the rDNA segments approximated the observed nucleolar volume. Chromosome 12, which contains the rDNA locus was therefore modeled as a copolymer (Wong et al., 2012).

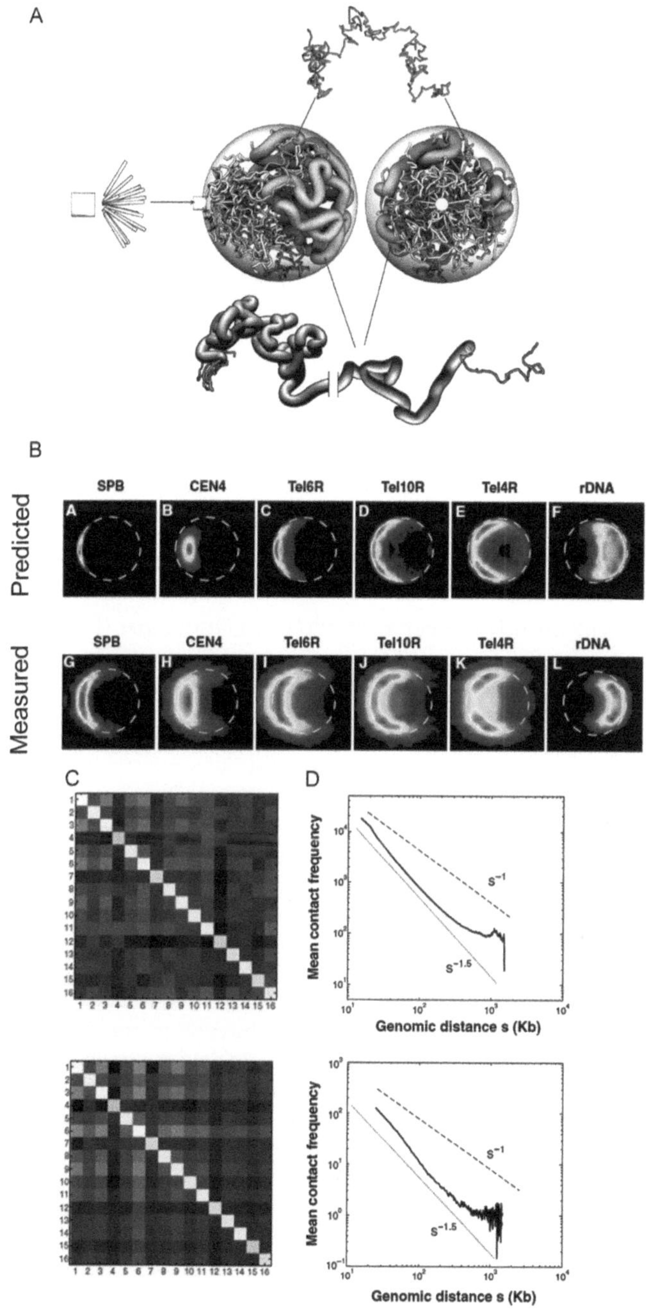

Figure 9.13 See figure caption on next page bottom.

Starting from an assumed initial configuration, the simulation was run until equilibrium, and then sampled at large time intervals. This enabled detailed comparisons of the model's predictions with quantitative data obtained from experiments on large cell populations. These data included the nuclear territories and distances between pairs of loci, as determined by live cell imaging, as well as genome-wide contact frequencies measured by high-throughput chromosome conformation capture (Berger et al., 2008; Duan et al., 2010; Thérizols et al., 2010). These comparisons showed a high degree of agreement, indicating that most of the variability observed experimentally could be accounted for by this simple model (Fig. 9.13B–D). Specifically, the model recapitulated the nuclear territories of selected loci previously obtained by imaging, most significantly that of the rDNA, which was predicted to segregate from the rest of the genome into a crescent-like compartment similar to that observed for the nucleolus (Fig. 9.13D); the model also explained much of the differences in absolute and relative positions of several tens of single loci and locus pairs (Wong et al., 2012, Fig. 2). Likewise, the model accounted for most of the differences in average contact frequencies within and between individual chromosomes (Fig. 9.13B and C) or chromosome arms. It also performed well in predicting contact features at smaller genomic scales, down to roughly 50 kbp (Wong et al., 2012, Figs. 4 and S4). The predicted average intrachromosomal contact frequencies $P_c(s)$

Figure 9.13—Cont'd Predictive polymer model of the yeast nucleus with minimal sequence specificity. *Reprinted from Wong et al. (2012) with permission from Elsevier.* (A) Simulation ingredients. In the middle, a snapshot of the dynamic simulation is shown from two view angles, with each of the 16 chromosomes depicted in a different color. Chromosomes are represented by self-avoiding* chains, as shown by the green chain on the top. The rDNA locus on chromosome 12 is assumed to have a larger diameter than the rest of the genome and is shown in pink. The centromere of each chromosome is tethered to the spindle pole body (SPB) by a single rigid microtubule, as highlighted on the left. (B–D) Comparison of model predictions (top) with experimental measurements obtained from imaging and genome-wide contact frequencies (bottom) (Berger et al., 2008; Duan et al., 2010; Thérizols et al., 2010). (B) Intranuclear territories of selected loci (three telomeres, a centromere, and the rDNA locus) and the SPB. (C) Average contact frequencies between pairs of chromosomes displayed as heat map (with the highest frequencies shown in white and the lowest in black). (D) Average intrachromosomal contact frequencies $P_c(s)$ as function of genomic* separation s. Note the strong agreement between model predictions and experimental measurements. See Wong et al. (2012) for more extensive and quantitative comparisons. (For interpretation of the references to color in this figure legend, the reader is referred to the online version of this chapter.)

decayed with genomic separation s approximately as $\sim s^{-3/2}$, as expected for equilibrated polymers, and in agreement with observations (Fig. 9.13D). This agrees with the earlier prediction that yeast chromosomes are near equilibrium, as discussed in Section 2.3.3 (Rosa and Everaers, 2008). Departures from this power law, evident at both small and large genomic separations, were also predicted by the model (Fig. 9.13D) (Wong et al., 2012).

Based on this evidence, it was concluded that yeast genome architecture can be understood largely as a consequence of the generic properties of confined polymers and does not appear to be dominated by sequence-specific effects—with the exception of the three sequences mentioned above.

3. INVERSE MODELS OF GENOME ARCHITECTURE

In Section 2, we discussed "direct" models of chromosomes and genomes based on polymer physics (Fig. 9.1, top). These models are characterized by a relatively limited set of assumptions and quantitative parameters, such as the persistence length of chromatin, or looping probabilities and accordingly have strong predictive power. However, the availability of increasingly large experimental data sets on genome architecture, generated by imaging methods and particularly by high-throughput 3C-based techniques, has prompted the development of "inverse" approaches, in which chromosome configurations are reconstructed from these large data sets (Fig. 9.1, bottom). In essence, these methods are not unlike those previously used to reconstruct atomic structures of molecules from nuclear magnetic resonance data, or molecular assemblies from protein interaction data (Alber et al., 2007; Rieping et al., 2005).

The distinction between "direct" and "inverse" models is not sharp: some experimental data usually underly key assumptions used in the polymer models (for example, the persistence* length of chromatin was often obtained by fitting theoretical predictions to distances between loci measured by imaging (Bystricky et al., 2004)). Conversely, the "inverse" data-driven reconstructions may also embody aspects of polymer physics, such as the quantitative relationship between genomic distance between two loci on a chromosome and their contact probability (see Section 3.1.1). In practice, however, a useful distinction can be made based on the amount of experimental data used in either type of model: typically a handful of parameters in the "direct" models versus up to thousands or more values from large, often genome-wide data, in the "inverse" models. In the following, we will review a selection of "inverse" models, starting with those that attempt to reconstruct a single configuration or arrangement of chromosome(s) (Section 3.1), then discussing models that produce multiple structures (Section 3.2).

3.1. Reconstruction of a single 3D structure

3.1.1 Reconstruction of a yeast chromosome from 3C data

An important example of reconstructing the configuration of a chromosome already accompanied the introduction of the 3C technique (Dekker et al., 2002). In this study, the authors quantified the amount of cross-linking events between 13 loci distributed along the ~320 kbp long yeast chromosome 3, resulting in measurements for 78 pairs. The cross-linking data are interpreted as contact frequencies between the corresponding segments of DNA. These frequencies were then transformed into average spatial distances. The conversion of cross-linking frequencies into spatial distances is an important ingredient of most computational reconstructions discussed below. Obviously, high contact frequencies are taken to indicate small spatial distances between the corresponding loci, while low frequencies indicate larger distances. How to map contact frequencies into distances quantitatively, however, is not straightforward. Accordingly, different approaches have been used, as will be apparent in the following sections.

In this study, the authors first plotted the cross-linking frequencies observed for all 78 pairs of loci as function of their genomic* distance s (Fig. 9.14A, dots). They then compared these data to a theoretical polymer of contact frequencies derived from the statistical properties of a WLC* (Rippe, 2001; Shimada and Yamakawa, 1984). This formula (equation in Fig. 9.14A) involved three unknown parameters, including the persistence* length L_p, a parameter c that reflects the circularization or near-circularization of a polymer (and a normalizing constant). These parameters were determined by fitting the model to the data. The authors reported a very good fit for $L_p = 56$ nm and $c = 363$ kbp ($R^2 = 0.86$) (red curve in Fig. 9.14A). A DNA compaction of $\beta_0 = 11.1$ nm/kbp was assumed when transforming genomic distances into spatial distances, in accordance with the assumption of a 30-nm chromatin fiber. According to the same polymer model, the average squared distances between two loci $\sqrt{\langle R^2 \rangle}$ are proportional to the contact frequency raised to the power $-2/3$, and a proportionality constant was obtained using the fitted parameters above, thus allowing to compute the distances $\sqrt{\langle R^2 \rangle}$ from the contact frequencies for all 78 pairs of loci (Fig. 9.14B).

The final step of the analysis was to determine the relative positions in 3D space of all 13 loci. If all loci had the exact same positions relative to each

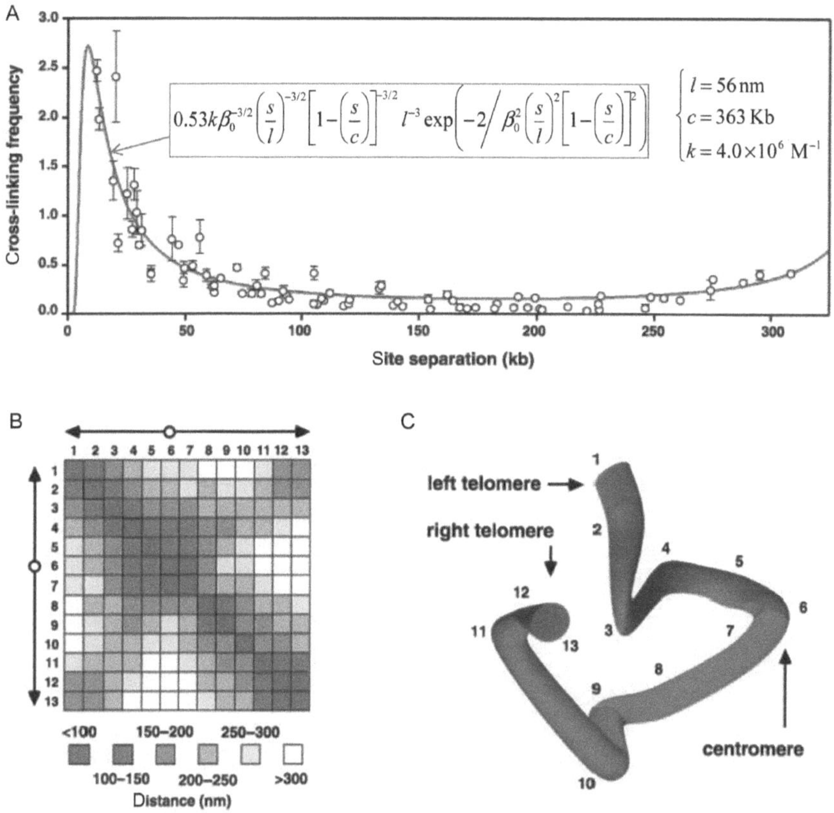

Figure 9.14 A 3D model of yeast chromosome 3 reconstructed from 3C data. *Adapted from Dekker et al. (2002). Reprinted with permission from AAAS.* (A) Cross-linking frequencies are plotted for all 78 pairs of loci against their genomic separation. Dots and error bars show the mean and standard error of the mean from triplicate experiments for each of the pairs. The red curve is a theoretical model given by the equation (red box). The values of the three parameters were obtained by fitting the model to the data. (B) Average distances between 13 loci along yeast chromosome 3, visualized as a heat map (see color legend on the bottom). These average distances were calculated from the cross-linking frequencies using the model shown in (A). The circle on the arrows indicates the location of the centromere; the arrowheads indicate the telomeres. (C) 3D model of the chromosome determined from the distance map in (B) (after approximating it by a matrix of rank 3 and performing Cholesky decomposition). (For interpretation of the references to color in this figure legend, the reader is referred to the online version of this chapter.)

other in all cells of the population, and in absence of measurement errors, the distance matrix (which has 78 distinct entries) would be entirely determined by the 3D coordinates of the 13 loci and could be converted to a 12×12 symmetric matrix (of inner vector products) containing only three independent columns (i.e., the matrix would have rank 3). This was indeed approximately the case, allowing the authors to use a standard mathematical matrix operation to retrieve the 3D coordinates of the 13 loci and hence a 3D model of the entire chromosome (Fig. 9.14C). Interestingly, the reconstructed configuration was nearly circular, with the two telomeric ends of the chromosome in close spatial vicinity, consistent with earlier observations from fluorescence microscopy.

3.1.2 Reconstruction of the immunoglobulin locus from FISH data

While most reconstructions of chromosome structure used data from 3C or its high-throughput variants (see below), it is also possible to use imaging data. A prominent example is a reconstruction of the immunoglobulin heavy-chain locus (which encodes the large parts of antibodies) based on single-cell FISH experiments (Jhunjhunwala et al., 2008). In this study, two colors were used to image many different pairs of ~10 kbp loci taken among 12 loci spanning the ~2.5 Mbp long genomic region. The authors used three sets of pairs, in each of which a single locus (called anchor) was paired with the 11 other loci. In addition, data for 11 pairs of consecutive loci were also included, thus totaling 47 pairs. For each of these pairs, the authors measured the distances between loci in 40–300 individual cells.

After this microscopy and image analysis step, the authors proceeded to reconstruct the average configuration of the chromosome region. In contrast to 3C, imaging allows to directly determine the distances between loci in each pair without requiring any particular assumptions (within experimental errors due to limited localization accuracy). For this purpose, the three anchor loci were placed in 3D space in such a way as to respect the mean distances measured between the corresponding pairs of anchors (trilateration). Then, the remaining eight loci were positioned, again using trilateration, based on average measured distances with the three anchors. Finally, an optimization* method (using a gradient-descent algorithm) was used to refine these positions in order to minimize an objective* function consisting of the (summed squared) error between all measured mean distances and their counterparts in the model. The authors used this approach to propose a 3D model of the immunoglobulin locus in B lymphocytes. During development of B cells, distinct segments of the locus undergo recombinatorial rearrangement to enhance antibody variability. The authors observed a much tighter folding of the reconstructed locus in pro-B cells (when genomic rearrangements occur) compared to cells at an earlier developmental stage (pre-pro-B cells), suggesting that the spatial configuration of this chromosome region facilitates genomic reshuffling (Jhunjhunwala et al., 2008).

3.1.3 Reconstruction of the budding yeast genome from Hi-C data

One year after the release of the first genome-wide contact data (Hi-C) from human cells (Lieberman-Aiden et al., 2009)—and following an earlier study (Rodley et al., 2009)—genome-wide contact frequencies were determined for the budding yeast genome (Duan et al., 2010). This data set was then used to reconstruct a 3D model of all 16 chromosomes in the nucleus (Fig. 9.15). In this model, each chromosome segment of 10 kbp was represented by a bead of initially unknown position. Next, the experimentally measured contact frequencies were transformed into desired spatial distances between pairs of beads. To perform this transformation, the authors proceeded as follows. First, they plotted the average measured intrachromosomal (*cis*) contact frequencies as function of genomic* separation *s* (see Fig. 9.13D). Then, they transformed the genomic separation *s* into spatial distances by assuming a DNA compaction in chromatin of 130 bp/nm (based on the 110–150 bp/nm range estimated previously (Bystricky et al., 2004)) and implicitly assuming a straight fiber geometry. Thus, the authors obtained a transformation of (intrachromosomal, *cis*) contact frequencies into spatial distances. This transformation was then also applied to interchromosomal

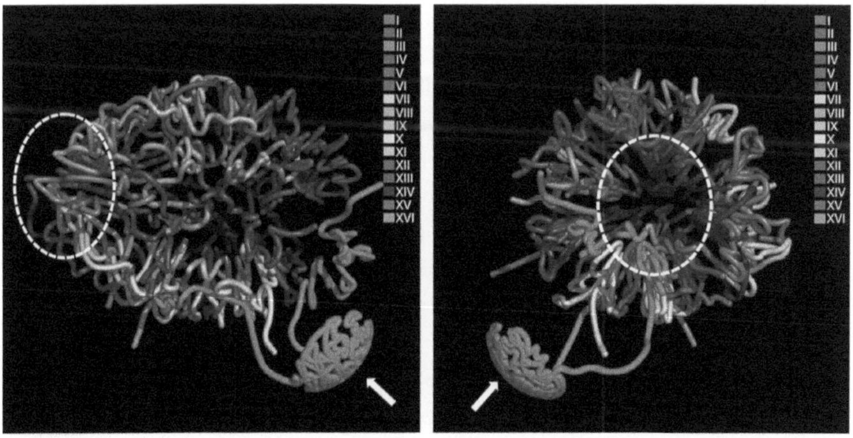

Figure 9.15 Static 3D model of the budding yeast nucleus reconstructed from genome-wide contact data. *Reprinted with permission from Duan et al. (2010) © (2010) Macmillan Publishers Ltd.* Each chromosome is shown by a different color as indicated in the legend. Left: view from the side, showing centromeres clustering in a small region (dashed oval). Right: view from the pole opposite the site of centromere clustering, showing a rosette-like arrangement of chromosomes. The white arrow points to the rDNA locus. (For color version of this figure, the reader is referred to the online version of this chapter.)

(*trans*) contact frequencies, thereby providing a desired distance for every pair of beads in the modeled genome.

In addition to these desired distances, the model also took into account other geometric constraints: all beads were confined to a 1 μm radius sphere; consecutive beads must lie within 66–91 nm (10 kbp divided by 110–150 bp/nm) from each other; two beads on the same chromosome must be at least 30 nm apart from each other (corresponding to the assumed diameter of the chromatin fiber), and two beads on separate chromosomes at least 75 nm apart; finally, based on imaging data (Berger et al., 2008), the beads corresponding to the rDNA locus on chromosome 12 (on which contacts could not be mapped because of its highly repetitive nature) were imposed to lie within a 300 nm radius sphere touching one pole of the nucleus, while the bead corresponding to the centromere was confined to a 100 nm radius sphere touching the opposite pole.

The entire set of desired distances and these additional geometric constraints cannot be satisfied simultaneously by any configuration of beads. Therefore, the authors resorted to an optimization* approach, in which the positions of the beads are moved until their relative distances most closely match the distances estimated from the contact frequencies (minimizing the sum of the squared differences between modeled distances and desired distances, as in the model of Section 3.1.2), while still satisfying the other constraints. The resulting configuration of chromosomes was described as a "water-lily," in which all 16 centromeres cluster near the same nuclear pole, chromosome arms extend away from this pole and the rDNA occupies the opposite pole (Fig. 9.15).

Although this model certainly reproduces some of the true features of yeast chromosome organization, it does not reflect the experimentally observed movement of chromatin loci nor the variability of their nuclear positioning observed in snapshots of cell populations (Berger et al., 2008; Heun et al., 2001). This limitation, in particular, has prompted several additional modeling studies, of which one was discussed above (Section 2.4.1) and others will be discussed below (Sections 3.2.4 and 3.2.7).

3.1.4 Reconstruction of fission yeast genome from Hi-C data

A similar experimental and computational approach was applied to reconstruct the architecture of the fission yeast genome, which totals ~14 Mbp and consists of only three chromosomes (Tanizawa et al., 2010). The conversion of contact frequencies into 3D distances, however, proceeded differently. Specifically, the authors used 3D FISH images to directly measure the spatial distances between 18 distinct pairs of loci (including 7 interchromosomal pairs), in 100 or more cells for each pair. They observed that the mean distances between loci were well approximated by an exponential function of the measured contact frequencies (with three fitted parameters). The authors then used this function to transform all measured contact frequencies into desired spatial distances. Chromosomes were modeled as chains of beads, each one corresponding to 20 kbp. The bead positions were determined by minimization of the same objective* function as in the previous model (Section 3.1.3), although only the 60% highest contact frequencies were used. Likewise, optimization* was performed under a few additional constraints. Specifically, all beads were constrained inside a 0.71 μm radius sphere (corresponding to the largest measured FISH distance); no pair of beads could come closer than 30 nm of each other; consecutive beads were located between 133 and 182 nm from each other; all three centromeres were forced to lie within a 30 nm radius sphere touching the spherical nuclear envelope; finally, telomeres were constrained to lie at the nuclear envelope. After reconstructing a 3D model of the three chromosomes, the authors proceeded to analyze the relative location of several genomic elements, finding in particular that co-regulated genes are frequently in close proximity, suggesting that they may share transcription factories (Tanizawa et al., 2010).

3.2. Reconstruction of multiple structures

The models described above generated a single 3D structure of one or multiple chromosomes. Clearly, these reconstructions do not account for the dynamics of the chromosomal polymers, nor for the fact that—even in absence of chromosome dynamics—distinct configurations may exist in different cells of a population. Furthermore, the optimization* procedures used in the methods above may yield quite different results depending on the initial configuration chosen, casting doubt on the meaning of the particular configurations obtained from a single optimization* run. The models described below address this limitation, at least to some extent.

3.2.1 Reconstruction of a human chromatin domain from 5C data

In 2011, Marti-Renom and colleagues described a reconstruction method adapted from a computational approach previously used to reconstruct large molecular assemblies (Alber et al., 2007). Their goal was to reconstruct the configuration of a 500 kbp chromatin domain on human chromosome

16 that contains the α-globin locus. It had previously been shown that expression of this gene required looping of the chromatin to bring the locus in contact with an enhancer sequence located 33–48 kbp upstream. To better characterize the spatial configuration of this chromatin domain in relation with α-globin gene expression, Baù et al. (2011) used 5C to measure the contact frequencies between 70 fragments distributed along this domain. The domain was modeled as 70 beads, each of which was linked to its two neighbors by springs. To model the interaction between consecutive fragments, the authors assumed a 30 nm chromatin fiber with a compaction of 100 bp/nm. In order to model the long-range interactions detected by 5C, additional springs were used to connect nonconsecutive beads. Each spring was given an equilibrium length (the length in absence of load) and a stiffness (which specifies the force exerted by the spring when compressed or extended). For springs connecting nonconsecutive beads, these two parameters were computed from the 5C data. To do this, the authors first normalized the data by computing the logarithm of the measured cross-linking frequencies and turning them into *Z*-scores (the number of standard deviations above or below the mean). The *Z*-scores were then transformed into spatial distances by assuming that these quantities are inversely proportional to each other. These distances were then used to define the equilibrium spring lengths. The spring stiffness was set as the square root of the *Z*-score's absolute value, such that springs corresponding to particularly low or high contact frequencies were stiffer than the average. Additional constraints were introduced to prevent pairs of beads with particularly high or low contact frequencies from moving too close to or too far away from each other. Furthermore, the model used spherical excluded volumes for each bead with radii proportional to the genomic length of the fragment.

Starting from an initial random placement in a cubic volume, the beads were then iteratively displaced using previously developed optimization* algorithms (Alber et al., 2007), until a configuration was obtained that minimized violation of all introduced constraints (Fig. 9.16A). This approach is similar to that discussed above (Sections 3.1.3 and 3.1.4) (Duan et al., 2010; Tanizawa et al., 2010). However, instead of optimizing a single 3D configuration, here the optimization was performed 50,000 times starting from as many random initial configurations (Baù et al., 2011). Because the initial structures were all different and the optimization procedure usually does not find the global minimum, but only a local minimum, the 50,000 optimized structures corresponding to these minima also differed, though often

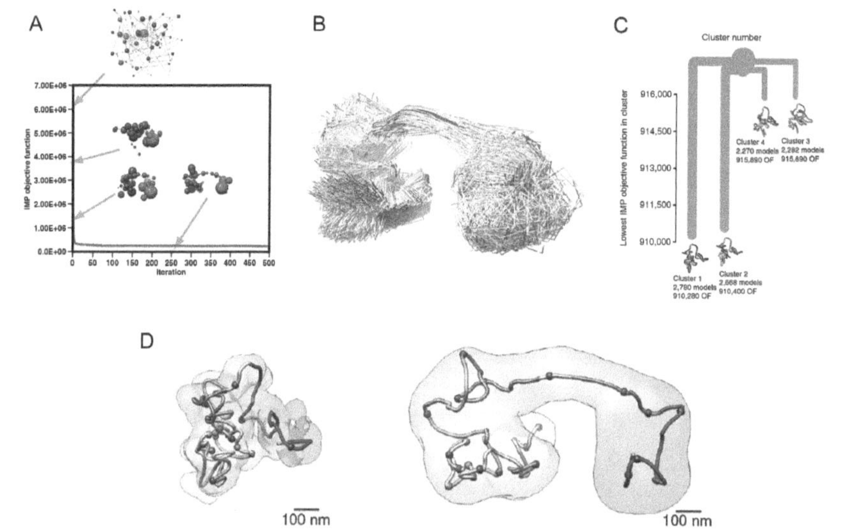

Figure 9.16 Reconstructing multiple configurations of a chromatin domain containing the α-globin locus from 5C data. *Reprinted with permission from Baù et al. (2011) © (2011) Macmillan Publishers Ltd.* (A) Iterative optimization* of a 3D model of the chromatin domain. The plot shows the objective* function (which expresses how strongly the constraints are violated) as function of iteration. The configuration is shown for four different iterations, with restriction fragments represented as balls connected by lines (balls have radii proportional to fragment lengths, and colors change progressively from red to blue along the chromatin fiber). The initial random configuration (structure on the top) corresponds to a high value of the objective* function, and rapidly evolves towards a configuration with much lower objective function characterized, in this case, by two distinct globular domains (rightmost structure). (B) Superposition of multiple aligned structures resulting from independent optimizations. (C) Clustering of structures obtained from independent optimizations yields four clusters for one of the cell lines analyzed. (D) Typical 3D structures of the chromatin domain in cell lines where the α-globin gene is silent (left) or expressed (right). The displayed structures correspond to the cluster centroid, that is, the average configuration of the largest cluster. The transparent "hull" refects the variability among all structures within the largest cluster. (For interpretation of the references to color in this figure legend, the reader is referred to the online version of this chapter.)

only slightly (Fig. 9.16B). The authors then retained only the 10,000 structures with the lowest minima (thus removing the optimized structures that most strongly violated the constraints) and proceeded to group them based on their structural similarity, using an unsupervised clustering algorithm. The parameters of the clustering algorithm were chosen such that the contact frequencies for the cluster containing most structures had the largest correlation with the experimental 5C data. Depending on the 5C data set, different numbers of clusters were obtained. Specifically, when applied to cell lines in which the α-globin gene was silent, this approach yielded only four clusters (Fig. 9.16C), the largest two corresponding approximately to mirror images of each other (mirror structures cannot be distinguished by contact frequency data because they yield identical distances). When applied to cell lines in which the α-globin gene was expressed, a much larger number of clusters were found. Analysis of the largest cluster indicated that the domain folded into two globules for the α-globin expressing cell line and into a single globule in the silent gene cell line (Fig. 9.16D). In the reconstructed models, the gene and the enhancer were in closer proximity for the α-globin expressing cell line, in agreement with previous 3C studies, and also exhibited a more extended domain than in the silent gene cell line. This prediction was confirmed by FISH experiments that measured the distance between the extremities of the chromatin domain (Baù et al., 2011).

3.2.2 Reconstruction of a bacterial genome from 5C data

A very similar computational approach was used by the same group to model the genome of the bacterium *Caulobacter crescentus*, which consists of a single circular ~4 Mbp long chromosome (Umbarger et al., 2011). The authors used 5C data for 339 fragments spanning this bacterial genome, thus providing a genomic resolution of ~12 kbp. Each fragment was represented by a bead whose size was determined based on the fragment's length and an assumed chromatin density of 0.3 bp/nm^3. As above, contact frequencies were normalized and used to compute the desired distances between bead pairs. To compute these distances, however, the authors took advantage of previous fluorescence microscopy measurements of distances between loci at different genomic separations (Viollier et al., 2004)—an approach similar to that discussed in Section 3.1.4 (Tanizawa et al., 2010). These data were combined with the average 5C contact frequencies as function of genomic distances in order to infer the relationship between contact frequencies and spatial distances. The authors approximated this relation by a fifth-order polynomial that mostly decreases with increasing spatial distances. (When contact frequencies could be mapped to more than one distance, the smallest distance was chosen.) The rigidity of springs was defined as the square of the normalized contact frequency as in Baù et al. (2011), with similar modifications for very low or very high contact frequencies or for neighboring fragments. The special treatments applied to high and low contact frequencies were necessary to prevent noisy reconstructed structures that did not fall into clear clusters. The authors also used a filtering approach to remove the effect of isolated outliers in the contact frequency maps. In addition, a large force was applied to maintain all beads inside a box of length 2.35 μm and width 0.6 μm, similar to the size of *C. crescentus* cells. The optimization* and clustering methods were similar to those used in Alber et al. (2007) and Baù et al. (2011), and resulted in two main configurations of the bacterial genome (that were not mirror images of each other). In both configurations, the circular genome was elongated as an ellipsoid and twisted ~1.5 times around its longest axis. Using genetic manipulations, the authors showed that the configuration of the chromosomes was determined by *parS*, a locus involved in chromosome segregation, which anchors the chromosome to one pole of the bacterium. However, the study also indicated that additional unknown factors might also influence chromosome positioning within the cell (Umbarger et al., 2011).

3.2.3 Reconstruction of structure probability distributions

In the reconstruction approaches discussed in Section 3.1, the experimental data were analyzed with an optimization* approach that determined a single 3D structure. By contrast, the two studies discussed just above (Sections 3.2.1 and 3.2.2) computed many thousands of structures by running the optimization algorithm independently starting from random initial configurations. The resulting structures were then grouped in clusters and further analyzed as described above. However, the variability observed in the obtained structures (partly reflected by the multiplicity of clusters) could not readily be interpreted in terms of either biological variability of structures in a cell population, or statistical uncertainties in determining a model from necessarily limited data. As a consequence it is unclear, for example, how to compute error bars on quantities predicted by the model, such as mean distances between loci.

This limitation motivated the development of a reconstruction method called MCMC5C, which allows to compute a probability distribution of structures from contact frequency data (Rousseau et al., 2011). In spirit, the method is quite similar to Bayesian approaches previously developed for protein structure determination (Rieping et al., 2005). The method uses the Bayes rule, which relates the probability of a model given the data (posterior probability) to the probability of the model in absence of data (prior) and the probability of the data given the model (likelihood). For the sake of description, we temporarily assume that the posterior probability can be calculated for a given model (here, a 3D structure)—how this was done will be summarized in the next paragraph. In order to determine the probability density of structures, the posterior probability should ideally be computed for all possible structures, but because the space of possible structures is enormous (in fact infinite), this is impossible. Therefore, the authors used a sampling approach, in which the posterior probability distribution was approximated using the posterior probabilities of a large number of samples, that is, individual 3D structures. For this approximation to be correct, the samples must be taken in a manner allowing to cover the regions of the (very high-dimensional) space of structures containing non-negligible posterior probabilities. To achieve this, the authors used a computational strategy known as Markov chain Monte Carlo (MCMC), in which an initial structure is iteratively modified using random "moves." The structure consisted of a single chain of beads, and each move consisted simply in a random displacement, within a sphere, of a randomly selected bead. This random move was either accepted or rejected depending on the ratio r of the posterior probability of the new structure divided by the posterior probability of the old structure: if $r > 1$ the move was always accepted, if $r < 1$, it was accepted with probability r and otherwise rejected. After several billions of iterations, the MCMC algorithm sampled a distribution of structures that did not depend on the arbitrary starting structure. By drawing a subset of the samples separated by a sufficiently large number of iterations (in this case 250–500), a good approximation of the true probability density of structures was obtained. This approach thus generated not just a single structure, but a family of structures with associated probabilities.

In order to compute the posterior probability of a structure from the Bayes rule, the authors used two main assumptions: (i) a flat prior, that is, the probability of any 3D structure in absence of data was assumed the same (thus effectively neglecting constraints such as volume exclusion or DNA compaction ratios); and (ii) a probabilistic relationship between the contact probabilities and the 3D structures. To determine the latter relationship, the authors assumed that the average contact frequency varied with a power law of the spatial distance. Rather than imposing a specific exponent for this power law, as the models discussed above, they determined it from the data using cross-validation: for a given exponent, they run the model using all of the contact data except for a specific pair of beads, then computed the chain structure with the highest probability and compared the distance between these two beads in the chain to the distance predicted from their contact frequency. After testing a range of values, the authors determined that an exponent of 2 provided the best cross-validation. Thus, the spatial distances were taken to be inversely proportional to the squared contact frequency. Finally, the model also included a probabilistic description of random noise in the measurement of contact frequencies, assuming a Gaussian with a variance estimated from the data.

The authors first validated their reconstruction method on contact frequencies computed from a simulated 3D structure (and assuming the above-mentioned relationship between distances and contact frequencies). They showed that the sampled structures aligned well with the simulated ground truth. They then applied the method to previously obtained 5C data of a 142 kbp region of chromosome 7 containing a cluster of *Hox* genes, which play a key role in development and cell differentiation (Fraser et al., 2009). The authors found that the chromosomal region in the differentiated cells was more compact than in undifferentiated cells and that this difference was statistically significant—a statistical test that was possible thanks to the availability of the structure probability distribution, as opposed to a single structure. They also applied MCMC5C to Hi-C data previously obtained for a 88.4 Mbp long arm of human chromosome 14 and showed that the distances predicted by their model for three pairs of loci correlated well with FISH data (Lieberman-Aiden et al., 2009). Extension of the analysis to the entire human genome was unfortunately not feasible because of the excessive computation time for reliable sampling.

Recently, another Bayesian method for chromosome reconstruction was proposed based on partly similar motivations (Hu et al., 2013) and using a similar, MCMC-based, sampling procedure. An important difference is the adoption of a mixture model, which partly relaxes the assumption underlying MCMC5C that all cells in the population of the Hi-C experiment display the same (or at least a similar) chromosome configuration (consensus structure). The new approach tests whether a mixture of different consensus structures can explain the Hi-C data significantly better than a single one. More precisely, this was done by dividing chromatin domains in two regions of equal genomic size, assuming that each can be described by a consensus structure, and asking if different relative orientations of the two regions are required to explain the data. A model selection approach (known as Akaike information criterion) was used to penalize this increase in model complexity. Another difference is the explicit inclusion in the model of various experimental biases known to affect the Hi-C data and which have been previously characterized statistically (Yaffe and Tanay, 2011). The authors also assumed a Poisson statistics of the contact frequencies that better reflects the counting process of Hi-C than a Gaussian distribution. As other methods described above, the authors assumed a power law relation between average distances and contact frequencies. The reconstruction method, called BACH-MIX, was then applied to determine chromosome structures within ~Mb long "topological domains" identified in a recent Hi-C data set on mouse embryonic stem cells (Dixon et al., 2012). Interestingly, the results suggested that most topological domains can be described by a single consensus structure or a mixture of two consensus structures in which one structure dominates over the other (i.e., is given a much higher weight, corresponding to a larger proportion of the cell population), and even that entire chromosomes may display a single dominant structure (Hu et al., 2013). The authors also argued that their method provided improved predictions of distances as validated with FISH data. As MCMC5C, the approach is currently limited to reconstructing a single chromosome.

3.2.4 Brownian dynamics simulation of yeast chromosomes biased by Hi-C data

As already mentioned, the static 3D model of budding yeast genome architecture described in Section 3.1.3 (Duan et al., 2010) was followed by several recent modeling studies aiming to account for the variability in chromosome configurations (Gehlen et al., 2012; Tjong et al., 2012; Tokuda et al., 2012; Wong et al., 2012). Two of these studies can be considered as hybrids between polymer models of the type described in Section 2 (and more specifically in Section 2.4.1) and data-based reconstructions (Gehlen et al., 2012; Tokuda et al., 2012). In the first of these models, chromosomes were represented as chains of beads (each of 3 kbp) undergoing motions which were simulated using Brownian* dynamics (Tokuda et al., 2012). The model included forces that prevented beads from penetrating each other (assuming a 30 nm chromatin fiber), spring forces that ensured a distance between consecutive beads consistent with a compaction of 130 bp/nm, and forces that resisted chain bending, assuming a persistence* length of $L_p = 170$–220 nm (Bystricky et al., 2004). In addition, the model included forces that maintained the beads inside a 1 μm radius sphere, the centromeres of all 16 chromosomes near the SPB, and sequestered exclusively the beads of the rDNA locus inside a cap on the opposite pole of the nucleus that represented the nucleolus. Furthermore, optional forces were introduced to maintain telomeres and/or rDNA loci at the nuclear periphery in accordance with observations (Mekhail et al., 2008; Taddei et al., 2010).

These model ingredients are quite similar to those of the polymer model with minimal sequence specificity discussed in Section 2.4.1. However, unlike Wong et al. (2012), the authors also included data-driven "forces" that were calculated from the genome-wide contact frequencies of Duan et al. (2010). These forces acted as springs between pairs of beads, with equilibrium lengths set to the distances previously estimated for the reconstruction discussed in Section 3.1.3. The weight of these data-driven forces relative to those mentioned above could be adjusted. From their dynamic simulation, the authors then computed the statistical distribution of distances between six pairs of telomeres and compared them to previously published experimental data from fluorescence imaging (Bystricky et al., 2005). According to their analysis, the model did not fit all measured distances well when the weight of the data-driven forces was too low or too weak. For an intermediate weight, however, a relatively good agreement between the model distances and the measurements could be obtained. This result led the authors to conclude that the dynamics of chromosomes is not entirely random, but is driven in part by specific interactions between chromatin segments as determined by the measured contact frequencies.

3.2.5 Reconstructing a population of genome structures

The "inverse models" summarized so far yielded interesting results, but face important limitations that are likely to restrict their generality applicability or may call into question the reliability of the obtained structures. One central limitation of these methods is their underlying assumption that the contact frequencies (or in one case, the average FISH distances (Jhunjhunwala et al., 2008)) reflect a single 3D structure. Although some models discussed above (Section 3.2) generate multiple structures, each of these structures results from constraints originating from the same contact frequency matrix (except for the recent model by Hu et al. (2013) mentioned in Section 3.2.3). Thus, all detected interactions simultaneously influence the final configuration of each computed structure. Such modeling approaches however do not account for the nature of the 3C-based experimental data, which are ensemble measurements that aggregate contact events over millions of cells, without the ability to trace back these contacts to individual cells. Thus, only a very small fraction of these contacts may actually occur in any given cell, and which set of contacts can occur in the same cell is unknown. For example, it is possible that very different genomic structures exist in distinct subpopulations of cells, and that these different structures give rise to nonoverlapping groups of contacts. Models that apply the entire contact frequency data to all structures can thus potentially generate structures very different from the real ones.

Another important (and partly related) limitation of the methods discussed above is their reliance on a transformation of contact frequencies into spatial distances. Although contacts detected by cross-linking events require physical proximity, it is not in general clear if and how a given contact frequency can be assigned to a spatial distance, or even a distribution of distances. In practice, the transformations used are based either on theoretical predictions from basic polymer physics (as in Section 3.1.1) or on empirical relations between contact frequencies and genomic separation combined with assumptions about the structure of the chromatin fiber (as in Section 3.1.3), sometimes supplemented by fluorescence microscopy data on spatial distances for selected loci (as in Section 3.2.2). The relationships between contact frequencies and spatial distances thus obtained from intrachromosomal (*cis*) contacts are then sometimes extrapolated to interchromosomal (*trans*) contacts. The validity of the underlying assumptions usually cannot be easily verified.

These two important limitations can be overcome with appropriate modeling approaches, as demonstrated by Kalhor et al. (2012). In this study, the authors modeled a population of 10,000 genome structures whose configurations were calculated based on constraints derived from genome-wide contact frequencies determined experimentally in human cells. The model employed random initial configurations and an optimization method using data-driven molecular* dynamics similar to that mentioned in Sections 3.2.1 and 3.2.2—but in contrast to these studies, the 10,000 structures were not optimized independently. Instead, a single optimization* problem was defined based on the entire population, that is, the objective function depended on all 10,000 structures simultaneously (Fig. 9.17D). Thus, the measured contacts did not need to be reproduced by individual structures, but only by their aggregation, in fine agreement with the population-based nature of the experiments. This approach also allowed the authors to dispense of the transformations of contact frequencies into spatial distances.

In this model, the entire diploid human genome was represented by a total of 2×428 spheres, each corresponding to a distinct chromosomal region. The genomic boundaries of these regions were determined from the 23 intrachromosomal contact frequency matrices (Fig. 9.17A) using a clustering algorithm that grouped consecutive bins into larger blocks such that the resulting block matrix provided a good approximation to the full-resolution matrix (Fig. 9.17B). Each block was represented by a sphere, the volume of which was chosen to be proportional to the corresponding nucleotide content—and assuming that 20% of the nuclear volume is occupied by genomic material (Fig. 9.17C). The spheres were forced inside a spherical nuclear volume of radius 5 μm and not allowed to penetrate each other (hard spheres). However, each sphere was given a soft outer shell allowed to overlap with that of other spheres. Overlaps between two spheres

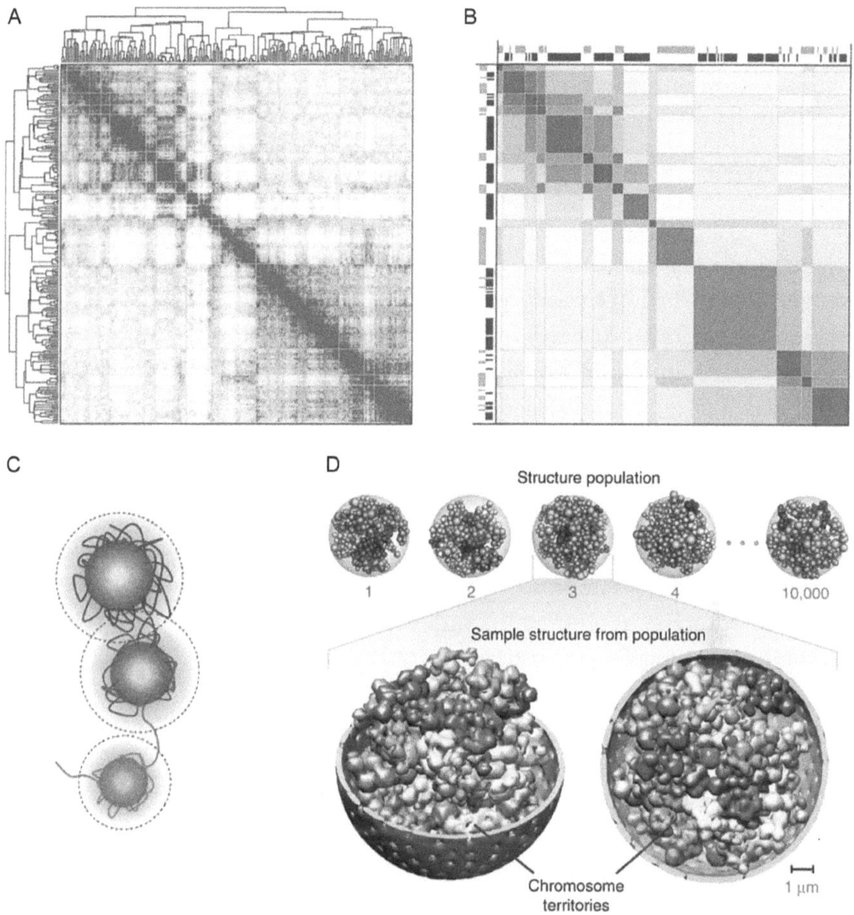

Figure 9.17 Ensemble of genome structures reconstructed from genome-wide contact frequencies in human cells. *Reprinted with permission from Kalhor et al. (2012) © (2012) Macmillan Publishers Ltd.* (A) Experimentally obtained contact frequency matrix for chromosome 11, of size 237 x 237. The dendrograms show the result of a hierarchical clustering procedure used to group rows (or columns) with similar contact frequency profiles. (B) Approximation of the contact matrix in (A) obtained by partitioning the chromosome based on an automated thresholding of the dendrogram. Consecutive bins within the same cluster are combined into a single larger bin. The resulting matrix is of size 15 x 15. (C) Each chromatin segment from this partition is represented by an impenetrable sphere with a volume proportional to the genomic content, surrounded by a soft shell of double radius, which is allowed to overlap with the outer shells from other chromatin segments. Such overlaps are used to define contact events. (D) Top: an ensemble of 10,000 genome structures (modeled using the genomic partition applied to all chromosomes) is optimized simultaneously to reproduce the entire genome-wide contact matrix. Bottom: zoom into a single genome structure from this ensemble, showing each of the 2 x 23 chromosomes in a different color (homologous chromosomes are given the same color). (For interpretation of the references to color in this figure legend, the reader is referred to the online version of this chapter.)

were scored as a contact. Such contacts were imposed between spheres corresponding to consecutive genomic blocks. For other pairs of spheres, overlaps were only imposed in a fraction of the structure population proportional to the experimentally measured contact frequency. This fraction was chosen heuristically as the subset of structures for which the spheres were already closest to each other in the initial random configurations. A special procedure was applied to account for the ambiguity posed by homologous chromosomes in this diploid genome.

Computational optimization of these constraints then produced a population of structures that was fully consistent with all these constraints (i.e., the objective function was minimized to zero, in contrast to models described in the previous sections). The modeling approach was applied to genome-wide contact data in human lymphoblasts obtained using an improved cross-linking protocol (in which cross-linked fragments were tethered to beads) (Kalhor et al., 2012). Importantly, their model was able to predict positional preferences of chromosome territories with respect to the nuclear center or periphery that agreed with those previously observed by FISH, thereby providing an independent validation of this powerful modeling strategy (Boyle, 2001).

3.2.6 Molecular dynamics biased by randomly subsampled contact data

In a partly similar effort, another group (Gehlen et al., 2012) developed a biased molecular* dynamics simulation of budding yeast chromosomes. This model used a previously obtained genome-wide contact data set (Rodley et al., 2009) to attract frequently interacting loci toward each other. As a key difference with the model of Tokuda et al. (2012) discussed in Section 3.2.4, the authors did not use the entire contact data simultaneously to bias chromosome dynamics. In line with the previous approach (Section 3.2.5), they considered instead that contacts detected between two genomic regions do not necessarily occur simultaneously in all cells, but only in subsets of the cell population. Because these subsets are unknown, they chose random subsets of contacts for each of 250 independent simulations and activated spring-like forces tending to push the corresponding pairs of chain segments toward each other.

Another distinguishing feature of this model is the assumption of two distinct chromatin structures, open or compact, with diameters of 10 and 30 nm, persistence* lengths of 30 and 200 nm, and DNA compactions of 22 and 130 bp/nm, respectively. Chain segments were assigned to either of the two chromatin states based on measured intrachromosomal contact frequencies, with the overall fraction of compact chromatin adjusted to 70%, a value for which the model agreed best with the experimental contact frequencies (Rodley et al., 2009). Much as in Tokuda et al. (2012), the simulation included forces to tether centromeres in the vicinity of the SPB, the telomeres near the nuclear envelope, and the chromosomes inside a 1 μm radius sphere. Similarly, the nucleolus was modeled as a spherical cap opposite the SPB (with a maximum thickness of 300 nm) sequestering the rDNA and excluding other chromosome regions.

By computationally aligning the configurations resulting from the independent simulations, the authors determined the nuclear space explored by each modeled chromosome. Although the tethering constraints by themselves led to a nonrandom positioning of chromosomes in nuclear space, incorporating the contact data led to a much tighter nuclear occupancy (i.e., aligned chromosomes were closer to each other). From this observation, the authors drew the conclusion, in line with Tokuda et al. (2012), that specific chromosomal interactions play a key role in chromosome positioning. However, the authors did not directly compare the predictions of their model in absence of contact data to experimental data on chromosome positioning or contact frequencies, thus leaving open whether such specific interactions were in fact required.

3.2.7 Geometric population modeling of the yeast genome

The structure ensemble model described in Section 3.2.5 (Kalhor et al., 2012) was adapted by the same group to the yeast genome (Tjong et al., 2012). However, in contrast to the study on human cells their yeast model did not incorporate any experimental data on contact frequencies. Instead, the authors only considered geometric constraints: chromosomes were modeled as chains of 30 nm beads, each containing 3.2 kbp of DNA, with mutually exclusive volumes. Much as in the other yeast models, the telomeres were constrained to lie within 50 nm of a 1 μm radius sphere; the centromere of each of the 16 chromosomes was confined to the vicinity of the SPB; finally, the rDNA locus was not explicitly modeled, but its two boundaries on chromosome 12 were constrained to lie on the surface of a crescent-shaped compartment opposite the SPB that approximated the observed nucleolus; all other loci were excluded from this nucleolar compartment. Starting from random initial distributions of the beads, 200,000 structures were optimized based on these geometric constraints, using algorithms similar to those in Alber et al. (2007) and Kalhor et al. (2012).

Because it did not directly use any experimental data as input, this model could actually predict such data, including contact frequencies across the genome and intranuclear distributions of loci. As such, and despite its origin in optimization*-based reconstruction rather than polymer physics, this model could arguably be classified among the direct models discussed in Section 2. Comparing the model's predictions to previously obtained experimental data, the authors found a remarkable good agreement for nuclear positions of loci and relative distances as well as genome-wide contact frequencies (Berger et al., 2008; Duan et al., 2010; Thérizols et al., 2010). In agreement with the study summarized in Section 2.4.1, the authors conclude that genome architecture in this organism results from a few tethering constraints and statistical properties of random chains, and does not require other sequence-specific interactions.

4. CONCLUDING REMARKS

We have presented a semi-historic overview of quantitative models developed over the last ~25 years to describe or explain genome architecture. For each model, we tried to emphasize the main assumptions, the targeted biological application, and the experimental data that were used to build or test it. We also discussed the similarities and differences between models and their main motivations. We hope that this will allow readers to better appreciate the power and limitations of individual models and their relevance for understanding genome architecture *in vivo*.

As detailed above, computational models of genome architecture roughly cluster into two groups: (i) direct models based on a small set of assumptions and physical parameters (Section 2); and (ii) inverse models, which use large experimental data and additional assumptions to reconstruct structures consistent with the data (Section 3) (Fig. 9.1) (Marti-Renom and Mirny, 2011). The models of the first group mainly describe chromosomes as long polymer fibers in solution. The heterogeneity of the fiber due to the DNA sequence, the chemical modifications of DNA or histones, or the presence of specific chromatin-binding proteins are mostly—if not entirely—ignored by these approaches. Thus, such models essentially remove the biological information and only retain physical assumptions. Despite this drastic simplification, we reviewed evidence that polymer physics models can in fact account for a surprisingly large set of observations. These include qualitative properties, such as spatial segregation of chromosomes into distinct territories, but also quantitative data such as the scaling laws of average contact frequencies with genomic distance (Lieberman-Aiden et al., 2009; Rosa and Everaers, 2008; Rosa et al., 2010; Wong et al., 2012). Nevertheless, generic polymer models also face limitations. These models cannot account for all features of rich experimental data sets, such as the observed partitioning of higher eukaryotic chromosomes into topological domains, the boundaries of which are associated with specific DNA-binding proteins (Dixon et al., 2012).

This of course is where the approaches of the second group come in. Inverse models can fully take into account rich experimental data sets such as genome-wide contact frequencies, because they use them as input to the generation of the model itself. Obviously, agreement of the models with the input data is required for internal consistency but has otherwise limited significance. The inverse models are still valuable, however, because they may yield more realistic 3D representations of chromosomes than generic polymer models and allow to predict other features of genome architecture that are not directly provided by the input experimental data, for example, the positions of chromosomes, chromatin domains or functionally related loci relative to each other or relative to nuclear landmarks such as the nuclear envelope. Such models also face limitations, however. Reconstruction of 3D structures from experimental data also requires the incorporation of assumptions. Some of these assumptions may be similar to those of the polymer models (e.g., an assumed persistence* length or chromatin occupancy volume), and as a result some inverse models can even be considered as extensions of polymer models (Meluzzi and Arya, 2013; Tokuda et al., 2012). Additional assumptions are necessary, however, in order to translate the experimental data into constraints for the model, and these are not always tested. For example, as we discussed in Section 3, many authors use different functions to transform cross-linking frequencies into preferred spatial distances—though some studies demonstrate that such assumptions can be dispensed of (Kalhor et al., 2012; Meluzzi and Arya, 2013). More generally, it often remains unclear how strongly the reconstructed models depend on the arbitrary parameters and assumptions. Thus, validations using independent data (e.g., imaging-based measurements for models reconstructed from cross-linking data) are often required to build confidence in the reconstructed models. Although most inverse models use optimization, probabilistic sampling approaches that can determine uncertainties in both reconstructed models and arbitrary parameters seem a promising strategy with the potential to providing a complete and less biased view of the ensemble of models that are consistent with the experimental data (Hu et al., 2013; Rieping et al., 2005; Rousseau et al., 2011). Of particular interest is the potential ability of such approaches to highlight how model uncertainties relate to the limitations of existing experimental data sets. This information could help guide the design of future experimental investigations aiming to decrease these uncertainties.

Notwithstanding such improvements, a key limitation of inverse models compared to direct models remains their relative lack of predictive power. Inverse models cannot predict, for example, how chromosome models will change as a result of a translocation or a change in gene expression, since—by definition—they require new data from such experiments as input for the reconstruction. Therefore, inverse models might be considered as a useful intermediate step to cope with the complexity of experimental observations on genome architecture until we can explain these observations using fully predictive models. An apparent avenue toward more realistic predictive models is to add functional biological information into generic polymer models. One example is provided by copolymer* models that partition chromosomes into distinct chromatin regions based on their genomic content, leading to phase separation and higher-order chromatin structures (Ostashevsky, 1998). Other examples are generic models of the yeast genome, where the introduction of a few sequence-specific spatial constraints was sufficient to account for the majority of experimental data on genome architecture (Tjong et al., 2012; Wong et al., 2012). Future studies might follow a similar path by identifying a limited set of specific elements that, combined with the generic physical properties of polymers, may allow not only to explain, but also to predict genomic architecture in more complex organisms.

Clearly, we are only at the beginning of this exciting journey. As new experiments produce increasingly complete and refined data, we will be able to better discriminate among competing modeling approaches and improve the most promising ones. The close interplay of experiments and computational modeling is poised to unveil the mechanisms of genome architecture and its connection with biological functions.

Box 9.1 Polymer physics: A brief introduction

Polymers are interesting physical systems that have stimulated intensive experimental and theoretical investigation. Nowadays, polymer physics is a very well understood topic, and several excellent textbooks are available (Doi and Edwards, 1988; De Gennes, 1979; Khokhlov and Grosberg, 1994; Rubinstein and Colby, 2003).

The predictive power of modern polymer theories stems from the fact that, for sufficiently long chains, the chemical specificity of the monomers composing the chain can usually be neglected. Each chain is modeled as a sequence of generic (abstract) monomers, which interact with each other through simple force fields. For example, in the case of chains with effects of volume exclusion, it is simply required that two monomers cannot occupy the same region of space. Since no specific chemistry is invoked, the range of applicability of modern polymer theories is enormous, ranging from synthetic polymers such as polystyrene and rubber to biological polymers such as DNA, RNA, and proteins (Harmandaris et al., 2007; Marko and Siggia, 1995; Micheletti, 2012; Rubinstein and Colby, 2003; Yoffe et al., 2008).

The reason why polymer models work so well despite neglecting chemical details can be understood by the closely related analogy with the statistical mechanics model of an *ideal gas*. As explained in introductory thermodynamics, most macroscopic properties of ideal gases, such as the famous law $PV=NK_BT$—which relates the pressure (P), and the volume (V) of a gas made of N identical particles to the temperature (T) and the Boltzmann constant (K_B)—can be derived from fundamental assumptions, modeling gases as *ensembles* of point-like particles inside a vessel, and in contact with a heat reservoir (Fig. 9.18A) (Huang, 1987). Thus, *macroscopic* observables such as temperature are derived as ensemble averages of specific *microscopic* observables (e.g., the particle velocity). Although it neglects chemical details such as the nature of the molecules composing the gas, the ideal gas model performs extremely well for dilute systems, and therefore plays a central role in many introductory courses in Physics (Huang, 1987).

Polymer physics shares many analogies with the thermodynamic theory of gases. Similarly to molecules in gases, polymer chain monomers move under the action of thermal agitation, and interact with other monomers (of the same chain or from other chains). However—at odds with the phenomenology of gases—monomers in a chain are connected to each other. This seriously limits the total number of configurations available to the chain,—that is, the entropy* of a polymer chain is significantly smaller than for a gas made of the same number of particles,—which in turn affects the thermodynamic properties of a polymer solution (Fig. 9.18B). In polymers, unlike in gases, particles are joined to each other and thus form chains, with no limit to the number of monomers contained in a single chain (Doi and Edwards, 1988; De Gennes, 1979; Khokhlov and Grosberg, 1994; Rubinstein and Colby, 2003). Moreover, chains can exist in different topologies, that is, one can have linear chains, circular chains (chains closed in a ring), or branched chains where different polymer branches can depart from the same monomer. Because of the chain constraint, the mathematical description of polymer solutions is—in general—more complicated than the "equivalent" theory of gases. As a result,

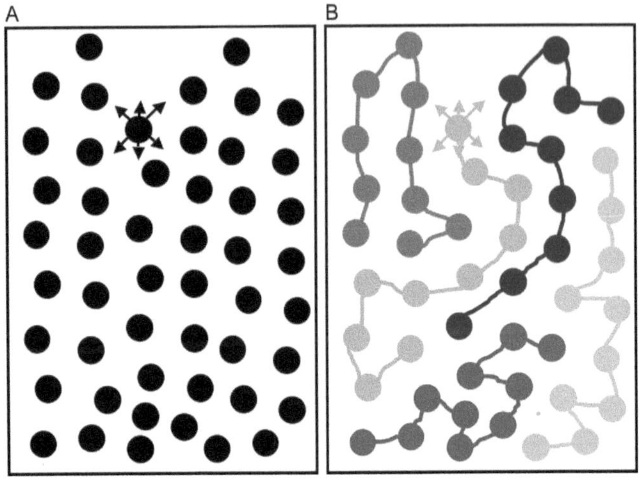

Figure 9.18 Physics of gases and polymers. (A) Illustration of point particles contained in a rectangular vessel, and driven by thermal motion (schematically described by arrows) as a model for an ideal gas (Huang, 1987). (B) The same molecules connected by strings constitute a model for a polymer solution. In this case, the particles have much less degrees of freedom compared to system (A), which is at the basis of many of the peculiar physical properties of polymer systems (Doi and Edwards, 1988; De Gennes, 1979; Khokhlov and Grosberg, 1994; Rubinstein and Colby, 2003). (For color version of this figure, the reader is referred to the online version of this chapter.)

analytical theories (using equations) must often be complemented by numerical methods, such as Monte Carlo* simulations (Allen and Tildesley, 1987) and molecular* dynamics (Kremer and Grest, 1990)—important tools to probe the equilibrium and dynamical properties of polymers under different experimental conditions.

Box 9.2 Essential Terminology

This section provides brief explanations of terms denoted by an asterisk (*) in the main text. The terms are listed here in alphabetical order.

Brownian dynamics: see molecular* dynamics.

Contour length, L: also called curvilinear length, is simply the distance covered if walking along the chain from one end to the other (units of nano- or micro-meters) (see Fig. 9.20). For DNA and chromatin fibers, it is directly related to the *genomic distance* s, which is naturally expressed in base pairs (bp). Using known or assumed geometric properties of DNA and chromatin, s can be converted into L. For example, a segment of $s=3$ bp of DNA in a normal double helix structure measures $L\sim 1$ nm.

Corresponding conversions for the chromatin fiber are less straightforward, since the structure of the fiber *in vivo* is still unclear and quite variable (Luger et al., 2012).

Copolymer: Copolymers (also known as heteropolymers) are polymer chains made of two (or, more) different species of monomers (Rubinstein and Colby, 2003) (Fig. 9.19). The properties of copolymers depend both on composition, and on the order in which these different monomers are distributed along the chain.

(Average squared) End-to-end distance, $\langle R^2(L)\rangle$: This quantity is often used as a measure of a polymer chain's average size and is defined as the squared distance between the chain ends (see Fig. 9.20), averaged over all possible conformations that the chain can assume in space due to random fluctuations (the average is denoted by the symbol $\langle\ \rangle$). It is often expressed as a function of the chain's contour* length L.

Entanglements: In polymer solutions, the dynamics of a chain is topologically constrained by the presence of surrounding chains because chains cannot cross (Fig. 9.21). These topological constraints result in a rich phenomenology best exemplified by the reptation* model.

Entropy: In Thermodynamics, entropy (traditionally indicated by the symbol S) is a measure of the total number of configurations (indicated by the symbol W) which can be ascribed to the system of interest (Huang, 1987).

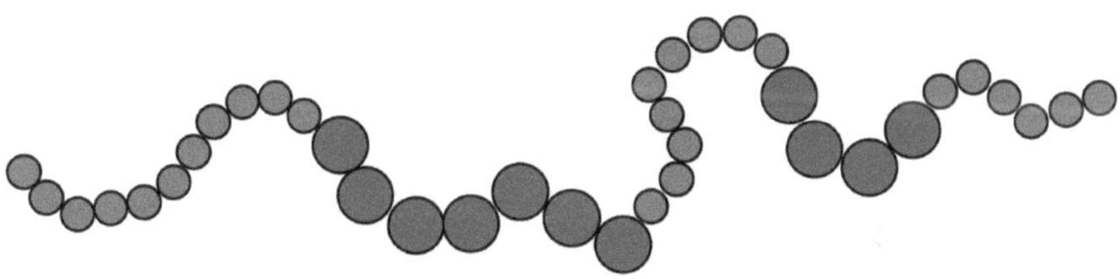

Figure 9.19 Copolymer consisting of two different species of monomers (blue and red). (For interpretation of the references to color in this figure legend, the reader is referred to the online version of this chapter.)

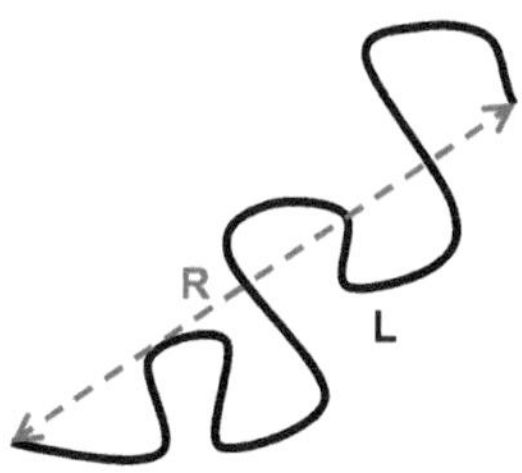

Figure 9.20 End-to-end distance R and contour length L of a polymer chain. (For color version of this figure, the reader is referred to the online version of this chapter.)

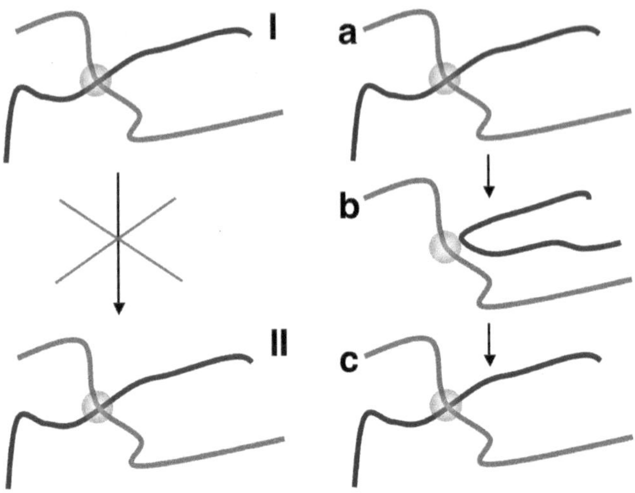

Figure 9.21 Entanglements in polymer solutions. (Left) The dynamics of the red chain in the yellow region is hindered by the presence of the blue chain. This prevents the chains from swapping their local positions (from I to II) by means of local "jumps." (Right) Chains can swap their local positions only by complete retraction of one end (from a to b), followed by chain rearrangements (from b to c). The slow dynamics of entangled chains in solution is described by the reptation* model. (For interpretation of the references to color in this figure legend, the reader is referred to the online version of this chapter.)

According to Boltzmann, $S = K_B \, log(W)$, where K_B is the Boltzmann constant. Entropy can be considered as a measure of the disorder of a system. The second law of thermodynamics states that the entropy of a closed system cannot decrease, but only increase with time.

Equilibrium: A system is said to be at equilibrium if its macroscopic physical properties, such as the average gyration* radius, or the average polymer dynamics, do not change over time. For dynamical quantities, such as the MSD, equilibrium means that the measured quantity shows the same time behavior, regardless of when we start observing the system. Thus, a system containing polymers with arbitrary configurations, for example linearly stretched out chains, is usually not at equilibrium, but will evolve towards an equilibrium state. This process is called relaxation.

Freely jointed chain (FJC): The FJC is the most simple polymer model. In this model, the polymer is represented by a chain of N rigid segments called Kuhn* segments. The length of each segment (Kuhn length), L_K, is a measure of the chain's rigidity (Rubinstein and Colby, 2003). The contour* length L is then simply $L = N \times L_K$. Each Kuhn segment can adopt a random orientation relative to its neighbors independently of the orientation of the other segments (Fig. 9.22).

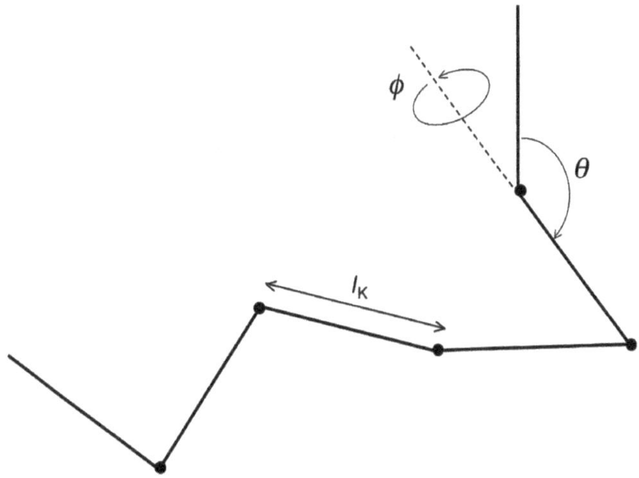

Figure 9.22 Schematic illustration of a freely jointed chain, composed of $N=6$ Kuhn* segments of length L_K. Each segment can be rotated by arbitrary polar (θ) and azimuthal (ϕ) angles around neighbor hinges (dotted line).

The average squared end-to-end* distance and the average squared gyration* radius are given by the following exact relations (Doi and Edwards, 1988):

$$\langle R^2(L)\rangle = L_K L = N L_K{}^2 \tag{9.1}$$

and

$$\langle R_g{}^2(L)\rangle = \frac{L_K L}{6} = \frac{1}{6} N L_K{}^2 = \frac{1}{6}\langle R^2(L)\rangle \tag{9.2}$$

Genomic distance, *s*: see contour* length.

(Average squared) Gyration radius, $\langle R_g{}^2(L)\rangle$: A quantity frequently employed to characterize the average polymer size, along with the average squared end-to-end* distance. It is defined as the average square distance between each monomer and the center of mass of the chain. As for the average squared end-to-end distance, this quantity is averaged over all possible conformations of the chain.

Heteropolymer: See copolymer*.

Kuhn length, L_K: See freely* jointed chain.

Melt of polymers: A polymer melt is a particular physical realization of a polymer solution, where solvent molecules consist of other polymer molecules. A simple physical picture of a melt is as follows. Consider an ensemble of identical polymer chains in solution at very dilute conditions, meaning that the average distance between chains is larger than the typical equilibrium size of the chains. By compressing the system, the chains will start approaching and interpenetrating each other. The onset of interpenetration defines the semi-dilute regime. Upon further compression, chains will eventually start to squeeze all the solvent out, such that the system will finally be composed of chains only. This is the melt state. For long-enough chains, the relaxation properties of a polymer melt are dominated by entanglement* effects. In a melt, according to the Flory theorem, excluded volume effects are screened and chains behave as ideal, FJCs (De Gennes, 1979; Doi and Edwards, 1988; Khokhlov and Grosberg, 1994; Rubinstein and Colby, 2003).

Molecular dynamics simulations: Molecular dynamics methods are computational techniques that simulate the dynamics of a physical system (Allen and Tildesley, 1987). Typically, the system is described as an ensemble of particles, each of which is subjected to forces coming from interactions with the other particles, plus, possibly, external forces. The trajectories of particles in space (and hence the configurations of polymers) are then obtained by numerical integration of Newton's law of motion. For polymer systems, a frequently adopted related technique is *Brownian dynamics*. This method simulates the behavior of chains subjected to inter-monomer forces and includes an external, stochastic term, which models the random thermal fluctuations of the chain due to collisions with particles in the surrounding medium.

Monte Carlo simulations: The Monte Carlo method is an efficient computational technique employed to obtain estimates of the statistical properties of a physical system. Many possible configurations of the system are simulated using an algorithm that applies random changes to the configurations according to an assumed probability distribution (Allen and Tildesley, 1987). Monte Carlo simulations can be used to provide statistical properties such as the average squared end-to-end* distance, but, unlike molecular* dynamics simulations, they usually do not provide dynamic information about the system.

Objective function: see optimization.

Optimization: The minimization (or maximization) of a mathematical function (often called objective function or cost function) that depends on several unknown parameters. In the context of this chapter, these parameters may typically consist of the 3D coordinates describing the conformation of chromosomes. The objective function typically measures the discrepancy between experimental data (such as distances between loci measured by microscopy) and corresponding quantities in the model (e.g., the distances between the loci in the computational chromosome model). The objective function usually also embodies assumed constraints that are independent of the experimental data set at hand and are based on prior knowledge, such as an assumed packing ratio of DNA in chromatin, the chromatin fiber persistence* length, nuclear tethering constraints, etc. Optimization is generally performed by an algorithm that iteratively changes the parameters so that the objective function diminishes until it can no longer be minimized further. The starting parameters (initialization) are often defined arbitrarily, for example, by random values. Different kinds of optimization algorithms exist. Gradient-descent methods try to determine how parameters should be changed to obtain the strongest reduction in the objective function at every iteration. These deterministic methods are often fast, but can easily get trapped in a local minimum. Alternative methods such as simulated annealing use a strategy that includes random changes of the parameters and occasionally allow for moves that temporarily increase the cost function. This allows to explore a larger space of possible configurations and to escape local minima, though with no absolute guarantee of reaching the global minimum and at the expense of much longer computation time.

Persistence length, L_p: The persistence length is a measure of the rigidity of a semi-flexible polymer chain. It corresponds to the length-scale along the polymer contour above which the chain looses the memory of its initial direction. Mathematically, it can be defined through the average value of the cosine of the angle θ between the tangents to the chain, as detailed in Fig. 9.23.

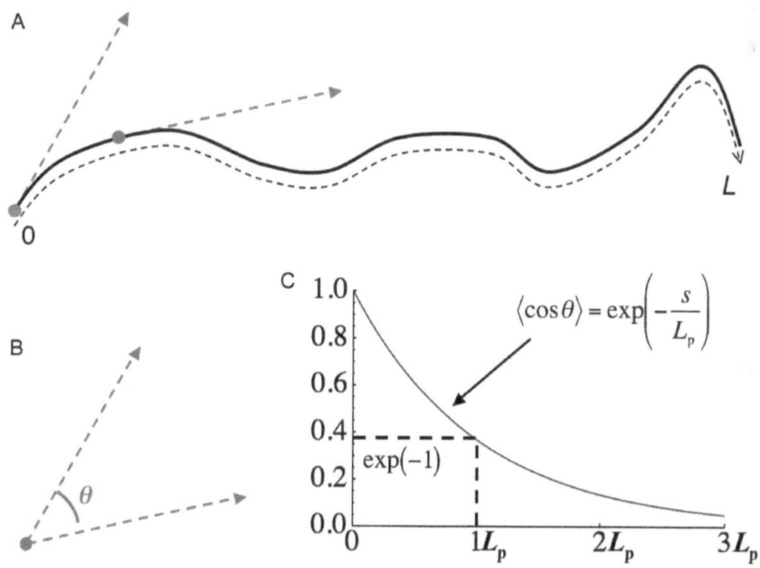

Figure 9.23 Persistence length. (A) Polymer chain modeled as a worm-like* chain (solid black line). The dashed line shows the chain profile along its contour* length, with curvilinear coordinate *s* going from 0 to *L*. Red arrows show the tangent to the chain at two selected points (curvilinear coordinates 0 and *s*). (B) θ is the angle between the two tangents. (C) The cosine of θ averaged over all possible polymer conformations decays exponentially with *s* (Rubinstein and Colby, 2003). The decay rate (coordinate *s* for which $\langle\cos(\theta)\rangle = \exp(-1)$) defines the persistence length, L_p. (For interpretation of the references to color in this figure legend, the reader is referred to the online version of this chapter.)

For sufficiently long chains (L much larger than L_p), the concepts of persistence length and Kuhn* length are equivalent to each other, since it can be proven that $L_K = 2L_p$ (Rubinstein and Colby, 2003).

Phantom chain: A phantom (or, ideal) chain is an abstract idealization of a polymer chain, where effects of volume exclusion from different monomers are neglected. In practice, two (or more) monomers are allowed to occupy the same place in space, as if the chain was transparent to itself (Fig. 9.24). Although this assumption is clearly not realistic, it underlies conceptually important models including the FJC and the WLC. The assumption is removed in more realistic models such as the self-avoiding* chain.

Relaxation: see Equilibrium*.

Rouse dynamics: The Rouse model is the "standard model" of polymer dynamics. Polymers are described as chains of beads connected by harmonic springs and subjected to the stochastic motion arising from random collisions with the surrounding medium. The Rouse model neglects otherwise important contributions arising from the subtle interplay between monomer motion and the surrounding medium (hydrodynamic effects). According to the Rouse model, chains relax* on a timescale, τ_R (the so-called Rouse time), given by:

$$\tau_R = \tau_R(L) \approx \frac{L^2}{D} \tag{9.3}$$

where L is the polymer contour* length and D is the diffusion coefficient of a Kuhn* segment (De Gennes, 1979; Doi and Edwards, 1988; Khokhlov and Grosberg, 1994; Rubinstein and Colby, 2003).
A quantity of notable experimental interest is the mean squared displacement (MSD) of a monomer as function of time, $\langle \delta r^2(t) \rangle$. For isolated diffusing monomers (not connected into a chain), the MSD is simply proportional to time:

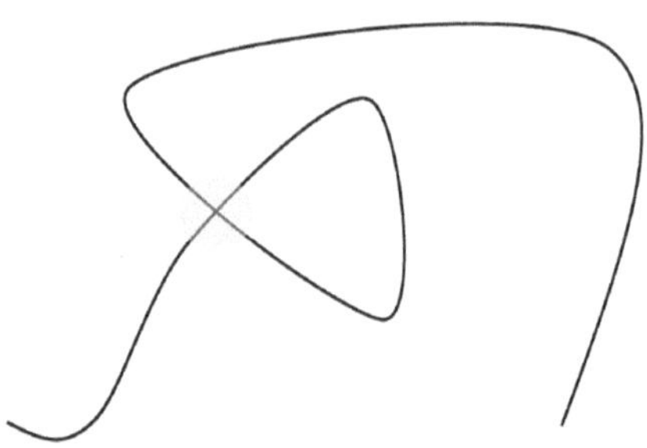

Figure 9.24 Phantom chain. Schematic illustration of a phantom chain, where two chain portions are allowed to overlap, that is, they can occupy the same region of space (yellow spot). (For interpretation of the references to color in this figure legend, the reader is referred to the online version of this chapter.)

$$\langle \delta r^2(t) \rangle = 6Dt$$

where D is the monomer diffusion coefficient. For monomers of a polymer chain, however, it can be shown that, at times t shorter than τ_R, the MSD is proportional to the square root of time (De Gennes, 1979; Doi and Edwards, 1988; Khokhlov and Grosberg, 1994; Rubinstein and Colby, 2003):

$$\langle \delta r^2(t) \rangle \approx \begin{cases} L_K (Dt)^{1/2}, & \text{for } t < \tau_R \\ \dfrac{DL_K}{L} t, & \text{for } t > \tau_R \end{cases} \tag{9.4}$$

This implies that at short time intervals the monomer behavior is always sub-diffusive, while at large times each monomer follows the motion of the rest of the chain. We stress the fact that short-time subdiffusive behavior is a general feature of polymer physics, and is not restricted to the Rouse model (see reptation* and (De Gennes, 1979; Doi and Edwards, 1988; Khokhlov and Grosberg, 1994; Rubinstein and Colby, 2003) for further details).

Reptation dynamics: The dynamics of entangled chains in concentrated solutions and melts (the physical state constituted only by polymer molecules, with no solvent) is richer than predictions from the Rouse* model. According to the Edwards-De Gennes reptation theory, chains are transiently confined to a tube-like region that roughly follows their contour and has a diameter $\sim \langle R^2(L_e) \rangle^{1/2}$, where R is the end-to-end distance (De Gennes, 1979; Doi and Edwards, 1988; Khokhlov and Grosberg, 1994; Rubinstein and Colby, 2003). L_e is defined as the entanglement length, and corresponds to the contour* length of chain segments between entanglements* (Fig. 9.25). In general, L_e is a nontrivial function of solution density, and persistence* length (Uchida et al., 2008).
Entanglements do not affect statistical properties of ensembles of chains at equilibrium (they are still described by the WLC model), but they do have severe consequences on chain dynamics. In particular, the relaxation of single chains is considerably slowed down. Because of entanglements, chains relax on a timescale called the reptation time, τ_{rept}, which is given by (De Gennes, 1979; Doi and Edwards, 1988; Khokhlov and Grosberg, 1994; Rubinstein and Colby, 2003):

$$\tau_{\text{rept}} = \tau_{\text{rept}}(L) \approx \tau_R(L_e) \left(\frac{L}{L_e} \right)^3 \tag{9.5}$$

From Eqs. (9.3) and (9.5), it can be seen that $\tau_{\text{rept}}(L)/\tau_R(L) = L/L_e$. Since in concentrated solutions the entanglement length L_e is much smaller than the contour length L ($L_e \ll L$), the reptation time τ_{rept} is much longer than the Rouse* time: $\tau_{\text{rept}} \gg \tau_R$.
Finally, the mean squared displacement (MSD) of a monomer at short time scales shows three well-distinct subdiffusive regimes:

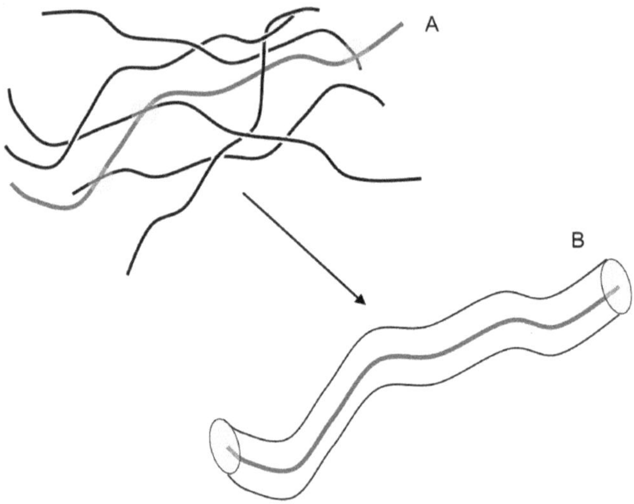

Figure 9.25 Reptation. (A) In polymer solutions, entanglements* (or topological constraints) strongly affect the dynamics of single chains: here, the red chain cannot cross other chains at yellow spots (see also Fig. 9.21). (B) Due to these constraints, chain dynamics is effectively confined to a tube-like region, whose center-line is given by the chain itself. Under these conditions, chains can only move large distances by snake-like displacements along the tube (hence the name "reptation"). These motions are characterized by extremely slow relaxation time (De Gennes, 1979; Doi and Edwards, 1988; Khokhlov and Grosberg, 1994; Rubinstein and Colby, 2003). (For interpretation of the references to color in this figure legend, the reader is referred to the online version of this chapter.)

$$\langle \delta r^2(t) \rangle \approx \begin{cases} L_K (Dt)^{1/2}, & \text{for } t < \tau_R(L_e) \\ \langle R^2(L_e) \rangle^{1/2} L_K{}^{1/2} (Dt)^{1/4}, & \text{for } \tau_R(L_e) < t < \tau_R(L) \\ \langle R^2(L) \rangle \left(t/\tau_{\text{rept}}(L) \right)^{1/2}, & \text{for } \tau_R(L) < t < \tau_{\text{rept}}(L) \\ \langle R^2(L) \rangle \left(t/\tau_{\text{rept}}(L) \right), & \text{for } t > \tau_{\text{rept}}(L) \end{cases} \tag{9.6}$$

where R denotes end-to-end* distance. In particular, the second regime, characterized by a time dependence with an exponent 0.25, trademarks the onset of entanglements effects and is a specific signature of reptation (De Gennes, 1979; Doi and Edwards, 1988; Khokhlov and Grosberg, 1994; Rubinstein and Colby, 2003). The complex behavior of Eq. (9.6) is summarized in Fig. 9.26.

Self-avoiding chain (SAC): In real chains, excluded volume effects often play an important role and tend to swell the chain. As a result, the typical chain size $\langle R^2(L) \rangle$ is given by:

$$\langle R^2(L) \rangle \approx L_K{}^2 \left(\frac{L}{L_K} \right)^{2\nu} \tag{9.7}$$

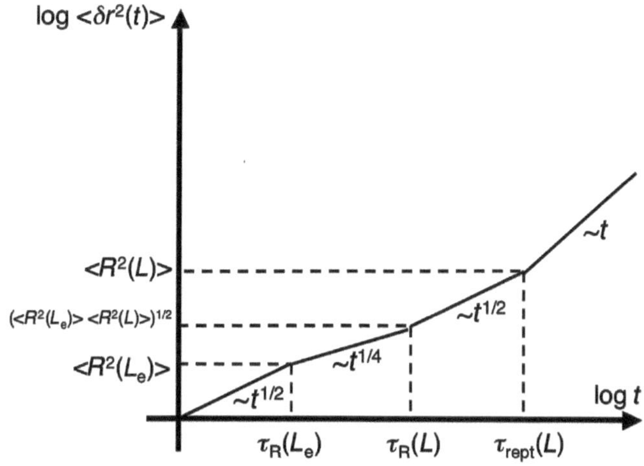

Figure 9.26 Mean-square displacement of a monomer of an entangled chain: schematic illustration of the different time regimes. The $\sim t^{1/4}$ behavior is typical of reptation.

The exponent ν is a measure of chain swelling, and for SAC's $\nu \approx 0.6$ (Doi and Edwards, 1988). For comparison, for a FJC or a WLC, $\nu = 0.5$.

Worm-like chain (WLC): In this model, semi-flexible chains are represented by a continuous rod of constant flexibility (as opposed to the FJC* model, in which only the hinges between the rigid Kuhn* segments are flexible). A WLC is described by its contour* length L, and persistence* length, L_P. For contour lengths $L \ll L_\text{P}$, thermal fluctuations have little effect and the chains are effectively rigid with mean squared end-to-end* distances $\langle R^2(L)\rangle = L^2$. For $L \gg L_\text{P}$, equilibrated linear chains exhibit random coil statistics with $\langle R^2(L)\rangle = 2L \times L_\text{P}$. The complete cross-over is described by the exact formula (Doi and Edwards, 1988):

$$\langle R^2(L)\rangle = 2{L_\text{P}}^2\left(\frac{L}{L_\text{P}} + \exp\left(-L/L_\text{P}\right) - 1\right) \quad (9.8)$$

The corresponding equation for the gyration* radius is given by (Benoit and Doty, 1953):

$$\langle {R_\text{g}}^2(L)\rangle = \frac{LL_\text{P}}{3} - {L_\text{P}}^2 + 2\frac{{L_\text{P}}^3}{L} - 2\frac{{L_\text{P}}^4}{L^2}\left[1 - \exp\left(-L/L_\text{P}\right)\right] \quad (9.9)$$

Alternative expressions as function of the Kuhn* length can be obtained using: $L_\text{P} = L_\text{K}/2$.

ACKNOWLEDGMENTS

We thank J-M. Arbona, H. Marie-Nelly, R. Koszul and O. Espeli for useful discussions. We apologize to colleagues whose work we did not discuss because of space limitations. AR acknowledges financial support from the Italian Ministry of Education, grant PRIN 2010HXAW77. C. Z. acknowledges financial support from Institut Pasteur, Fondation pour la Recherche Médicale (Equipe FRM) and Agence Nationale de la Recherche (specifically grants ANR-09-PIRI-0024 and ANR-11-MONU-020-02).

REFERENCES

Alber, F., Dokudovskaya, S., Veenhoff, L.M., Zhang, W., Kipper, J., Devos, D., Suprapto, A., Karni-Schmidt, O., Williams, R., Chait, B.T., Rout, M.P., Sali, A., 2007. Determining the architectures of macromolecular assemblies. Nature 450, 683–694.

Alberts, B., Johnson, A., Lewis, J., Raff, M., Roberts, K., Walter, P., 2008. Molecular Biology of the Cell: Reference Edition, vol. 1. Garland Publishing, New York.

Allen, M.P., Tildesley, D.J., 1987. Computer Simulation of Liquids. Oxford University Press, Oxford.

Barbieri, M., Chotalia, M., Fraser, J., Lavitas, L.-M., Dostie, J., Pombo, A., Nicodemi, M., 2012. Complexity of chromatin folding is captured by the strings and binders switch model. Proc. Natl. Acad. Sci. U. S. A. 109, 16173–16178.

Baù, D., Sanyal, A., Lajoie, B.R., Capriotti, E., Byron, M., Lawrence, J.B., Dekker, J., Marti-Renom, M.A., 2011. The three-dimensional folding of the α-globin gene domain reveals formation of chromatin globules. Nat. Struct. Mol. Biol. 18, 107–114.

Belmont, A.S., 2001. Visualizing chromosome dynamics with GFP. Trends Cell Biol. 11, 250–257.

Benoit, H., Doty, P.M., 1953. Light scattering from non-Gaussian chains. J. Phys. Chem. 57, 958–963.

Berger, A.B., Cabal, G.G., Fabre, E., Duong, T., Buc, H., Nehrbass, U., Olivo-Marin, J.C., Gadal, O., Zimmer, C., 2008. High-resolution statistical mapping reveals gene territories in live yeast. Nat. Methods 5, 1031–1037.

Bickmore, W.A., van Steensel, B., 2013. Genome architecture: domain organization of interphase chromosomes. Cell 152, 1270–1284.

Bohn, Manfred, Heermann, D., Van Driel, R., 2007. Random loop model for long polymers. Phys. Rev. E 76, 051805.

Boyle, S., 2001. The spatial organization of human chromosomes within the nuclei of normal and emerin-mutant cells. Hum. Mol. Genet. 10, 211–219.

Bystricky, K., Heun, P., Gehlen, L., Langowski, J., Gasser, S.M., 2004. Long-range compaction and flexibility of interphase chromatin in budding yeast analyzed by high-resolution imaging techniques. Proc. Natl. Acad. Sci. U. S. A. 101, 16495–16500.

Bystricky, K., Laroche, T., Van Houwe, G., Blaszczyk, M., Gasser, S.M., 2005. Chromosome looping in yeast: telomere pairing and coordinated movement reflect anchoring efficiency and territorial organization. J. Cell Biol. 168, 375–387.

Cates, M.E., Deutsch, J., 1986. Conjectures on the statistics of ring polymers. J. Phys. France 47, 2121–2128.

Cavalli, G., 2007. Chromosome kissing. Curr. Opin. Genet. Dev. 17, 443–450.

Cavalli, G., Misteli, T., 2013. Functional implications of genome topology. Nat. Struct. Mol. Biol. 20, 290–299.

Cook, P.R., 1999. The organization of replication and transcription. Science 284, 1790–1795.

Cremer, T., Cremer, C., 2001. Chromosome territories, nuclear architecture and gene regulation in mammalian cells. Nat. Rev. Genet. 2, 292–301.
Cremer, T., Cremer, M., 2010. Chromosome territories. Cold Spring Harb. Perspect. Biol. 2, a003889.
De Gennes, P.G., 1979. Scaling Concepts in Polymer Physics. Cornell University Press, Ithaca.
De Gennes, P.G., 1971. Reptation of a polymer chain in the presence of fixed obstacles. J. Chem. Phys. 55, 572.
De Laat, W., Dekker, J., 2012. 3C-based technologies to study the shape of the genome. Methods 58, 189–191.
De Wit, E., De Laat, W., 2012. A decade of 3C technologies: insights into nuclear organization. Genes Dev. 26, 11–24.
Dekker, Job, Marti-Renom, Marc A., Mirny, L.A., 2013. Exploring the three-dimensional organization of genomes: interpreting chromatin interaction data. Nat. Rev. Genet. 14, 390–403.
Dekker, J., Rippe, K., Dekker, M., Kleckner, N., 2002. Capturing chromosome conformation. Science 295, 1306–1311.
Di Stefano, M., Rosa, A., Belcastro, V., Di Bernardo, D., Micheletti, C., 2013. Colocalization of coregulated genes: a steered molecular dynamics study of human chromosome 19. PLoS Comput. Biol. 9, e1003019.
Dixon, J.R., Selvaraj, S., Yue, F., Kim, A., Li, Y., Shen, Y., Hu, M., Liu, J.S., Ren, B., 2012. Topological domains in mammalian genomes identified by analysis of chromatin interactions. Nature 485, 376–380.
Doi, M., Edwards, A.M., 1988. The Theory of Polymer Dynamics. Oxford University Press, Oxford.
Dostie, J., Bickmore, W.A., 2012. Chromosome organization in the nucleus—charting new territory across the Hi-Cs. Curr. Opin. Genet. Dev. 22, 125–131.
Duan, Z., Andronescu, M., Schutz, K., McIlwain, S., Kim, Y.J., Lee, C., Shendure, J., Fields, S., Blau, C.A., Noble, W.S., 2010. A three-dimensional model of the yeast genome. Nature 465, 363–367.
Edelman, L.B., Fraser, P., 2012. Transcription factories: genetic programming in three dimensions. Curr. Opin. Genet. Dev. 22, 110–114.
Eltsov, M., MacLellan, K.M., Maeshima, K., Frangakis, A.S., Dubochet, J., 2008. Analysis of cryo-electron microscopy images does not support the existence of 30-nm chromatin fibers in mitotic chromosomes in situ. Proc. Natl. Acad. Sci. U. S. A. 105, 19732.
Flors, C., 2011. DNA and chromatin imaging with super-resolution fluorescence microscopy based on single-molecule localization. Biopolymers 95, 290–297.
Fraser, J., Rousseau, M., Shenker, S., Ferraiuolo, M.A., Hayashizaki, Y., Blanchette, M., Dostie, J., 2009. Chromatin conformation signatures of cellular differentiation. Genome Biol. 10, R37.
Gasser, S.M., 2002. Visualizing chromatin dynamics in interphase nuclei. Science (New York, NY) 296, 1412–1416.
Gehlen, L.R., Gruenert, G., Jones, M.B., Rodley, C.D., Langowski, Jörg, O'Sullivan, J.M., 2012. Chromosome positioning and the clustering of functionally related loci in yeast is driven by chromosomal interactions. Nucleus 3 (4), 370–383.
Grosberg, A.Y., Nechaev, S.K., Shakhnovich, E.I., 1988. The role of topological constraints in the kinetics of collapse of macromolecules. J. Phys. France 49, 2095–2100.
Grosberg, A.Y., Rabin, Y., Havlin, S., Neer, A., 1993. Crumpled globule model of the three-dimensional structure of DNA. Europhys. Lett. 23, 373–378.
Hadizadeh Yazdi, N., Guet, C.C., Johnson, R.C., Marko, J.F., 2012. Variation of the folding and dynamics of the Escherichia coli chromosome with growth conditions. Mol. Microbiol. 86, 1318–1333.
Halperin, A., 1991. On the collapse of multiblock copolymers. Macromolecules 24, 1418–1419.

Harmandaris, V.A., Reith, D., Van der Vegt, N.F.A., Kremer, Kurt, 2007. Comparison between coarse-graining models for polymer systems: two mapping schemes for polystyrene. Macromol. Chem. Phys. 208, 2109–2120.
Heun, P., Laroche, T., Shimada, K., Furrer, P., Gasser, S.M., 2001. Chromosome dynamics in the yeast interphase nucleus. Science 294, 2181–2186.
Hu, M., Deng, K., Qin, Z., Dixon, J., Selvaraj, S., Fang, J., Ren, B., Liu, J.S., 2013. Bayesian inference of spatial organizations of chromosomes. PLoS Comput. Biol. 9, e1002893.
Huang, K., 1987. Statistical Mechanics, second ed. Wiley, New York.
Israelachvili, J.N., 2011. Intermolecular and Surface Forces (Google eBook), Revised third ed. Academic Press, London.
Iyer, B.V.S., Arya, G., 2012. Lattice animal model of chromosome organization. Phys. Rev. E 86, 011911.
Jackson, D.A., 1998. Replicon clusters are stable units of chromosome structure: evidence that nuclear organization contributes to the efficient activation and propagation of S phase in human cells. J. Cell Biol. 140, 1285–1295.
Jhunjhunwala, S., Van Zelm, M.C., Peak, M.M., Cutchin, S., Riblet, R., Van Dongen, J.J.M., Grosveld, F.G., Knoch, T.A., Murre, C., 2008. The 3D structure of the immunoglobulin heavy-chain locus: implications for long-range genomic interactions. Cell 133, 265–279.
Jun, S., Mulder, B., 2006. Entropy-driven spatial organization of highly confined polymers: lessons for the bacterial chromosome. Proc. Natl. Acad. Sci. U. S. A. 103, 12388–12393.
Kalhor, R., Tjong, H., Jayathilaka, N., Alber, F., Chen, L., 2012. Genome architectures revealed by tethered chromosome conformation capture and population-based modeling. Nat. Biotechnol. 30, 90–98.
Khokhlov, A., Grosberg, A., 1994. Statistical Physics of Macromolecules. American Institute of Physics, New York.
Kim, J.S., Pradhan, P., Backman, V., Szleifer, I., 2011. The influence of chromosome density variations on the increase in nuclear disorder strength in carcinogenesis. Phys. Biol. 8, 015004.
Kornberg, R.D., 1974. Chromatin structure: a repeating unit of histones and DNA. Science 184, 868–871.
Kouzarides, T., 2007. Chromatin modifications and their function. Cell 128, 693–705.
Kremer, K., Grest, G.S., 1990. Dynamics of entangled linear polymer melts: a molecular-dynamics simulation. J. Chem. Phys. 92, 5057–5086.
Kreth, G., Finsterle, J., Von Hase, J., Cremer, M., Cremer, C., 2004. Radial arrangement of chromosome territories in human cell nuclei: a computer model approach based on gene density indicates a probabilistic global positioning code. Biophys. J. 86, 2803–2812.
Lieberman-Aiden, E., Van Berkum, N.L., Williams, L., Imakaev, M., Ragoczy, T., Telling, A., Amit, I., Lajoie, B.R., Sabo, P.J., Dorschner, M.O., Sandstrom, R., Bernstein, B., Bender, M.A., Groudine, M., Gnirke, A., Stamatoyannopoulos, J., Mirny, L.A., Lander, E.S., Dekker, Job, 2009. Comprehensive mapping of long-range interactions reveals folding principles of the human genome. Science (New York, NY) 326, 289–293.
Luger, K., Dechassa, M.L., Tremethick, D.J., 2012. New insights into nucleosome and chromatin structure: an ordered state or a disordered affair? Nat. Rev. Mol. Cell Biol. 13, 436–447.
Maeshima, K., Hihara, S., Eltsov, M., 2010. Chromatin structure: does the 30-nm fibre exist in vivo? Curr. Opin. Cell Biol. 22, 291–297.
Marenduzzo, D., Micheletti, C., Cook, P.R., 2006. Entropy-driven genome organization. Biophys. J. 90, 3712–3721.
Markaki, Y., Smeets, D., Fiedler, S., Schmid, V.J., Schermelleh, L., Cremer, T., Cremer, M., 2012. The potential of 3D-FISH and super-resolution structured illumination

microscopy for studies of 3D nuclear architecture: 3D structured illumination microscopy of defined chromosomal structures visualized by 3D (immuno)-FISH opens new perspectives for stud. BioEssays 34, 412–426.

Marko, J.F., Siggia, E.D., 1995. Stretching DNA. Macromolecules 28, 8759–8770.

Marti-Renom, M.A., Mirny, L.A., 2011. Bridging the resolution gap in structural modeling of 3D genome organization. PLoS Comput. Biol. 7, e1002125.

Mateos-Langerak, J., Bohn, M., De Leeuw, W., Giromus, O., Manders, E.M.M., Verschure, P.J., Indemans, M.H.G., Gierman, H.J., Heermann, D.W., Van Driel, R., 2009. Spatially confined folding of chromatin in the interphase nucleus. Proc. Natl. Acad. Sci. U. S. A. 106, 3812.

Mekhail, K., Seebacher, J., Gygi, S.P., Moazed, D., 2008. Role for perinuclear chromosome tethering in maintenance of genome stability. Nature 456, 667–670.

Meluzzi, D., Arya, G., 2013. Recovering ensembles of chromatin conformations from contact probabilities. Nucleic Acids Res. 41, 63–75.

Micheletti, C., 2012. Comparing proteins by their internal dynamics: exploring structure-function relationships beyond static structural alignments. Phys. Life Rev. 10, 1–26.

Misteli, T., 2010. Higher-order genome organization in human disease. Cold Spring Harb. Perspect. Biol. 2, a000794.

Müller, M., Wittmer, J., Cates, M., 1996. Topological effects in ring polymers: a computer simulation study. Phys. Rev. E 53, 5063–5074.

Münkel, C., Eils, R., Dietzel, S., Zink, D., Mehring, C., Wedemann, G., Cremer, T., Munkel, C., Langowski, J., 1999. Compartmentalization of interphase chromosomes observed in simulation and experiment. J. Mol. Biol. 285, 1053–1065.

Münkel, C., Langowski, J., 1998. Chromosome structure predicted by a polymer model. Phys. Rev. E 57, 5888–5896.

Ostashevsky, J., 1998. A polymer model for the structural organization of chromatin loops and minibands in interphase. Mol. Biol. Cell 9, 3031–3040.

Pelletier, J., Halvorsen, K., Ha, B.-Y., Paparcone, R., Sandler, S.J., Woldringh, C.L., Wong, W.P., Jun, S., 2012. Physical manipulation of the Escherichia coli chromosome reveals its soft nature. Proc. Natl. Acad. Sci. U. S. A. 109, E2649–E2656.

Possoz, C., Junier, I., Espeli, O., 2012. Bacterial chromosome segregation. Front. Biosci. 17, 1020–1034.

Rieping, W., Habeck, M., Nilges, M., 2005. Inferential structure determination. Science 309, 303.

Rippe, K., 2001. Making contacts on a nucleic acid polymer. Trends Biochem. Sci. 26, 733–740.

Rodley, C.D.M., Bertels, F., Jones, B., O'Sullivan, J.M., 2009. Global identification of yeast chromosome interactions using Genome conformation capture. Fungal Genet. Biol. 46, 879–886.

Rosa, A., Becker, N.B., Everaers, R., 2010. Looping probabilities in model interphase chromosomes. Biophys. J. 98, 2410–2419.

Rosa, A., Everaers, R., 2008. Structure and dynamics of interphase chromosomes. PLoS Comput. Biol. 4.

Rouquette, J., Cremer, C., Cremer, T., Fakan, S., 2010. Functional nuclear architecture studied by microscopy: present and future. Int. Rev. Cell Mol. Biol. 282, 1–90.

Rousseau, M., Fraser, J., Ferraiuolo, M.A., Dostie, J., Blanchette, M., 2011. Three-dimensional modeling of chromatin structure from interaction frequency data using Markov chain Monte Carlo sampling. BMC Bioinformat. 12, 414.

Rubinstein, M., Colby, R., 2003. Polymer Physics. Oxford University Press, New York.

Sachs, R.K., Van den Engh, G., Trask, B., Yokota, H., Hearst, J.E., 1995. A random-walk/giant-loop model for interphase chromosomes. Proc. Natl. Acad. Sci. U. S. A. 92, 2710–2714.

Schram, R.D., Barkema, G.T., Schiessel, H., 2013. On the stability of fractal globules. J. Chem. Phys. 138, 224901.
Shimada, J., Yamakawa, H., 1984. Ring-closure probabilities for twisted wormlike chains. Application to DNA. Macromolecules 17, 689–698.
Sikorav, J.L., Jannink, G., 1994. Kinetics of chromosome condensation in the presence of topoisomerases: a phantom chain model. Biophys. J. 66, 827–837.
Sumners, D.W., Whittington, S.G., 1988. Knots in self-avoiding walks. J. Phys. A Math. Genet. 21, 1689–1694.
Taddei, A., Schober, H., Gasser, S.M., 2010. The budding yeast nucleus. Cold Spring Harb. Perspect. Biol. 2, a000612.
Tanizawa, H., Iwasaki, O., Tanaka, A., Capizzi, J.R., Wickramasinghe, P., Lee, M., Fu, Z., Noma, K., 2010. Mapping of long-range associations throughout the fission yeast genome reveals global genome organization linked to transcriptional regulation. Nucleic Acids Res. 38, 8164–8177.
Thérizols, P., Duong, T., Dujon, B., Zimmer, C., Fabre, E., 2010. Chromosome arm length and nuclear constraints determine the dynamic relationship of yeast subtelomeres. Proc. Natl. Acad. Sci. U. S. A. 107, 2025.
Tjong, H., Gong, K., Chen, L., Alber, F., 2012. Physical tethering and volume exclusion determine higher-order genome organization in budding yeast. Genome Res. 22, 1295–1305.
Toan, N., Marenduzzo, D., Cook, P., Micheletti, C., 2006. Depletion effects and loop formation in self-avoiding polymers. Phys. Rev. Lett. 97, 178302.
Tokuda, N., Terada, T.P., Sasai, M., 2012. Dynamical modeling of three-dimensional genome organization in interphase budding yeast. Biophys. J. 102, 296–304.
Uchida, N., Grest, G.S., Everaers, R., 2008. Viscoelasticity and primitive path analysis of entangled polymer liquids: from F-actin to polyethylene. J. Chem. Phys. 128, 044902.
Umbarger, M.A., Toro, E., Wright, M.A., Porreca, G.J., Bau, D., Hong, S.H., Fero, M.J., Zhu, L.J., Marti-Renom, M.A., McAdams, H.H., Shapiro, L., Dekker, J., Church, G.M., 2011. The three-dimensional architecture of a bacterial genome and its alteration by genetic perturbation. Mol. Cell 44, 252–264.
Van Rensburg, E.J.J., Madras, N., 1992. A nonlocal Monte Carlo algorithm for lattice trees. J. Phys. A Math. Genet. 25, 303–333.
Vanderzande, C., 1998. Lattice Models of Polymers. Cambridge University Press, Cambridge.
Vettorel, T., Grosberg, A.Y., Kremer, K., 2009. Statistics of polymer rings in the melt: a numerical simulation study. Phys. Biol. 6, 025013.
Viollier, P.H., Thanbichler, M., McGrath, P.T., West, L., Meewan, M., McAdams, H.H., Shapiro, L., 2004. Rapid and sequential movement of individual chromosomal loci to specific subcellular locations during bacterial DNA replication. Proc. Natl. Acad. Sci. U. S. A. 101, 9257–9262.
Wang, X., Montero Llopis, P., Rudner, D.Z., 2013. Organization and segregation of bacterial chromosomes. Nat. Rev. Genet. 14, 191–203.
Wiggins, P.A., Cheveralls, K.C., Martin, J.S., Lintner, R., Kondev, J., 2010. Strong intranucleoid interactions organize the Escherichia coli chromosome into a nucleoid filament. Proc. Natl. Acad. Sci. U. S. A. 107, 4991–4995.
Wong, H., Marie-Nelly, H., Herbert, S., Carrivain, P., Blanc, H., Koszul, R., Fabre, E., Zimmer, C., 2012. A predictive computational model of the dynamic 3D interphase yeast nucleus. Curr. Biol. 22, 1881–1890.
Yaffe, E., Tanay, A., 2011. Probabilistic modeling of Hi-C contact maps eliminates systematic biases to characterize global chromosomal architecture. Nat. Genet. 43, 1059–1065.
Yoffe, A.M., Prinsen, P., Gopal, A., Knobler, C.M., Gelbart, W.M., Ben-Shaul, A., 2008. Predicting the sizes of large RNA molecules. Proc. Natl. Acad. Sci. U. S. A. 105, 16153–16158.

Yokota, H., Van den Engh, G., Hearst, J.E., Sachs, R.K., Trask, B.J., 1995. Evidence for the organization of chromatin in megabase pair-sized loops arranged along a random walk path in the human G0/G1 interphase nucleus. J. Cell Biol. 130, 1239–1249.
Zhang, Y., Heermann, D.W., 2011. Loops determine the mechanical properties of mitotic chromosomes. PLoS One 6, e29225.
Zimmer, C., Fabre, E., 2011. Principles of chromosomal organization: lessons from yeast. J. Cell Biol. 192, 723–733.

CHAPTER TEN

How Chromatin Looping and Nuclear Envelope Attachment Affect Genome Organization in Eukaryotic Cell Nuclei

Hansjoerg Jerabek[*,†], **Dieter W. Heermann**[*,†,‡,§,1]
[*]Institute for Theoretical Physics, Heidelberg, Germany
[†]Interdisciplinary Center for Scientific Computing (IWR), Heidelberg, Germany
[‡]The Jackson Laboratory, Bar Harbor, Maine, USA
[§]Shanghai Center for Bioinformation Technology (SCBIT), Shanghai, PR China
[1]Corresponding author: e-mail address: heermann@tphys.uni-heidelberg.de

Contents

International Review of Cell and Molecular Biology, Volume 307
ISSN 1937-6448
http://dx.doi.org/10.1016/B978-0-12-800046-5.00010-2

Abstract

To understand how interphase chromatin is organized in eukaryotic cell nuclei, it is essential to understand what kind of interactions influence the nuclear architecture and to what extent. Using a mesoscale model that incorporates chromatin–chromatin interactions as well as binding of chromatin to the nuclear envelope, we can show that chromatin loops and envelope bonds are major players in genome organization because they largely affect the entropy of the chromatin fibres. The model allows us to consistently reproduce multiple characteristic chromatin parameters in agreement with experimental data. We focus on the question of how an inversion of the nuclear architecture, in the course of which the highly active euchromatin changes its preferential position from the nuclear center to the periphery, can be achieved. We find that the transition between the common and inverted organization is driven by the strength of the envelope interaction and the nuclear chromatin density.

1. INTRODUCTION

1.1. Chromatin folding and chromosome organization in eukaryotes

The question of how the genome of eukaryotic cells is organized during interphase, that is, the main part of the cell cycle where the cell expresses and copies its DNA, has been subject to many experimental and simulational studies. Up to date, we know that eukaryotic DNA is wrapped around histone proteins. A histone molecule together with the DNA that is bound to it is called a nucleosome. The nucleosomes and the DNA connecting them form a beads-on-a-string-like filament with a diameter of $\approx$11 nm (Schiessel et al., 2001). This filament is called chromatin and represents a more compact form of DNA. In eukaryotes, the DNA must be compacted because of the large number of base pairs the genomes consist of. In humans, for example, the over 3 billion base pairs in each cell form a DNA strand with a length of more than 2 m, yet the diameter of the cell nucleus is only of the order of a few μm. As the chromatin fiber itself is still not compact enough to fit into the nucleus, there must also be multiple higher levels of compaction. In this context, *in vitro* experiments have shown that chromatin can condense to a filament with a diameter of $\approx$30 nm, which consists of nucleosome arrays. However, there is no proof for the existence of this filament in living cells, and on the scale above 30 nm, even less is known about how the genome is structured (Diesinger and Heermann, 2009; Maeshima et al., 2010).

1.2. Experimental methods to investigate chromatin structure

The main reason why there is no detailed structural information about chromatin in interphase is that there are no direct experimental methods to investigate the structure. While light microscopes suffice to visualize full cell nuclei and total chromosomes, they are limited in spatial resolution and do not provide the accuracy required to exhibit the chromatin fiber itself. Because of these factors, reliable structural data about chromatin can only be obtained using indirect experimental methods. The two most frequently used methods are, on the one hand, fluorescence *in situ* hybridization (FISH) (Goetze et al., 2007; Mateos-Langerak et al., 2009) and, on the other hand, chromosome conformation capture (3C) and its derivatives (4C, 5C, Hi-C) (Lieberman-Aiden et al., 2009; Simonis et al., 2006, 2007; Vassetzky et al., 2009; Zhao et al., 2006).

In FISH experiments, fluorescent probes are utilized to bind to specific DNA sequences. These markers bind only to those parts of the genome with which they show a high degree of sequence complementarity. Hence, it is possible to measure the spatial position of certain genomic regions. By using different fluorescence colors, FISH furthermore allows determining the relationship between spatial distance (R) and genomic distance (g). More precisely, the measure of interest is the mean square distance (MSD) $\langle R^2(g) \rangle$. From polymer physics we know that the characteristics of the MSD allow making conclusions about the underlying type of polymer. Hence, the experimentally measured MSD can be used to determine what kind of biopolymer chromatin is.

In contrast to FISH, chromosome conformation capture experiments do not provide absolute spatial information, but they allow us to measure whether or not different genomic regions are in spatial proximity. In 3C-type experiments, formaldehyde is added to living cells, which leads to cross-linking of proteins to DNA as well as to each other. As a consequence, DNA strands that are in spatial proximity are also cross-linked. Making use of restriction enzymes, the cross-linked parts of DNA can be separated from the rest. By analyzing the cross-linked sequences, it is possible to reconstruct which parts of the chromatin fiber were involved, that is, were spatially close when the formaldehyde was added. The resulting contact map or contact pattern provides information about the local chromatin structure as well as about genome-wide interactions (Hi-C) (Lieberman-Aiden et al., 2009). The most important measure in 3C-like experiments is the contact/loop distribution $\langle P_l(g) \rangle$, which exhibits how frequently two genomic regions contact each other and form a loop, respectively, when there is a genomic distance g between them. In order to get a better insight into what type of polymer chromatin is, this distribution can be compared to those generated by different polymer models.

1.3. Basic polymer models for chromatin

Because of the fact that chromatin is a biopolymer, one should think that it will be easily possible to find a polymer model that describes the folding of the fiber as well as all of its structural and topological properties. However, there are mainly two problems. On the one hand, the chromatin fiber does not consist only of DNA and histones, and there are a huge number of other proteins bound to the DNA that are essential for the overall organization of the genome. The immense complexity of the real system does not allow incorporating all of its features into a single model. As a consequence, all polymer models and their results are solely approximations. On the other hand, the accuracy of the experimental data often does not allow making definite conclusions about which polymer model provides the best agreement.

Despite these drawbacks, developing polymer models that incorporate new features is essential because it allows conclusions to be made about which of these features govern the highly complex process of chromatin folding. The simplest polymer model for chromatin is the freely jointed chain model in which the smallest subunits of the polymer, that is, its monomers, are connected by links of fixed length. The monomers as well as the links are allowed to overlap and there are no constraints for the angle between the links. Thus, the resulting chain is also called a random walk (RW) polymer. Owing to the fact that chemical bonds are not totally rigid but possess a certain flexibility in length, it is logical to incorporate fluctuating bond distances into a next-level polymer model. A frequently used model of this kind is, for example, the Gaussian chain (GC) model where the bond distances follow a Gaussian distribution (Bohn, 2010).

The most important polymer features that the RW and GC models neglect are excluded volume interactions. In principle, these interactions prevent the polymer from overlapping with itself because the same position in space cannot be occupied by different molecules. For this reason, chains incorporating excluded volume are called self-avoiding walk (SAW) polymers. As the links cannot point to the opposite direction of the previous links as a result of the excluded volume constraint, the direction of adjacent links is correlated. In polymer physics, the general measure of this kind of angular correlation is the persistence length l_p (De Gennes, 1979; Hsu et al., 2010). The models mentioned up to now are for free polymers; they neglect the influence of the solvent in which they are immersed, yet the solvent can strongly influence the structure of a polymer. For example, a hydrophobic polymer in a hydrophilic solvent will try to minimize the interaction with its polar environment by reducing the polymer–solvent surface. As a sphere is the shape that has the smallest surface in relation to its volume, the polymer will collapse to a more compact spherical conformation, a so-called globular state (GS). While only the total polymer would have a globular structure in the GS model, recent papers have proposed a version of the GS in which subparts of the fiber also exhibit the formation of globules. The resulting chain of globules can again form globules, such that the final incrementally folded conformation is denoted as a fractal globule (FG), originally proposed by Grosberg (Grosberg et al., 1988; Mirny, 2011). The word fractal is used because the structure of the FG looks similar at all scales, an aspect called self-similarity, which is the characteristic feature of fractals.

As a first test for how well a given polymer model can reproduce the structure of real chromatin fibers, the MSD model is compared to the corresponding results from FISH experiments. The RW and FG models, for example, predict that

$$\left\langle R^2(g)\right\rangle \propto g^{2\upsilon}, \tag{10.1}$$

with υ being a model-specific parameter (see Fig. 10.1).

For a SAW polymer also, a similar power-law is observed (Bohn, 2010). Yet the FISH experiments show that *in vivo,* the MSD does not exhibit this behavior. Instead, it reaches a plateau level above a certain genomic distance, that is, it remains almost constant. This allows us to conclude that the

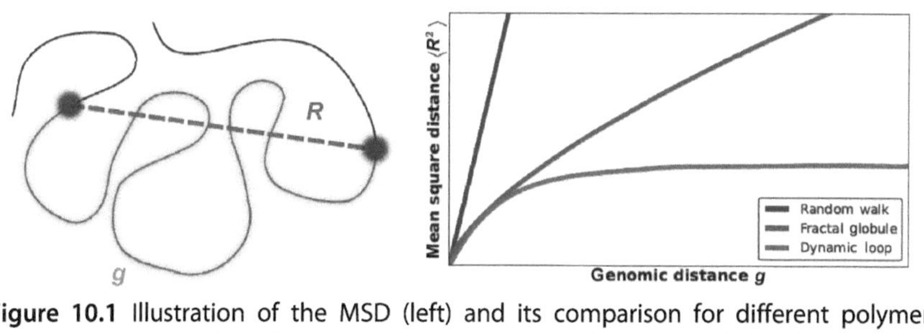

Figure 10.1 Illustration of the MSD (left) and its comparison for different polymer models (right). (For the color version of this figure, the reader is referred to the online version of this chapter.)

chromosomal fibers are confined inside certain volumes (Goetze et al., 2007; Mateos-Langerak et al., 2009). The fact that the plateau distance is much smaller than the nuclear radius indicates that these volumes are much smaller than the nuclear volume, that is, that the plateau does not occur simply because the chromosomes are confined inside the nucleus. In contrast to the RW, SAW, and FG models, the GS model allows for reproducing the leveling-off of the MSD with a polymer length-dependent plateau level much lower than the surrounding boundary. However, this model assumes that the chromatin fiber is surrounded by a "bad solvent" that makes the polymer try to reduce the surface exposed to the solvent. In addition, the GS cannot explain another important feature of chromatin, namely, its distance fluctuations. A measure for these fluctuations are the higher-order moment ratios of distances, which can be calculated from simulation as well as from FISH data (Bohn, 2010). It has been shown that the experimental fluctuations are much larger than those obtained in the RW, SAW, and GS model. This is especially remarkable considering that the RW model naturally exhibits quite large fluctuations. Of the given models, the GS has the lowest fluctuations and shows even less agreement with the experimental data than does the SAW model.

1.4. Importance of chromatin loops and loop entropy for folding of chromatin

A few years back, the aspect of chromatin looping received only little attention when it came to modelling chromatin. However, since the invention of the 3C technique and its derivatives, a lot of evidence has come up suggesting that the formation of chromatin loops could be essential for genome organization and gene regulation. The experimental data show that loops of all sizes can be found *in vivo*, that is, that chromatin contacts can be established even between loci that are genomically separated by hundreds of megabase pairs (Mb). The formation of loops is important for the organization of chromosomes for multiple reasons: primarily, they lead to compaction of a chromosome because they force different parts of the fiber to stick together. Yet, this at first sight simple way of compacting a polymer differs from, for example, a compaction induced by external forces. Without loops, the polymer chain would strongly intermingle with itself and form a lot of knots while being compacted. This is not the case for a looped polymer because the loops repel each other (Bohn and Heermann, 2011; Marenduzzo and Orlandini, 2009). As a result of this entropic repulsion, different parts of the polymer can intermingle only to a limited extent.

Preventing full intermingling is especially important because having to disentangle the chromatin fibers during each mitosis or meiosis would strongly complicate and slow down cell division. In contrast to this unspecific looping, chromatin loops have also been reported to play a role in gene regulation. Regulatory genomic elements such as enhancers or silencers, for example, can strongly alter gene transcription by forming functional contacts with the promotor regions of the corresponding genes (Kurukuti et al., 2006; Valenzuela et al., 2008). Hence, the formation of loops can also modify a cell's expression levels. For the beta-globin gene locus, for instance, experiments have demonstrated that these functional contacts are indeed able to influence transcription (Dekker, 2003).

1.5. Overall genome architecture in eukaryotic cell nuclei

As eukaryotic genomes consist of multiple chromosomes, not just the folding of single chromatin fibers but also the interplay between different chromosomes and their interaction with the environment largely influence nuclear architecture. From FISH experiments, we know that single chromosomes do not spread over the total volume of the cell nucleus, and that each of them is confined to a subnuclear volume much smaller than the nuclear volume. These compartments are called chromosome territories (Bolzer et al., 2005; Cremer and Cremer, 2001; Kosak and Groudine, 2004). Even though recent studies have shown that the territories may intermingle to a certain extent (Branco and Pombo, 2006), they remain well segregated during interphase. As for the case of single chromosomes, the loops are responsible for this separation because the entropic loop repulsion prevents the fibers from interfusing (Bohn and Heermann, 2011; Marenduzzo and Orlandini, 2009). One of the main aspects of genome organization is the positioning of genomic regions depending on their transcriptional activity. In this context, positioning can have two meanings: on the one hand, the spatial location relative to the center of the chromosome territory, and on the other hand, the position relative to the center of the nucleus. As far as the first case is concerned, multiple studies have shown that chromosomes exhibit an expression-dependent topology, with frequently transcribed genes preferentially located at the periphery of their chromosome territories and weakly expressed loci situated closer to the center (Clemson et al., 2006; Cremer et al., 2004; Kosak and Groudine, 2004; Scheuermann et al., 2004). These experimental findings are also in agreement with recent computational data suggesting that entropic forces are responsible for this kind of chromosome topology (Jerabek and Heermann, 2012). However, there are also studies stating that the periphery is on the average as transcriptionally active as the whole chromosome territory (Branco and Pombo, 2006) and that transcription sites are basically uniformly distributed throughout the nucleoplasm and the territories (Abranches et al., 1998; Verschure et al., 1999).

Despite these opposing opinions about the territory-related positioning, scientists agree that the expression-dependent location of chromatin relative to the nuclear center is the same in almost all eukaryotes. Highly active chromatin (euchromatin) is mainly located close to the center of the cell's nucleus, while the majority of the lowly active parts of the chromosomes (heterochromatin) preferentially reside at the nuclear periphery (see Fig. 10.2A).

Yet there are exceptions to this rule: a recent experiment (Solovei et al., 2009) has shown that a nonconventional architecture can be found in rod photoreceptor cells of nocturnal mammals, while rod cells in diurnal mammals exhibit the common organization pattern. As illustrated in Fig. 10.2B, the nocturnal cells have an inverse architecture, with euchromatin near the envelope and heterochromatin in the center of the nucleus. This inversion is thought to increase light transmission to photoreceptors in nocturnal mammals. However, as the progenitors of the rod cells exhibit the conventional organization, the architectural inversion must be a result of the specialization that comes with cell differentiation.

What drives this inversion of nuclear architecture is a key question we are trying to address in this work. Therefore, we make use of a recently developed model based on loops and entropy, two concepts that have

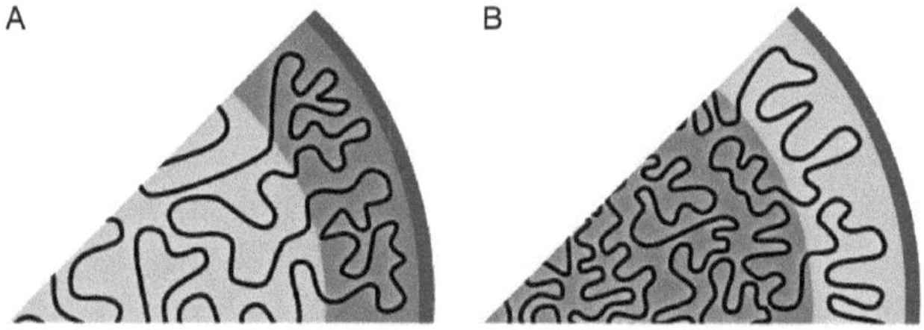

Figure 10.2 Illustration of differences in nuclear architecture. (A) Usually, chromatin is organized such that the highly active euchromatin (light grey) is located in the nuclear center while the compact heterochromatin (dark grey) resides mainly near the nuclear envelope. (B) In rod photoreceptor cells of nocturnal mammals, this organization is inversed (Solovei et al., 2009). (For the color version of this figure, the reader is referred to the online version of this chapter.)

emerged as the major players in the organization of the nucleus (Bohn and Heermann, 2011; Hancock, 2007; Heermann, 2011; Jun and Mulder, 2006; Marenduzzo et al., 2006). It is natural to ask whether the inverted structure can be derived from the common structure. This has been investigated in the light of the interplay between intra-chromosomal, inter-chromosomal and chromatin–envelope interactions.

2. THE MODEL

The model that we propose and investigate consists of a set of polymers representing coarse-grained versions of the chromosomes. Coarse-grained means that the monomers and links of the simulated polymers do not represent single base pairs, but long stretches of DNA. Even though this simplification ignores most of the rich diversity in the distribution of chromatin modifiers on the short scale, it needs to be introduced in order to reduce the complexity of the system and, consequently, the computational effort to a practical level.

The simulated polymers interact via intra- and inter-chromosomal interactions (see later) as well as with the boundary that encloses the polymers, represented by the nuclear envelope. Since these interactions are mediated by proteins in a nondeterministic fashion, the interaction is not set up as potentials but rather by binding probabilities (see later). Thus, two averaging procedures apply to all quantities that we calculate: first, the configurational averaging where excluded volume applies, and second, the averaging over the "disorder" that results from the stochastic binding between monomers and monomers with the wall.

2.1. Chromatin model

For the intra- and inter-chromosomal interactions, we use the Dynamic Loop (DL) model (Bohn and Heermann, 2010). In this model, two monomers that are nonadjacent along the polymer are allowed to interact only if they are in spatial proximity, that is, if the distance between them is below a certain cutoff. Thus, when two monomers approach each other because of diffusional motion and the cutoff condition is fulfilled, a bond can be formed between them with a certain probability. This bond is established with a probability p_b, which is determined by the interaction affinities a of the two monomers involved:

$$p_{\mathrm{b},ij} = \frac{a_i + a_j}{2} \tag{10.2}$$

Here, the indices i and j correspond to the different monomers. In case the bond is formed, a lifetime τ_b chosen from a Poisson distribution with mean T_b is assigned to it, determining when the bond dissociates. Hence, the bonds can change frequently, which mimics the effect of the highly dynamic chromatin–chromatin and chromatin–protein interactions. In contrast, the bonds along the backbone of the polymer are fixed. In the model, the proteins that are bound to the DNA and are expected to facilitate the binding of the fiber are not modeled explicitly. This means that the simulated polymer represents the chromatin fiber in conjunction with the proteins bound to it. To take the local binding affinities of the proteins into account, one has to adapt the local interactions affinities that are used to calculate the binding probabilities of the monomers. Even though there are other chromatin models that explicitly incorporate binding ligands such as proteins (Barbieri et al., 2012), the advantage of this model's implicit approach is that the computational effort of the simulations is reduced strongly. Otherwise, it would be practically impossible to finish the simulations of a system as complex as a cell nucleus in a manageable time frame. Since the binding affinities can be varied along the chromosome, it is possible to obtain largely heterogeneous chromosomal conformations where subparts strongly differ in structure. For the DL model, it has been shown that this also allows reproducing the structural properties of real eu- and hetero-chromatic regions, in agreement with experimental data (Jerabek and Heermann, 2012). It has to be mentioned that *in vivo*, the structural differences between these regions are a result of numerous factors such as, for example, helical unwinding of DNA and binding of large multi-protein complexes. However, it is not possible to incorporate all these complex processes into a single chromatin model. Yet, at the level of coarse-graining, which we use in our simulations with one monomer representing a few hundred kb, the average chromatin properties can be reproduced consistently with the DL model by using only one parameter, namely, the interaction affinity.

2.2. Nuclear envelope attachment

In order to incorporate interactions with the nuclear envelope into our model, we confine the polymer chains in a sphere. For the attachment, we used a similar approach as for the chromosome bonds: as soon as a monomer comes close enough to the envelope and is closer than a certain cutoff distance, the monomer can attach to the envelope with a certain probability p_m. For each attachment, a lifetime τ_m chosen from a Poisson distribution with mean T_m is set to determine how long the attachment is retained. The envelope attachment probability p_m of the monomers was set to be proportional to the chromatin binding affinity:

$$p_{m,i} = f_m \times a_i, \tag{10.3}$$

with f_m being a scaling factor that determines how strong the chromatin–envelope interaction is in relation to the chromosome–chromosome interaction. In this study, f_m was varied between 0.0 and 1.2, which corresponds to situations where no envelope interaction is present and where chromosome–envelope interaction is 20% stronger than intra- and interchromosomal interaction, respectively. We did not investigate cases where $f_m > 1.2$ because for all simulated nuclear densities, the effect of interest, namely, the transition between the common and the inverted architecture, already occurred at lower values.

2.3. Model parameters

In this study, the model nucleus contains 20 chromosomes and each chromosome-representing polymer consists of 400 monomers. The boundary of the nucleus is spherical and the size of the sphere is set such that the confined chromosomes account for a certain percentage of the total volume. Thus, three parameters can be varied:

- ρ, the chromosome density inside the nucleus
- a, the chromosome interaction affinity
- f_m, the strength of chromatin–envelope interaction relative to chromatin–chromatin interaction

For the nuclear density, we used values of 20%, 30% and 40%. Chromatin usually occupies ≈10% of the nuclear volume, yet it is expected that many other biomolecules, for example, ribonucleoproteins, are also packed into nuclei. Hence we considered densities ≥20% a reasonable choice. The average chromosome binding affinity, that is, the probability of bond formation between the monomers, was varied between 0.06 and 0.18. These values were set because the level of compaction of the chromosomes changes significantly in that range. In order to investigate how the fiber heterogeneity, for example, because of the presence of euchromatic and heterochromatic parts, affects the interplay between the different kinds of interactions, the

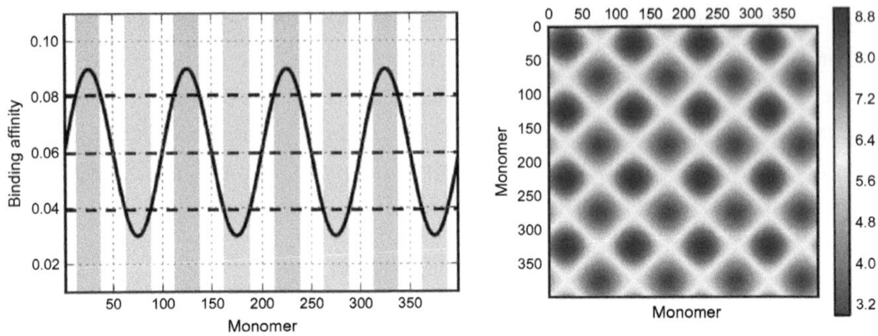

Figure 10.3 Sinusoidal binding affinity profile that was used as input for the simulation (left) and the resulting binding probability map for all pairs of monomers (right). Grey regions represent low affinity (euchromatin); black ones have high affinity (heterochromatin). (For interpretation of the references to color in this figure legend, the reader is referred to the online version of this chapter.)

strength of the affinity was varied along the fibers. For simplicity, we added a sinusoidal shift to the average affinity (see Fig. 10.3):

$$a_i = \bar{a}\left(1 + f_{\text{amp}} \sin\left(2\pi\, n_{\text{osc}} \frac{i}{n-1}\right)\right), \quad \text{for all } i \in [0, n-1]. \qquad (10.4)$$

Here, $\bar{a}$ is the average affinity, i is the index of the monomer in the polymer, and n is the length of the polymer. f_{amp} is the amplitude of the sinusoidal shift relative to $\bar{a}$ and was set to 50% for all simulations. n_{osc} determines the number of affinity changes along each polymer. We arbitrarily set this parameter to 4, so that each model chromosome has 4 less compact (euchromatic) and 4 more compact (heterochromatic) regions. The heterochromatic regions have a higher affinity than the euchromatic ones because for the DL model, this has been shown to provide the best agreement between simulation and experimental data (Jerabek and Heermann, 2012).

2.4. Model implementation

The model was implemented using the bond-fluctuation model (Carmesin and Kremer, 1988), which allows bond and bond-angle fluctuations and in which the consistency with the dynamic properties for the polymers is ensured automatically. To generate conformations of interphase chromosomes in nuclei, lattice Monte-Carlo simulations with spherical boundaries were performed. Because of the fact that eukaryotic genomes, such as the human one, often consist of billions of base pairs, it is not possible to treat full sets of chromosomes on the single base pair level. Hence, the chromosomes are coarse-grained versions of the real chromosomes and each monomer represents a long stretch of chromatin. If the length of the stretch is much larger than the persistence length of the chromatin fiber, we can expect the polymer to be totally flexible and can neglect the influence of the local bending rigidity. With a persistence length below 250 nm (Langowski, 2006) and a fiber packing where a length of 10 nm $\approx$ 1 kilobase pair (kb), the monomers should represent chromatin stretches with a length of at least 50 kb. In this study, each monomer corresponds to a genomic content of at least 200 kb. The initial configuration is created by generating SAW polymers in a sphere whose radius is set such that the total monomer density has a given value. Each polymer represents a single interphase chromosome. At first, an equilibration run of 2×10^8 Monte Carlo Steps (MCS) where only intra-chromosomal interactions are allowed is performed. This is done because it is known that interphase chromosomes in real nuclei are not totally intermingled but tend to be confined to certain sub-nuclear volumes, the chromosome territories (Cremer and Cremer, 2001). Since the SAW polymers are strongly intermingling, turning on all interactions right from the start would lead to biased conformations. Therefore, by allowing only intrachromosomal interactions, a segregation of the polymer chains is achieved. This is followed by a second equilibration run of 10^8 MCS where we additionally allow interchromosomal interactions. In a third equilibration run of 10^8 MCS, we start to allow the attachment of the monomers to the surface of the spherical boundary, meaning the nuclear envelope. When all three equilibration runs are finished, we start the main simulations where we save the total nuclear conformation every 2×10^6 MCS.

2.5. Mean square distance

The MSD provides the average spatial distance by which two monomers or loci are separated as a function of the contour length or genomic distance between them, respectively. With $\mathbf{r}_{j,i}$ being the coordinate vector of the ith monomer of the jth conformation and n being the contour length along the polymer chain, the MSD is calculated as follows:

$$\left\langle R_n^2 \right\rangle = \frac{1}{(N-n)N_c} \sum_{j=0}^{N_c-1} \sum_{i=0}^{N-n-1} \left| \mathbf{r}_{j,i} - \mathbf{r}_{j,i+n} \right|^2 \quad (10.5)$$

Here, N is the number of monomers the polymer consists of and N_c is the number of polymer conformations generated. As the simulations are carried out on a lattice, the coordinates $\mathbf{r}_{j,i}$ and, consequently, the MSD are given in lattice units. The scaling factor that relates lattice units to real units is obtained by scaling the computed to the experimental MSD.

2.6. Volume ratio

The volume ratios of different genomic regions are calculated from the corresponding monomer densities of the simulated polymer chains. These densities are obtained using three-dimensional Gaussian kernel density estimation (Jones et al., 2001). The volume ratios are indirectly proportional to the obtained density ratios:

$$\phi_{ij} = \frac{V_i}{V_j} = \frac{\langle \rho \rangle_j}{\langle \rho \rangle_i}. \tag{10.6}$$

Here, V_i is the volume and $\langle \rho \rangle_i$ is the average density of all monomers in the ith genomic region.

2.7. Polymer intermingling

A monomer of a polymer is considered to be intermingling with a group of other polymers when we find more, or the same, number of monomers from neighboring polymers than from the polymer it belongs to inside a sphere around this primary monomer. Here, the maximum bond length from the bond fluctuation model (Carmesin and Kremer, 1988) was used as the sphere radius.

3. RESULTS

The results presented here are obtained by calculating the average of the respective observables over thousands of uncorrelated nuclear conformations, which corresponds to averaging over the same number of cells. Hence, our conclusions always refer to the ensemble of possible conformations, which does not mean that every single conformation would allow deriving exactly the same conclusions. The structural and topological variations between the single conformations can be seen to reflect the cell-to-cell variations.

3.1. Formation of chromosome territories and volume ratios between eu- and hetero-chromatin

The results of MC simulations of our model show that chromosome separation and intermingling depend strongly on the binding affinity of the chromosomes. For low affinities, the chromosomes have a very open structure because the number of established intra-chain bonds is too small to maintain a compact conformation. With increasing affinity, the fibers become more compact because the higher number of bonds prevents subparts of the chain from diffusing away from each other. As a measure for the degree of compaction, we primarily used the mean square spatial distance between two monomers as a function of the contour length *g* by which they are separated. In living cells, this measure can be obtained in FISH experiments (Goetze et al., 2007; Mateos-Langerak et al., 2009). Just as *in vivo*, we obtain a leveling-off above a certain genomic distance (see Fig. 10.4), indicating that the polymer is confined inside a certain volume. If the fibers would simply be confined inside the nuclear volume, the height of the MSD plateau level would roughly equal the squared radius of the nucleus. However, the calculated plateau level is much smaller, that is, the simulated chromosomes are confined to subnuclear volumes. This means that the model allows reproducing the formation of chromosome territories.

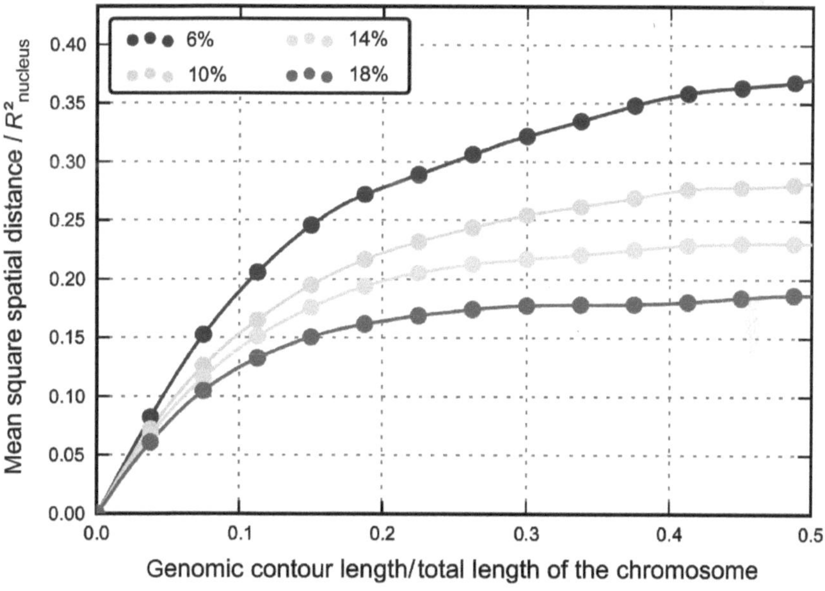

Figure 10.4 The mean square spatial distance as a function of the contour length along the fiber for different binding affinities. With increasing binding affinity, the plateau level decreases because the chromosomes become more compact. Parameters: $\rho = 20\%$, $f_m = 70\%$. (For the color version of this figure, the reader is referred to the online version of this chapter.)

Figure 10.4 shows that the size of these territories (their volume) decreases with increasing binding affinity.

Besides the chromatin–chromatin interactions, we expected the envelope attachment also to influence chromosome compaction. However, the results do not show that the interaction of the chromosomes with the envelope significantly influences the MSD plateau level. As the MSD is calculated over the total chromosome, it does not provide information on the difference in compaction between hetero- and eu-chromatic regions. Yet, the volume ratio between these two kinds of regions can be measured experimentally (Goetze et al., 2007), which gives us the opportunity to check whether the fluctuations in binding affinity that we input into our model fit to the *in vivo* observations. To calculate the volume ratios from our simulation data, we first determined the local chromatin density along the fiber using Gaussian kernel density estimation. As one can see in Fig. 10.5, the

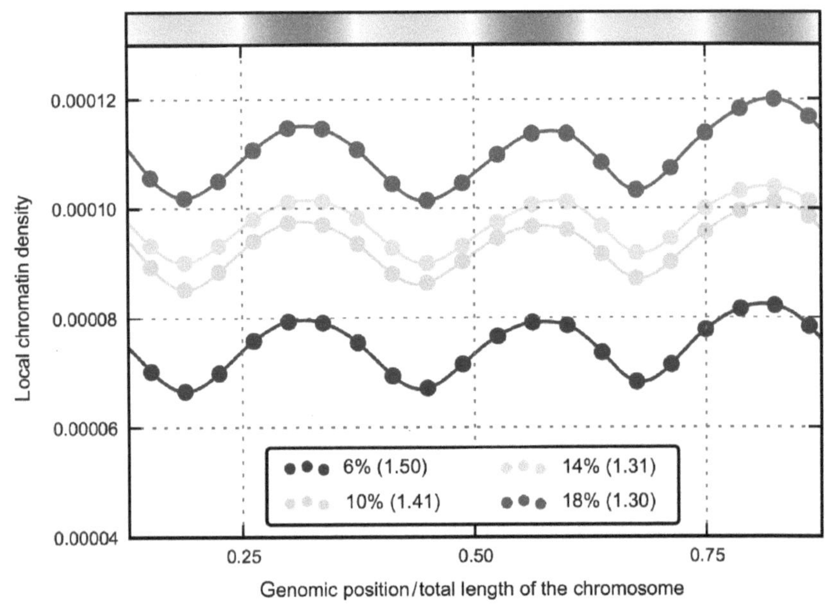

Figure 10.5 Local chromatin density determined with Gaussian kernel density estimation (Jones et al., 2001). With increasing binding affinity, the local chromatin density increases, as the chromosomes are compacted more strongly. The bracketed values in the legend are the volume ratios between hetero- and euchromatic regions. The color bar represents the affinity along the fiber, with red regions (heterochromatin) having high affinity and green regions (euchromatin) having low affinity. Parameters: $\rho = 20\%$, $f_m = 70\%$. (For interpretation of the references to color in this figure legend, the reader is referred to the online version of this chapter.)

heterochromatic regions are much denser because they have a higher chromatin affinity. The volume ratios were calculated between the 25% of the monomers with the lowest affinity and the 25% with the highest affinity (see green and red areas in Fig. 10.3). For all the simulations, we obtained volume ratios in the range between 1.27 and 1.61, which is around the experimental value of $\approx$1.4 (Goetze et al., 2007). Here, the volume ratios increase with increasing average chromatin affinity because the absolute difference between the highest and lowest input affinities also becomes larger. However, for all simulations, the calculated volume ratios deviate from the experimental value by less than 15%. This allows us to conclude that nuclear conformations with realistic activity-dependent chromatin density fluctuations were obtained for all the parameter combinations we explored.

3.2. Chromosome intermingling

Since chromosomes are not hard spheres, they can intermingle up to a certain extent. In the current model, the degree of intermingling depends on the chromatin affinity, yet there are two effects influencing it in opposing ways. On the one hand, an increase in intra-chain bonds leads to a compaction of the polymer and an increase in the number of loops. As the entropic repulsion between polymers is known to become stronger with an increase in the number of loops (Bohn and Heermann, 2011; Marenduzzo and Orlandini, 2009), the repulsion between the chromosomes increases and, consequently, the intermingling is reduced. On the other hand, establishing more interchromosomal contacts leads to an attraction of the chromosomes, disturbing chromosome compaction and increasing intermingling (intra- and inter-chromosomal contact maps are shown in Figs. 10.6 and 10.7).

Here, these opposing effects do not neutralize each other. We find that the intermingling increases toward higher binding affinities. For 20% nuclear density, for example, the intermingling changes from $\approx$28% to $\approx$34% when increasing the chromatin affinity from 0.06 to 0.18 (see Fig. 10.8). However, as the standard deviation of the degree of intermingling is $\approx$6%, the affinity-induced change is rather small compared to the observed variations. As far as the influence of nuclear chromatin density is concerned, we find that with increasing density, the intermingling becomes significantly stronger. This is obviously an effect of the reduction in free volume that the chromosomes could use to avoid each other in order to increase the entropy. As a consequence, the number of interchromosomal contacts increases and the compaction of the chromosomes is more

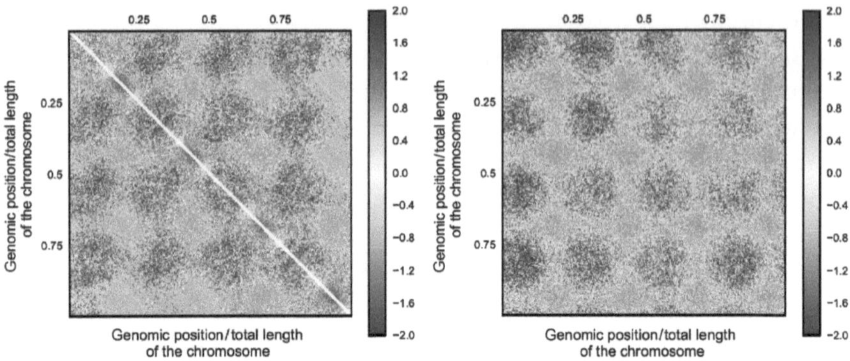

Figure 10.6 Normalized contact maps for intra- (left) and inter-chromosomal (right) contacts. Normalization was done by subtracting the mean and dividing by the standard deviation. Red regions frequently contact each other; green ones less often form chromatin bonds. Parameters: $\rho = 20\%$, $a = 6\%$, $f_m = 10\%$. (For interpretation of the references to color in this figure legend, the reader is referred to the online version of this chapter.)

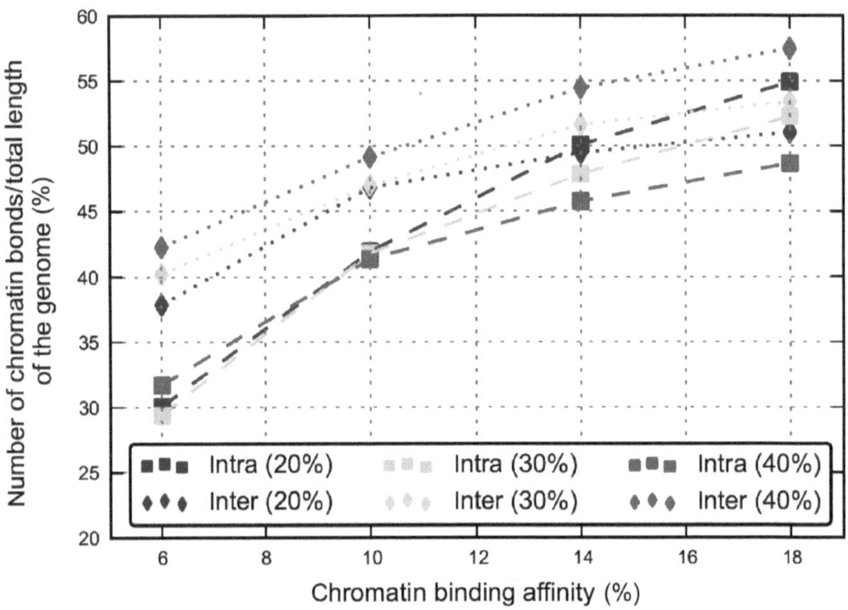

Figure 10.7 The relative number of intra- (square markers) and inter-chromosomal (diamond markers) contacts as a function of the chromatin-binding affinity for different nuclear densities. The number of contacts increases monotonically with increasing binding affinity. While the number of inter-chromosomal contacts is generally higher than the number of intra-chromosomal contacts, it is the other way round for the simulation settings with the lowest density and the highest chromatin-binding affinity. (For the color version of this figure, the reader is referred to the online version of this chapter.)

disturbed. Even though we also expected the envelope interaction to influence the intermingling, we did not observe significant changes when increasing the affinity to the envelope.

For all simulation settings, the calculated mean of the degree of intermingling was between ≈25% and ≈43%, with an average standard deviation of ≈6%. The intermingling of a chromosome with the rest of the genome has also been measured in an experimental study (Branco and Pombo, 2006): the obtained value of ≈40% is in agreement with our findings, at least for nuclear densities of 30% and 40%. However, for the lowest chromatin density of 20% (see Fig. 10.8), the intermingling of ≈32% is much lower than the experimental one, indicating that real nuclei might have higher chromatin densities.

3.3. Chromosome topology

A frequently studied feature of interphase chromosomes is the position of hetero- and eu-chromatin relative to the center of the corresponding

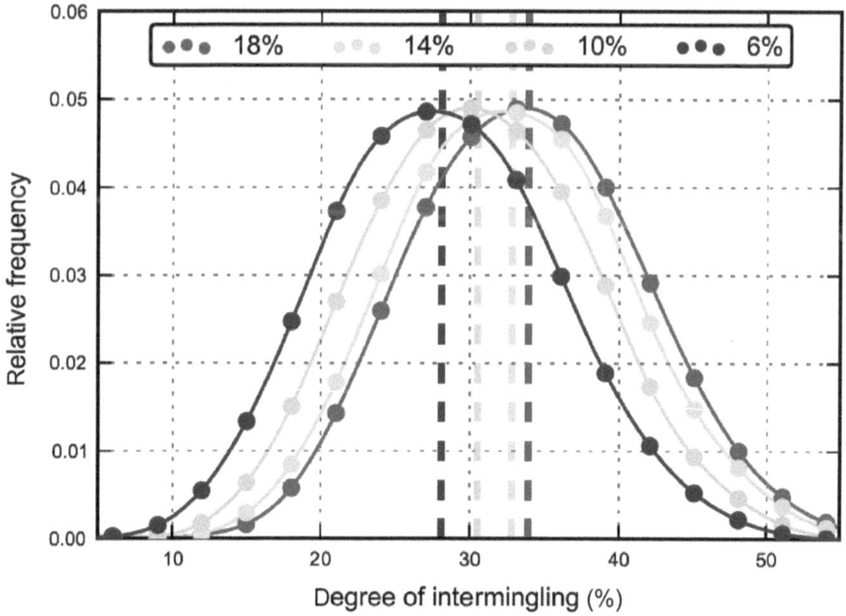

Figure 10.8 The distribution of the degree of intermingling of single chromosomes with the rest of the genome for different chromatin–chromatin binding affinities. The higher the affinity, the more the chromosomes intermingle with each other. Parameters: $\rho = 20\%$. (For the color version of this figure, the reader is referred to the online version of this chapter.)

chromosome territory. While most studies state that there is an expression-dependent topology with euchromatin being preferentially located closer to the surface of the territories (Clemson et al., 2006; Cremer et al., 2004; Kosak and Groudine, 2004; Scheuermann et al., 2004), others state that the surface is on average as transcriptionally active as the whole chromosome (Branco and Pombo, 2006) and that transcription sites are basically uniformly distributed throughout the nucleoplasm and the territories (Abranches et al., 1998; Verschure et al., 1999). Our Monte Carlo simulations show that whether an expression-dependent topology is established or not depends on how strong the chromosome binding affinity is. Figure 10.9 depicts that as long as there is little chromatin–chromatin interaction, the euchromatin is more likely to be found near the surface of the chromosome

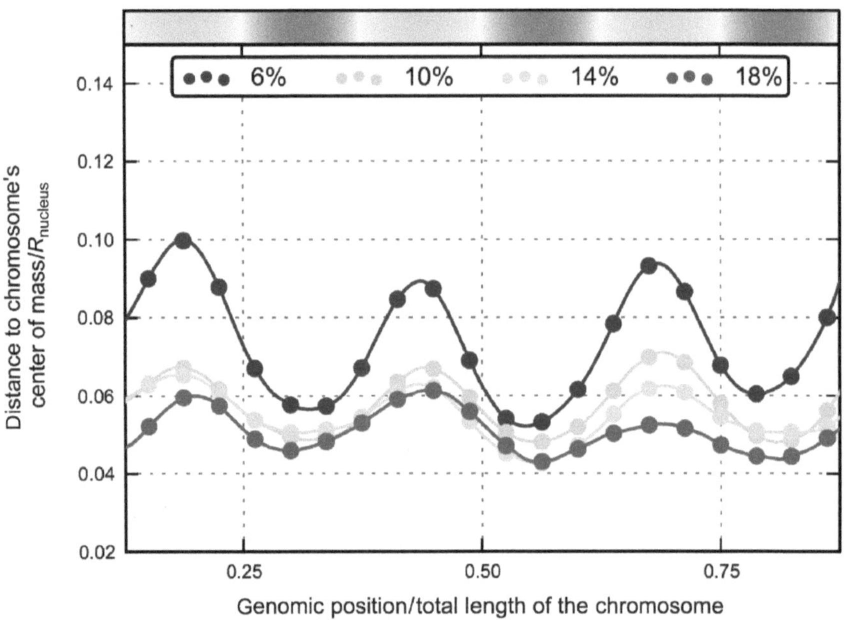

Figure 10.9 Spatial distance of each locus to the center of mass of the chromosome territory to which it belongs. In general, the heterochromatic regions can be found closer to the territory's center, while euchromatin is more likely to be located near the surface. However, this effect becomes less pronounced with increasing binding affinity because the number of interchromosomal contacts between heterochromatic parts increases, thereby forcing these regions to get closer to the surface of the corresponding chromosomes. The color bar represents the affinity along the fiber, with red regions (heterochromatin) having high affinityand green regions (euchromatin) having low affinity. Parameters: $\rho = 20\%$, $f_m = 10\%$. (For interpretation of the references to color in this figure legend, the reader is referred to the online version of this chapter.)

territory while the dense heterochromatic parts preferentially reside closer the territory's center. This is because at the surface the open euchromatic parts of the fiber can move easier and effectively have more degrees of freedom than in the dense center, which increases the entropy of the chain. When the chromosome affinity increases, the expression-dependent positioning relative to the chromosome center of mass gets more and more lost: the more the number of inter-chromosomal contacts established, the smaller the inter-chromosomal space becomes. Hence, the gain of entropy becomes smaller because the degrees of freedom for chromatin regions at the territory's periphery approach the value of those in the center and the entropy-driven preference becomes less pronounced. For the same reason, increasing the nuclear chromatin density also changes the positioning because of the reduction of free space. As a consequence, the higher the density, the more uniform a chromosome territory's topology becomes. Despite the density- and chromosome affinity-induced topological changes, the results obtained show that the envelope attachment does not seem to significantly affect chromosome topology.

3.4. Inversion of expression-dependent nuclear organization

The main goal of this study is to investigate how the interplay between intra-chromosomal, inter-chromosomal, and envelope interactions influences the organization of nuclei and how an inversion of the common architecture, where heterochromatin is mainly located at the nuclear periphery, can be achieved. To determine whether the simulated conformations exhibit different nuclear architectures, we first calculated the average spatial distance of each monomer to the center of the nucleus. This measure provides information about the difference between hetero- and eu-chromatin as far as their preferential radial position is concerned.

The results (Fig. 10.10) show that the expression-dependent nuclear positioning depends on the strength of the envelope interaction. When there is little envelope attachment, heterochromatin can be found close to the center while euchromatin preferentially resides closer to the envelope (=inverse organization). As for the case of single chromosomes, this positioning increases the entropy because the euchromatic parts have more degrees of freedom near the periphery, as only a small fraction of chromatin is bound to the envelope. By increasing the overall envelope affinity above a certain threshold, the radial positioning becomes inverted, and the heterochromatic regions move to the periphery and the euchromatin gets

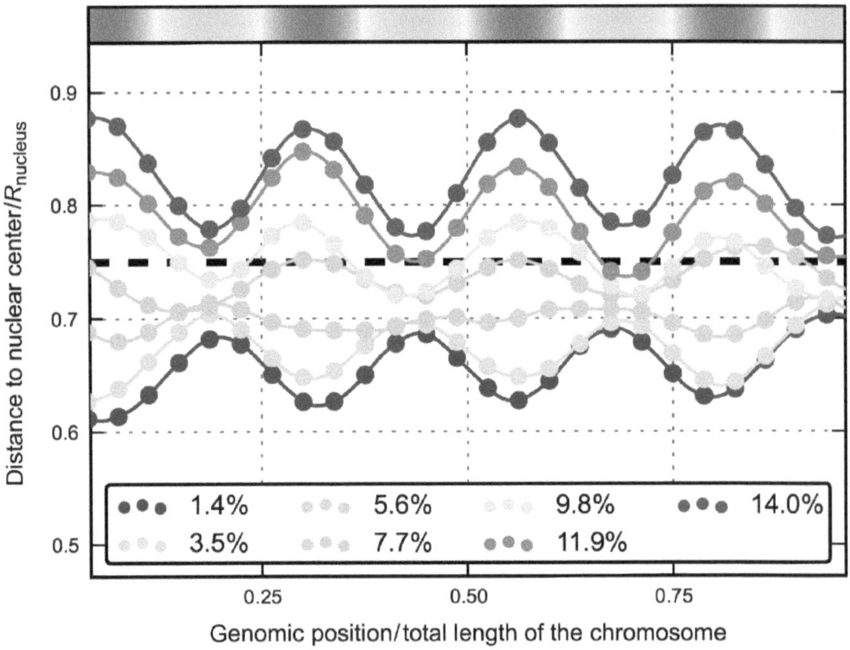

Figure 10.10 Spatial distance of genomic regions to the center of the nucleus for different nuclear envelope-binding affinities. For a weak envelope interaction (blue), euchromatin is more likely to be found closer to the nuclear periphery. With increasing envelope affinity, the positioning is inverted and euchromatin prefers to reside closer to the nuclear center (red). The dashed black line gives the average distance to the center for a sphere with homogenous density. The color bar represents the affinity along the fiber, with red regions (heterochromatin) having high affinity and green regions (euchromatin) having low affinity. Parameters: $\rho = 30\%$, $a = 14\%$. (For interpretation of the references to color in this figure legend, the reader is referred to the online version of this chapter.)

pushed to the center (=common organization). The inversion occurs because with higher envelope affinity, the number of envelope bonds increases and these bonds increasingly constrain the movement of regions that are located near the surface. Consequently, the entropy gain of the euchromatic regions strongly decreases when they reside near the envelope; hence, they will try to avoid the envelope. Since the heterochromatic regions, furthermore, have a higher chromatin–chromatin as well as envelope–chromatin affinity, they often start binding to the envelope and, thus, additionally push the euchromatin away to the center. As the observed inversion is strongly entropy-driven, we also investigated the effect of nuclear chromatin density on this process. Therefore, we calculated the

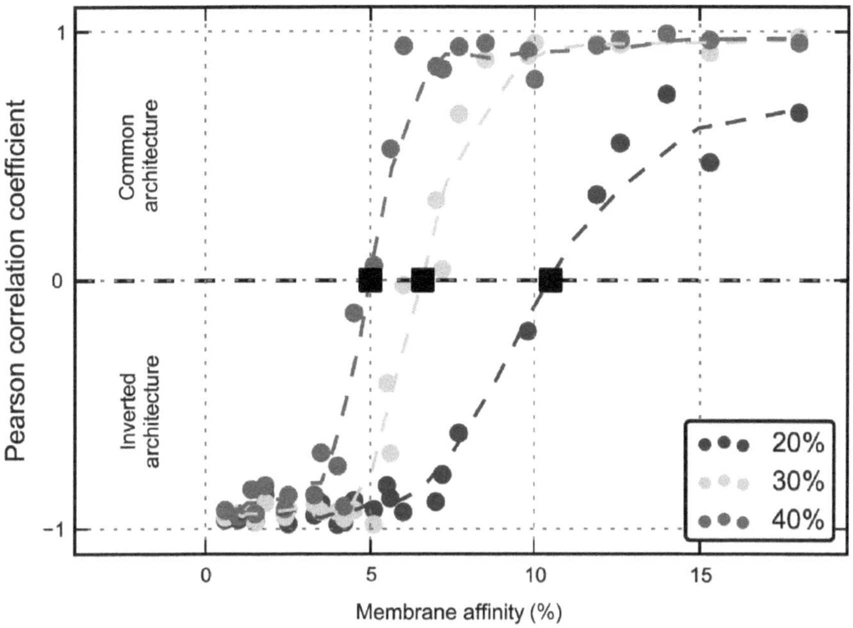

Figure 10.11 Pearson correlation coefficient between spatial distance to center (Fig. 10.10) and affinity profile (Fig. 10.3) for different nuclear densities as a function of the envelope affinity. Around certain threshold affinities (black squares), the correlation coefficient changes from −1 to +1 because of the transition from the common to the inverted nuclear organization. The slope of the transition increases and the threshold decreases with higher nuclear density. (For the color version of this figure, the reader is referred to the online version of this chapter.)

Pearson correlation between the affinity profiles (Fig. 10.3) and the profiles of the distance to the nuclear center (Fig. 10.10) as a measure for the type of architecture: if euchromatin has a smaller distance to the center, we obtain a positive correlation because euchromatin also has a low affinity. Consequently, we calculate a negative correlation if euchromatin is preferentially located to the periphery. Figure 10.11 exhibits the correlation coefficients for different envelope affinities and nuclear chromatin densities.

As described earlier, we find that when the strength of the envelope interaction exceeds a certain threshold, the correlation changes from −1 to +1 because of the inversion of the nuclear organization. However, the threshold envelope affinity as well as the slope from −1 to +1 changes: the higher the nuclear density, the steeper the slope and the lower the threshold becomes. This is because the free volume is smaller at higher densities and hence there is less entropy gain for euchromatin at the periphery. Consequently, euchromatin tries to avoid the envelope at even lower envelope affinities. In addition, the change of envelope affinity that is necessary to achieve a total inversion of the architecture, meaning from correlation −1 to +1, is also lower. In order to get information about how much chromatin is attached to the envelope, we calculated the average fraction of monomers that are simultaneously bound to the envelope (see Fig. 10.12). We found that the number of envelope bonds increases linearly with the envelope affinity. For the different nuclear densities, we also determined the number of bonds that are present at the corresponding threshold affinities where the inversion between the two architectures occurs.

The results of our model system show that the transition from the common to the inverse organization happens when ≈4.7% of the genome is bound to the envelope. This critical fraction is the same for all densities

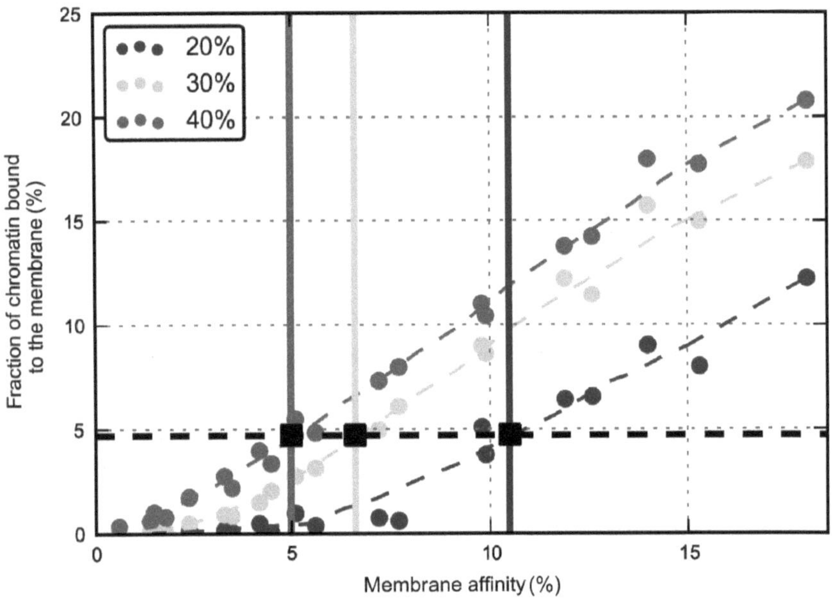

Figure 10.12 Fraction of chromatin that is bound to the envelope for different nuclear densities. The number of envelope bonds increases linearly with the envelope-binding affinity. All affinity thresholds for the transition from the common to the inverted architecture (vertical continuous lines) intersect with the corresponding graphs at ≈4.7% (black dashed line, black squares), meaning that the transition occurs when this fraction of the genome is bound to the nuclear envelope, independent of the overall chromatin density. (For thecolor version of this figure, the reader is referred to the online version of this chapter.)

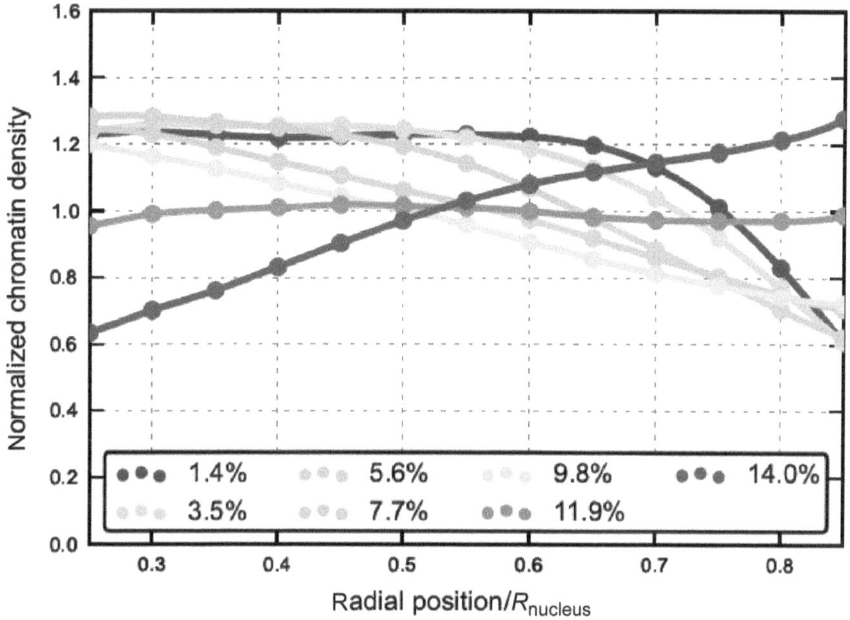

Figure 10.13 Radial chromatin density distribution relative to the center of the nucleus. Because of the inversion of nuclear architecture, the density pattern changes strongly. For the common organization, it increases monotonically toward the envelope while for the inverted case, it decreases monotonically. Parameters: $\rho = 30\%$, $a = 14\%$. (For the color version of this figure, the reader is referred to the online version of this chapter.)

we simulated and hence appears to be the best measure of which of these two genome architectures is present. To relate the alterations of the preferential positioning of hetero- and eu-chromatin to the changes in the density distribution of chromatin inside the nucleus, we additionally calculated the radial distribution $g(r)$ of chromatin relative to the center of the nucleus (see Fig. 10.13):

$$g(r) = 4\pi r^2 \rho(r) \mathrm{d}r \tag{10.7}$$

Here r is the distance to the nuclear center, ρ is the local density, and $\mathrm{d}r$ is an infinitesimal radial change. $\rho(r)$ is normalized such that the integral over g from 0 to R_{nucleus} is 1. For low envelope binding, the distribution has the same plateau level for all densities and decays when approaching the envelope as the less dense euchromatin is preferentially located there. The gradient becomes steeper with increasing density because the euchromatic regions at the surface are the first to be squeezed when the nuclear radius is reduced. For strong envelope interaction, when the euchromatin is preferentially located in the nuclear center, the characteristics of the distributions change strongly because the easily squeezed open chromatin is no longer located at the periphery. In the simulations, we observe that at 20% overall density, the radial distribution decreases monotonically toward the surface, while for 40%, the distribution increases and has a maximum close to the envelope.

3.5. Chromatin diffusion

To measure chromatin diffusion, we calculated the mean square displacement of every monomer inside the nucleus as a function of time. We averaged the displacement of 25% of the monomers with the highest and lowest affinity in order to compare the diffusion of hetero- and eu-chromatic regions, respectively. The simulations clearly show that euchromatin is much more dynamic than heterochromatin (see Fig. 10.14), mainly because

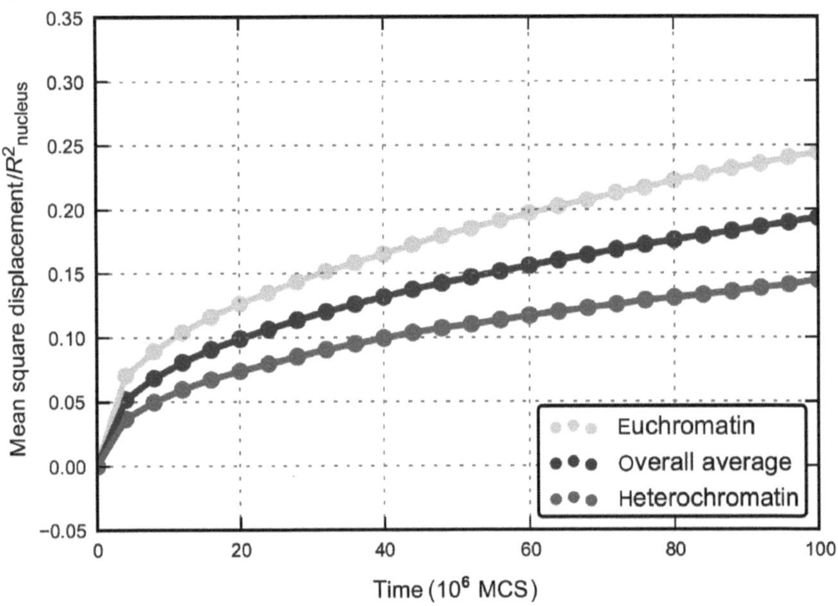

Figure 10.14 Mean square displacement of genomic loci with different chromatin affinities. Euchromatic regions are much more dynamic than heterochromatic ones because they are less dense and, thus, effectively have more degrees of freedom. Parameters: $\rho = 20\%$, $a = 10\%$, $f_m = 10\%$. (For the color version of this figure, the reader is referred to the online version of this chapter.)

euchromatic regions are less dense and, thus, effectively have more degrees of freedom.

This is in agreement with studies suggesting that the chromatin fiber rather than the transcription machinery is the mobile partner during gene transcription; when euchromatin is more dynamic, it is more likely that these active regions find a transcription factory (Jackson, 2005; Schoenfelder et al., 2010; Sexton et al., 2007). It is also of interest how the total chromosomes diffuse depending on the parameters. Therefore, we calculated the mean square displacement of the center of mass of each chromosome (see Fig. 10.15). We found that with increasing chromatin–chromatin and chromatin–envelope interaction, the speed of chromosome diffusion is reduced. This is because at higher binding affinities, the chromosomes often bind to each other or to the envelope, which strongly hinders them from diffusing inside the nucleus. Obviously, an increase in the nuclear chromatin density also reduces the mean square displacement as there is less

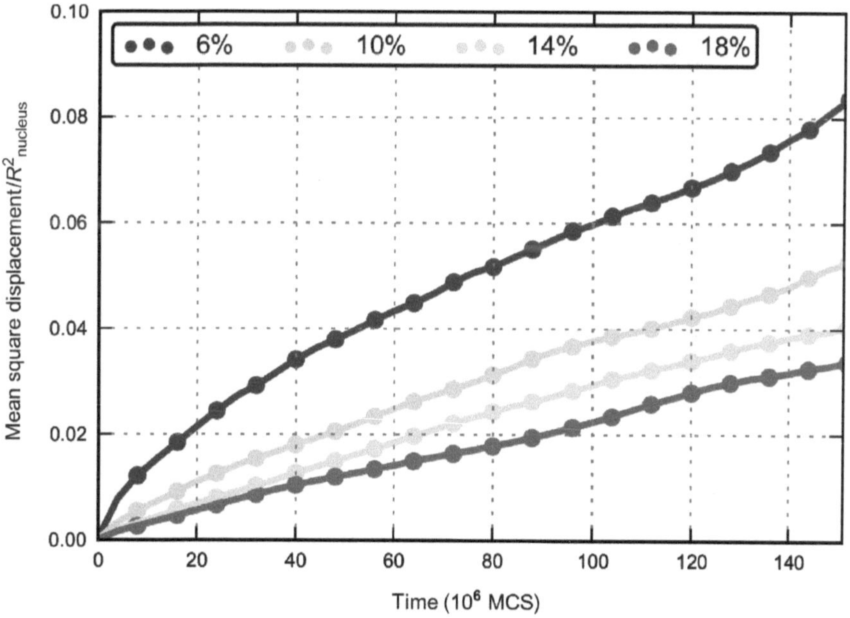

Figure 10.15 Mean square displacement of the chromosomes' center of mass for different chromatin–chromatin binding affinities. With increasing strength of the chromatin–chromatin interactions, the chromosomes stick together more and more. This strongly reduces their movement because the frequent binding to neighboring chromosomes hinders their diffusion. Parameters: $\rho = 20\%$, $f_m = 10\%$. (For the color version of this figure, the reader is referred to the online version of this chapter.)

free volume that the chromosomes could use to avoid each other. This is in agreement with the idea that cell differentiation, and the subsequent reduction in nuclear volume, reduces the mobility of chromatin (Thakar and Csink, 2005).

4. PERSPECTIVES

The results obtained have shown that most chromosome parameters, for example, size and intermingling of chromosome territories, local chromatin density, and chromosome topology, depend mainly on the strength of the chromatin–chromatin interactions. However, it turned out that the interplay of chromatin with the nuclear envelope is essential for the overall organization of interphase cell nuclei. We found that there are two possible ways to reproduce the transition from the common to the inverted nuclear architecture, namely, by changing the nuclear envelope affinity or by altering the chromatin density inside the nucleus. As a consequence, there are two explanations why the rod photoreceptor cells of nocturnal mammals exhibit the inverted pattern, while their progenitor cells have the common one. On the one hand, the differentiation of the cells and the change of gene expression that comes along with it could alter the production of proteins that are responsible for chromatin–envelope interactions. In this context, our data show that already small changes in the envelope affinity can induce a transition of the architecture, especially when being close to the threshold affinity (see Fig. 10.11). On the other hand, as the differentiation of cells leads to a reduction in their volume, this process effectively increases the nuclear chromatin density. Since the threshold envelope affinity that is needed for the transition to happen decreases with increasing density, the threshold could drop below the present envelope affinity. Thus, the differentiation-induced transition could be density-based without a change of envelope affinity being necessary. Despite the fact that the current nuclear model allows us to reproduce most aspects of chromosomal and nuclear organization, it has to be mentioned that the simulated systems are only coarse-grained representations of real systems. At the moment, the simplifications incorporated into the model are necessary to reduce the computational effort to a practical level. However, the more we learn about the complex mechanisms responsible for nuclear organization, the better the simulation techniques and computers will become, and the greater the number of features that can be incorporated into future models will be. Up to date, for example, only a few proteins, such as CTCF and cohesin, have been reported to be directly involved in chromatin looping (Handoko et al., 2011; Parelho et al., 2008; Phillips and Corces, 2009; Rubio et al., 2008; Wendt et al., 2008). Yet there are a huge number of unknown proteins that can be expected to contribute to the folding of the fiber and the binding to the nuclear envelope. Since we have shown how important the formation of chromatin loops and the related entropic changes are for the nuclear organization, next-generation models that are going to investigate the cell nucleus at a higher level of detail will have to additionally incorporate protein–chromatin as well as protein–protein interactions.

ACKNOWLEDGMENTS

We would very much like to thank Roel van Driel and Mariliis Tark-Dame for the stimulating discussions. Furthermore, Hansjoerg Jerabek would like to thank the Heidelberg Graduate School of Mathematical and Computational Methods for the Sciences and the Institute for Theoretical Physics for funding. Dieter W. Heermann would like to thank Yixue Li and Lei Liu as well as the SIBS for the discussions and the hospitality.

REFERENCES

Abranches, R., Beven, A.F., Aragn-Alcaide, L., Shaw, P.J., 1998. Transcription sites are not correlated with chromosome territories in wheat nuclei. J. Cell Biol. 143, 5–12.

Barbieri, M., Chotalia, M., Fraser, J., Lavitas, L.-M., Dostie, J., Pombo, A., Nicodemi, M., 2012. Complexity of chromatin folding is captured by the strings and binders switch model. Proc. Natl. Acad. Sci. U. S. A. 109, 16173–16178.

Bohn, M., 2010. Modelling of interphase chromosomes: From genome function to spatial organization. Ph. D. thesis, University of Heidelberg.

Bohn, M., Heermann, D.W., 2010. Diffusion-driven looping provides a consistent framework for chromatin organization. PLoS One 5, e12218.

Bohn, M., Heermann, D.W., 2011. Repulsive forces between looping chromosomes induce entropy-driven segregation. PLoS One 6, e14428.

Bolzer, A., Kreth, G., Solovei, I., Koehler, D., Saracoglu, K., Fauth, C., Mller, S., Eils, R., Cremer, C., Speicher, M.R., Cremer, T., 2005. Three-dimensional maps of all chromosomes in human male fibroblast nuclei and prometaphase rosettes. PLoS Biol. 3, e157.

Branco, M.R., Pombo, A., 2006. Intermingling of chromosome territories in interphase suggests a role in translocations and transcription-dependent associations. PLoS Biol. 4, e138.

Carmesin, I., Kremer, K., 1988. The bond fluctuation method: a new effective algorithm for the dynamics of polymers in all spatial dimensions. Macromolecules 21, 2819–2823.

Clemson, C.M., Hall, L.L., Byron, M., McNeil, J., Lawrence, J.B., 2006. The X chromosome is organized into a gene-rich outer rim and an internal core containing silenced nongenic sequences. Proc. Natl. Acad. Sci. U. S. A. 103, 7688–7693.

Cremer, T., Cremer, C., 2001. Chromosome territories, nuclear architecture and gene regulation in mammalian cells. Nat. Rev. Genet. 2, 292–301.

Cremer, T., Küpper, K., Dietzel, S., Fakan, S., 2004. Higher order chromatin architecture in the cell nucleus: on the way from structure to function. Biol. Cell. 96, 555–567.

De Gennes, P.G., 1979. Scaling Concepts in Polymer Physics. Cornell University Press, Ithaca, NY.

Dekker, J., 2003. A closer look at long-range chromosomal interactions. Trends Biochem. Sci. 28, 277–280.

Diesinger, P.M., Heermann, D.W., 2009. Depletion effects massively change chromatin properties and influence genome folding. Biophys. J. 97, 2146–2153.

Goetze, S., Mateos-Langerak, J., Gierman, H.J., de Leeuw, W., Giromus, O., Indemans, M.H.G., Koster, J., Ondrej, V., Versteeg, R., van Driel, R., 2007. The three-dimensional structure of human interphase chromosomes is related to the transcriptome map. Mol. Cell. Biol. 27, 4475–4487.

Grosberg, A.I., Nechaev, S.K., Shakhnovich, E.I., 1988. The role of topological limitations in the kinetics of homopolymer collapse and self-assembly of biopolymers. Biofizika 33, 247–253.

Hancock, R., 2007. Packing of the polynucleosome chain in interphase chromosomes: evidence for a contribution of crowding and entropic forces. Semin. Cell Dev. Biol. 18, 668–675.

Handoko, L., Xu, H., Li, G., Ngan, C.Y., Chew, E., Schnapp, M., Lee, C.W.H., Ye, C., Ping, J.L.H., Mulawadi, F., Wong, E., Sheng, J., Zhang, Y., Poh, T., Chan, C.S., Kunarso, G., Shahab, A., Bourque, G., Cacheux-Rataboul, V., Sung, W.-K., Ruan, Y., Wei, C.-L., 2011. Ctcf-mediated functional chromatin interactome in pluripotent cells. Nat. Genet. 43, 630–638.

Heermann, D.W., 2011. Physical nuclear organization: loops and entropy. Curr. Opin. Cell Biol. 23, 332–337.

Hsu, H., Paul, W., Binder, K., 2010. Standard definitions of persistence length do not describe the local intrinsic stiffness of real polymer chains. Macromolecules 43, 3094–3102.

Jackson, D.A., 2005. The amazing complexity of transcription factories. Brief Funct. Genomic. Proteomic. 4, 143–157.

Jerabek, H., Heermann, D.W., 2012. Expression-dependent folding of interphase chromatin. PLoS One 7, e37525.

Jones, E., Oliphant, T., Peterson, P., 2001. SciPy: Open Source Scientific Tools for Python. http://www.scipy.org/.

Jun, S., Mulder, B., 2006. Entropy-driven spatial organization of highly confined polymers: lessons for the bacterial chromosome. Proc. Natl. Acad. Sci. U. S. A. 103, 12388–12393.

Kosak, S.T., Groudine, M., 2004. Gene order and dynamic domains. Science 306, 644–647.

Kurukuti, S., Tiwari, V.K., Tavoosidana, G., Pugacheva, E., Murrell, A., Zhao, Z., Lobanenkov, V., Reik, W., Ohlsson, R., 2006. Ctcf binding at the h19 imprinting control region mediates maternally inherited higher-order chromatin conformation to restrict enhancer access to igf2. Proc. Natl. Acad. Sci. U. S. A. 103, 10684–10689.

Langowski, J., 2006. Polymer chain models of DNA and chromatin. Eur. Phys. J. E Soft Matter 19, 241–249.

Lieberman-Aiden, E., van Berkum, N.L., Williams, L., Imakaev, M., Ragoczy, T., Telling, A., Amit, I., Lajoie, B.R., Sabo, P.J., Dorschner, M.O., Sandstrom, R., Bernstein, B., Bender, M.A., Groudine, M., Gnirke, A., Stamatoyannopoulos, J., Mirny, L.A., Lander, E.S., Dekker, J., 2009. Comprehensive mapping of long-range interactions reveals folding principles of the human genome. Science 326, 289–293.

Maeshima, K., Hihara, S., Eltsov, M., 2010. Chromatin structure: does the 30-nm fibre exist *in vivo*? Curr. Opin. Cell Biol. 22, 291–297.

Marenduzzo, D., Orlandini, E., 2009. Topological and entropic repulsion in biopolymers. J. Stat. Mech. L09002.

Marenduzzo, D., Micheletti, C., Cook, P.R., 2006. Entropy-driven genome organization. Biophys. J. 90, 3712–3721.

Mateos-Langerak, J., Bohn, M., de Leeuw, W., Giromus, O., Manders, E.M.M., Verschure, P.J., Indemans, M.H.G., Gierman, H.J., Heermann, D.W., van Driel, R., Goetze, S., 2009. Spatially confined folding of chromatin in the interphase nucleus. Proc. Natl. Acad. Sci. U. S. A. 106, 3812–3817.

Mirny, L.A., 2011. The fractal globule as a model of chromatin architecture in the cell. Chromosome Res. 19, 37–51.
Parelho, V., Hadjur, S., Spivakov, M., Leleu, M., Sauer, S., Gregson, H.C., Jarmuz, A., Canzonetta, C., Webster, Z., Nesterova, T., Cobb, B.S., Yokomori, K., Dillon, N., Aragon, L., Fisher, A.G., Merkenschlager, M., 2008. Cohesins functionally associate with ctcf on mammalian chromosome arms. Cell 132, 422–433.
Phillips, J.E., Corces, V.G., 2009. Ctcf: master weaver of the genome. Cell 137, 1194–1211.
Rubio, E.D., Reiss, D.J., Welcsh, P.L., Disteche, C.M., Filippova, G.N., Baliga, N.S., Aebersold, R., Ranish, J.A., Krumm, A., 2008. Ctcf physically links cohesin to chromatin. Proc. Natl. Acad. Sci. U. S. A. 105, 8309–8314.
Scheuermann, M.O., Tajbakhsh, J., Kurz, A., Saracoglu, K., Eils, R., Lichter, P., 2004. Topology of genes and nontranscribed sequences in human interphase nuclei. Exp. Cell Res. 301, 266–279.
Schiessel, H., Gelbart, W.M., Bruinsma, R., 2001. DNA folding: structural and mechanical properties of the two angle model for chromatin. Biophys. J. 80, 1940–1956.
Schoenfelder, S., Sexton, T., Chakalova, L., Cope, N.F., Horton, A., Andrews, S., Kurukuti, S., Mitchell, J.A., Umlauf, D., Dimitrova, D.S., Eskiw, C.H., Luo, Y., Wei, C.-L., Ruan, Y., Bieker, J.J., Fraser, P., 2010. Preferential associations between co-regulated genes reveal a transcriptional interactome in erythroid cells. Nat. Genet. 42, 53–61.
Sexton, T., Umlauf, D., Kurukuti, S., Fraser, P., 2007. The role of transcription factories in large-scale structure and dynamics of interphase chromatin. Semin. Cell Dev. Biol. 18, 691–697.
Simonis, M., Klous, P., Splinter, E., Moshkin, Y., Willemsen, R., de Wit, E., van Steensel, B., de Laat, W., 2006. Nuclear organization of active and inactive chromatin domains uncovered by chromosome conformation capture-on chip (4C). Nat. Genet. 38, 1348–1354.
Simonis, M., Kooren, J., de Laat, W., 2007. An evaluation of 3C-based methods to capture DNA interactions. Nat. Methods 4, 895–901.
Solovei, I., Kreysing, M., Lanctt, C., Ksem, S., Peichl, L., Cremer, T., Guck, J., Joffe, B., 2009. Nuclear architecture of rod photoreceptor cells adapts to vision in mammalian evolution. Cell 137, 356–368.
Thakar, R., Csink, A.K., 2005. Changing chromatin dynamics and nuclear organization during differentiation in *Drosophila* larval tissue. J. Cell Sci. 118, 951–960.
Valenzuela, L., Dhillon, N., Dubey, R.N., Gartenberg, M.R., Kamakaka, R.T., 2008. Long-range communication between the silencers of HMR. Mol. Cell. Biol. 28, 1924–1935.
Vassetzky, Y., Gavrilov, A., Eivazova, E., Priozhkova, I., Lipinski, M., Razin, S., 2009. Chromosome conformation capture (from 3C to 5C) and its chip-based modification. Methods Mol. Biol. 567, 171–188.
Verschure, P.J., van Der Kraan, I., Manders, E.M., van Driel, R., 1999. Spatial relationship between transcription sites and chromosome territories. J. Cell Biol. 147, 13–24.
Wendt, K.S., Yoshida, K., Itoh, T., Bando, M., Koch, B., Schirghuber, E., Tsutsumi, S., Nagae, G., Ishihara, K., Mishiro, T., Yahata, K., Imamoto, F., Aburatani, H., Nakao, M., Imamoto, N., Maeshima, K., Shirahige, K., Peters, J.-M., 2008. Cohesin mediates transcriptional insulation by ccctc-binding factor. Nature 451, 796–801.
Zhao, Z., Tavoosidana, G., Sjlinder, M., Gndr, A., Mariano, P., Wang, S., Kanduri, C., Lezcano, M., Sandhu, K.S., Singh, U., Pant, V., Tiwari, V., Kurukuti, S., Ohlsson, R., 2006. Circular chromosome conformation capture (4C) uncovers extensive networks of epigenetically regulated intra- and interchromosomal interactions. Nat. Genet. 38, 1341–1347.

CHAPTER ELEVEN

Crowding, Diffusion, and Biochemical Reactions

Matthias Weiss[1]

Department of Experimental Physics I, University of Bayreuth, Bayreuth, Germany

[1]Corresponding author: e-mail address: matthias.weiss@uni-bayreuth.de

Contents

Abstract

Diffusion is the basic mode of transport for molecules in living cells. Diffusion leads to dispersion of individual molecules, but it is also the driving force behind biochemical reactions and pattern formation as diffusional motion mediates reactant encounters. Owing to macromolecular crowding in all cellular fluids and biomembranes, diffusion of molecules in cells is quite different from the motion observed in dilute solutions in a test tube. Hindered and anomalous diffusion are seen in cells, and biochemical reactions are affected by these. This review is intended to give an introduction and a brief overview about causes and consequences of crowding-induced diffusion anomalies and their impact on biochemical reactions.

International Review of Cell and Molecular Biology, Volume 307

ISSN 1937-6448

http://dx.doi.org/10.1016/B978-0-12-800046-5.00011-4

1. INTRODUCTION

Cellular fluids are crowded with macromolecules. Indeed, the concentration of macromolecules in cytoplasm reaches up to 400 mg/ml (Ellis and Minton, 2003). Also cellular membranes, that is, two-dimensional fluids, are highly crowded (Engelman, 2005). Here, depending on the organelle, about 50,000 transmembrane proteins have been reported (Griffiths et al., 1984; Quinn et al., 1984), and peripheral membrane proteins make this crowded condition even worse. Proteins are therefore tightly packed into the cell with little recreation space. Recent computer simulations nicely illustrate this fact (McGuffee and Elcock, 2010). Although crowding is a hallmark of living matter, typical biochemical assays rather work at low protein concentrations in dilute solution. Yet, given the extraordinary dense and crowded environment in cells, it is questionable how much these assays can report on cellular biochemistry. One may even wonder if and how proteins can actually move and perform their function in living cells.

This chapter is intended to give a brief overview of phenomena associated with diffusion in crowded media, for example, diffusion-limited reactions and anomalous diffusion. Focusing on these dynamic aspects, some steady-state considerations associated with crowding and excluded-volume interactions can only be touched on briefly. For a more extensive review on these aspects we refer the reader to (Zhou et al., 2008). Also, in an attempt to not overload the reader with mathematics on (anomalous) diffusion, several details have not been included here. The interested reader will find a comprehensive review on these aspects in Hoefling and Franosch (2013) and Klafter and Sokolov (2012).

The following part of this chapter is divided into three main sections. In Section 2, different mathematical descriptions of diffusional motion are recapitulated, and subsequently the concept of anomalous diffusion is introduced in some detail. While some mathematics is certainly needed for that, the intention here was to reduce formulas to an absolute minimum and rather describe the basic ideas. Section 3 is devoted to state-of-the-art approaches to quantify diffusional motion. Here, experimental findings on crowding and (anomalous) diffusion are also summarized. In Section 4, an introduction to diffusion-limited reactions is given and shortcomings of these classical considerations in the context of crowded systems are discussed, and recent results on crowding and diffusion-limited reactions are highlighted.

2. DIFFUSION AND RANDOM WALKS

Any (bio)chemical reaction requires the educts to meet in order to form a product. While one may facilitate the reactants' encounter in a test tube by using magnetic stir bars, virtually all reactions in living cells rely on diffusion as a means of transport. Sometimes, diffusion only has to bridge short distances after molecular motors have supported traffic on larger length scales, yet the majority of cellular reactions depend entirely on diffusional motion. Elucidating cellular reactions therefore requires some knowledge about diffusion. The theoretical background of diffusion, however, is taught very differently in the different disciplines of natural sciences. This section is intended to review the basic concepts and mathematical framework of normal and anomalous diffusion in a way that diffusion-driven reactions in crowded media, for example, the cell's cytoplasm, can also be understood. The mathematics used is restricted to the minimum needed for a basic description to not overload the reader. More detailed and mathematically deeper introductions may be found, for example, in Hoefling and Franosch (2013).

The first section focuses on the macroscopic, phenomenological theory of diffusion that was established in the nineteenth century. In this context, diffusion is viewed as a transport mechanism that leads to the equilibration of concentration gradients. This view is extended in Section 2, where a microscopic interpretation of diffusion as an unbiased, independent random walk of molecules is discussed. While these two sections only discuss normal Brownian diffusion, Section 2.3 introduces the framework of anomalous diffusion in a broader context.

2.1. Diffusion as macroscopic transport mechanism

Diffusion is a phenomenon that can be observed in everyday life. For example, if a drop of ink is carefully placed in water, one can observe a slow spreading of ink molecules until the entire water assumes a homogenous bluish color (cf. sketch in Fig. 11.1A). Performing such an experiment under well-controlled conditions, and evaluating the spatiotemporal evolution of the ink concentration, $c(\mathbf{r},t)$, yields a number of insights: (i) ink molecules seem to prefer leaving regions of high concentrations, (ii) concentration gradients vanish over time, (iii) equilibration becomes faster for higher temperatures, (iv) equilibration becomes slower for more viscous fluids, that is, ink spreads slower in glycerol than in water.

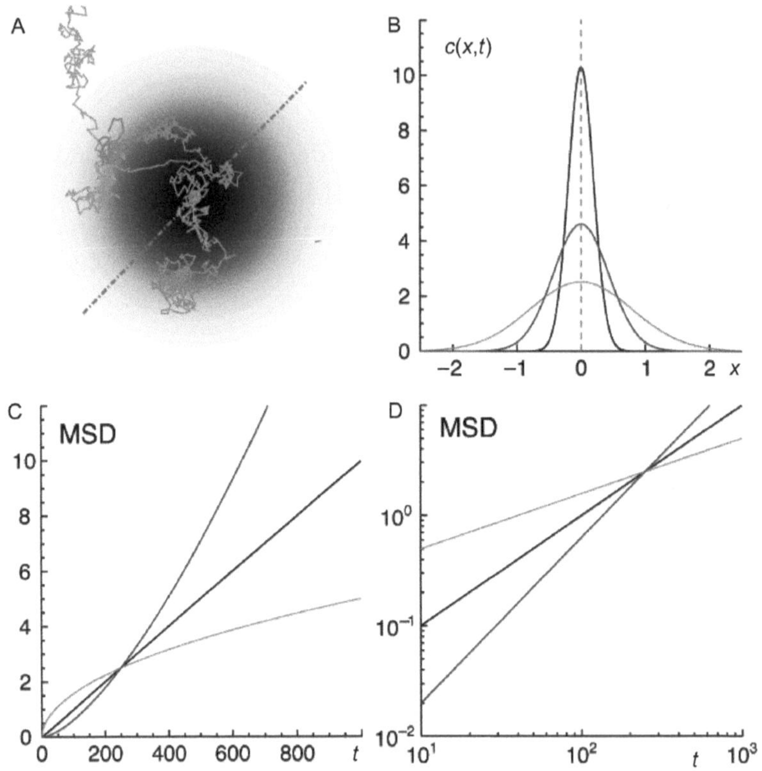

Figure 11.1 Diffusion in a nutshell. (A) Diffusively spreading ink droplet at some instant of time seen from the top. While the darkish center of the bluish contour plot indicates a still high concentration at the initial point of injection, remote regions have also assumed a light blue color. Three representative random walks of individual ink molecules are superimposed in red, green, and orange. (B) Gaussian concentration profiles of the ink, $c(x,t)$, along the dash-dotted line shown in (A). The more time has elapsed since placing the ink droplet in water (black to blue to green curves), the wider is the Gaussian and the lower is the ink concentration at the injection point (dashed red line). (C) The width of the Gaussian distribution shown in (B) is characterized by the mean square displacement (MSD). Normal Brownian diffusion shows a linear scaling MSD$\sim t$ (black line), whereas sub- and super-diffusion (green and blue curves, respectively) show a power-law increase MSD$\sim t^{\alpha}$ with $\alpha<1$ and $\alpha>1$, respectively. (D) The power-law increase shown in (B) is highlighted in a double-logarithmic plot. (For interpretation of the references to color in this figure legend, the reader is referred to the online version of this chapter.)

The first observation states that the flux of ink, $\mathbf{J}$, points downhill of the concentration gradient, that is, $\mathbf{J}=-D\nabla c(\mathbf{r},t)$. Here, bold letters indicate three-dimensional vectors and ∇ denotes the first spatial derivative along x-, y-, and z-directions of space, respectively; the action of ∇ therefore creates a three-dimensional vector out of the scalar concentration profile $c(\mathbf{r},t)$.

The prefactor D has dimensions of an area per time, for example, $\mu m^2/s$, and is called the "diffusion constant." The linear relation between flux and concentration gradients is known as Fick's first law and, as it stands, it is solely justified by its excellent description of experimental data. Owing to observations (iii) and (iv), one anticipates a relation $D \sim T/\eta$, where T is the (absolute) temperature and η is the viscosity of the fluid in which the ink spreads. Again, this is a purely empirical law that fits experimental observations; a justification for this relation will be given in the next section.

In addition to Fick's first law, we know that ink molecules are not destroyed in this experiment but only move to a different location. This can be stated formally by the so-called continuity equation $\frac{\partial}{\partial t}c(r,t) = -\nabla J$, which states that the concentration's temporal variation (left-hand side) is solely given by sources and sinks of the diffusional flux (right-hand side). Inserting now Fick's first law into the continuity equation yields the famous diffusion equation (also called Fick's second law):

$$\frac{\partial}{\partial t}c(\mathbf{r}, t) = D\nabla^2 c(\mathbf{r}, t) \tag{11.1}$$

This equation states that the temporal evolution of the concentration (left-hand side) equals the second spatial derivative, that is, the curvature, of the concentration profile $c(\mathbf{r},t)$. Interestingly, this equation was known already in the early nineteenth century as Laplace's heat equation.

To understand the contents of the diffusion equation, we will restrict the discussion for simplicity to one dimension, say, along a coordinate x (cf. Fig. 11.1A). The diffusion equation then reads $\frac{\partial}{\partial t}c(x,t) = D\frac{\partial^2}{\partial x^2}c(x,t)$. We start now with an infinitely thin but very concentrated ink droplet at $x=0$, that is, $c(x, t=0) = c_0\delta(x)$ in mathematical terms. Plugging this initial concentration into the diffusion equation (Eq. (11.1)) yields for all times $t>0$ a Gaussian concentration profile:

$c(x,t) = c_0 \exp(-(x^2/4Dt))/\sqrt{4\pi Dt}$ (cf. Fig. 11.1B). The peak in the concentration profile at $x=0$ does not move in time, yet it becomes less pronounced as the width of the concentration profile grows with time. The squared width is nothing else but the Gaussian's variance, $\mathrm{MSD}(t) = \int_{-\infty}^{\infty} x^2 c(x,t)\mathrm{d}x = 2Dt$. The variance in this context is also called the *mean square displacement* (MSD) of the ink molecules and is illustrated in two ways in Fig. 11.1C and D.

The corresponding result for arbitrary dimensions is easily obtained in a very general form:

$$c(\mathbf{r}, t) = \exp\left(-\frac{\mathbf{r}^2}{4Dt}\right)/(4\pi Dt)^{\dim/2} \tag{11.2}$$

and

$$\mathrm{MSD}(t) = \int_{-\infty}^{\infty} \mathbf{r}^2 c(\mathbf{r}, t)\mathrm{d}^{\dim}\mathbf{r} = 2Dt \times \dim \tag{11.3}$$

Here, bold letters indicate vectors, for example, in a one-, two-, or three-dimensional space for which $\dim = 1, 2, 3$. With these equations, macroscopic diffusion is well described and a material-specific diffusion constant D that describes the speed of gradient equilibration can be quantified.

While it is clear that the diffusion equation seems to act on behalf of the second law of thermodynamics (maximizing entropy by spreading/mixing the ink), the microscopic driving force that seems to fuel the motion of ink particles away from areas with high concentrations remains elusive in this picture. In order to overcome this limitation, one needs to take on a statistical perspective for single particles that was pioneered by Boltzmann.

2.2. Diffusion as random walk of molecules

In the late nineteenth century, Ludwig Boltzmann introduced a more atomistic view to the natural sciences that aimed at explaining thermodynamic phenomena via the statistics of individual particles instead of continuous fields or concentrations. Against all odds and severe criticism of famous colleagues, he formulated the basis for what is known today as statistical physics. Seeking for microscopic interpretations of thermodynamic laws also brought diffusion back into focus. In one of his seminal papers in 1905, Albert Einstein gave the first consistent microscopic interpretation of Fick's laws including a microscopic derivation of the diffusion constant (Einstein, 1905). The major step in this article was the assumption that small particles in or near thermodynamic equilibrium will perform a random walk due to a finite temperature. Before diving into details, let us fill the concept of a random walk with some life.

Let us consider a drunkard on the way home after a pub-crawl. Having had too many beers, the person will stagger from lamppost to lamppost, forgetting instantly where he/she came from. The chances to go left or right therefore are equal, namely $p = 1 - p = 0.5$. Having visited N lampposts in total, the distance that the person has walked to the right (measured in lampposts k) is given by a simple Binomial distribution:

$$P(N,k) = \binom{N}{k} p^k (1-p)^{N-k} = \frac{N!}{(N-k)!k!} p^N.$$

Replacing k by $(N-k)$, one obtains the Binomial for the number of steps to the left. Since both expressions are identical for $p=0.5$, and as steps to the left have to be counted negative (going to the right being defined as the positive direction) the net movement of the drunkard is strictly zero. This is indeed a result one might have guessed already. Nevertheless, the drunkard has moved while aiming to get back home. This (unproductive) movement is captured by the second moment of the Binomial, that is, the width of the distribution aka the MSD: $\mathrm{MSD}(N) = 2Np(1-p) = N/2$. The squared excursion length of the drunkard therefore grows linearly with the number of visited lampposts, N. In this example, N reflects the time that has elapsed after the random walk had started. If the street is a long one, that is, if the number of visited lampposts is very large, the Binomial distribution for the actual position of the drunkard converges toward a Gaussian due to the central limit theorem. It can therefore be written in the form $p(x,t) = \exp(-(x^2/4Dt))/\sqrt{4\pi Dt}$ with x denoting the distance from the starting point of the random walk (=pub), D being a microscopic diffusion coefficient, and t counting the elapsed time. The function $p(x,t)$ is called a probability distribution function (PDF), and it states the probability of finding a particle (=the drunkard) at a locus between x and $x + dx$ at any given time t. Here, dx is an infinitesimal small step. Certainly, we can now extend this consideration to any dimension dim = 1, 2, 3...:

$$p(\mathbf{r},t) = \exp\left(-\frac{\mathbf{r}^2}{4Dt}\right) / (4\pi Dt)^{\mathrm{dim}/2} \tag{11.4}$$

Looking at the PDF in Eq. (11.4), we see that it is identical to the solution of the diffusion equation in Eq. (11.2)! Hence, the PDF for a single particle that undergoes an unbiased random walk has the same shape as a spreading concentration profile $c(\mathbf{r},t)$. As a consequence, the equation of motion for both quantities is the diffusion equation, Eq. (11.1). This fairly deep result states that the random walk of a single particle is representative for an ensemble of identical, noninteracting particles. In other words, if we follow a single representative particle over a long time, we get the same information about an ensemble as if we had monitored the ensemble itself. Many experimental approaches to quantify diffusion, for example, single-particle tracking microscopy (cf. Section 3.2), rely on this equivalence.

We can now state that the equilibration of concentration gradients is not a driven process, but it simply reflects a large collection of particles that undergo a random walk with the same starting point. Moreover, even after equilibration the motion of individual particles will not stop. But due to their homogenous distribution, the same number of particles will move on average to the right as those going to the left.

In the above reasoning, we have tacitly expanded the drunkard's motion from a regular grid of lampposts to continuous space and time variables. To be more explicit on that, we could write down Newton's equation of motion for a single particle that is subject only to a randomly fluctuating force. This random force is analogous to the wish of the drunkard to move in a random direction. The choice of a random force that causes the random walk is justified by arguing that the diffusing particle is hit by thermally excited, much smaller solvent molecules (e.g., water molecules or lipids around a protein). Since there is many of them, these impacts will be uncorrelated and the net force is a random kick. Moreover, for objects smaller than a micrometer the fluid's Reynolds number is very small (even for water), that is, all inertia terms can be neglected. The particle's motion is then given by the overdamped Langevin equation $d\mathbf{r}/dt = \boldsymbol{\xi}(\mathbf{r},t)$, where $\boldsymbol{\xi}$ describes the random force that may depend on time and the position of the particle. As a result of this stochastic differential equation, one obtains single trajectories of a diffusing particle and the set of all these trajectories is described statistically via the diffusion equation. For illustration, representative random walks from a Langevin equation are superimposed to the density profile in Fig. 11.1A.

Due to the link between Langevin and diffusion equations, the strength of the random force ξ needs to be related somehow to the diffusion constant. Moreover, the strength of the random kick needs to be related to temperature as higher temperatures excite the solvent molecules more. As a consequence, the diffusion constant has to be proportional to temperature. Indeed, Einstein first derived in 1905 that the diffusion coefficient can generally be written as $D = k_B T/\gamma$ (Einstein, 1905) with k_B being Boltzmann's constant, T the absolute temperature, and γ being the friction coefficient of the particle in the fluid. This equation is the basic version of a fluctuation–dissipation theorem, that is, thermally induced fluctuations (described by D) are related to frictional forces that moving particles experience (friction γ). For a spherical particle, the well-known Stokes–Einstein relation is recovered, that is, from $\gamma = 6\pi\eta R$ one obtains $D = k_B T/6\pi\eta R$. Here, η is the fluid's viscosity, and R is the radius of the particle. With these insights, the experimental observation in the previous section, $D \sim T/\eta$, is totally explained in terms of fundamental quantities. With that, diffusional motion has a proper basis that is fueled by first principles rather than empirical laws. In addition, the concept of a random walk is an intuitive picture for Brownian motion of individual molecules.

2.3. Anomalous diffusion

Having constructed a link between Fick's diffusion equation and random walks of individual molecules, one might ask now which types of random walks are compatible with the above theoretical construction. Clearly, we have assumed an unbiased movement, that is, going left and right should be equally probable. Beyond that, one can show with mathematical rigor that normal Brownian diffusion is obtained when the (symmetric) random walk has the following ingredients:

- Successive steps are uncorrelated (this is called a Markovian random walk)
- The PDF of step lengths, $p(\Delta x)$, has finite mean and variance, that is, $\int_{-\infty}^{\infty} x^n p(\Delta x) dx < \infty$ for $n < 3$.
- The PDF of waiting times, $p(\tau)$, between successive steps of length Δx has a finite mean and variance, that is, $\int_{-\infty}^{\infty} \tau^n p(\tau) d\tau < \infty$ for $n < 3$.
- Under these conditions, the famous central limit theorem holds and the motion will be eventually normal Brownian diffusion. Breaking one of these rules typically yields diffusion anomalies that are the topic of this section. Deeper mathematical details may be found, for example, in Klafter and Sokolov (2012).

One of the most straightforward extensions to a Brownian random walk is a violation of the Markovian assumption, that is, successive steps are chosen to be correlated. If this is the case, the random walk is said to have a memory and the random force ξ in the Langevin equation "remembers" its action at previous time steps. This might influence the random walk in two ways: If successive steps are positively correlated, the random walk becomes persistent, whereas for anticorrelated steps an antipersistent motion emerges. In the picture of our drunkard, the persistent random walk means that having gone to the right will bias the drunkard's decision to go to the right again in the next move. For an antipersistent walk, the bias is just the opposite, that is, having stepped left, the more likely decision of the drunkard is to take a step to the right at the next instant of time. In both cases, a few more beers may restore Markovian behavior...

Memory effects, however, massively impede finding a particle's PDF, that is, the analogue to Eq. (11.4). To avoid nasty mathematical details, we will not dive into this realm. However, the width of the PDF, that is, the MSD(t), is a pretty straightforward quantity that can already report on the characteristics of the diffusional motion. Depending on the type of memory, the MSD can show an anomalous power-law increase: $\mathrm{MSD}(t) = Kt^{\alpha}$ with $0 < \alpha \leq 2$. Here, K is a generalized diffusion coefficient with fractional units. The cases $\alpha = 1$ (normal diffusion with $K = D$, cf. previous Section 2.2) and $\alpha = 2$ (directed ballistic motion with velocity $K = v$) are trivial special cases. If α deviates from unity but is lower than two, one speaks of anomalous diffusion to highlight the difference in the power-law increase with time. Here we can distinguish two regimes: if $\alpha < 1$ the motion is called *subdiffusion*, whereas for $\alpha > 1$ it is called *superdiffusion*.

For a correlated random walk, the most popular model is fractional Brownian motion that has been introduced by Mandelbrot in the 1960s (Mandelbrot and van Ness, 1968). The anomaly of fractional Brownian motion (from sub- to superdiffusion) can be tuned at will via the so-called Hurst coefficient $H = 1 - \alpha/2$, that is, $0 < H < 1$. The model of fractional Brownian motion is particularly relevant for the case of subdiffusion ($\alpha < 1$, $H > 0.5$), where the antipersistent, anticorrelated random walk can be related to the viscoelasticity of the fluid in which the particles diffuse (Ernst et al., 2012; Mason and Weitz, 1995).

Instead of introducing correlations in successive steps of the random walk, the step length distribution and/or the waiting time distribution can also be altered in such a way that an anomalous diffusion emerges. Letting the distribution of waiting times between successive steps compatible with normal diffusion but changing the step length distribution may lead to superdiffusion. If the variance and/or the mean of the step length distribution diverges, the distribution is said to be of Levy-type. Levy-stable distributions, introduced by Paul Levy in the early twentieth century, are proper distribution functions that do not obey the central limit theorem, that is, the sum of Levy-distributed random numbers is not a new random number that asymptotically follows a Gaussian distribution. A particular realization of a Levy distribution is the Cauchy distribution $p(\Delta x) = 1/(\pi(1 + \Delta x^2))$ that is a properly normalized distribution of step lengths with diverging moments (infinite mean and variance). This type of step length distributions yields so-called Levy flights and the particle trajectories have superdiffusive characteristics (Klafter et al., 1996). Superdiffusive behavior has been observed, for example, for diving whales/fish

(Humphries et al., 2010; Sims et al., 2008) or in the context of human mobility (Brockmann et al., 2006; Gonzalez et al., 2008). Yet, in the context of cells, superdiffusion is not a tremendously relevant concept. Superficially, it might be observed as a consequence of transient ballistic motion on cytoskeletal filaments with intermittent periods of free diffusion. Taking a closer look reveals, however, that this example is simply the superposition of normal diffusion and ballistic motion. Only the crossover region between a linear (diffusive) and a quadratic (ballistic) increase of the MSD in time mimics a nontrivial power-law on certain time scales.

More relevant in the context of cell biology are Levy distributions for the waiting times $p(\tau)$ between two steps that are drawn from a well-behaved step length distribution (e.g., a Gaussian). If already the mean of the waiting time distribution $p(\tau)$ diverges (e.g., when having a Cauchy-like distribution for waiting times), very long residence times also become very probable and the diffusive spreading of molecules seems to become slower and slower as time progresses; the whole system shows aging. The asymptotic power-law behavior of $p(\tau)$ for large residence times enforces a so-called continuous time random walk (CTRW) for individual particles. As a consequence, the trajectory of the diffusing particle looks like that of a normally diffusing particle yet the time spent at each location can be extremely long. For an ensemble of particles, the CTRW induces a MSD that follows a subdiffusive characteristics, MSD $\sim t^{\alpha}$, and the choice of the waiting time distribution determines the anomaly degree α. Yet, compared to fractional Brownian motion the CTRW is a particular model. As it conflicts with the central limit theorem, a single particle's trajectory is not representative any more for an ensemble. In fact, inspecting the MSD for a single particle yields normal diffusion (MSD $\sim t$) whereas the ensemble shows subdiffusion. This phenomenon, that is, time-averaged and ensemble-averaged MSD being qualitatively and quantitatively different, is called weak ergodicity breaking (He et al., 2008; Lubelski et al., 2008). Hence, depending on the type of experiment, grossly different results may be observed. Some aspects of this peculiar phenomenon will be discussed in more detail in the context of experimental findings in Section 3.

Last but not least in a set of models that are associated with anomalous diffusion are obstructed random walks. If a particle diffuses in a maze of immobile obstacles, and if the density of obstacles is sufficiently high, subdiffusion can be observed over extended time scales (Havlin and Benavraham, 1987). If obstacles are also mobile the anomaly may vanish (see, e.g., discussion in Malchus and Weiss, 2010). The reason for anomalous diffusion in a course of dense obstacles is the geometric shape of the still accessible space: if obstacles are placed randomly, a connected cluster of non-accessible sites will emerge at a critical density ρ_c the so-called percolation threshold (Havlin and Benavraham, 1987). At (and near) to this concentration, the still accessible space has a fractal geometry, that is, it looks like a self-similar arrangement of paths on smaller and smaller length scales (Bouchaud and Georges, 1990). Some more details of fractals will be discussed in Section 4 and Fig. 11.4 in the context of fractal kinetics. Normal diffusion on a fractal has all signatures of anomalous diffusion, that is, tracer particles will be seen to display subdiffusion while exploring the fractal. Yet, only at the percolation threshold will an anomaly be observed for all time scales. Below ρ_c a crossover to normal diffusion will be seen for large times (Bouchaud and Georges, 1990; Havlin and Benavraham, 1987) and the anomaly exponent α will eventually converge toward unity for normal diffusion. By the very nature of the obstruction process, α will depend on the dimensionality of the space in which the obstructed diffusion is taking place. For membranes (two dimensions), $\alpha \approx 0.69$, whereas in cytoplasm (three dimensions) $\alpha \approx 0.55$ at the percolation threshold (Bouchaud and Georges, 1990). It is also worth noting that α does not smoothly interpolate between unity and the stated lower bounds. Strictly speaking, the anomaly is only well defined at the percolation threshold and only the transient nature of the subdiffusion in case of an obstacle density below ρ_c imitates a power-law increase of the MSD with an intermediate value of α for finite time scales (Hofling et al., 2006).

Having introduced different models for subdiffusion, one may ask now how they are related to biological samples. Obstructed diffusion, for example, might be expected if larger membrane structures like vesicles and endosomes try to diffuse through the cell's interior that is heavily decorated with cytoskeletal elements and intermediate filaments. On membranes, almost immobile protein complexes and adsorbed filaments may act as obstacles for small proteins. Fractional Brownian motion will reflect a viscoelastic characteristic of fluids due to its antipersistent character: any movement of a macromolecule in crowded fluids results in squeezing nearby macromolecules. These, however, will respond analogous to a compressed polymer, that is, an entropic spring behavior will aim at restoring the old positions. Such a local elastic response will enforce a memory and hence the desired antipersistence. A CTRW, on the other hand, may be envisaged as a diffusive motion in an environment with many traps (Saxton, 1996). These traps will stop the diffusion intermittently and hence spread the waiting times between periods of free diffusional motion, yet a few conceptual problems make this scenario rather unlikely for biological probes (Saxton, 2007). The following section will pick up these considerations in the context of experimental approaches to quantify diffusion *in vivo*.

3. QUANTIFYING DIFFUSION IN LIVING CELLS

This section is intended to give an overview of light microscopy methods that are frequently used to assess transport in living cells. Here, the more theoretical concepts of the preceding section will be put into an experimental context. Sections 3.1 and 3.2 focus on ensemble-based and single-particle-based experimental approaches, respectively. Important insights on diffusional motion derived from these approaches are summarized in Section 3.3.

3.1. Ensemble methods

Probably the most popular technique for quantifying diffusive transport in living cells is fluorescence recovery after photobleaching (FRAP). Since its first realization in the late 1970s (Axelrod et al., 1976a,b), a number of variants have been developed to match particular scientific problems, especially in cell biology (Lippincott-Schwartz et al., 2001).

Before explaining FRAP in some detail, let us briefly recall how fluorescent molecules like GFP can be excited, how they emit their fluorescence light, and how they can be bleached. In its most basic form, a fluorescent molecule can be described as a three-level system (Fig. 11.2). An incoming photon, from an arc lamp or a laser, can excite an electron from the dye's ground state (S_0) to the first excited state (S_1) if its energy fits the energy gap between both states. The photon's energy is determined by its wavelength, $E \sim 1/\lambda$. Owing to the more complex structure of dyes as compared to single atoms, each of the states S_0 and S_1 is subdivided into many vibrational levels. Therefore, the absorption spectrum of a dye is not sharply peaked at a particular wavelength λ, but rather covers a certain range of wavelengths (470–500 nm for GFP). After having been shifted to the excited state, the electron relaxes quickly via vibrational modes to the lowest excited state S_1. From here, the electron can decay back after some nanoseconds to the sublevels of the ground state S_0 via spontaneous emission of a fluorescence photon. Due to the energy loss when sliding down vibrational modes in the excited state, the emitted photon has a lower energy than the exciting photon;

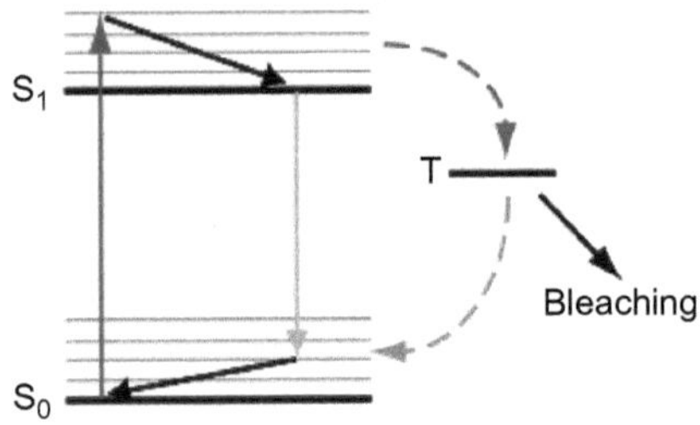

Figure 11.2 Fundamentals of photophysics. A fluorescent dye, for example, GFP, can be excited by a photon with the appropriate wavelength, that is, an electron is shifted from the ground state S_0 to the excited state S_1 (blue arrow). In the excited state, the electron can relax via vibrational sublevels to the edge of the excited state (black arrow). From there, the electron can decay back to the ground state by emitting a red-shifted photon (green arrow). Alternatively, the electron may reach the triplet state T from the excited state S_1 (red arrow). From this state the electron cannot relax back to the ground state by sending out a fluorescence photon any more; instead, it needs to relax via collisions (orange arrow). From the triplet state, the dye may also get irreversibly bleached. While this is unfavorable for FCS, it is the basis for FRAP experiments. (For interpretation of the references to color in this figure legend, the reader is referred to the online version of this chapter.)

the fluorescence is red-shifted (Stokes-shift). GFP, for example, is excited by blue light ($\lambda = 490$ nm) and fluoresces in green ($\lambda = 515$ nm).

In addition to these two so-called singlet levels, a third state is important: the triplet state. This state is different in its total spin configuration and as dipole-transitions like $S_0 \leftrightarrow S_1$ do not flip spins, the excited electron should never reach the triplet level. Collisions with surrounding molecules and the complex internal modes of fluorescent dyes, however, enable the excited electron to fall into this unfavorable state when the dye has absorbed a photon. As a transition back to S_0 is quantum-mechanically forbidden, this triplet state can only relax via collisions with surrounding molecules. The lifetime of the triplet state is therefore much longer than that of the S_1 state. In combination with a collision, the triplet state will decay radiation less after 1–100 μs to the ground state S_0. As a consequence, fluorescent dyes blink on short time scales. While being in the triplet state, the dye also can undergo an irreversible transition to a nonexcitable state—the dye bleaches and remains dark. This irreversible bleaching is the basis for FRAP.

The concept of FRAP is straightforward (cf. Fig. 11.3A): given a spatial distribution of fluorescently labeled particles, for example, GFP-tagged proteins in the cytoplasm, one selects a region of interest (ROI) in which dye molecules are irreversible bleached by a short but intense illumination with

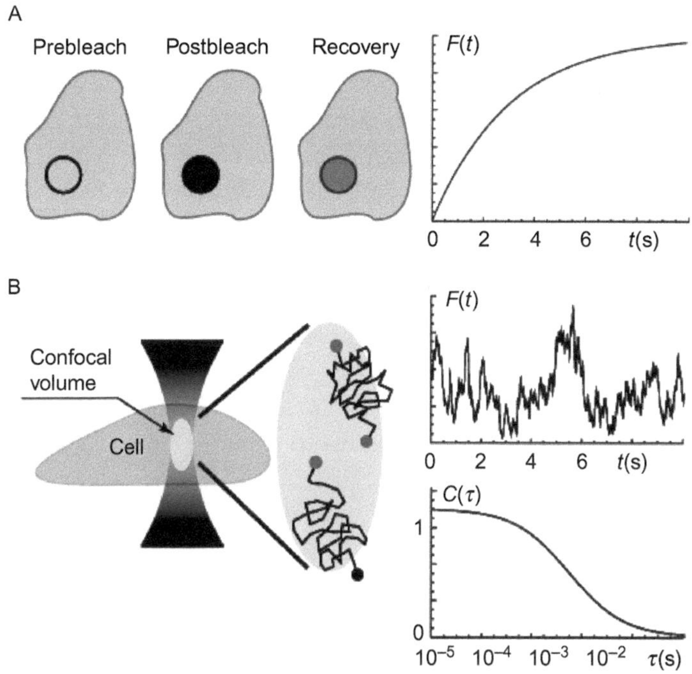

Figure 11.3 Principle of FRAP and FCS experiments. (A) Sketch of a prebleach and a postbleach image of a GFP-expressing cell with a circular region of interest (ROI). After bleaching, the diffusion-driven recovery from unbleached regions to the bleached ROI is monitored with low illumination intensities (third sketch). The total fluorescence $F(t)$ of the ROI starting at the first postbleach image rises and eventually saturates to a steady state. Fitting the curve with theoretical expressions yields the diffusion coefficient. (B) In FCS, only the fluctuating fluorescence $F(t)$ in a confocal volume is monitored as fluorescent particles go in and out of the illuminated focus. The autocorrelation function of these fluctuations decays sigmoidally in a semilogarithmic plot, and the half-time of this decay reports on the diffusion coefficient and the type of diffusion. See text for details. (For color version of this figure, the reader is referred to the online version of this chapter.)

the appropriate wavelength. The intense illumination will force the dyes to undergo many excitation/emission cycles (each taking a few nanoseconds). At each cycle the triplet state, from where the dye can bleach, will be reached with a certain probability, and thanks to the vast quantity of cycles many dyes in the ROI will be bleached. The emerging dark region is subsequently monitored with a much lower illumination intensity that does not lead to bleaching any more. If the bleached ROI recovers at least partially, one knows that an exchange with nonbleached regions has occurred. For example, the observed recovery can be due to diffusion, directed transport, or binding events. For the sake of brevity we will focus only on diffusion-driven recoveries and neglect the latter two possibilities.

Typical FRAP experiments require to choose a ROI, to take at least one prebleach image of it with low illumination power, to bleach it rapidly with very strong illumination power, and to monitor then the postbleach images with low illumination settings again (cf. Fig. 11.3A). The integral fluorescence of the ROI is the experimental quantity that reports on the diffusion-driven exchange of bleached and nonbleached molecules. Since the diffusion constant D has dimensions of an area per time (see Section 2.1), the spatial form of the bleached ROI is of crucial importance in this protocol. Knowing the shape of the ROI, one can analytically derive fitting functions for FRAP curves via the diffusion equation (Eq. (11.1)), and fitting these functions to experimental data allow one to determine D. A key step in deriving these fitting functions is an assumption about the fluorescence profile right after the bleaching, that is, the profile seen in the first postbleach image. Certain assumptions for the postbleach profile may lead to systematic errors (see below).

The most basic variant of FRAP utilizes a stationary Gaussian-shaped laser beam that illuminates a roughly cylindrical ROI throughout the thickness of the entire specimen, for example, a cell. The diameter of the beam (typically a few 100 nm) is dictated by the optical setup. The fitting function in this case can be calculated analytically and by varying parameters to match experimental data, one may determine the diffusion coefficient that has carried the observed recovery. Due to the small beam diameter, transport properties can be probed locally in cells and even binding reactions can be monitored with some accuracy (Berkovich et al., 2011).

Today, confocal laser scanning microscopes (LSMs) are commonly used for FRAP as they allow for choosing arbitrary ROI geometries. With an LSM, one obtains raster images of a specimen by moving the laser beam in a rectangular grid over the sample. Selecting a set of pixels that define the ROI and scanning them with much higher laser power allows for bleaching an (almost) arbitrary ROI. While the optical resolution is certainly still diffraction-limited (typically to roughly half of the wavelength), the distance between two pixels can be chosen as low as 10 nm. Hence, an image (and also a ROI) can be massively oversampled to enhance the signal and/or the bleaching.

To obtain a reasonably simple fitting function for diffusion measurements, one typically restricts FRAP experiments to bleaching a line, a square, or a circle. Circular ROIs are very convenient for diffusion measurements, and an analytical fitting function for the fluorescence recovery is available (Soumpasis, 1983):

$$F(t) = \exp\left(-\frac{w^2}{2Dt}\right) \times \left\{ I_0\left(\frac{w^2}{2Dt}\right) + I_1\left(\frac{w^2}{2Dt}\right)\right\}.$$

Here, w denotes the radius of the ROI (typically some μm), and I_0 and I_1 are modified Bessel functions (Abramowitz and Stegun, 1970). This expression not only is a valuable one for fitting FRAP data of normal diffusion (cf. Fig. 11.3), but it can also be extended to report on subdiffusion. To this end, one simply replaces Dt in the above equation by Kt^{α}, where K is a generalized diffusion coefficient with units of an area per fractional time (Saxton, 2001).

A word of caution is needed, however, when fitting experimental data. While one assumes almost always a sharp, step-like transition of the fluorescence from the ROI to the rest of the specimen when deriving a fitting function, the finite time needed for bleaching invalidates this assumption. The bleach profile looks much smoother than step-like. To put it simply: Bleaching takes a finite time during which the diffusion does not stop, and the motion of dye molecules that have been bleached early will render the ROI somewhat wider than intended. Fitting experimental recovery data now with an analytical expression that does not take this into account will lead to a systematic, two- to fivefold underestimation of the diffusion constant (Weiss, 2004). To cope with this problem, more sophisticated fitting approaches have been developed, for example, by taking into account the actual fluorescence profile in the first postbleach image (Goehring et al., 2010). Despite this source of uncertainty, FRAP has been employed successfully in many studies to estimate the diffusional mobility of proteins and larger structures (Lippincott-Schwartz et al., 2001).

An alternative to FRAP when aiming and quantifying diffusion coefficients is the use of fluorescence fluctuation techniques. Presumably the most prominent of these fluctuations techniques is fluorescence correlation spectroscopy (FCS) that was invented even before FRAP (Magde et al., 1972). Being halfway between ensemble and single-particle approaches, the concept of fluctuation spectroscopy experiments will be outlined by using FCS as an example. For more recent developments related to FCS, we would like to refer the reader to Digman and Gratton (2011).

While FCS was invented already in the early 1970s (Magde et al., 1972), the technique only became popular with the advent of modern confocal microscopy in the mid-1990s (Rigler and Elson, 2001). The prime reason for this late application was the need for small optical volumes to make FCS robust and feasible for cell biological applications. Technically, this could be achieved with the advent of robust confocal microscopes. In a nutshell, modern FCS setups monitor the mean dwell time τ_D of fluorescent particles in a femtoliter-sized confocal volume (Fig. 11.3B). The faster the diffusion (high D), the lower the measured mean dwell time in the focus. At the heart of FCS lies a noise analysis of fluorescence fluctuations: whenever a particle enters the confocal volume it will contribute to the fluorescence, whereas the fluorescence drops slightly if a particle leaves the confocal volume. From the mean dwell time (typically about 500 μs) and the radius of the confocal volume (typically 250 nm, depending on the optics used in the setup), one can easily build a quantity that yields a rough estimate for the diffusion constant: $(250\ \text{nm})^2/500\ \mu\text{s} = 125\ \mu\text{m}^2/\text{s}$. Going through some more mathematical details (cf. below) reveals that this estimate for the diffusion coefficient is correct up to a factor four that is missing in the denominator. The requirements for an FCS setup are hence just a confocal microscope and a fast, sensitive detector that counts the fluorescence photons that are emitted from particles in the focus (Fig. 11.3B).

To obtain quantitative insights into the diffusion coefficient and a potential anomaly of the diffusion, the crude estimate above is clearly insufficient. The fluorescence monitored by the detector can be split into a constant average fluorescence and fluctuations around this mean value: $F(t) = \langle F \rangle + f(t)$. Here, brackets denote a temporal average. When taking confocal images, one clearly seeks to maximize the mean so that the fluctuations $f(t)$ become negligible and do not distort the quality of the image. For FCS, however, the strategy is opposite as one needs the fluctuations for evaluating the particles' diffusion. Therefore, FCS works best at low dye concentrations and in principle a single dye per confocal volume (i.e., a concentration of 1 nM) is sufficient to obtain reliable results. Since particles are not tracked individually, FCS may be regarded as an on-average-single-molecule technique. The noise analysis is done by calculating the so-called autocorrelation function $C(\tau) = \langle F(t)F(t+\tau) \rangle / \langle F \rangle^2 - 1 = \langle f(t)f(t+\tau) \rangle / \langle F \rangle^2$. This definition states that one should calculate the average of all possible products $F(t)F(t+\tau)$ along the recorded time series (i.e., for all t) for a set of time lags τ. These average values depend by definition only on the lag time, τ. Dividing by the squared average fluorescence and subtracting unity ensures that the autocorrelation function will decay to zero for large lag times and that the offset will be related to the number of particles in the focus. Essentially, $C(\tau)$ states the probability that a photon that has been recorded at time $t+\tau$ comes from the same dye as a photon that has been recorded at time t. Hence, if τ is extremely small, the photons will always come from the same dye molecules and $C(\tau)$ is larger than zero. In contrast, for large lag times all dye molecules that contributed to the fluorescence at time t have already diffused out of the confocal volume and another set of dyes sends out the photons. These photons are totally uncorrelated and thus $C(\tau)$ is zero for large τ. The interpolation between these two extremes, that is, the typical time at which the correlation has dropped by 50%, yields the desired information about the type and kinetics of the diffusion.

Fitting curves can be obtained by bearing in mind that the fluorescence is simply the sum of all fluorescence contributions from individual dyes in a given illumination profile $I(\mathbf{r})$, that is, $F(t)=\int I(\mathbf{r})c(\mathbf{r},t)\mathrm{d}V$. Here, the spatiotemporally-varying concentration of particles $c(\mathbf{r},t)$ is included that evolves in time according to the diffusion equation; integration runs over the entire space. In a confocal microscope, the illumination profile (more correctly, the superposition of illumination and detection profiles) is a Gaussian of the form $I(\mathbf{r})=I_0\exp(-(x^2+y^2+z^2/S^2)/r_0{}^2)$. Typically, the beam radius is connected to the illumination wavelength as $r_0\approx\lambda/2$. The elongation of the confocal volume by a factor $S=3$–5 along the optical axis is a feature of the usual optical path, in which only fluorescence in a half-space is detected while the fluorescence in the opposite half-space is lost. Inserting this expression into the definition of the autocorrelation function and splitting the concentration into a mean and a fluctuating part yields a double integral expression for $C(\tau)$. The argument of the integration is essentially the product of the Gaussian illumination profile and a correlator of the concentration fluctuations. The latter is nothing else but the propagator of the diffusion equation, namely the Gaussian PDF of Eq. (11.4) in the previous section! The whole calculation now reduces to evaluating two Gaussian integrals. The result for three dimensions finally reads:

$$C(\tau)=\frac{1/N}{1+\tau/\tau_{\mathrm{D}}}\cdot\frac{1}{\sqrt{1+\tau/(S^2\tau_{\mathrm{D}})}};\quad \tau_{\mathrm{D}}=\frac{r_0{}^2}{4D}.$$

The offset of the correlation function thus reports on the inverse average number of particles in the focus, N, while the typical decay time τ_{D} reports on the inverse diffusion constant. Comparing this expression to the above crude estimate, only a factor four and a proper definition of the focus width have led to a slight update. Subdiffusion can be easily implemented into the above equation when replacing τ/τ_D by $(\tau/\tau_D)^\alpha$ (Weiss et al., 2004). While one may derive more elaborate fitting functions for anomalous diffusion (Lubelski and Klafter, 2009), the improvement in describing experimental data is typically smaller than the measurement error (Weiss et al., 2004). Hence, the simple exchange method (τ/τ_D replaced by $(\tau/\tau_D)^\alpha$ in the fit equation) yields a straightforward and easy procedure that captures the essence of the potential subdiffusion.

Using FCS one can determine diffusion constants on a local scale (within the focus) without perturbing or bleaching the cell. Unlike FRAP, where bleaching occurs along the optical axis throughout the whole cell, FCS also probes diffusion along the optical axis, albeit not very sensitively. Being an almost-single-molecule technique, the overexpression of proteins can also be held low in FCS experiments to not perturb the cell's native state. Yet, quantifying long-range diffusion beyond the focus is not possible. This can be achieved, however, with techniques that use two foci (Dertinger et al., 2007) or whole images (Kolin and Wiseman, 2007) and still evaluate the fluctuations due to diffusion in a very similar manner as the approach for FCS.

3.2. Single-particle methods

A drawback with FRAP and even fluctuation methods like FCS is the lack of information about the individual trajectory of a diffusing particle. To put it differently, FRAP and FCS only yield the particles' MSD. For subdiffusion, however, we have discussed in the previous chapter three very different models that can lead to the same nonlinear scaling of the MSD. Therefore, FCS and FRAP can hardly distinguish between such models. Tracking individual particles (even individual molecules) provides the maximum information as the whole trajectory is available after the measurement and not just its temporally changing variance.

In principle, tracking fluorescently labeled particles like quantum dots in two dimensions, for example, on the cell's plasma membrane, only requires an epifluorescence microscope with a stable illumination source and a fast and sensitive camera for image acquisition. Often, an EM-CCD or a sCMOS camera are used, depending on the strength of the fluorescence signal and the required speed of acquisition. If the number of labeled particles is low and if the fluorescence background is not too strong, individual nanometer-sized particles can be detected easily with this setup. Although the particles are much smaller than the diffraction limit (i.e., the resolution limit of the optics), their position can be determined with nanometer precision. Simplifying the particle to a point-like source of fluorescence light, the particle will emit a characteristic pattern of light, the so-called point-spread function (PSF). The PSF states the probability that a fluorescence photon will be sent from the particle to a particular position on the detector (=the camera chip). For point-like particles, the PSF image has almost a Gaussian shape. Fitting a Gaussian to the camera image of the tracked particle can be done with an accuracy that only depends on the number of photons that have been captured to image the PSF (Brauchle et al., 2009). For a high frame rate, only few photons contribute whereas lower frame rates yield a smoother image and therefore a better fit. The peak position of the Gaussian can be extracted via that approach with a precision of a few nanometers. In fact, the use of this simple trick also underlies modern superresolution microscopy techniques like PALM or STORM (Thompson et al., 2012).

A number of problems, however, make single particle tracking in cells a challenging project. First, GFP is prone to rapid bleaching, that is, the number of positions in the tracking data is fairly limited. Therefore, one has to average the MSDs of many individual particles or one needs to employ more stable dyes. Second, cells show a strong fluorescent background that may mask individual particles. This is particularly problematic for blue/green variants of GFP; more red-shifted variants suffer less from background issues. Third, particles move while acquiring the photon for fitting the PSF image. This leads to an apparent superdiffusive behavior that may obscure the data evaluation (Goulian and Simon, 2000). Fourth, tracking in three dimensions, for example, in cytoplasm, is challenging as the positional accuracy along the optical axis can be estimated less well. Moreover, one also needs to compensate for the z-motion to not lose the particle.

Despite these challenges, single-particle tracking has been frequently used to assess diffusion in cells. Owing to the full trajectory, one can also inspect additional quantities that go beyond the MSD. A particularly interesting quantity is the probability distribution of first passage times, that is, the individual times that particles need to bridge a certain distance. These times are predicted to follow an inverse Gaussian law for normal diffusion (Klafter and Sokolov, 2012), yet to probe or exploit this distribution a very large number of transit times need to be recorded. A more robust measure is the asphericity of the diffusion track of individual particles. Counterintuitive to a naïve expectation, individual random walks are not spherical on average but assume an elliptical shape (Rudnick and Gaspari, 1987). The asphericity of this ellipsoid changes with the type of random walk and also with the dimension of the environment in which the diffusion takes place (e.g., bulk or on a surface). Since individual particles do not align their ellipsoidal walks, an ink droplet composed of many particles spreads symmetrically, that is, in the ensemble our assumption of an isotropic random walk is fine again. As we shall see in Section 3.3, (sub)diffusion can be characterized with the asphericity in much more detail as compared to FCS or FRAP experiments.

3.3. Experimental results on diffusion in crowded media

Using the above techniques, the diffusive motion of cytoplasmic and membrane-bound proteins has been investigated in great detail. For the sake of brevity, we will restrict this section to only those phenomena that are associated with crowding.

In general, small cytosolic or nucleoplasmic proteins have been seen to diffuse with typical diffusion constants of about 20 $\mu m^2/s$, whereas membrane proteins are about 50-fold less mobile (Dix and Verkman, 2008). Complex formation and/or interactions with larger structures certainly may slow down the apparent diffusion even more. However, from basic hydrodynamics and thermodynamics (Einstein–Stokes equation, cf. Section 2.2), one can estimate the diffusion constant of small globular proteins (say, with a mass of 25 kDa) to be in the range of 80 $\mu m^2/s$ in water or buffer solution. Using the Saffman–Delbruck relation (Saffman and Delbruck, 1975) one obtains for single-pass transmembrane proteins approximately 3 $\mu m^2/s$. The strongly reduced mobility observed in cellular environments as compared to these first-principle estimates is a first signature of crowding (Dix and Verkman, 2008). A plethora of dissolved macromolecules increases the apparent viscosity η of cellular fluids and membranes, which leads to a significant reduction of the diffusion coefficient. The change of viscosity from water to cytoplasm has been reported for many different cell types and conditions to be roughly three- to fivefold (see, e.g., Elsner et al., 2003; Seksek et al., 1997). Hence, due to macromolecular crowding diffusion is about three- to fivefold slower in cytoplasm (and nucleoplasm) as compared to water or buffer solution. For diffusion on membranes, similarly strong reductions have been observed (Frick et al., 2007). As an additional note, we would like to highlight that diffusion constants on membranes in general have a peculiar size-dependence that strongly deviates from the commonly envisaged Einstein–Stokes law $D \sim 1/R$ in bulk fluids (R denoting the protein's hydrodynamic radius). For details, we refer the reader to the extensive literature on this aspect.

Starting points may be, for example, Guigas and Weiss (2006), Hughes et al. (1982), and Petrov and Schwille (2008).

Going beyond a mere reduction of diffusion constants, the emergence of anomalous diffusion has been observed for protein-sized particles in the cytoplasm (Guigas et al., 2007a,b; Tolic-Norrelykke et al., 2004; Weiss et al., 2004) and nucleoplasm (Guigas et al., 2007a,b; Wachsmuth et al., 2000) of eukaryotes, in the cytoplasm of bacteria (Golding and Cox, 2006; Weber et al., 2010), and on membranes (Weigel et al., 2011; Weiss et al., 2003). Even cell extracts (Guigas et al., 2007b) and biomimetic artificial solutions (Banks and Fradin, 2005; Pan et al., 2009; Szymanski and Weiss, 2009) with different crowding agents have shown subdiffusion. Typically, the anomaly in cellular fluids was seen in the range $0.5 \leq \alpha \leq 0.7$ whereas artificial fluids and membranes rather showed $0.7 \leq \alpha \leq 0.9$.

It should be noted at this point that the observed subdiffusion is always a transient phenomenon that asymptotically converges toward normal diffusion. On length and time scales that are large enough, normal diffusion ought to be restored. Indeed, all subdiffusion models in Section 2.2 are just mathematical constructs that assume, for example, that proteins can get trapped for an infinitely long time. Certainly this is unphysical, that is, in all of these models exists a smallest and largest length/time scale beyond which normal diffusion has to emerge. So far, all observations of subdiffusion in cells are linked to length and time scales below a few microns and a few seconds. Probing diffusion on larger scales, for example, by large area FRAP, reveals normal Brownian motion yet with a much smaller diffusion coefficient than anticipated.

Linking experimental observations of subdiffusion to any of the above models is not easy, as typically only the MSD can be evaluated, for example, in FCS and FRAP experiments. Yet, this does not discriminate which type of random walk one has observed in the microscope. Several reports for subdiffusion on membranes are most consistent with a model of obstructed diffusion (percolation). Proteins and larger complexes are "sieved" by more immobile structures, for example, translocons (Malchus and Weiss, 2010) or membrane-adsorbed cytoskeletal filaments (Fujiwara et al., 2002; Tomishige and Kusumi, 1999). For subdiffusion in artificial solutions and the cytoplasm, fractional Brownian motion seems to be the most reasonable explanation. Indeed, single particle tracking has revealed a distinct asphericity of the subdiffusive random walk that is most consistent with fractional Brownian motion (Ernst et al., 2012; Weiss, 2013). While a normal Brownian motion ($\alpha = 1$) with the theoretically predicted asphericity 4/7 was observed for purely viscous fluids, crowded fluids showed a significantly decreased anomaly value ($\alpha = 0.8$) and a lower asphericity. In comparison to computer simulations of CTRW, obstructed diffusion, and fractional Brownian motion, it turned out that this result was only compatible with fractional Brownian motion. In line with this result, the cytoplasm appears viscoelastic when shearing it with high frequencies, that is, the complex shear modulus has equally strong viscous and elastic contributions (Guigas et al., 2007a,b; Tseng et al., 2002; Yamada et al., 2000). Viscoelasticity of the crowded cytoplasm therefore can be seen to induce anomalous diffusion and the deformation memory induced by the elastic part is the cause for proteins to undertake a fractional Brownian walk. This very feature can dramatically alter biochemical reactions, as we shall see in Section 4.

4. DIFFUSION AS A DRIVING FORCE FOR BIOCHEMICAL REACTIONS

This section is intended to give an overview of the impact of diffusion on the progress of biochemical reactions. In the first part, classical results on diffusion-driven reactions and limitations in these approaches are reviewed. In this context, the impact of the reaction space's dimensionality on the progress and outcome of (bio)chemical reactions will also be discussed. The second part summarizes recent findings on biochemical reactions in crowded systems and the role of (anomalous) diffusion in this context.

4.1. Diffusion-controlled reactions

Virtually all biochemical reactions in a cell rely on diffusion. Reactions may be dissected into two phases with distinct time scales, and in each of these phases random walks are needed to drive molecular association. First, reactants need to undergo a diffusive search to come into close contact (encounter phase). If two cytosolic molecules with diffusion constants D_1 and D_2 that wish to react with each other are separated by a distance s, the typical time for their encounter is determined by the MSD, that is, $\tau = s^2/[6(D_1 + D_2)]$. The sum of diffusion constants appears here since the particles need to move relative to each other and not just with respect to an immobile reference point for which the individual diffusion constant is the appropriate quantity. Diffusion coefficients therefore determine the time for complex formation of two (or more) molecules. In fact, as diffusion is a random walk the typical encounter time τ only is the first moment (i.e., the mean) of a distribution of encounter times that is given by an inverse Gaussian. Once two molecules are in close proximity, the second phase of a reaction can proceed. Molecules reorient by rotational diffusion until the binding sites of both reactants face each other. At this point a small lateral diffusion step downhill the binding potential finalizes the binding event and leads to a lock-key type of arrangement.

Let us focus in the remainder to reactions that are limited by the first phase, that is, by the encounter time. These reactions are named diffusion-limited since the search process dominates the time needed for a reaction. For the case of two molecules that wish to undergo a reaction in bulk fluid but are present only in minor amounts, Smoluchowski derived already in 1916 an analytical expression for the reaction constant in terms of the diffusion constants (Smoluchowski, 1916): let us consider the basic reaction $A+B \rightarrow C$ between two species, where the reaction rate is k. For simplicity, we assume that the reaction occurs rapidly after the molecules have met each other. If the two reactants A and B have radii r_A and r_B, and diffusion coefficients D_A and D_B, respectively, one can analytically calculate via the diffusion equation that $k=4\pi(D_A+D_B)(r_A+r_B)$. Again, the sum of both diffusion constants reflects the relative motion of two diffusing particles. On can extend this result to include also the second phase of binding, that is, when reactants need some additional time to bind after having encountered each other:

$1/k=1/\{4p(D_A+D_B)(r_A+r_B)\}+1/\kappa$, with κ being the second order rate constant. The total reaction rate is hence the inverse sum of the times needed for encountering and reacting.

Smoluchowski's construction seems to yield an elegant solution for considering diffusional motion in reactions. However, it relies critically on the assumption that molecules move in a three-dimensional fluid. If molecules move in one or two dimensions, for example, on a membrane or on a cytoskeletal filament, the construction fails and an analytical expression for a reaction constant cannot be derived any more. Indeed, the situation is even worse: the reaction coefficient is not a constant any more but depends on time! The reason for this counterintuitive phenomenon is based on the geometrical properties of the molecules' random walk as compared to the space in which they move. In brief, a random walk is a two-dimensional object by its very construction, and Smoluchowski's construction is only valid when reactants perform a random walk in an environment that has a dimension larger than two (e.g., in a three-dimensional bulk fluid). The next paragraph will give a more concise explanation of this statement.

In order to understand why a random walk's geometry is important, we need to briefly reflect on the definition of the term "dimension." Our

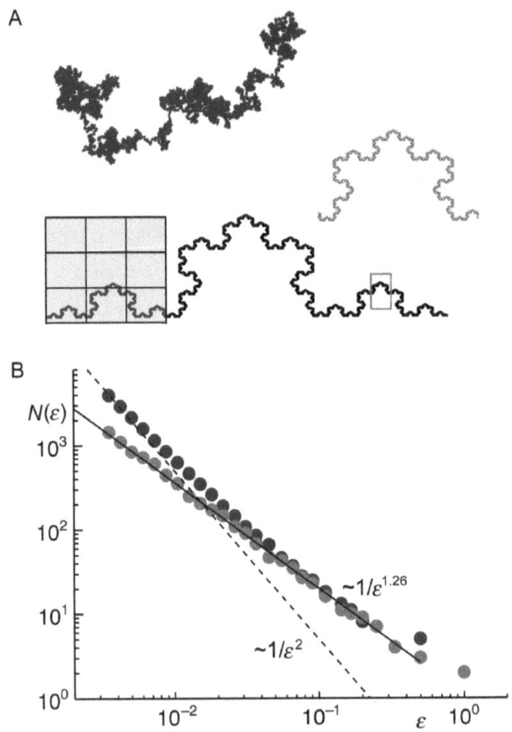

Figure 11.4 Fractal dimension of random walks. (A) Representative Brownian random walk (blue) as compared to the famous fractal Koch curve (black). The strict self-similarity of the Koch curve is highlighted by zooming into the red region (magnification of the curve shown in red). Superimposing quadratic grids of different sizes ε (shown here as a grey box with a refined black mesh of nine boxes) allows for quantifying the fractal dimension d_0 of both curves. (B) The number of boxes that touch one of the curves, $N(\varepsilon)$, shows a power-law decay $N(\varepsilon) \sim 1/\varepsilon^{d_0}$. For small box sizes, an exponent $d_0 = 1.26$ is found for the Koch curve and $d_0 = 2$ for the Brownian random walk. Hence, a random walk is plane-filling while the Koch curve is something between a line and a surface. (For interpretation of the references to color in this figure legend, the reader is referred to the online version of this chapter.)

intuition is compatible with Euclid's definition of dimension. Points, lines, surfaces, and volumes are zero-, one-, two-, and three-dimensional objects, respectively. Yet, in the 1960s a new definition of so-called "fractal" dimensions was introduced (Mandelbrot, 1983) that triggered not only a heap of scientific work but also inspired artists (just search the internet for a visualization of the Mandelbrot set). To understand the concept of a fractal dimension, let us imagine an arbitrary object whose dimensions we wish to determine, say, a wiggly line on a flat piece of paper (cf. Fig. 11.4A). If we superimpose a square grid with mesh length ε onto the paper, we can determine the number of tiles that are visited by the line, $N(\varepsilon)$. We then iteratively refine the mesh by decreasing ε and count again. Having done so many times, the number of tiles that touch the line for a given mesh size ε shows a scaling $N(\varepsilon) \sim I/\varepsilon^{d_0}$ for small box sizes ε (Fig. 11.4B). The exponent d_0 is called the fractal dimension (more precisely, it is named the box-counting dimension), and it agrees with our intuitive Euclidian definition: for points $d_0 = 0$, for lines $d_0 = 1$, and so on. The value of d_0, however, does not have to assume integer values. If the wiggly line is the so-called Koch curve (cf. Fig. 11.4), the fractal dimension is $d_0 \approx 1.26$. This curve is therefore neither a line nor a surface; it is something in between. Indeed, zooming into the curve, it seems to repeat itself on smaller and smaller length scales. This self-similarity is the hallmark of fractals and one measure for the self-similarity is the fractal dimension d_0.

A box-counting procedure for a protein's random walk in buffer solution yields the fractal dimension of normal Brownian diffusion: $d_0 = 2$. Normal diffusion is therefore filling a surface but not a three-dimensional volume. This property of "filling" an environment or not is captured by the term "compact" versus "noncompact" random walk. Diffusion is a compact random walk on a membrane but not in three-dimensional cytoplasm. With this insight, we can now see a deeper reason why Smoluchowski's construction does not work in one or two dimensions: the random walk is compact. One can only define a reaction constant if diffusion is noncompact, otherwise the reaction rate k depends on time (Benichou et al., 2010). This is particularly important for diffusion in two dimensions (e.g., for proteins in the plasma membrane) but also for subdiffusion in cytoplasm. If proteins are subject to fractional Brownian motion, that is, if they show crowding-induced subdiffusion in a viscoelastic fluid (cf. Section 3), the random walk takes on a fractal dimension $d_0 = 2/\alpha$. Subdiffusion therefore can be a compact random walk in three dimensions (for $\alpha < 2/3$, i.e., for values that have been experimentally observed, cf. Section 3). In this case Smoluchowski's approach fails even for reactions in bulk fluids, and one cannot define a proper reaction constant any more. These reactions are said to be fractal or fractional and the consequences for biology will be discussed in Section 4.2.

4.2. Target finding, rebinding, and enzymatic reactions in crowded media

From the above it is clear that understanding biochemical reactions in crowded cellular fluids requires some knowledge about diffusive transport and its potential anomalies. Especially reactions that rely on few molecules (say, a couple of thousands per cell, i.e., at nanomolar concentrations and below) are particularly affected by properties of diffusive transport, as these reactions are diffusion-limited.

General approaches to describe biochemical reactions, for example, in many systems biology approaches (Kholodenko, 2006), are based on the law of mass action and sets of ordinary differential equations (ODEs). A prominent example is the Michaelis–Menten reaction $E+S \leftrightarrow C \rightarrow E+P$ in which an enzyme E can reversibly form a complex C with its substrate (S). Alternatively to falling apart, the substrate/enzyme complex can progress to a product P and the unconsumed enzyme. This chemical equation can be translated to a set of ODEs for the reactants' concentration. In this approach, well-mixed conditions are assumed for all times, and reaction coefficients include diffusional motion of molecules only via the Smoluchowski construction (cf. Section 4.1). Such an approach, however, may be problematic. First, if the number of particles involved is small, a concentration may not be a good description in the first place: if there are just 1000 substrate molecules to be modified in a cell of volume 1000 μm^3, the concentration of unmodified substrates can only assume discrete values between 1.6 nM (max) and 0.0016 nM (min). Any value below the lower bound is strictly zero, that is, there is simply no particle left to be modified. Yet, in an ODE system this lower bound is typically not considered. Second, assuming well-mixed conditions implies that substrate and enzyme will immediately take on a nonzero equilibrium distance when dissociating from each other without product formation. In better words, rebinding of the same enzyme molecule to the substrate is as likely as binding a completely different enzyme molecule. This can only hold true for high enzyme concentrations, say, above $\sim$100 μM. Third, the dimension in which particles have to search for each other matters (cf. above), that is, the Smoluchowski construction may not be valid at all. Finally, an ODE formulation implies that diffusion can be described by a single time scale, namely the ratio of the squared typical distance between reactants and the sum of diffusion constants. Yet, in the presence of anomalous diffusion, multiple time scales are involved in the diffusional motion.

Several lines of evidence have recently supported these fairly general caveats. The first two criticisms are highlighted, for example, in a particle-based simulation of the MAPK signaling pathway (Takahashi et al., 2010). Unlike ODEs, these stochastic simulations include natural rebinding events of just-released enzyme molecules via an event-based algorithm. With this approach, the authors were able to show that the commonly assumed bistability of the MAPK pathway (Markevich et al., 2004) is an artifact of the widely used ODE description. For physiological concentrations of MAPK and its kinase, rebinding is of major importance. This result indeed challenges a fair number of ODE-based systems biology studies.

When considering subdiffusion of the fractional Brownian motion type, simulations also showed that product formation in multiphosphorylation cascades can be massively enhanced as compared to normal diffusion (Hellmann et al., 2012). In terms of the MAPK example, this means that subdiffusion strongly enhances rebinding of MAPK kinases and therefore increases the yield of double-phosphorylated MAPK. Indeed, crowding-induced subdiffusion may allow for a continuous transition between a processive and a distributive kinase action: while a distributive kinase will dissociate from its substrate after having performed the first phosphorylation, subdiffusion strongly counteracts a diffusion-driven separation of the reactants and leads to a fast rebinding. Therefore it seems as if the kinase is processive, that is, it stays bound to the substrate until all phosphorylations have been done. Thus, naming a kinase processive or distributive depends on the viscoelastic properties of the surrounding fluid. Indeed, strong changes of MAPK phosphorylation have been found when comparing crowded and noncrowded buffer conditions (Aokia et al., 2011).

Single-step Michaelis–Menten kinetics has also been studied numerically for obstructed diffusion in two dimensions (Berry, 2002). Here, enzyme and substrate were assumed to perform subdiffusion due to a percolation cluster of obstacles. In line with previous reports on fractal substrates (Kopelman, 1988), a fractal/fractional kinetics has been observed in these cases. A description via heuristic ODEs was only possible for certain time scales. These observations are in line with simple reactions on a percolation cluster, where encounter and reaction have been probed separately (Saxton, 2002). Again, enhanced rebinding due to subdiffusion has been found in this case.

Simulations of subdiffusion on membranes also revealed reactant segregation (Hellmann et al., 2011). Since normal diffusion and subdiffusion are compact random walks in two-dimensional environments, reactant segregation (the so-called Zeldovich effect) is expected (Kopelman, 1988). In this case, reactions only occur at the interface between regions that are enriched in complementary reactants and a chemical structuring of the surface is observed. Also, the formation of Turing patterns, which have been implicated in many cell biological and developmental processes (Meinhardt, 1982), is strongly facilitated by subdiffusion (Weiss, 2003). Since the emergence of Turing patterns critically relies on strongly different diffusional mobilities of two counteracting reactants (typically named "activator" and "inhibitor"), subdiffusion can massively enhance the stability of emerging patterns. In both of these situations, (sub)diffusion is a bad mixing strategy and domain formation may occur from local heterogeneities. Membrane domain formation in cell biology therefore may be just a reflection of the particularly bad diffusive mixing, and cells may have turned this into an advantageous strategy to foster signaling and mass exchange at particular spots.

While the above examples have been focused on certain classes of reactions, for example, of the Michaelis–Menten type, the general problem of finding a target (say, a repressor aiming to bind at a certain DNA operon) is also strongly affected by the properties of diffusion. A classical result of this is the so-called "reduction of dimension mechanism" for the Lac repressor (Berg et al., 1981). Given that there are only about ten copies of the repressor in each *E. coli* cell, one may ask how these few molecules can search and bind a single DNA operon in a short time. Diffusion through the cell's cytoplasm and once in a while hitting the DNA at a random location is surely a bad search strategy. On the other hand, diffusing only along the DNA is by far too slow. Switching between these two extremes, however, massively improves the searching. Diffusion through cytoplasm and binding to a random piece of the DNA followed by a brief sliding along the DNA in the neighborhood of the attachment point yields a fast and reliable search strategy (Berg et al., 1981). When two soluble proteins need to find each other in cytoplasm, however, reduction of dimension is in most cases not an option. Yet, if the particles are subject to subdiffusion (fractional Brownian motion) their random walk is a fractal with dimensions $d_0 = 2/\alpha$. If the anomaly is below 2/3, subdiffusion becomes a compact random walk and all points in space will be visited repeatedly. Hence, proteins will be able to find each other with a high probability. Indeed, simulations have shown that the probability of finding a target is massively enhanced for a subdiffusive search strategy although the diffusion process is slower than normal diffusion (Guigas and Weiss, 2008). Subdiffusion therefore does not speed up a search process but it makes it more reliable, that is, proteins find their binding partners with a higher probability. If a cellular response does not have to occur too rapidly but requires a high degree of reliability, subdiffusive searching may be advantageous for cells.

Most of the above-mentioned results have been found in numerical approaches. Experiments that probe specifically the influence of (sub) diffusion on biochemical reactions are still rare. The formation of complexes in crowded environments, for example, amyloid fiber formation, has been studied in some detail. However, most of the measurements and interpretations have been focused on the effect of excluded volume in crowded solutions (Zhou et al., 2008). Owing to the greater complexity of experiments, the contributions of (sub)diffusion have been studied much less. In a very recent study, we have been able to show that subdiffusion indeed can facilitate complex formation (Stiehl and Weiss, 2013). Here, the opening/closing of a ssDNA hairpin in artificial crowded solutions was studied via FCS. As a result of crowding and subdiffusion, an enhanced fraction of closed ssDNA hairpins was observed which matches well to the predictions of numerical simulations. It will be interesting to further extend these approaches to more complex biochemical reactions, for example, signaling cascades in bulk and on membranes.

5. CONCLUDING REMARKS

Despite the fact that diffusion has been studied intensively for more than 100 years, its impact on the secret life of a cell is still far from being well understood. This review has only highlighted a few ideas and problems in this context that hopefully will stimulate other researchers to consider transport properties more thoroughly when aiming at an understanding of complex biochemical reactions. Having identified important players of cell biology via many -omics approaches in the last decade, it is now about time to dive into the realm of quantitative biology that uses these molecular players together with concepts from physics in order to understand cellular organization and dynamics in detail.

REFERENCES

Abramowitz, M., Stegun, I.A., 1970. Handbook of Mathematical Functions: With Formulas, Graphs, and Mathematical Tables. Dover, New York.

Aokia, K., Yamada, M., Kunida, K., Yasuda, S., Matsuda, M., 2011. Processive phosphorylation of ERK MAP kinase in mammalian cells. Proc. Natl. Acad. Sci. U.S.A. 108, 12675–12680.

Axelrod, D., Koppel, D.E., Schlessinger, J., Elson, E., Webb, W.W., 1976a. Mobility measurement by analysis of fluorescence photobleaching recovery kinetics. Biophys. J. 16, 1055–1069.

Axelrod, D., Ravdin, P., Koppel, D.E., Schlessinger, J., Webb, W.W., Elson, E.L., Podleski, T.R., 1976b. Lateral motion of fluorescently labeled acetylcholine receptors in membranes of developing muscle-fibers. Proc. Natl. Acad. Sci. U.S.A. 73, 4594–4598.

Banks, D.S., Fradin, C., 2005. Anomalous diffusion of proteins due to molecular crowding. Biophys. J. 89, 2960–2971.
Benichou, O., Chevalier, C., Klafter, J., Meyer, B., Voituriez, R., 2010. Geometry-controlled kinetics. Nat. Chem. 2, 472–477.
Berg, O.G., Winter, R.B., von Hippel, P.H., 1981. Diffusion-driven mechanisms of protein translocation on nucleic acids. 1. Models and theory. Biochemistry 20, 6929–6948.
Berkovich, R., Wolfenson, H., Eisenberg, S., Ehrlich, M., Weiss, M., Klafter, J., Henis, Y.I., Urbakh, M., 2011. Accurate quantification of diffusion and binding kinetics of non-integral membrane proteins by FRAP. Traffic 12, 1648–1657.
Berry, H., 2002. Monte Carlo simulations of enzyme reactions in two dimensions: fractal kinetics and spatial segregation. Biophys. J. 83, 1891–1901.
Bouchaud, J.P., Georges, A., 1990. Anomalous diffusion in disordered media—statistical mechanisms, models and physical applications. Phys. Rep. 195, 127–293.
Brauchle, C., Lamb, D.C., Michaelis, J., 2009. Single Particle Tracking and Single Molecule Energy Transfer. Wiley-VCH, Weinheim.
Brockmann, D., Hufnagel, L., Geisel, T., 2006. The scaling laws of human travel. Nature 439, 462–465.
Dertinger, T., Pacheco, V., von der Hocht, I., Hartmann, R., Gregor, I., Enderlein, J., 2007. Two-focus fluorescence correlation spectroscopy: a new tool for accurate and absolute diffusion measurements. ChemPhysChem 8, 433–443.
Digman, M.A., Gratton, E., 2011. Lessons in fluctuation correlation spectroscopy. Annu. Rev. Phys. Chem. 62, 645–668.
Dix, J.A., Verkman, A.S., 2008. Crowding effects on diffusion in solutions and cells. Annu. Rev. Biophys. 37, 247–263.
Einstein, A., 1905. Über die von der molekularkinetischen Theorie der Wärme geforderte Bewegung von in ruhenden Flüssigkeiten suspendierten Teilchen. Ann. Phys. 322, 549–560.
Ellis, R.J., Minton, A.P., 2003. Cell biology: join the crowd. Nature 425, 27–28.
Elsner, M., Hashimoto, H., Simpson, J.C., Cassel, D., Nilsson, T., Weiss, M., 2003. Spatiotemporal dynamics of the COPI vesicle machinery. EMBO Rep. 4, 1000–1005.
Engelman, D.M., 2005. Membranes are more mosaic than fluid. Nature 438, 578–580.
Ernst, D., Hellmann, M., Kohler, J., Weiss, M., 2012. Fractional Brownian motion in crowded fluids. Soft Matter 8, 4886–4889.
Frick, M., Schmidt, K., Nichols, B.J., 2007. Modulation of lateral diffusion in the plasma membrane by protein density. Curr. Biol. 17, 462–467.
Fujiwara, T., Ritchie, K., Murakoshi, H., Jacobson, K., Kusumi, A., 2002. Phospholipids undergo hop diffusion in compartmentalized cell membrane. J. Cell Biol. 157, 1071–1081.
Goehring, N.W., Chowdhury, D., Hyman, A.A., Grill, S.W., 2010. FRAP analysis of membrane-associated proteins: lateral diffusion and membrane-cytoplasmic exchange. Biophys. J. 99, 2443–2452.
Golding, I., Cox, E.C., 2006. Physical nature of bacterial cytoplasm. Phys. Rev. Lett. 96, 098102.
Gonzalez, M.C., Hidalgo, C.A., Barabasi, A.L., 2008. Understanding individual human mobility patterns. Nature 453, 779–782.
Goulian, M., Simon, S.M., 2000. Tracking single proteins within cells. Biophys. J. 79, 2188–2198.
Griffiths, G., Warren, G., Quinn, P., Mathieu-Costello, O., Hoppeler, H., 1984. Density of newly synthesized plasma membrane proteins in intracellular membranes. I. Stereological studies. J. Cell Biol. 98, 2133–2141.
Guigas, G., Weiss, M., 2006. Size-dependent diffusion of membrane inclusions. Biophys. J. 91, 2393–2398.

Guigas, G., Weiss, M., 2008. Sampling the cell with anomalous diffusion—the discovery of slowness. Biophys. J. 94, 90–94.
Guigas, G., Kalla, C., Weiss, M., 2007a. The degree of macromolecular crowding in the cytoplasm and nucleoplasm of mammalian cells is conserved. FEBS Lett. 581, 5094–5098.
Guigas, G., Kalla, C., Weiss, M., 2007b. Probing the nanoscale viscoelasticity of intracellular fluids in living cells. Biophys. J. 93, 316–323.
Havlin, S., Benavraham, D., 1987. Diffusion in disordered media. Adv. Phys. 36, 695–798.
He, Y., Burov, S., Metzler, R., Barkai, E., 2008. Random time-scale invariant diffusion and transport coefficients. Phys. Rev. Lett. 101, 058101.
Hellmann, M., Heermann, D.W., Weiss, M., 2011. Anomalous reaction kinetics and domain formation on crowded membranes. EPL 94, 5.
Hellmann, M., Heermann, D.W., Weiss, M., 2012. Enhancing phosphorylation cascades by anomalous diffusion. EPL 97, 58004.
Hoefling, F., Franosch, T., 2013. Anomalous transport in the crowded world of biological cells. Rep. Prog. Phys. 76, 046602.
Hofling, F., Franosch, T., Frey, E., 2006. Localization transition of the three-dimensional lorentz model and continuum percolation. Phys. Rev. Lett. 96, 165901.
Hughes, B.D., Pailthorpe, B.A., White, L.R., Sawyer, W.H., 1982. Extraction of membrane microviscosity from translational and rotational diffusion coefficients. Biophys. J. 37, 673–676.
Humphries, N.E., Queiroz, N., Dyer, J.R.M., Pade, N.G., Musyl, M.K., Schaefer, K.M., Fuller, D.W., Brunnschweiler, J.M., Doyle, T.K., Houghton, J.D.R., Hays, G.C., Jones, C.S., Noble, L.R., Wearmouth, V.J., Southall, E.J., Sims, D.W., 2010. Environmental context explains Levy and Brownian movement patterns of marine predators. Nature 465, 1066–1069.
Kholodenko, B.N., 2006. Cell-signalling dynamics in time and space. Nat. Rev. Mol. Cell Biol. 7, 165–176.
Klafter, J., Sokolov, I.M., 2012. First Steps in Random Walks: From Tools to Applications. Oxford University Press, Oxford/New York.
Klafter, J., Shlesinger, M.F., Zumofen, G., 1996. Beyond Brownian motion. Phys. Today 49, 33–39.
Kolin, D.L., Wiseman, P.W., 2007. Advances in image correlation spectroscopy: measuring number densities, aggregation states, and dynamics of fluorescently labeled macromolecules in cells. Cell Biochem. Biophys. 49, 141–164.
Kopelman, R., 1988. Fractal reaction-kinetics. Science 241, 1620–1626.
Lippincott-Schwartz, J., Snapp, E., Kenworthy, A., 2001. Studying protein dynamics in living cells. Nat. Rev. Mol. Cell Biol. 2, 444–456.
Lubelski, A., Klafter, J., 2009. Fluorescence correlation spectroscopy: the case of subdiffusion. Biophys. J. 96, 2055–2063.
Lubelski, A., Sokolov, I.M., Klafter, J., 2008. Nonergodicity mimics inhomogeneity in single particle tracking. Phys. Rev. Lett. 100, 250602.
Magde, D., Webb, W.W., Elson, E., 1972. Thermodynamic fluctuations in a reacting system—measurement by fluorescence correlation spectroscopy. Phys. Rev. Lett. 29, 705–708.
Malchus, N., Weiss, M., 2010. Anomalous diffusion reports on the interaction of misfolded proteins with the quality control machinery in the endoplasmic reticulum. Biophys. J. 99, 1321–1328.
Mandelbrot, B.B., 1983. The Fractal Geometry of Nature: Updated and Augmented. W.H. Freeman, San Francisco.
Mandelbrot, B., van Ness, J.W., 1968. Fractional Brownian motions, fractional noises and applications. SIAM Rev. 10, 422–437.

Markevich, N.I., Hoek, J.B., Kholodenko, B.N., 2004. Signaling switches and bistability arising from multisite phosphorylation in protein kinase cascades. J. Cell Biol. 164, 353–359.

Mason, T.G., Weitz, D.A., 1995. Optical measurements of frequency-dependent linear viscoelastic moduli of complex fluids. Phys. Rev. Lett. 74, 1250–1253.

McGuffee, S.R., Elcock, A.H., 2010. Diffusion, crowding and protein stability in a dynamic molecular model of the bacterial cytoplasm. PLoS Comp. Biol. 6, e1000694.

Meinhardt, H., 1982. Models of Biological Pattern Formation. Academic Press, London/New York.

Pan, W.C., Filobelo, L., Pham, N.D.Q., Galkin, O., Uzunova, V.V., Vekilov, P.G., 2009. Viscoelasticity in homogeneous protein solutions. Phys. Rev. Lett. 102, 058101.

Petrov, E.P., Schwille, P., 2008. Translational diffusion in lipid membranes beyond the Saffman-Delbruck approximation. Biophys. J. 94, L41–L43.

Quinn, P., Griffiths, G., Warren, G., 1984. Density of newly synthesized plasma membrane proteins in intracellular membranes. II. Biochemical studies. J. Cell Biol. 98, 2142–2147.

Rigler, R., Elson, E., 2001. Fluorescence Correlation Spectroscopy: Theory and Applications. Springer, Berlin/New York.

Rudnick, J., Gaspari, G., 1987. The shapes of random-walks. Science 237, 384–389.

Saffman, P.G., Delbruck, M., 1975. Brownian motion in biological membranes. Proc. Natl. Acad. Sci. U.S.A. 72, 3111–3113.

Saxton, M.J., 1996. Anomalous diffusion due to binding: a Monte Carlo study. Biophys. J. 70, 1250–1262.

Saxton, M.J., 2001. Anomalous subdiffusion in fluorescence photobleaching recovery: a Monte Carlo study. Biophys. J. 81, 2226–2240.

Saxton, M.J., 2002. Chemically limited reactions on a percolation cluster. J. Chem. Phys. 116, 203–208.

Saxton, M.J., 2007. A biological interpretation of transient anomalous subdiffusion. I. Qualitative model. Biophys. J. 92, 1178–1191.

Seksek, O., Biwersi, J., Verkman, A.S., 1997. Translational diffusion of macromolecule-sized solutes in cytoplasm and nucleus. J. Cell Biol. 138, 131–142.

Sims, D.W., Southall, E.J., Humphries, N.E., Hays, G.C., Bradshaw, C.J.A., Pitchford, J.W., James, A., Ahmed, M.Z., Brierley, A.S., Hindell, M.A., Morritt, D., Musyl, M.K., Righton, D., Shepard, E.L.C., Wearmouth, V.J., Wilson, R.P., Witt, M.J., Metcalfe, J.D., 2008. Scaling laws of marine predator search behaviour. Nature 451, 1098–1102.

Smoluchowski, M.V., 1916. Drei Vorträge über Diffusion, Brownsche Molekularbewegung und Koagulation von Kolloidteilchen. Phys. Z. 17, 557–571.

Soumpasis, D.M., 1983. Theoretical analysis of fluorescence photobleaching recovery experiments. Biophys. J. 41, 95–97.

Stiehl, O., Weidner-Hertrampf, K., Weiss, M., 2013. Kinetics of conformational fluctuations in DNA hairpin loops in crowded fluids. New J. Phys. 15, 113010.

Szymanski, J., Weiss, M., 2009. Elucidating the origin of anomalous diffusion in crowded fluids. Phys. Rev. Lett. 103, 4.

Takahashi, K., Tanase-Nicola, S., ten Wolde, P.R., 2010. Spatio-temporal correlations can drastically change the response of a MAPK pathway. Proc. Natl. Acad. Sci. U.S.A. 107, 2473–2478.

Thompson, M.A., Lew, M.D., Moerner, W.E., 2012. Extending microscopic resolution with single-molecule imaging and active control. Annu. Rev. Biophys. 41, 321–342.

Tolic-Norrelykke, I.M., Munteanu, E.L., Thon, G., Oddershede, L., Berg-Sorensen, K., 2004. Anomalous diffusion in living yeast cells. Phys. Rev. Lett. 93, 078102.

Tomishige, M., Kusumi, A., 1999. Compartmentalization of the erythrocyte membrane by the membrane skeleton: intercompartmental hop diffusion of band 3. Mol. Biol. Cell 10, 2475–2479.

Tseng, Y., Kole, T.P., Wirtz, D., 2002. Micromechanical mapping of live cells by multiple-particle-tracking microrheology. Biophys. J. 83, 3162–3176.

Wachsmuth, M., Waldeck, W., Langowski, J., 2000. Anomalous diffusion of fluorescent probes inside living cell nuclei investigated by spatially-resolved fluorescence correlation spectroscopy. J. Mol. Biol. 298, 677–689.

Weber, S.C., Spakowitz, A.J., Theriot, J.A., 2010. Bacterial chromosomal loci move subdiffusively through a viscoelastic cytoplasm. Phys. Rev. Lett. 104, 238102.

Weigel, A.V., Simon, B., Tamkun, M.M., Krapf, D., 2011. Ergodic and nonergodic processes coexist in the plasma membrane as observed by single-molecule tracking. Proc. Natl. Acad. Sci. U.S.A. 108, 6438–6443.

Weiss, M., 2003. Stabilizing Turing patterns with subdiffusion in systems with low particle numbers. Phys. Rev. E. 68, 036213.

Weiss, M., 2004. Challenges and artifacts in quantitative photobleaching experiments. Traffic 5, 662–671.

Weiss, M., Hashimoto, H., Nilsson, T., 2003. Anomalous protein diffusion in living cells as seen by fluorescence correlation spectroscopy. Biophys. J. 84, 4043–4052.

Weiss, M., Elsner, M., Kartberg, F., Nilsson, T., 2004. Anomalous subdiffusion is a measure for cytoplasmic crowding in living cells. Biophys. J. 87, 3518–3524.

Weiss, M., 2013. Single-particle tracking data reveal anticorrelated fractional Brownian motion in crowded fluids. Phys. Rev. E. 88, 010101(R).

Yamada, S., Wirtz, D., Kuo, S.C., 2000. Mechanics of living cells measured by laser tracking microrheology. Biophys. J. 78, 1736–1747.

Zhou, H.X., Rivas, G., Minton, A.P., 2008. Macromolecular crowding and confinement: biochemical, biophysical, and potential physiological consequences. Annu. Rev. Biophys. 37, 375–397.

CHAPTER TWELVE

Importance of Crowding in Signaling, Genetic, and Metabolic Networks

Pieter Rein ten Wolde[1], Andrew Mugler
FOM Institute AMOLF, Amsterdam, The Netherlands
[1]Corresponding author: e-mail address: tenwolde@amolf.nl

Contents

Abstract

It is now well established that the cell is a highly crowded environment. Yet, the effects of crowding on the dynamics of signaling pathways, gene regulation networks, and metabolic networks are still largely unknown. Crowding can alter both molecular diffusion and the equilibria of biomolecular reactions. In this chapter, we first discuss how diffusion can affect biochemical networks. Diffusion of transcription factors can increase noise in gene expression, while diffusion of proteins between intracellular compartments or between cells can reduce concentration fluctuations. In push–pull networks diffusion can impede information transmission, while in multisite protein modification networks diffusion can qualitatively change the macroscopic response of the system, such as the loss or emergence of bistability. Moreover, diffusion can directly change the metabolic flux. We describe how crowding affects diffusion, and thus how all these phenomena are influenced by crowding. Yet, a potentially more important effect of crowding on biochemical networks is mediated via the shift in the equilibria of

International Review of Cell and Molecular Biology, Volume 307
ISSN 1937-6448
http://dx.doi.org/10.1016/B978-0-12-800046-5.00012-6

bimolecular reactions, and we provide computational evidence that supports this idea. Finally, we discuss how the effects of crowding can be incorporated in models of biochemical networks.

1. INTRODUCTION

Cellular function arises at the level of the network of interactions between proteins and DNA inside the cell. Whether a cell differentiates or proliferates, stays put or moves, is not determined by a single component or reaction, but rather by the collective dynamics of signaling pathways and gene regulation networks. These biochemical networks are often modeled as well-stirred chemical reactors. However, cellular networks are fundamentally different from well-stirred chemical reactors. The copy numbers of the components are often low, and, as a result, biochemical networks can be highly stochastic (Elowitz, 2002; Ozbudak et al., 2002; Paulsson, 2004; Taniguchi et al., 2010; Yu, 2006). Moreover, the interior of the cell is a highly crowded environment, with proteins and DNA occupying up to 20–30% of the cellular volume (Ellis, 2001; Zimmerman and Trach, 1991). These crowded conditions strongly affect the chemical reactions and physical interactions that form the biochemical network. Yet, while the effects of crowding on the dynamics of individual molecules and reactions have been investigated extensively (Muramatsu and Minton, 1988), how crowding affects the collective dynamics of biochemical networks and thereby cellular function has remained largely unexplored.

Crowding can affect biochemical networks in two principal ways. First, crowding shifts the equilibrium of association–dissociation reactions to the associated state. Molecules in their associated state take up less volume, thereby creating more space for the other molecules in the system. This increases the entropy of the system and leads to an effective attraction between the molecules which is often called a depletion interaction or an excluded volume interaction (Asakura and Oosawa, 1954); this depletion interaction effectively increases the binding affinity between two molecules. Second, crowding changes the diffusive motion of the molecules. At long length and timescales, diffusion appears normal, which means that the mean-squared displacement increases linearly with time, just as in a dilute system without crowding agents; however, the slope of this linear relation, which gives the effective diffusion constant, can be orders of magnitude lower because the crowders obstruct molecular motion (Bancaud et al., 2009; Elf et al., 2007; Muramatsu and Minton, 1988). This lower diffusion constant decreases the association rate of diffusion-limited bimolecular reactions. On short length and timescales, on the other hand, crowding can lead to anomalous diffusion, or, more precisely, subdiffusion, which means that the mean-squared displacement grows sublinearly with time (Banks and Fradin, 2005; Bénichou et al., 2010; Golding and Cox, 2006; Weiss et al., 2004). This implies that it takes longer for initial positions to be forgotten, and that molecules that have come together, stay together longer.

While the effects of crowding on binding affinities and the diffusion of individual molecules have been studied in considerable detail both theoretically and experimentally, the role of crowding in the behavior of biochemical networks has only begun to be addressed. Yet, a number of studies in recent years have predicted that diffusion can have dramatic and often counterintuitive effects on the dynamics of signaling pathways, gene regulation networks, and metabolic networks. In this review, we will summarize these studies. We hope to explain the phenomena using simple, intuitive arguments, and we will discuss how the crowded environment of the cell is expected to affect them. These studies also shed light on the question how crowding should be described in models of biochemical networks. Do the crowding agents have to be described explicitly or can the effect of their presence be captured by renormalizing the association and dissociation rates, thus allowing for a simpler model? Finally, we address the question whether the dominant effect of crowding on the dynamics of biochemical networks is the change in the diffusion dynamics or the shift in the binding equilibria.

2. DIFFUSION OF TRANSCRIPTION FACTORS CAN INCREASE NOISE IN GENE EXPRESSION

Experiments in recent years have vividly demonstrated that gene expression can be highly stochastic (Elowitz, 2002; Ozbudak et al., 2002; Paulsson, 2004; Yu, 2006). In particular, both in prokaryotes and eukaryotes, genes are often expressed in bursts. The burst-like nature of gene expression can emerge from the fact that one mRNA transcript can give rise to more than one protein copy—translation then occurs in bursts (Ozbudak et al., 2002). Alternatively, the production of the mRNA itself can occur in bursts, as has been observed for prokaryotes (Chubb et al., 2006; Golding et al., 2005; Raj et al., 2006; So et al., 2011). This can happen when the promoter switches between an "inactive" and an "active" state at timescales that are slower than the timescale of transcription initiation when the promoter is in the active state: then, more than one mRNA transcript will be produced from the promoter before the promoter switches back to the inactive state. While detailed time-series of mRNA production is consistent with such a two-state switching model of the promoter (So et al., 2011), the origin of the slow switching timescales that are necessary for bursty transcription is still unknown. One hypothesis is that it is due to the diffusion of transcription factors (TFs).

Many genes are regulated by TFs. The transcription-factor concentrations can be as low as a nanomolar, or even lower. At such low concentrations, the stochastic arrival of TFs at the promoter can be a major source of noise in gene expression. In recent years, several groups have begun to address this question by theory and computer simulations (Bialek and Setayeshgar, 2005; Meyer et al., 2012; Tkacik and Bialek, 2009; van Zon et al., 2006). These studies revealed that below a micromolar, the diffusion of TFs can be the dominant source of noise in gene expression (van Zon et al., 2006). These modeling studies also provide an explanation for why diffusion can contribute to the noise in transcription (Fig. 12.1).

When a TF dissociates from the promoter, it has a high probability of rebinding the promoter instead of diffusing into the bulk of other molecules. As a consequence, a TF will undergo many rounds of rebinding and dissociation before it diffuses away from the promoter and escapes into the bulk. These rounds of rebinding and dissociation increase the effective active and inactive times of the promoter, which means that the mRNA transcripts are

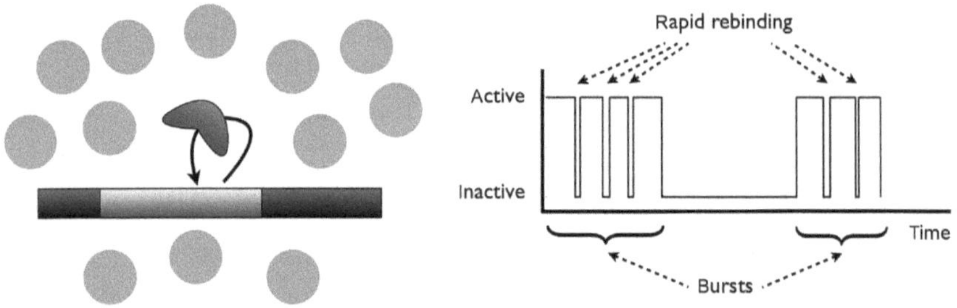

Figure 12.1 A transcription factor (red) diffusively interacting with the promoter region of DNA will undergo several rounds of dissociation and rapid rebinding before ultimately escaping to the bulk. As a result, gene expression will occur in bursts—alternating periods of high and low transcription. Crowders (gray) enhance bursts by promoting rebinding and slowing diffusion. (For interpretation of the references to color in this figure legend, the reader is referred to the online version of this chapter.)

more likely to be produced in bursts—long periods in which multiple mRNA copies are produced, alternating with long periods in which hardly any mRNA copies are produced. Crowding will decrease the diffusion constant and thereby decrease the association rate of the TF with the DNA. It will also increase the number of rebindings and thereby decrease the effective dissociation rate (Bénichou et al., 2010; Lomholt et al., 2007a,b; Meyer et al., 2012; Zaid et al., 2009). Both effects will slow down the dynamics of the promoter state, with slower switching between the active and the inactive state of the promoter. This will lead to more bursts in transcription. As such, crowding has been predicted to increase the noise in gene expression (Meyer et al., 2012; van Zon et al., 2006).

3. DIFFUSION BETWEEN COMPARTMENTS CAN REDUCE PROTEIN CONCENTRATION FLUCTUATIONS

Diffusion not only underlies bursts in gene expression, it can also act to "wash out" bursts across multiple compartments. Arguably the first hint of this effect was an experimental study on the precision of the *hunchback* (*hb*) expression domain in the developing embryo of the fruitfly Drosophila. Shortly after fertilization, the morphogen protein Bicoid (Bcd) forms an exponential concentration gradient along the anterior–posterior axis of the embryo, which provides positional information to the differentiating nuclei. One of the target genes of Bcd is *hb*, which is expressed in the anterior half of the embryo. The posterior boundary of the Hb expression domain is very sharp: by cell cycle 13, the position of the boundary varies only by about one nuclear spacing (Gregor et al., 2007; He et al., 2008; Houchmandzadeh et al., 2002). This precision is higher than the best achievable precision for a time-averaging-based readout mechanism of the Bcd gradient (Gregor et al., 2007). Intriguingly, the study of Gregor et al. revealed that the Hb concentrations in neighboring nuclei exhibit spatial correlations, and the authors suggest that this implies a form of spatial averaging, which enhances the precision of the posterior Hb boundary. However, the mechanism for spatial averaging remained unclear.

To elucidate the mechanism of spatial averaging, Erdmann et al. (2009) and independently Okabe-Oho et al. (2009) performed a computational study of the Bcd–Hb system. Their analysis revealed a simple but robust mechanism for spatial averaging, which is based on the diffusion of Hb itself. The expression of Hb in the respective nuclei is, according to their analysis,

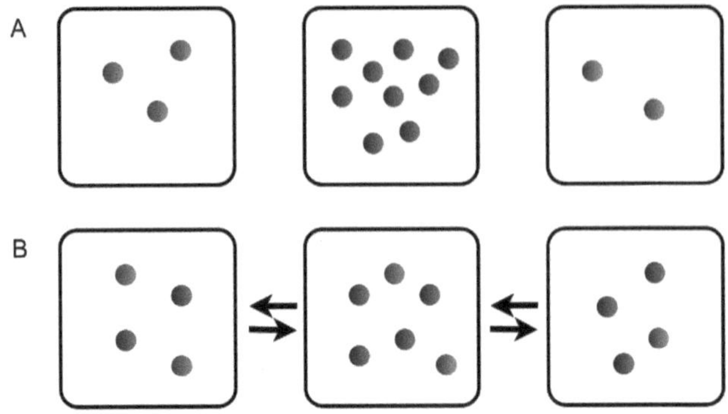

Figure 12.2 (A) In a developing Drosophila embryo, proteins in each cell nucleus are produced in bursts, meaning that some nuclei transiently have more proteins than others. (B) Diffusion of proteins between neighboring nuclei "washes out" these bursts, reducing transient fluctuations in protein number, a mechanism known as spatial averaging. Crowding will slow down diffusion and thus decrease the strength of spatial averaging. In both A and B, proteins are identical but are colored according to the compartment where they have been produced. (For color version of this figure, the reader is referred to the online version of this chapter.)

produced in bursts. The diffusion of Hb between neighboring nuclei can, however, wash out these bursts in gene expression by rapidly distributing the protein copies that are produced during one burst in a nucleus over the neighboring nuclei (Fig. 12.2). Concomitantly, the fluctuations in the Hb concentration in a given nucleus depend not only on the expression of Hb in that nucleus, but, due to the diffusive exchange, also on the expression of Hb in the neighboring nuclei. Since Hb expression in the different nuclei is, to a good approximation, uncorrelated, this means that the concentration in a nucleus represents an average over the fluctuations in that nucleus and the neighboring nuclei. This is the mechanism of spatial averaging. Its strength increases with increasing diffusion constant, because the number of neighboring nuclei that contribute to the average concentration in a nucleus increases with the diffusion constant.

Interestingly, the computational study by Erdmann et al. (2009) also suggested that there exists an optimal diffusion constant that maximizes the precision of the boundary. On the one hand, a higher diffusion constant reduces the fluctuations in the Hb concentration in each nucleus via the mechanism of spatial averaging. On the other hand, a higher diffusion constant also lessens the steepness of the boundaries of the Hb expression domain. The trade-off between these two effects yields an optimal diffusion constant that maximizes the sharpness or precision of the Hb expression domain.

The spatial averaging mechanism revealed by Erdmann et al. (2009) and Okabe-Oho et al. (2009) is very generic, and applies not only to any developmental system, but also to any biochemical network in general. For example, if a messenger protein is activated at one end of the cell and then has to diffuse to another place to activate another system, then the bursts in the activation of the messenger protein will be tempered by its diffusion. For this reason, it may be beneficial to spatially separate the input and output of the signaling pathway. Similarly, the diffusive exchange of signaling proteins between different intracellular compartments or between neighboring cells in a tissue will also tend to reduce the fluctuations in their concentration. A similar mechanism has also been implied in enhancing the fidelity of quorum sensing (Weber and Buceta, 2011). Crowding will slow down diffusion and thus decrease the strength of spatial averaging.

4. CROWDING CAN ENHANCE INFORMATION TRANSMISSION BY REMOVING CORRELATIONS

Since the activation and expression of proteins is often irregular or bursty, the argument of spatial averaging may seem to suggest that diffusion of the activated or expressed protein is always beneficial. However, this is not necessarily the case, as the phenomenon of membrane protein clustering shows.

Membrane-bound proteins are often clustered into lipid domains or partitioned via cytoskeleton-induced corralling. Intriguingly, experiments have revealed that these clusters have a characteristic size of only a few molecules. For example, the GPI-anchored receptor CD59 is observed to form clusters of 3–9 molecules upon interaction with the cytoskeleton and lipid rafts (Suzuki et al., 2007a,b). Similarly, the well-studied membrane-bound GTPase Ras forms clusters of 6–8 molecules, which also depend on interactions with the cytoskeleton and rafts (Plowman et al., 2005; Prior, 2003). While several theoretical studies had predicted that clustering modulates signaling (Gurry et al., 2009; Kalay et al., 2012; Mugler et al., 2012; Tian et al., 2007), the origin of the characteristic cluster size remained unknown.

To elucidate the origin of the characteristic cluster size, Mugler et al. (2013) performed a computational study of a common motif in membrane signaling. It consists of two layers of a push–pull network, in which the active component of the top layer activates the component of the bottom layer (Fig. 12.3). This motif can be found, for example, in CD59 and

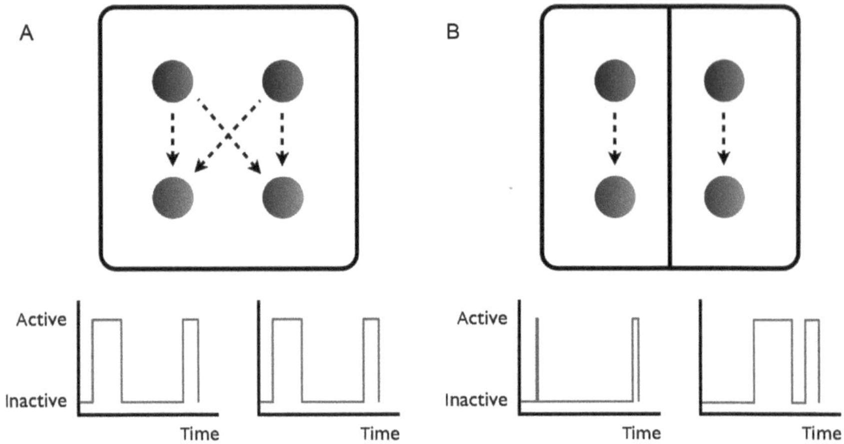

Figure 12.3 (A) In a typical signaling module, an upstream component, blue activates a downstream component (red). Since both downstream molecules respond to both upstream molecules, the activation states of the two downstream molecules will be correlated with each other in time. (B) Spatial partitioning, which can be induced or enhanced by crowding, isolates molecules, thereby reducing these correlations and lowering the noise in the average activation across all downstream molecules. (For interpretation of the references to color in this figure legend, the reader is referred to the online version of this chapter.)

Ras membrane signaling. Stimulated CD59 receptors—the component of the top layer—induce the switching of several Src family kinases from an unphosphorylated to a phosphorylated state (Suzuki et al., 2007a,b)—the component of the bottom layer. Similarly, stimulated EGF receptors (top component) induce the switching of Ras proteins from an inactive GDP-loaded state to an active GTP-loaded state (bottom component) (Tian et al., 2007).

The authors found that spatially partitioning the components into domains or compartments not only linearizes the input–output relation, but also reduces the noise in the total concentration of the output. This is perhaps surprising, because partitioning the components reduces the size of the reaction volume, the partition, which tends to increase the fluctuations. Indeed, the fluctuations per partition do increase, as compared to a scenario without compartments and all molecules well mixed. Yet, the fluctuations in the output summed over all partitions decrease. The reason is that partitioning isolates molecules, as a result of which the molecules in the different compartments are activated independently. This reduces correlations in the switching behavior between their active and inactive state, and it is this removal of correlations that lowers the output noise (Fig. 12.3).

While these results show that it can be beneficial to spatially separate molecules, it does not explain the characteristic cluster size. The characteristic size suggests that there exists a trade-off between two competing effects. But what is the effect that competes with the beneficial effect of noise reduction? The work of Mugler et al. (2013) suggests it is signal attenuation: making the partitions finer not only reduces noise, but also increases the probability that in a partition no signal can be propagated because only molecules of the top layer or only molecules of the bottom layer are present in that partition. This trade-off between noise reduction and linearization on the one hand, and signal attenuation on the other hand, naturally leads to an optimal cluster size that maximizes signaling fidelity and information transmission (Mugler et al., 2013). Interestingly, the predicted optimal cluster size is in very good agreement with the characteristic cluster sizes that have been found experimentally for the Ras system and the CD59 system.

Spatial partitioning of membrane signaling proteins can be induced by the underlying cytoskeletal network or by membrane domains such as lipid rafts. However, even in a spatially homogeneous system, the finite speed of diffusion can lead to an effective partitioning of the molecules, which can enhance information transmission. The effect of partitioning is also not limited to membrane signaling. Partitioning may also be induced by scaffold proteins or by large macromolecular complexes in the cytoplasm. Each scaffold or macromolecular protein complex would act as an independent reaction compartment, and the exchange between these compartments would be mediated by rare dissociation events followed by rapid cytoplasmic diffusion and fast association with other complexes in the cytoplasm.

Crowding may play a key role in all these scenarios. Crowding can enhance the residence time of molecules on scaffolds and in macromolecular complexes, thereby allowing for the effective spatial isolation of molecules on the timescale of biochemical signaling. Even in the absence of scaffolds or macromolecular complexes, crowding is expected to be important. Crowding slows down diffusion, which leads to a finer effective partitioning of the signaling molecules. This may allow for the more effective removal of correlations between the signaling molecules, enhancing the fidelity of signaling.

5. CROWDING CAN PROMOTE MEMBRANE REBINDING, WHICH CAN ENHANCE DOWNSTREAM SIGNAL PROPAGATION

Clustering of membrane-bound proteins can not only enhance signal transmission in the membrane, but also change signal propagation from the membrane into the cytoplasm. A canonical motif in cell signaling is one in which a signaling protein in the cytoplasm is chemically modified by two antagonistic enzymes, with one enzyme, for example, a kinase, residing at the membrane and the other, the phosphatase, in the cytoplasm. In some systems, such as the bacterial chemotaxis system, the cytoplasmic protein needs to be modified at only one site to become active, while in other systems, such as the Ras–Raf–Mek–Erk system and other MAPK signaling pathways (Chang and Karin, 2001), the cytoplasmic protein needs to be modified at multiple sites for full activation. It turns out that the effect of protein clustering on signal propagation strongly depends on the activation scenario (Mugler et al., 2012).

When the cytoplasmic protein has to be modified at only a single site, then clustering of the activating, membrane-bound enzyme reduces the response, taken to be the concentration of the activated protein in the cytoplasm. This is a bulk effect: the clustering of the enzyme molecules at the membrane reduces the effective target size for the substrate molecules in the bulk (Fig. 12.4). This slows down the diffusive search, reducing the response. However, when the cytoplasmic molecules have to be modified at more than one site to become active, then clustering of the membrane-bound enzyme molecules can enhance the response. This is a rebinding effect: a cytoplasmic molecule that has been modified at it its first site, may rapidly rebind a signaling molecule in the cluster before it diffuses into the bulk where it is most likely deactivated by the deactivating enzyme. In a multisite protein modification network, clustering thus allows for rapid rebindings, which may overcome the detrimental effect of the reduced target size. Crowding will, because of subdiffusion, promote rebinding

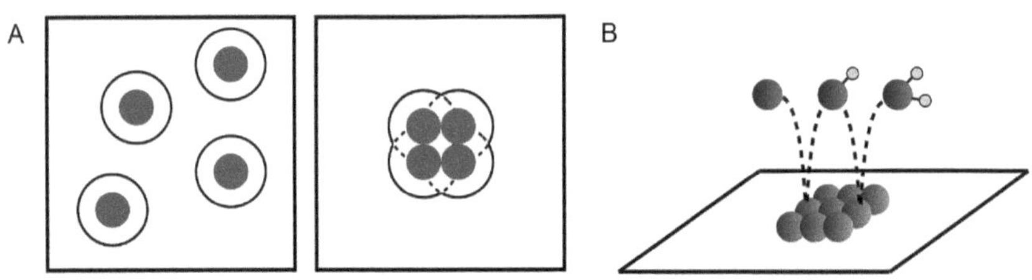

Figure 12.4 (A) Clustering of enzyme molecules reduces their effective target size for substrate molecules in the bulk, which decreases substrate activation. (B) On the other hand, clustering also enhances rapid substrate rebinding, which increases substrate activation if multisite modification is required. Crowding further promotes rebinding, and thus may allow the latter effect to overcome the former. (For color version of this figure, the reader is referred to the online version of this chapter.)

(Lomholt et al., 2007a,b). These observations suggest that crowding can enhance the output of multisite protein modification networks.

6. CROWDING CAN QUALITATIVELY CHANGE THE RESPONSE OF BIOCHEMICAL NETWORKS

Cellular systems that exhibit macroscopic concentration gradients, such as those occurring during embryonic development or macroscopic spatiotemporal oscillations as in the Min system of *Escherichia coli*, are typically treated as reaction–diffusion systems. In contrast, cellular systems that appear to be homogeneous at the cellular scale—that is, they do not exhibit concentration gradients across the cell—are typically modeled as well-stirred chemical reactors, with the Gillespie algorithm being the most powerful numerical technique to simulate these models (Gillespie, 1977). In these well-stirred descriptions, it is assumed that at each instant in time the molecules are randomly distributed in space. This assumption seems reasonable when there are no macroscopic concentration gradients at the cellular scale. While it is commonly appreciated that many cellular systems exhibit spatial structures such as scaffold proteins that tend to localize near the membrane, in many signaling pathways and gene regulatory networks the components do, in fact, have a uniform concentration profile on average. It then only seems natural to simulate these systems as well-stirred chemical reactors.

However, this approach ignores the fact that even in a system that does not exhibit macroscopic concentration gradients, spatiotemporal correlations between the molecules may develop. These correlations mean that while each component individually may on average be uniformly distributed in space, some molecules tend to be close to other molecules, perhaps only transiently. Recent experimental and computational studies have shown that such spatiotemporal correlations at the molecular scale can drastically change the macroscopic behavior of the system at the macroscopic cellular scale (Abel et al., 2012; Aoki et al., 2011, 2013; Hellmann et al., 2012; Takahashi et al., 2010). This phenomenon can arise in multisite protein modification networks.

Multisite protein modification networks are omnipresent in living cells. Well-known examples are the Kai system (Zwicker et al., 2010), the cyclin-dependent kinase inhibitor Sic1 (Nash et al., 2001), the nuclear factor of activated T cells system (Crabtree and Olson, 2002), and the CaM kinase II system (Miller et al., 2005). Arguably the best-known example is the MAPK pathway. MAPK pathways are involved in cell differentiation, cell proliferation, and apoptosis (Chang and Karin, 2001), and they exhibit very rich dynamics. It has been predicted mathematically and shown experimentally that they can generate an ultrasensitive response (Ferrell, 1996; Ferrell and Bhatt, 1997; Huang and Ferrell, 1996) and exhibit bistability via positive feedback (Ferrell and Machleder, 1998). It has also been predicted that they can generate oscillations (Kholodenko, 2000; Wang et al., 2006), amplify weak but attenuate strong signals (Locasale et al., 2007), and give rise to bistability due to enzyme sequestration (Elf and Ehrenberg, 2004; Markevich, 2004).

A typical MAPK pathway consists of three layers where in each layer a kinase activates the kinase of the next layer. Importantly, full activation of the kinase requires that it becomes doubly phosphorylated. This is regulated via a dual phosphorylation cycle in which the upstream kinase and a phosphatase control the modification state of the two sites of the kinase (the substrate) in an antagonistic fashion. It had long been believed that the response of the system strongly depends on whether the enzymes act in a processive or in a distributive manner (Ferrell, 1996; Ferrell and Bhatt, 1997; Huang and Ferrell, 1996). In a distributive mechanism, the enzyme has to release the substrate after it has modified the first site, before it can rebind and modify the second site. In contrast, in a processive mechanism, the enzyme remains bound to the substrate in between the modification of the two sites. Whereas a processive mechanism requires only a single enzyme–substrate encounter for the modification of both sites, a distributive mechanism requires at least two enzyme–substrate encounters. This may seem a small difference, but it can have marked implications. While a processive scheme gives rise to a hyperbolic response, a distributive scheme may give rise to a sigmoidal, ultrasensitive response (Ferrell, 1996; Ferrell and Bhatt, 1997; Huang and Ferrell, 1996). Moreover, a distributive scheme can, in contrast to a processive scheme, give rise to bistability, meaning that the distribution of phosphoforms becomes bimodal (Markevich, 2004). However, the studies that predicted all these behaviors assumed that the system can be described as a well-stirred chemical reactor. They thus ignored spatiotemporal correlations, which, as it turns out, can drastically change the behavior of the system.

To assess the importance of spatiotemporal correlations, Takahashi et al. (2010) performed a computational study of one layer of a MAPK cascade. In their model, the enzymes operate according to a distributive scheme. They found, however, that spatiotemporal correlations can reduce the response of the system to that of a system in which the enzymes act according to a

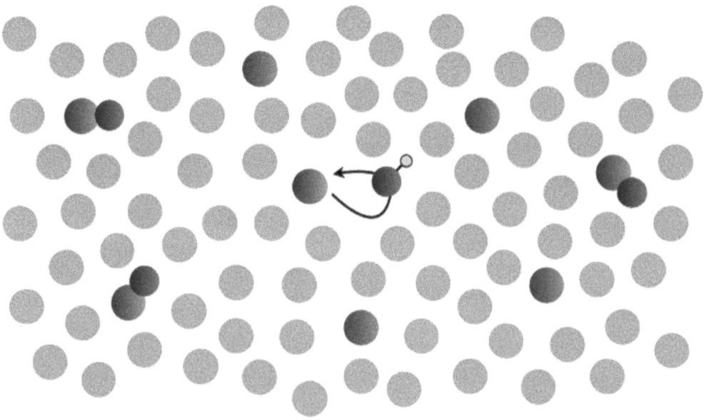

Figure 12.5 A substrate molecule (blue) has just been phosphorylated by a kinase (green) in a system where most phosphatases (red) are free and most kinases are sequestered by other substrate molecules. In a well-stirred model, the substrate molecule is more likely to bind a phosphatase because there are more free phosphatases than free kinases. In contrast, in a spatially resolved model, the substrate molecule is more likely to rapidly rebind the kinase because it remains nearby. These rebindings can lead to the loss of bistability, that is, the loss of bimodality in the distribution of substrate phosphoforms. Crowders (gray) promote rebinding and, as such, they can qualitatively change the macroscopic response of the system. (For interpretation of the references to color in this figure legend, the reader is referred to the online version of this chapter.)

processive scheme: ultrasensitivity and bistability may be lost in regimes where the well-stirred model predicts that they should be present. The origin of this effect is rapid enzyme–substrate rebinding (Fig. 12.5).

After a kinase has phosphorylated a substrate molecule at its first site, the two molecules will dissociate. The kinase then has to exchange ADP for ATP before it can bind the substrate molecule again to phosphorylate the second site. During this refractory time, the kinase and the substrate molecule will diffuse away from each other. When the refractory time is long and/or the diffusion constant is high, the kinase and the substrate molecule will have diffused away from each other into the bulk by the time the kinase becomes active again. In this case, the second site of the substrate molecule will most likely be phosphorylated by another kinase molecule. This is the scenario that is captured by well-stirred chemical models. However, when the refractory time is short and/or the diffusion constant is low, then the kinase and the substrate molecule will probably still be in close physical proximity to each other when the kinase becomes active again (Fig. 12.5). The kinase and the substrate molecule can then rebind. Importantly, this means that both sites are phosphorylated by the same kinase molecule. In this case, the response of the system becomes that of a processive scheme and ultrasensitivity and bistability are lost. The distinction between a processive scheme and a distributive one is thus not whether the enzyme and the substrate molecule remain physically connected to each other in between the modification of the two sites, but rather whether the two sites are phosphorylated by the same enzyme molecule—the enzyme and the substrate molecule may diffuse away from each other and back in between the modification steps.

The loss of bistability can be understood at a mechanistic level. In a well-stirred system, bistability arises due to enzyme sequestration. Imagine, for example, that the system is in the state in which most substrate molecules are unphosphorylated; in this state, the kinase molecules are mostly bound to substrate molecules, in the process of phosphorylating them, while the phosphatase molecules are freely diffusing (Fig. 12.5). The question then is what will happen to a substrate molecule that has just been phosphorylated by a kinase at its first site: will it bind a kinase that will phosphorylate it further or a phosphatase that will dephosphorylate it, driving it back to its original state? In the well-stirred model, the spatial distribution of the molecules is not important, and one only needs to consider the number of kinase and phosphatase molecules that are available for binding. Clearly, in the well-stirred model, the singly phosphorylated substrate molecule will most likely bind a phosphatase, because the phosphatase molecules are free while the kinase molecules are sequestered by the other substrate molecules. In the spatially resolved model, however, it may be more likely that the singly phosphorylated molecule binds a kinase molecule—indeed the kinase molecule that just phosphorylated it and is still nearby—even though there are many more phosphatase molecules that it could bind to.

These results show that spatiotemporal correlations due to rapid rebindings can drastically change the macroscopic behavior of a multisite protein modification network. Crowding can enhance rebindings (Lomholt et al., 2007a,b), which means that crowding can change the response of these multisite protein modification networks (Aoki et al., 2011, 2013; Hellmann et al., 2012; Takahashi et al., 2010). Recently, this prediction was confirmed experimentally by Aoki et al. (2011, 2013). They studied the Raf–Mek–Erk MAPK pathway in HeLa cells and found the response to be processive. To elucidate the effect of crowding, they also studied the same pathway in a test tube, both in the presence and in the absence of the crowding agent polyethylene glycol (PEG). In the absence of PEG, the response was observed to be distributive. However, in the presence of PEG, the response became processive, precisely as predicted by Takahashi et al. (2010). These beautiful experiments unambiguously demonstrate that crowding can have a dramatic effect on the response of biochemical networks.

7. DIFFUSION CAN AFFECT METABOLIC FLUX

Crowding is expected to not only affect cell signaling, but also the metabolic flux. It has long been thought that the spatial colocalization of metabolic enzymes might increase the metabolic flux, minimize the pool of unwanted intermediate products, and reduce transient timescales (Gaertner, 1978; Heinrich et al., 1991). New experimental techniques have now made it possible to also study these effects quantitatively by experiment. Advances in single-molecule techniques enable the investigation (Xie, 2001) and manipulation (Gumpp et al., 2009) of the dynamics of individual enzyme molecules. In parallel, molecules can now be positioned with nanometer precision in artificial systems, along one-dimensional channels, on two-dimensional surfaces (Kufer et al., 2008), and with DNA origami even in three dimensions (Fu et al., 2012; Müller and Niemeyer, 2008). These experimental developments now allow the detailed characterization of how metabolic flux depends on the spatial distribution of the enzymes and how this is affected by crowding.

Recently, Buchner et al. (2013) began to address this question using a simple theoretical one-dimensional model of a metabolic network. In this model, a metabolite enters the system at one end; the metabolite then diffuses through the system until it either leaves the other end or is converted by an enzyme into the desired product. The simplicity of the model permitted a detailed analysis, which revealed that the effect of enzyme clustering can be highly nontrivial. In particular, they found that the optimal arrangement of the enzyme molecules depends on a parameter that characterizes the probability that the enzyme reacts with the metabolite versus the probability that the metabolite is lost by diffusion. When this parameter is high, meaning that the catalytic rate of the enzyme is high, then product formation is maximized by uniformly distributing the enzyme molecules over the system. However, when this parameter is low, then it is advantageous to cluster the enzyme molecules near the origin where the metabolite enters the system. The authors explain these observations via the concept of enzyme exposure, which quantifies how much the metabolite molecules are exposed to the enzyme molecules (Fig. 12.6).

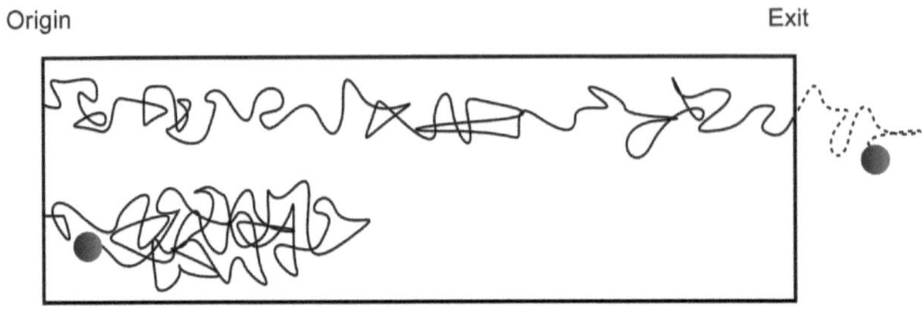

Figure 12.6 Metabolite molecules (blue) enter the left end of a one-dimensional channel and diffuse until either exiting the right end or being converted by an enzyme into a product. Most metabolite molecules reach the exit (top trajectory). Therefore when conversion is fast, product formation is optimized by a uniform enzyme distribution. When conversion is slow, however, only those metabolites that happen to stay long in the system have any chance of being converted. The molecules that stay long in the system are those that happen to spend much time near the origin (bottom trajectory). Hence, when conversion is slow, it becomes optimal to put the enzymes near the origin. Diffusive residence times in this and other geometries are significantly enhanced by crowding. (For interpretation of the references to color in this figure legend, the reader is referred to the online version of this chapter.)

Most molecules only stay a relatively short time in the system, because they quickly leave the system at the other end; this means that most metabolite molecules cover the entire space of the system. Hence, when the catalytic rate is high, and metabolites can be quickly converted into product, it becomes beneficial to spread the enzyme molecules over the system. In contrast, when the catalytic rate is low, metabolite and enzyme interact a large number of times before a reaction takes place. Metabolite molecules that quickly leave the system at the other end will thus not be converted; only those metabolite molecules that happen to stay in the system for a long time have a significant chance to react with the enzyme. The metabolite molecules that happen to stay long in the system are those that by chance remain close to the origin. It then becomes beneficial to cluster the enzymes close to the origin. The authors also found that for intermediate values of the control parameter—the catalytic rate over the diffusion speed—the optimal enzyme profile is an arrangement in which a fraction of the enzyme molecules is clustered near the origin, while the other fraction is approximately uniformly distributed over the system.

Related work shows that similar effects also occur in systems with different geometries, as well as higher dimensions (Buchner et al., 2013). What is of particular interest from the perspective of crowding is that recurrence or residence time is a key concept—this was indeed precisely the reason why the clustered configuration is optimal when the catalytic rate is low. Crowding can enhance the residence time, which means that crowding is expected to also affect how the flux in metabolic networks depends on the spatial distribution of the enzymes.

8. HOW TO MODEL BIOCHEMICAL NETWORKS IN THE PRESENCE OF CROWDING

The above results show that diffusion and crowding can significantly change the behavior of biochemical networks. This suggests that biochemical networks cannot be treated as well-stirred chemical reactors and that spatially resolved models are necessary. However, while in recent years efficient numerical techniques have been developed to simulate spatially resolved models (Elf and Ehrenberg, 2004; Erban and Chapman, 2009; Hattne et al., 2005; Takahashi et al., 2010; van Zon and ten Wolde, 2005a,b), simulating these models remains computationally challenging. The question arises whether the effects of diffusion cannot be captured by renormalizing the association and dissociation rates and then use these effective rates in a well-stirred model, which could then be simulated using the powerful Gillespie algorithm (Gillespie, 1977).

To answer this question we need to inspect the collective dynamics of the network on short and long timescales and compare this to the dynamics of the diffusive motion of the network components on these timescales. With regard to diffusion, recent theoretical (Kim and Yethiraj, 2009) and experimental (Bancaud et al., 2012; Elf et al., 2007) work shows that while the diffusion constant is decreased by the presence of the crowders, the molecules still move by normal diffusion on long length and timescales. Elf et al. (2007) showed that in *E. coli* the Lac repressor moves by normal diffusion on timescales longer than 10 ms and Bancaud et al. (2012) showed that also in eukaryotic nuclei the proteins move by normal diffusion on timescales longer than 20 ms. At these long length and timescales, the molecules have lost memory of where they came from, and to a good approximation they have distributed themselves randomly over the system, just as in a well-stirred chemical reactor. Consequently, we expect that the distribution of association times is exponential at long times even in the crowded environment of the cell (Takahashi et al., 2010; van Zon et al., 2006). On short length and timescales, however, crowding will enhance the probability of rebinding (Bénichou et al., 2010; Lomholt et al., 2007a; Meyer et al., 2012). Such rapid rebindings lead to an algebraic distribution of association times on short timescales

(Takahashi et al., 2010; van Zon et al., 2006); moreover, the probability of rebinding does not depend on the concentration. In contrast, in a well-stirred model, all association reactions depend on the concentration and are exponentially distributed on all timescales. The question then becomes whether the algebraic diffusion dynamics on these short timescales, originating from rebindings, can qualitatively change the behavior of the network.

As we have seen above, rebindings can qualitatively change the macroscopic behavior of the system in multisite protein modification networks (Abel et al., 2012; Hellmann et al., 2012; Takahashi et al., 2010). Here, a single rebinding event can cause irreversible modification of a species, which then takes the system from one state to another, effectively putting the system onto a different dynamical pathway (Mugler et al., 2013). Also for the metabolic network problem discussed above, the recurrence or return of trajectories to the origin (the source) which underlies the benefit of clustering the enzyme molecules, cannot be captured in a well-stirred model (Buchner et al., 2013). These systems have to be treated as reaction–diffusion systems, even though they may not exhibit concentration gradients at the cellular scale.

However, in many biochemical networks, the state of the system is the same before and after rebinding. Moreover, the timescale of rebinding is typically much faster than the timescale that governs the system dynamics. If both conditions are met, it is possible to treat a group of rebinding events as a single effective event and to renormalize the association and dissociation rates by the number of rebindings (Mugler et al., 2013; van Zon et al., 2006). An example is the binding of a TF to DNA: the TF–DNA complex is identical before and after a rebinding event and the rebinding time is much shorter than the protein lifetime, making it possible to integrate out the rebindings. Indeed, a spatially resolved model of gene expression that explicitly describes the diffusive motion of a TF in space gave essentially identical results to a well-stirred model with renormalized TF–DNA association and dissociation rates (van Zon et al., 2006). Hence, when rebindings do not put the system onto a different dynamical pathway, then the effect of diffusion and crowding may be captured in the association and dissociation rates in a zero-dimensional model, which can be simulated much more efficiently than a spatially resolved model.

9. EFFECT OF CROWDING: DIFFUSION OR ENTROPY?

The previous sections discussed how crowding could influence biochemical networks by introducing spatiotemporal correlations in the dynamics or by simply changing the magnitude of the diffusion constant. However, it is well-known that crowding also shifts the equilibrium of association–dissociation reactions toward the associated state. This is largely because the associated state occupies less volume than the dissociated molecules. In the associated state, more volume is therefore available for the surrounding crowder molecules, thereby increasing the total entropy of the system (Asakura and Oosawa, 1954). Importantly, most if not all, biochemical networks involve bimolecular reactions. This means that the shift in equilibria can have a pronounced effect on the behavior of biochemical networks. The question is whether this effect outweighs the effect of diffusion.

To address this question, Morelli et al. (2011) developed a simple scheme to model the effects of crowding on biochemical networks. It is based on the observation discussed above that even in the presence of crowding, the distribution of association and dissociation is expected to remain exponential on the relevant timescales of the network. This observation makes it possible to use a well-stirred model and capture the effect of crowding on both diffusion and the equilibrium shift by appropriately scaling the association and dissociation rates. They used this approach in combination with kinetic Monte Carlo simulations to analyze the effect of crowding on a constitutively expressed gene, a repressed gene, and a model for the bacteriophage λ switch. They found that the effects of crowding are mainly caused by the shift of association–dissociation equilibria rather than the slowing down of protein diffusion. While slower diffusion of the regulatory proteins increases the noise in gene expression (van Zon et al., 2006), it does not significantly change the mean expression level, in contrast to the shift in the equilibrium of the binding of the RNA polymerase (RNAP) and the regulatory proteins to the promoter. The results on the constitutively expressed gene and the self-repressed gene are therefore not very surprising. The results on the phage switch λ are more surprising, because previous work has shown that the stability of the switch is very sensitive to the noise in gene expression (Morelli et al., 2008; Warren and ten Wolde, 2005), which is increased by the slowdown in diffusion. However, it turns out that the behavior of the switch is much more affected by the shift in the equilibria of the binding of the RNAP and the regulatory proteins to the promoter. The study also revealed that the result of these equilibrium shifts can be highly counterintuitive, because of their competing effects on the flipping behavior of the switch. For example, the authors found that the switching rate depends nonmonotonically on the concentration of crowder. This observation is potentially much more generic: crowding will affect the equilibria of all bimolecular reactions in any biochemical network, but how this plays out in the collective dynamics of the network is difficult to predict by intuition.

10. PERSPECTIVES

While the effect of crowding on the diffusion of individual molecules has been characterized extensively in the past decades, its importance for the dynamics of biochemical networks remains a wide open question. We hope that this review shows that crowding is expected to significantly alter the dynamics of signal transduction pathways, gene regulation networks, and metabolic networks. As such, we hope that this review stimulates further research in this direction. A particularly promising approach is to reconstitute the biochemical network of interest in the test tube and to elucidate the importance of crowding by systematically varying the concentration of a crowding agent, such as PEG (Aoki et al., 2011). Such *in vitro* studies have provided tremendous insight in the diffusion dynamics of individual molecules, and we expect that they will be similarly fruitful in unraveling the role of crowding in biochemical networks. Finally, in this review we have mostly focused on cellular systems that are homogeneous at the cellular scale. Yet, in many signaling pathways cellular concentration gradients exist because proteins are activated at one position but deactivated at another (Kholodenko, 2006; van Albada and ten Wolde, 2007); the range of these gradients is set by the diffusion constant over the deactivation rate, which means that crowding will also change the dynamics of these networks. Moreover, systems that exhibit spatiotemporal oscillations, such as the Min system (de Boer et al., 1989), are also sensitive to the diffusion dynamics of the components and hence to crowding (Weiss, 2003). Indeed, we expect that many interesting crowding-induced phenomena will be uncovered in the coming years.

ACKNOWLEDGMENTS

This work is part of the research programme of the Foundation for Fundamental Research on Matter (FOM), which is part of the Netherlands Organisation for Scientific Research (NWO).

REFERENCES

Abel, S.M., Roose, J.P., Groves, J.T., Weiss, A., Chakraborty, A.K., 2012. The membrane environment can promote or suppress bistability in cell signaling networks. J. Phys. Chem. B 116, 3630–3640.

Aoki, K., Yamada, M., Kunida, K., Yasuda, S., Matsuda, M., 2011. Processive phosphorylation of ERK MAP kinase in mammalian cells. Proc. Natl. Acad. Sci. U.S.A. 108, 12675–12680.

Aoki, K., Takahashi, K., Kaizu, K., Matsuda, M., 2013. A quantitative model of ERK MAP kinase phosphorylation in crowded media. Sci. Rep. 3, 1541.

Asakura, S., Oosawa, F., 1954. On interaction between two bodies immersed in a solution of macromolecules. J. Chem. Phys. 22, 1255.
Bancaud, A., Huet, S., Daigle, N., Mozziconacci, J., Beaudouin, J., Ellenberg, J., 2009. Molecular crowding affects diffusion and binding of nuclear proteins in heterochromatin and reveals the fractal organization of chromatin. EMBO J. 28, 3785–3798.
Bancaud, A., Lavelle, C., Huet, S., Ellenberg, J., 2012. A fractal model for nuclear organization: current evidence and biological implications. Nucleic Acids Res. 40, 8783–8792.
Banks, D.S., Fradin, C., 2005. Anomalous diffusion of proteins due to molecular crowding. Biophys. J. 89, 2960–2971.
Bénichou, O., Chevalier, C., Klafter, J., Meyer, B., Voituriez, R., 2010. Geometry-controlled kinetics. Nat. Chem. 2, 472–477.
Bialek, W., Setayeshgar, S., 2005. Physical limits to biochemical signaling. Proc. Natl. Acad. Sci. U.S.A. 102, 10040–10045.
Buchner, A., Tostevin, F., Gerland, U., 2013. Clustering and optimal arrangement of enzymes in reaction-diffusion systems. Phys. Rev. Lett. 110, 208104.
Chang, L., Karin, M., 2001. Mammalian MAP kinase signalling cascades. Nature 410, 37–40.
Chubb, J.R., Trcek, T., Shenoy, S.M., Singer, R.H., 2006. Transcriptional pulsing of a developmental gene. Curr. Biol. 16, 1018–1025.
Crabtree, G.R., Olson, E.N., 2002. NFAT signaling: choreographing the social lives of cells. Cell 109 (Suppl.), S67–S79.
de Boer, P.A., Crossley, R.E., Rothfield, L.I., 1989. A division inhibitor and a topological specificity factor coded for by the minicell locus determine proper placement of the division septum in E. coli. Cell 56, 641–649.
Elf, J., Ehrenberg, M., 2004. Spontaneous separation of bi-stable biochemical systems into spatial domains of opposite phases. Syst. Biol. (Stevenage) 1, 230–236.
Elf, J., Li, G.W., Xie, X.S., 2007. Probing transcription factor dynamics at the single-molecule level in a living cell. Science 316, 1191.
Ellis, R.J., 2001. Macromolecular crowding: an important but neglected aspect of the intracellular environment. Curr. Opin. Struct. Biol. 11, 114–119.
Elowitz, M.B., 2002. Stochastic gene expression in a single cell. Science 297, 1183–1186.
Erban, R., Chapman, S.J., 2009. Stochastic modelling of reaction–diffusion processes: algorithms for bimolecular reactions. Phys. Biol. 6, 046001.
Erdmann, T., Howard, M., ten Wolde, P.R., 2009. Role of spatial averaging in the precision of gene expression patterns. Phys. Rev. Lett. 103, 258101.
Ferrell, J.E., 1996. Tripping the switch fantastic: how a protein kinase cascade can convert graded inputs into switch-like outputs. Trends Biochem. Sci. 21, 460–466.
Ferrell, J.E., Bhatt, R.R., 1997. Mechanistic studies of the dual phosphorylation of mitogen-activated protein kinase. J. Biol. Chem. 272, 19008–19016.
Ferrell, J.E., Machleder, E.M., 1998. The biochemical basis of an all-or-none cell fate switch in Xenopus oocytes. Science 280, 895–898.
Fu, J., Liu, M., Liu, Y., Woodbury, N.W., Yan, H., 2012. Interenzyme substrate diffusion for an enzyme cascade organized on spatially addressable DNA nanostructures. J. Am. Chem. Soc. 134, 5516–5519.
Gaertner, F.H., 1978. Unique catalytic properties of enzyme clusters. Trends Biochem. Sci. 3, 63–65.
Gillespie, D.T., 1977. Exact stochastic simulation of coupled chemical reactions. J. Phys. Chem. 81, 2340–2361.
Golding, I., Cox, E., 2006. Physical nature of bacterial cytoplasm. Phys. Rev. Lett. 96, 098102.
Golding, I., Paulsson, J., Zawilski, S.M., Cox, E.C., 2005. Real-time kinetics of gene activity in individual bacteria. Cell 123, 1025–1036.

Gregor, T., Tank, D.W., Wieschaus, E.F., Bialek, W., 2007. Probing the limits to positional information. Cell 130, 153–164.
Gumpp, H., Puchner, E.M., Zimmermann, J.L., Gerland, U., Gaub, H.E., Blank, K., 2009. Triggering enzymatic activity with force. Triggering enzymatic activity with force. Nano Lett. 9, 3290–3295.
Gurry, T., Kahramanogullari, O., Endres, R.G., 2009. Biophysical mechanism for Ras-nanocluster formation and signaling in plasma membrane. PLoS One 4, e6148.
Hattne, J., Fange, D., Elf, J., 2005. Stochastic reaction-diffusion simulation with MesoRD. Bioinformatics 21, 2923–2924.
He, F., Wen, Y., Deng, J., Lin, X., Lu, L.J., Jiao, R., Ma, J., 2008. Probing intrinsic properties of a robust morphogen gradient in Drosophila. Dev. Cell 15, 558–567.
Heinrich, R., Schuster, S., Holzhütter, H.G., 1991. Mathematical analysis of enzymic reaction systems using optimization principles. Eur. J. Biochem. 201, 1–21.
Hellmann, M., Heermann, D.W., Weiss, M., 2012. Enhancing phosphorylation cascades by anomalous diffusion. EPL (Europhysics Letters) 97, 58004.
Houchmandzadeh, B., Wieschaus, E., Leibler, S., 2002. Establishment of developmental precision and proportions in the early Drosophila embryo. Nature 415, 798–802.
Huang, C., Ferrell, J.E., 1996. Ultrasensitivity in the mitogen-activated protein kinase cascade. Proc. Natl. Acad. Sci. U.S.A. 93, 10078–10083.
Kalay, Z., Fujiwara, T.K., Kusumi, A., 2012. Confining domains lead to reaction bursts: reaction kinetics in the plasma membrane. PLoS One 7, e32948.
Kholodenko, B.N., 2000. Negative feedback and ultrasensitivity can bring about oscillations in the mitogen-activated protein kinase cascades. Eur. J. Biochem. 267, 1583–1588.
Kholodenko, B.N., 2006. Cell-signalling dynamics in time and space. Nat. Rev. Mol. Cell Biol. 7, 165–176.
Kim, J.S., Yethiraj, A., 2009. Effect of macromolecular crowding on reaction rates: a computational and theoretical study. Biophys. J. 96, 1333–1340.
Kufer, S.K., Puchner, E.M., Gumpp, H., Liedl, T., Gaub, H.E., 2008. Single-molecule cut-and-paste surface assembly. Science 319, 594–596.
Locasale, J.W., Shaw, A.S., Chakraborty, A.K., 2007. Scaffold proteins confer diverse regulatory properties to protein kinase cascades. Proc. Natl. Acad. Sci. U.S.A. 104, 13307–13312.
Lomholt, M.A., Zaid, I.M., Metzler, R., 2007a. Subdiffusion and weak ergodicity breaking in the presence of a reactive boundary. Phys. Rev. Lett. 98, 200603.
Lomholt, M.A., Zaid, I.M., Metzler, R., 2007b. How subdiffusion changes the kinetics of binding to a surface. Biophys. J. 97, 710–721.
Markevich, N.I., 2004. Signaling switches and bistability arising from multisite phosphorylation in protein kinase cascades. J. Cell Biol. 164, 353–359.
Meyer, B., Bénichou, O., Kafri, Y., Voituriez, R., 2012. Geometry-induced bursting dynamics in gene expression. Biophys. J. 102, 2186–2191.
Miller, P., Zhabotinsky, A.M., Lisman, J.E., Wang, X.-J., 2005. The stability of a stochastic CaMKII switch: dependence on the number of enzyme molecules and protein turnover. PLoS Biol. 3, e107.
Morelli, M.J., Tanase-Nicola, S., Allen, R.J., ten Wolde, P.R., 2008. Reaction coordinates for the flipping of genetic switches. Biophys. J. 94, 3413–3423.
Morelli, M.J., Allen, R.J., ten Wolde, P.R., 2011. Effects of macromolecular crowding on genetic networks. Biophys. J. 101, 2882–2891.
Mugler, A., Bailey, A.G., Takahashi, K., ten Wolde, P.R., 2012. Membrane clustering and the role of rebinding in biochemical signaling. Biophys. J. 102, 1069–1078.
Mugler, A., ten Wolde, P.R., 2013. The macroscopic effects of microscopic heterogeneity in cell signaling. In: Rice, S.A., Dinner, A.R. (Eds.). Adv. Chem. Phys. 153, 373–396.
Mugler, A., Tostevin, F., ten Wolde, P.R., 2013. Spatial partitioning improves the reliability of biochemical signaling. Proc. Natl. Acad. Sci. U.S.A. 110, 5927–5932.

Müller, J., Niemeyer, C.M., 2008. DNA-directed assembly of artificial multienzyme complexes. Biochem. Biophys. Res. Commun. 377, 62–67.
Muramatsu, N., Minton, A.P., 1988. Tracer diffusion of globular proteins in concentrated protein solutions. Proc. Natl. Acad. Sci. U.S.A. 85, 2984–2988.
Nash, P., Tang, X., Orlicky, S., Chen, Q., Gertler, F.B., Mendenhall, M.D., Sicheri, F., Pawson, T., Tyers, M., 2001. Multisite phosphorylation of a CDK inhibitor sets a threshold for the onset of DNA replication. Nature 414, 514–521.
Okabe-Oho, Y., Murakami, H., Oho, S., 2009. Stable, precise, and reproducible patterning of bicoid and hunchback molecules in the early Drosophila embryo. PLoS Comput. Biol. 5, e1000486.
Ozbudak, E.M., Thattai, M., Kurtser, I., Grossman, A.D., 2002. Regulation of noise in the expression of a single gene. Nat. Genet. 31, 69–73.
Paulsson, J., 2004. Summing up the noise in gene networks. Nat. Cell Biol. 427, 415–418.
Plowman, S.J., Muncke, C., Parton, R.G., Hancock, J.F., 2005. H-ras, K-ras, and inner plasma membrane raft proteins operate in nanoclusters with differential dependence on the actin cytoskeleton. Proc. Natl. Acad. Sci. U.S.A. 102, 15500–155005.
Prior, I.A., 2003. Direct visualization of Ras proteins in spatially distinct cell surface microdomains. J. Cell Biol. 160, 165–170.
Raj, A., Peskin, C.S., Tranchina, D., Vargas, D.Y., Tyagi, S., 2006. Stochastic mRNA synthesis in mammalian cells. PLoS Biol. 4, e309.
So, L.H., Ghosh, A., Zong, C., Sepúlveda, L.A., Segev, R., Golding, I., 2011. General properties of transcriptional time series in Escherichia coli. Nat. Genet. 43, 554–560.
Suzuki, K.G., Fujiwara, T.K., Edidin, M., 2007a. Dynamic recruitment of phospholipase Cγ at transiently immobilized GPI-anchored receptor clusters induces IP3–Ca2+ signaling: single-molecule tracking study 2. J. Cell Biol. 177, 731–742.
Suzuki, K.G., Fujiwara, T.K., Sanematsu, F., Iino, R., Edidin, M., Kusumi, A., 2007b. GPI-anchored receptor clusters transiently recruit Lyn and G for temporary cluster immobilization and Lyn activation: single-molecule tracking study 1. J. Cell Biol. 177, 717–730.
Takahashi, K., Tanase-Nicola, S., ten Wolde, P.R., 2010. Spatio-temporal correlations can drastically change the response of a MAPK pathway. Proc. Natl. Acad. Sci. U.S.A. 107, 2473–2478.
Taniguchi, Y., Choi, P.J., Li, G.-W., Chen, H., Babu, M., Hearn, J., Emili, A., Xie, X.S., 2010. Quantifying E. coli proteome and transcriptome with single-molecule sensitivity in single cells. Science 329, 533–538.
Tian, T., Harding, A., Inder, K., Plowman, S., Parton, R.G., Hancock, J.F., 2007. Plasma membrane nanoswitches generate high-fidelity Ras signal transduction. Nat. Cell Biol. 9, 905–914.
Tkacik, G., Bialek, W., 2009. Diffusion, dimensionality, and noise in transcriptional regulation. Phys. Rev. E. 79, 051901.
van Albada, S.B., ten Wolde, P.R., 2007. Enzyme localization can drastically affect signal amplification in signal transduction pathways. PLoS Comput. Biol. 3, e195.
van Zon, J.S., ten Wolde, P.R., 2005a. Simulating biochemical networks at the particle level and in time and space: Green's function reaction dynamics. Phys. Rev. Lett. 94, 128103.
van Zon, J.S., ten Wolde, P.R., 2005b. Green's-function reaction dynamics: a particle-based approach for simulating biochemical networks in time and space. J. Chem. Phys. 123, 234910.
van Zon, J.S., Morelli, M.J., ten Wolde, P.R., 2006. Diffusion of transcription factors can drastically enhance the noise in gene expression. Biophys. J. 91, 4350–4367.
Wang, X., Hao, N., Dohlman, H.G., Elston, T.C., 2006. Bistability, stochasticity, and oscillations in the mitogen-activated protein kinase cascade. Biophys. J. 90, 1961–1978.
Warren, P.B., ten Wolde, P.R., 2005. Chemical models of genetic toggle switches. J. Phys. Chem. B 109, 6812–6823.

Weber, M., Buceta, J., 2011. Noise regulation by quorum sensing in low mRNA copy number systems. BMC Syst. Biol. 5, 11.
Weiss, M., 2003. Stabilizing turing patterns with subdiffusion in systems with low particle numbers. Phys. Rev. E Stat. Nonlin. Soft Matter Phys. 68, 036213.
Weiss, M., Elsner, M., Kartberg, F., Nilsson, T., 2004. Anomalous subdiffusion is a measure for cytoplasmic crowding in living cells. Biophys. J. 87, 3518–3524.
Xie, S., 2001. Single-molecule approach to enzymology. Single Mol. 2, 229–236.
Yu, J., 2006. Probing gene expression in live cells, one protein molecule at a time. Science 311, 1600–1603.
Zaid, I.M., Lomholt, M.A., Metzler, R., 2009. How Subdiffusion Changes the Kinetics of Binding to a Surface. Biophysical Journal 97 (3), 710–721.
Zimmerman, S.B., Trach, S.O., 1991. Estimation of macromolecule concentrations and excluded volume effects for the cytoplasm of Escherichia coli. J. Mol. Biol. 222, 599–620.
Zwicker, D., Lubensky, D.K., ten Wolde, P.R., 2010. Robust circadian clocks from coupled protein-modification and transcription–translation cycles. Proc. Natl. Acad. Sci. U.S.A. 107, 22540–22545.

CHAPTER THIRTEEN

Relevance and Limitations of Crowding, Fractal, and Polymer Models to Describe Nuclear Architecture

Sébastien Huet[*,†,‡], Christophe Lavelle[‡,§,¶,||], Hubert Ranchon[‡,,††], Pascal Carrivain[‡,#], Jean-Marc Victor[‡,#], Aurélien Bancaud[‡,**,††,1]**

[*]CNRS, UMR 6290, Institut Génétique et Développement de Rennes, Rennes, France
[†]Université Rennes 1, Université Européenne de Bretagne, Structure fédérative de recherche, Biosit, Faculté de Médecine, Rennes, France
[‡]CNRS GDR 3536, University Pierre et Marie Curie-Paris 6, Paris, France
[§]National Museum of Natural History, Paris, France
[¶]CNRS UMR 7196, Paris, France
[||]INSERM U565, Paris, France
[#]LPTMC UMR 7600, CNRS, Université Pierre et Marie Curie-Paris 6, Paris, France
[**]CNRS, LAAS, Toulouse, France
[††]Université de Toulouse, LAAS, Toulouse, France
[1]Corresponding author: e-mail address: abancaud@laas.fr

Contents

International Review of Cell and Molecular Biology, Volume 307
ISSN 1937-6448
http://dx.doi.org/10.1016/B978-0-12-800046-5.00013-8

Abstract

Chromosome architecture plays an essential role for all nuclear functions, and its physical description has attracted considerable interest over the last few years among the biophysics community. These researches at the frontiers of physics and biology have been stimulated by the demand for quantitative analysis of molecular biology experiments, which provide comprehensive data on chromosome folding, or of live cell imaging experiments that enable researchers to visualize selected chromosome loci in living or fixed cells. In this review our goal is to survey several nonmutually exclusive models that have emerged to describe the folding of DNA in the nucleus, the dynamics of proteins in the nucleoplasm, or the movements of chromosome loci. We focus on three classes of models, namely molecular crowding, fractal, and polymer models, draw comparisons, and discuss their merits and limitations in the context of chromosome structure and dynamics, or nuclear protein navigation in the nucleoplasm. Finally, we identify future challenges in the roadmap to a unified model of the nuclear environment.

1. INTRODUCTION

Chromosome architecture plays an essential role for transcription, replication, recombination, and repair, but the molecular mechanisms of these transactions as well as the properties of chromosomes remain major unsolved problems in biology. The best-studied chromosome component is the nucleosome, which is its constitutive element composed of ~150 bp of deoxyribonucleic acid (DNA) wrapped around an octamer of histone proteins, as inferred at atomic precision by X-ray diffraction (Harp et al., 2000; Luger et al., 1997). Beyond the nucleosome level, chromatin structure is far less characterized. Condensed nucleosome arrays that fold into a "30 nm fiber" have been described using electron microscopy of chromatin spread at moderate ionic strength (van Holde, 1989), yet their exact structure still remains strongly debated (van Holde and Zlatanova, 2007), and recent cryo-electron microscopy imaging performed *in situ* does not even support the existence of the 30-nm fiber (Maeshima et al., 2010; Woodcock and Ghosh, 2010). For genomic distances higher than ~100 kb, long-distance interactions along the chromatin fiber were detected long ago using optical imaging (Mateos-Langerak et al., 2009; Yokota et al., 1995), and more recently by the molecular biology technique called Hi-C, which enables detection of distant chromosome contacts in *cis* and in *trans* at the genome wide level (Dekker, 2006).

Topological domains, which are stable across different cell types and highly conserved across species, have recently been identified

(Dixon et al., 2012; Nora et al., 2012; Sexton et al., 2012), allowing for cross talk between distant chromatin loci in three dimensions (Göndör and Ohlsson, 2009) and participating in the regulation of expression (Göndör and Ohlsson, 2009; Krivega and Dean, 2012). The largest level of organization is the chromosome territory, referring to the folding of chromosomes in compact and self-segregating entities (Bolzer et al., 2005). Territories seem to be conserved in daughter cells after mitosis (Gerlich et al., 2003; Walter et al., 2003), following a "space memory" mechanism still under debate (Orlova et al., 2012). Some authors have proposed the existence of a compartment complementary to the chromosome territory, as defined by the interchromatin space (Cremer and Cremer, 2010). This structure, which is mainly devoid of chromatin, forms a porous network between and within chromosome territories. However the validity of this model is still controversial, in particular due the observation of significant chromatin intermingling at the periphery of the chromatin territories (Branco and Pombo, 2006). Overall, the architecture of chromosomes is defined by a series of structural elements including DNA, the nucleosome, the chromatin fiber, higher-order chromatin loops, coils or folds, chromosome territories and their complement, and the nucleus itself, which are associated with different spatial scales of about 2, 10, 30, 200, 1000, and 20,000 nm, respectively, that is, extending over four orders of magnitude. Whereas adequate space exists in the nucleus such that DNA compaction rate is not so a concern (Carrivain et al., 2012), functional packaging, which consists in providing the genome with the plasticity in the right place and at the right time, remains an open problem that requires proposition and validation of new models.

Chromosomes were viewed as static entities in the early history of the field, but this description has been considerably revisited over the last decade with emerging light microscopy techniques. For instance, fluorescence recovery after photobleaching (FRAP) or any variant of this technique (Bancaud et al., 2011; Hinde and Cardarelli, 2011) to investigate molecular interactions *in vivo* showed that fluorescence redistribution occurred very rapidly for nearly every chromatin-interacting protein, leading to a dynamic picture for molecular interactions in the chromatin context. Further, the examination of single nucleosome motion by single-particle tracking indicated the local dynamics of these particles in interphase (Hihara et al., 2012). These results are consistent with the spatiotemporal fluctuations of chromosome segments, that is, tandemly repeated bacterial operator sequences inserted in the genome of living cells, which showed random

(Heun et al., 2001; Marshall et al., 1997) and infrequent directional movements (Levi and Gratton, 2008) associated with transcription activation (Chuang et al., 2006). Finally, chromosome reorganization events have been detected during development (Avivi et al., 2004), and they have been correlated to transcription activation (Chambeyron and Bickmore, 2004). The dynamics of chromosomes illustrated in these three examples cover a temporal range of ~5 orders of magnitudes from seconds to days, and models to describe the underlying movements are still severely limited. Thus, the dynamic nature and the complex topology of chromosomes appear as an intractable puzzle.

Over the last few years however the efforts of the community to propose physical descriptions of the nuclear environment have intensified to fulfill the demand for quantitative analysis of molecular biology experiments, which provide comprehensive data on chromosome folding, or of live cell imaging experiments that enable researchers to visualize selected chromosome loci in living or fixed cells. Several nonmutually exclusive models have thus emerged to describe the folding of DNA in the nucleus, the dynamics of proteins in the nucleoplasm, or the movements of chromosome loci. These models rely on different assumptions depending on the spatial precision of the experiment: conventional optics, for instance, gathers information at spatial scales larger than ~300 nm so that structural details of chromatin folding can be omitted. On the contrary, the folding principles of chromosomes, as inferred from molecular biology techniques, which now reach 100 kb resolution, can only be analyzed with structural models based on polymer physics. Our goal is to describe three classes of models of chromosomes, namely molecular crowding, fractal, and polymer models (Fig. 13.1), draw comparisons, and discuss their merits and limitations in the context of chromosome structure and dynamics, or nuclear protein navigation in the nucleoplasm. We first focus on molecular crowding, which has proven to be useful to interpret experimental findings on the dynamics of nuclear proteins and the biogenesis of nuclear organelles. We then explain that this model is insufficient to make precise predictions about the architecture of the chromosomes, and we will go beyond this theory, focusing on fractal and polymer models to describe the organization of chromosomes and the nucleoplasm. Finally, we identify future challenges in the roadmap to a unified model of the nuclear environment, in which the spatial organization of chromosomes and reaction–diffusion dynamics of nuclear proteins are accounted for with minimal structural parameters.

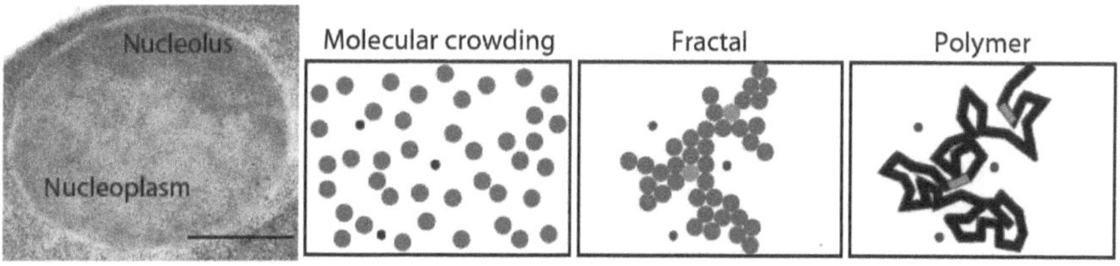

Figure 13.1 Schematic overview of crowding, fractal, and polymer models. The left panel is an electron micrograph of ultra-thin sections of cryo-fixed, cryo-substituted yeast contrasted with uranyl acetate and lead citrate. The rim defines the nuclear periphery, and two compartments can be detected based on their difference in contrast, the nucleoplasm and the nucleolus. The scale bar corresponds to 1 μm. Three main models have been proposed to describe the nuclear environment, as schematized in the cartoons in the right. Molecular crowding assumes a random distribution of co-solutes (gray circles). The fractal model implies that co-solutes are ordered, following a spatial repartition that can be repeatedly detected at any zoom factor. Note that the number of structural elements is equal in the molecular crowding and fractal model cartoons. The polymer model is based on the connectivity of the crowding molecule: the path between the two green elements is essential, whereas the distribution of structuring objects is characteristic of fractals. The orange circles represent tracers diffusing in the nucleoplasm. (For interpretation of the references to color in this figure legend, the reader is referred to the online version of this chapter.)

2. FIRST GLANCE AT METHODS TO INVESTIGATE NUCLEAR ORGANIZATION

Our current knowledge of DNA organization in the nucleus is the result of a century and a half of various experimental breakthroughs to visualize chromatin and analyze its biochemical content. From the first observations of chromosomes (Flemming, 1878) in an organelle (Brown, 1831) to the evidence for chromatin heterogeneous distribution (Baccarini, 1908; Boveri, 1914; Heitz, 1928; Rabl, 1885), and the striking link between gene position and expression (with the discovery of position effect variegation (Muller, 1930), considered for many as the true "birth" of epigenetics), scientists from all kind of backgrounds have struggled to decipher the nature and properties of our genome. Technical progresses in visualization techniques were first reported with electron microscopy of the whole cell (Porter et al., 1945), followed by specific developments for nuclear visualization (Bernhard, 1969). Fluorescence microscopy gained popularity in the 1980s, first in fixed cells (Cremer et al., 2006) and then, in the 1990s, in living cells using fluorescent proteins, which greatly improved our view of the complex nuclear architecture, as well as of the dynamics of molecular interactions. Fluorescence microscopy techniques recently received renewed interest as their spatial resolution increased with super resolution microscopy methods (Hell, 2003). At the same time, biochemical techniques, which rely on the cross-linking of DNA protein interactions as initially proposed in 1978 (Jackson, 1978), can now provide genome wide contact maps of chromosome interactions (Dekker, 2006; Dekker et al., 2002; van Steensel and Dekker, 2010), as well as the genomic distribution of virtually every chromatin-interacting protein (Mardis, 2007), giving access to an additional perspective on DNA folding in the nucleus.

3. MOLECULAR CROWDING

3.1. Definition of molecular crowding

The textbook picture of cells, as popularized by illustrations of David S. Goodsell (Goodsell, 2005), consists of a variety of biomolecules, including DNA, RNA, and proteins, at high concentrations, sometimes assembled into large-scale structures. This representation emphasizes the degree of crowding in the cell, and hints at the fact that biomolecules cannot be considered as isolated objects flowing in an obstacle-free environment. Conversely, conventional molecular biology techniques aim to characterize interaction properties of purified biomolecules, in order to help the interpretation of *in vivo* measurements after an extrapolation from *in vitro* to *in vivo* conditions. Although this extrapolation has met considerable success to characterize molecular reactions, it disregards the impact of crowding, which is expected to introduce a systematic bias through generic mechanisms, including, among others, molecular crowding or phase separation (Ellis, 2001; Hyman and Brangwynne, 2011). For example, a crowded and inhomogeneous environment may induce local onsets in viscosity and/or repulsions of large complexes, modulate stereo-specific interactions, and macromolecular complexes may spontaneously segregate due to phase separation processes (Keating, 2012). Although the impact of these mechanisms remains largely elusive, they are generally thought to have a great impact on molecular interactions in the nucleus (Iborra, 2007). This argument has been indirectly supported by the observation that polynucleosomes extracted from interphase cells self-assemble into compact arrays *in vitro* in the presence of crowding agents (Hancock, 2008), suggesting that crowding is necessary to reproduce *in vivo* conditions.

The concept of macromolecular crowding applies to continuous media in which a fraction of the volume, typically more than 20% (Ellis, 2001), is occupied by background molecules and, thus, is not accessible to other molecules. The cell nucleus definitely meets the concentration criteria because chromatin alone, which is the main crowding agent in the nucleus, occupies 30–50% of the nuclear volume depending on the cell type, as inferred from contrast analysis of 2D or 3D electron micrographs (Lopez-Velazquez et al., 1996; Rouquette et al., 2009). We propose to overview the consequences of molecular crowding *in vitro* in the following section.

3.2. Consequences of molecular crowding *in vitro*

The consequences of molecular crowding have been extensively documented by adding inert co-solutes such as synthetic polymers (dextran, ficoll, polyethylene glycol) or serum proteins (BSA) in test tubes, and by monitoring reaction kinetics as a function of the background molecule concentration. This research has been supported by numerous analytical models (Zimmerman and Minton, 1993) and Brownian dynamics simulations (Schnell and Turner, 2004) which described three main effects associated with molecular crowding (Fig. 13.2).

First, volume exclusion is associated with the inaccessible volume fraction occupied by background species, which is defined by the combination of hard steric repulsion between macromolecules and by soft nonspecific interactions such as electrostatic, hydrophobic or Van der Waal's interactions (Hall and Minton, 2003) that may reinforce or instead partially cancel steric repulsion (Jiao et al., 2010). In addition, volume exclusion is dependent on the size and shape of probe molecules: small molecules diffuse throughout crowded environments in the interstices between co-solutes, whereas large ones are confined in narrow regions (Minton, 2006).

Second, the presence of high concentrations of co-solutes slows down diffusion, because frequent collisions with crowding species impede diffusion-driven motions. The resulting diffusive hindrance, which is defined as the ratio of the diffusion coefficient in a crowded versus constraint-free space, increases exponentially with the concentration of crowding co-solutes, and also increases with the size of the tracer (Muramatsu and Minton, 1988).

Last but not least, volume exclusion alters equilibrium constants and reaction rates of biochemical reactions (Ellis, 2001). For concentrations of co-solutes of ~100 g/L or more (Hancock, 2004), recent studies have shown nonlinear relationships between the concentration of macromolecules and

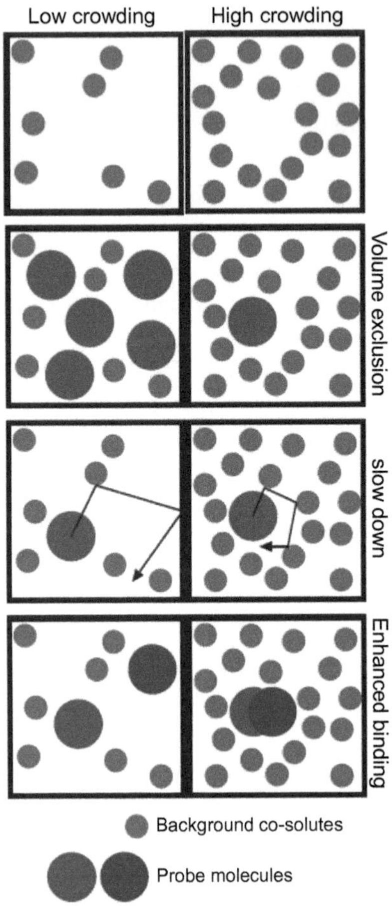

Figure 13.2 Consequences of molecular crowding. The presence of high amounts of background (right panel) species reduces the volume accessible to additional macromolecules, inducing volume exclusion (second row). Background species also act as obstacles impeding the diffusion of the molecules (third row), and they alter the thermodynamics of binding reactions, which tend to be shifted toward the bound state (fourth row). (For color version of this figure, the reader is referred to the online version of this chapter.)

equilibrium constants and reaction rates in different contexts (Zhou et al., 2008). The free energy pathway of a reaction involving two reactants, which form an activated state and then a product, can be described in a crowded environment by summing (i) the free energy in dilute conditions, with (ii) the free energy required to place the system in a crowded medium. The latter term depends on the interaction properties between these components (reactant, activated state, and product) with the background species (Minton, 2006), potentially dramatically affecting the reaction properties and dynamics. In the limit of steric repulsion, the inaccessible volume fractions can be used to determine the constants of the reaction (Minton, 1983), showing that molecular crowding tends to stabilize the reaction product, which minimizes the entropic contribution of the free energy. Kinetics are determined by the transition rate to the activated state, as well as the encounter rates between diffusing molecules. Thus, crowding has two antagonistic effects: it stabilizes the activated state and hence accelerates the formation of this activated state, but it lowers the chances that reactants encounter due to diffusion slow down. In fact, it is generally admitted that increasing the degree of crowding ultimately leads to reaction inhibition because reactants can no longer diffuse (Ellis, 2001).

While molecular crowding was beautifully illustrated by *in vitro* experiments and convincing evidence based on diffusion hindrance or aggregation reaction kinetics was reported in prokaryotes (de Vries, 2010; Foley et al., 2010; Ignatova and Gierasch, 2004; Konopka et al., 2006), its relevance for eukaryotes remains less investigated. In the following section, we review recent experimental studies indicating that molecular crowding is a consistent ingredient for modeling the nucleus of eukaryotes.

3.3. Confirmations of molecular crowding predictions in eukaryotes

Confocal imaging of living cells showed that the steady-state distribution of inert fluorescent tracers, such as microinjected dextrans or GFP arrays, was heterogeneous in the nucleus, and that these probes were excluded from regions enriched in chromatin, that is, heterochromatin foci (Gorisch et al., 2003; Verschure et al., 2003). This effect appeared to be enhanced as the tracer size increased, as inferred from the simultaneous observation of 25 and 500 kDa dextrans labeled with two different fluorophores in heterochromatin foci (Bancaud et al., 2009). Volume exclusion was also detected in nucleoli, which are dense nuclear compartments containing rDNA sequences, specific transcription machinery, ribosomal subunits, as well as different nucleolus-associated proteins (Handwerger et al., 2005). Furthermore, modifications of chromatin compaction by histone hyperacetylation, which induces chromatin decondensation through electrostatic repulsion (Tóth et al., 2004), or by overexpression of the histone methyltransferase Suv39H1 (Bancaud et al., 2009) that imprints epigenetic marks characteristic of heterochromatin modulated the degree of exclusion of fluorescent probes. Chromatin therefore appears to be a crowding component, which induces volume exclusion of physical tracers.

The mobility of fluorescent tracers was also monitored by FRAP, showing that their diffusion throughout the nucleus is rapid so long as their molecular weight is smaller than ~500 kDa (Görisch et al., 2005; Seksek et al., 1997). Their mobility is moderately slowed down by a factor of 3–5 in comparison to free solution (Beaudouin et al., 2006; Seksek et al., 1997; Wachsmuth et al., 2000), and the diffusive hindrance is constant with the tracer size (Bancaud et al., 2009; Pack et al., 2006; Seksek et al., 1997). Analyzing further the dynamics of the tracers showed an enhanced diffusive hindrance in nucleoli and heterochromatin in comparison to euchromatin (Bancaud et al., 2009; Pack et al., 2006), in agreement with the enhanced degree of volume exclusion and diffusion hindrance in these compartments.

The effect of crowding on reaction equilibria and rate constants was also investigated in heterochromatin versus euchromatin by performing FRAP on three generic chromatin-interacting proteins, namely the guanine nucleotide exchange factor RCC1, the linker histone isoform H1.1, and the C-terminal tail of H1.1 (H1t) (Bancaud et al., 2009). Fluorescence redistribution was slower in heterochromatin for the three proteins, suggesting that their association with chromatin was enhanced in this dense compartment. The same effect was observed by comparing redistribution kinetics before and after chromatin hypercondensation by osmotic shocks (Martin and Cardoso, 2010). These observations suggest a generic enhancement of the binding to chromatin in dense chromatin regions, in line with the theoretical predictions. Moreover, a recent study monitoring the folding kinetics of the phosphoglycerate kinase showed that this protein adopts a more compact configuration in the nucleus of living yeasts than in dilute *in vitro* media (Dhar et al., 2011).

3.4. Relevance of molecular crowding for nuclear compartmentalization

The enhanced binding reactions in dense environments have been often speculated to play a role for the formation of nuclear compartments, which are dynamic and self-organizing structures defined by specific and transient protein–protein, chromatin epigenetic status, and/or probably protein–RNA interactions (Dundr, 2012; Meldi and Brickner, 2011; Rajapakse and Groudine, 2011; Tripathi and Prasanth, 2011). The molecular mechanism driving their formation remains elusive, and it has been suggested that molecular crowding plays a role for the formation and maintenance of these macromolecular assemblies (Cho and Kim, 2012; Richter et al., 2007, 2008), in addition to biochemical cues delivered during interphase (Lavau et al., 1995).

Molecular crowding may act as a self-reinforcing mechanism, stabilizing the interaction of proteins in defined nuclear regions and the initial specific nucleation event between pioneering scaffolding molecules (Dundr and Misteli, 2010), as was suggested by Marenduzzo and colleagues for the clustering of RNA and DNA polymerases into factories (Marenduzzo et al., 2006a,b). The segregation mechanism may be spontaneous (Hancock, 2004; Iborra, 2007), facilitating the binding of additional diffusing scaffolding proteins to the nucleator until the compartment reaches a steady-state size (Bancaud et al., 2009). This model has been supported by the evidence that molecular crowding was necessary for the maintenance of nucleoli and nuclear bodies in isolated nuclei (Hancock, 2004). Notably, this research involved changes in osmotic pressure, as was performed in numerous studies showing that reorganizations of chromatin compaction occurred after hyper- or hypo-osmotic treatment (Brasch et al., 1971; Delpire et al., 1985) (Fig. 13.3). However, this method induces adverse changes in ionic conditions (Martin and Cardoso, 2010), and the role of crowding remains a topic of discussion because both effects alter chromatin compaction *in vitro* (Hancock, 2007; Maeshima et al., 2010). In line with this discussion, our observations suggest that crowding is neither sufficient to induce the formation of heterochromatin nor necessary for its maintenance (Walter et al., 2013). The driving force generating distinct compartments may then arise from chromatin polymeric properties, because polymers with distinct physical properties spontaneously segregate in crowded media (Keating, 2012), a scenario recently proposed to account for the formation of the nucleolus in budding yeast (Wong et al., 2012). Note that this mechanism could be enhanced by changes in crowding and/or ionic strength. Thus, the generic role of molecular crowding in nuclear compartmentalization remains unclear and more studies are still needed to clarify the relative contribution of the crowding as compared to other biophysical processes such as phase separation or colloidal behaviors (Iborra, 2007). We propose that this research could benefit from the possibility of tuning the amount of DNA in the nucleus of fission yeast up to 16-fold with minimal change in nuclear volume (Neumann and Nurse, 2007), hence circumventing the artifacts of osmotic stress.

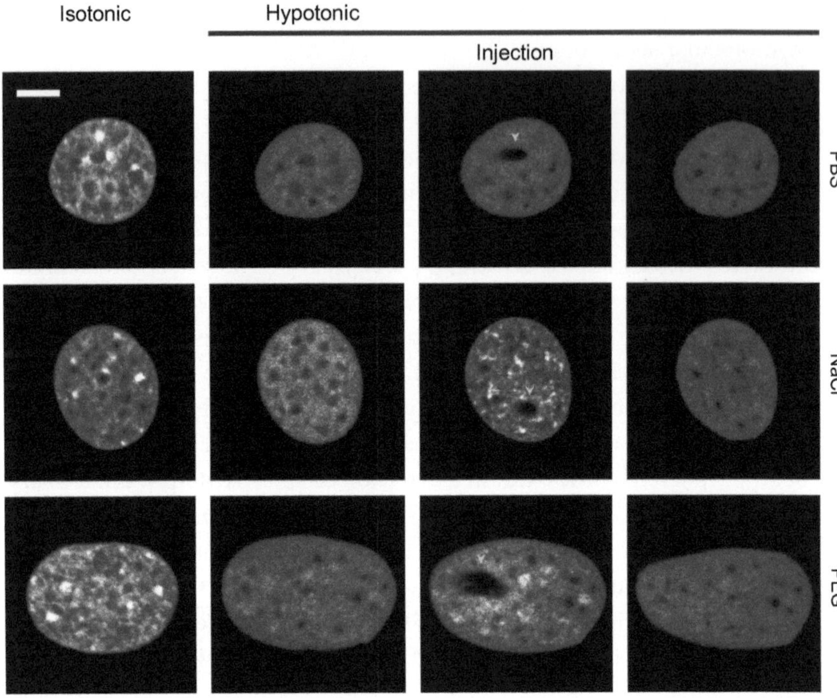

Figure 13.3 Chromatin architecture and hypotonic shocks. The different panels represent confocal image sequences of living 3T3 cells expressing the core histone H2B fused to EGFP submitted to hypotonic shocks, which are induced by dilution of the imaging medium with distilled water (35:65%, v/v, respectively). To rescue the isotonic chromatin compaction level, cells were microinjected with PBS (upper row), a mixture of PBS and water 40:60 (v/v) supplemented with NaCl at a final concentration of 320 mM (middle row), or with polyethylene glycol (MW = 1500 g/mol) at a dilution of 25% (w/v) (bottom row). While the addition of PBS alone did not induce changes in chromatin compaction, chromatin rich compartments could be detected upon addition of salt or polyethylene glycol. The position of the tip of the microinjection needle is indicated by green arrowheads. The scale bar represents 5 μm. (For interpretation of the references to color in this figure legend, the reader is referred to the online version of this chapter.)

3.5. Limitations of molecular crowding model

Despite the oversimplification of molecular crowding, which assumes a random distribution of background co-solutes, this model met some success in accounting for the dynamics and the assembly properties of nuclear proteins and macromolecular assemblies. However, it fails to precisely describe the properties of nuclear diffusion. Several authors have reported, for instance, that the diffusive hindrance is independent of the molecular weight of the probe in the nucleus (Pack et al., 2006; Seksek et al., 1997), in contradiction with several *in vitro* studies that indicated an increase of the hindrance with the size of the probe in random crowded media (Han and Herzfeld, 1993; Muramatsu and Minton, 1988). A second discrepancy is related to the anomaly of diffusion in the nuclear environment. Monte-Carlo simulations as well as *in vitro* studies indicate that, except at very short timescales, tracers moving within a crowded medium composed of mobile co-solutes usually display pure Brownian diffusion (Dauty and Verkman, 2004; Saxton, 1994). This result is essentially explained by the fact that the crowding co-solutes define an unstructured medium after a travel displacement larger than the average distance between obstacles, so that the crowding medium appears as homogeneous and normal diffusion occurs. Anomalous diffusion persists in the case of a structured pool of obstacles, that is, a polymer melt obtained with polymer solutions at high concentrations (Banks and Fradin, 2005). This structuration of co-solutes is consonant with the observation of an anomalous diffusion for GFP multimers containing one, two, five, and ten GFPs at length scales smaller than ~300 nm in the nucleus of living cells, as inferred from fluorescence correlation spectroscopy, which consists in measuring fluorescence intensity fluctuations in the confocal volume of an epifluorescence microscope, and from single-particle tracking (Bancaud et al., 2009; Grünwald et al., 2008; Pack et al., 2006; Wachsmuth et al., 2000). Anomalous diffusion is characterized by the anomaly parameter α, which is defined using the power-law scaling of the temporal evolution of the mean square displacement (MSD) with the time t:

$$\mathrm{MSD}(\Delta t) \propto \Delta t^{a} \tag{13.1}$$

α is equal to 1 for normal diffusion in free space, but its value somewhat varied between 0.5 and 0.8 in the nucleus. Although the molecular mechanism accounting for anomalous diffusion remains unclear (Saxton, 2012; Szymanski and Weiss, 2009), these results do not support a conventional molecular crowding picture. Furthermore, the analysis of the spatial redistribution of the three chromatin-interacting proteins RCC1, H1.1, and H1t (Bancaud et al., 2009), the navigation of which is determined by diffusion and interaction dynamics in the nucleus, showed fractal-like dynamics meaning that the association constant of these proteins to chromatin appeared to be time-dependent (Kopelman, 1988). This result could not be reproduced assuming a homogeneous repartition of chromatin obstacles in the nucleus, and rather hinted to the fractal organization of the nucleoplasm, as will be discussed in the following section.

Overall, the assumption that the nuclear environment can be modeled with a homogeneous network of obstacles appears questionable, supporting the modern view of chromatin as a highly ordered structure (van Steensel, 2011). In the meantime the fractal model has gained popularity to describe chromatin architecture (Mazza and Mc Nally, 2010; Mirny, 2011; Sanyal et al., 2011), and we recently surveyed some of the main experimental evidence supporting its validity (Bancaud et al., 2012). We will now briefly overview the physical description and consequences of the fractal model.

4. FRACTAL MODELS

4.1. The fractal hypothesis for nuclear diffusion

Fractal objects were initially imagined to describe geometrical objects with irregular appearances (Mandelbrot, 1982), which cannot be mapped with a finite number of geometrical elements. The evaluation of the contour of a coastline is, for instance, characterized by an infinite length l because the measured length increases as the resolution to sample the contour increases. The morphology of this object can nevertheless be characterized by the fractal dimension f, which is determined by assessing the mass distribution (i.e., the number of structural elements N) at a given zoom factor R according to the following relationship:

$$f = \frac{\ln(N)}{\ln(R)} \qquad (13.2)$$

The fractal dimension is a noninteger number, which varies somewhat between 1 and 2 for a line in 2D: its value is 1.25 for the British coastline (Mandelbrot, 1967). Mathematical fractals are self-similar over an unlimited range of scales, but natural fractals are self-similar only within a spatial domain with upper and lower scaling limits of ~20 μm and ~2 nm in the nuclear context, corresponding to the lateral dimension of the nucleus and the DNA diameter, respectively.

Let us note that the assumption of a fractal repartition of obstacles has been investigated in the seminal contributions of M.J. Saxton on molecular diffusion in the presence of obstacles, which were randomly placed on a lattice (Saxton, 1994) or in percolation or diffusion-limited clusters (Saxton, 1993), which exhibit self-similar properties characterized by fractal dimensions of 1.9 and 1.7 in 2D, respectively. These molecular dynamics simulations showed that diffusion was anomalous over a broad spatial domain, and this result suggested that the anomalous diffusive response observed in living cells originated from the fractality of the nucleoplasm (Münkel et al., 1999). The fractal dimension of the accessible nucleoplasm has then been derived by modeling interaction kinetics of chromatin-interacting proteins, revealing that the fractal architecture of the nucleoplasm of euchromatin and heterochromatin were markedly distinct and associated with fractal dimensions of $f \sim 2.6$ and 2.2, respectively, in the ~2–100 nm space domain (Bancaud et al., 2009). These values appear to be consistent with the idea that heterochromatin is more condensed than euchromatin because the topography of the complement of heterochromatin is smooth and leaves a smaller part of chromatin surface for transcription factors to scan for target sites. By contrast, nucleoplasm in euchromatin is defined by a larger fractal dimension, occupying more of the total volume, and therefore giving access to a larger part of the rough chromatin surface for scanning chromatin-interacting proteins.

Moreover, it was recently shown that the comparison of the fractal dimension f to the anomaly parameter α provides general predictions regarding the target-search mechanism in fractal environments (Bénichou et al., 2010; Condamin et al., 2007, 2008). Exploration by diffusion is compact when $\alpha > f$, meaning that diffusing molecules systematically visit their neighboring sites, and oversample their surrounding environment. On the contrary, the time to reach a target locus is independent of the distance from the initial position for $f \geq \alpha$, so diffusing molecules rapidly travel over large distances while overlooking nearby sites. This second regime is referred to as non-compact exploration. Because the fractal dimension of the heterochromatin nucleoplasm is lower than in euchromatin but the anomaly parameter is similar (Bancaud et al., 2009), exploration is more compact in heterochromatin, and chromatin-interacting proteins should systematically bind to all of their available binding sites in heterochromatin, in agreement with the long residence of generic chromatin binding proteins observed in heterochromatin (Bancaud et al., 2009) and with the higher frequency of transient protein trapping in heterochromatin (Grünwald et al., 2008). On the contrary, the target-search strategy in euchromatin likely favors a faster exploration, which is presumably adapted to scan for comparatively rare regulatory elements in the genome. Notably, it was recently shown that the facilitated diffusion model, which describes the search for a target site based on alternating phases of free diffusion in the bulk and sliding diffusion of a bound complex, was a robust mechanism for a fractal template such as chromatin (Bénichou et al., 2011). In turn, this study indicated that the exploration strategy could be finely tuned by adjusting the molecular interactions on chromatin. This hypothesis could be tested by tracking the motion of, for example, transcription factors at the single-molecule level (Normanno et al., 2012), potentially shedding new light on the regulation of transcription activation; chromatin might be able to switch the expression of different loci by altering its fractal structure, as was suggested by theoretical models (Meyer et al., 2012).

4.2. Fractal hypothesis for chromosome architecture

The fractal model has also been applied to describe the structure of chromatin by image analysis of fluorescently labeled DNA, showing a fractal topography characterized by a fractal dimension of $f \sim 2.5$ in the 0.15–2.7 μm spatial range (Einstein et al., 1998; Huisman et al., 2005; Tóth et al., 2004). Furthermore, textural analysis of transmission electron micrographs of cell nuclei stained with uranyl acetate as a contrast agent indicated that chromatin organization was fractal (Castelli and Rosa, 2001). Neutron scattering, which is presumably the optimal method to probe fractal architecture over a broad spatial range of 15 nm to 10 μm (Lebedev et al., 2005, 2008), has been applied to isolated chicken erythrocyte nuclei, showing a power-law scaling relationship between the scattering intensity and the scattering vector which is characteristic of a fractal organization. A biphasic response with a fractal dimension of $f \sim 2.4$ in the length spectrum 15–400 nm and $f \sim 2.9$ for larger length scales was detected. The response could then be studied in finer details by assaying the contribution of DNA architecture that also exhibited two different regimes of fractality with a fractal dimension of $f \sim 2.2$ in the 15–400 nm space domain, and a $f \sim 3.2$ exponent for larger length scales ((Lebedev et al., 2005), see table in (Bancaud et al., 2012) for a classification of measured fractal dimensions). Last but not least, the fractal model gained additional support from Hi-C (Lieberman-Aiden et al., 2009), which unraveled a power-law scaling associated with a slope of -1.08 in the range 0.5–7 Mb, which corresponds to a spatial range of ~0.5–2 μm. This structural property appeared to be consistent with a fractal organization of DNA characterized by a fractal dimension of $f \sim 3$. This folding can be assigned to the folding of DNA into a crumpled globule conformation, as defined by Grosberg and coworkers (Grosberg et al., 1993). The crumpled globule is not entangled, and large-scale loops should be reorganized at a low energetic cost with no need to break physical contacts to liberate genomic sequences.

Consequently, several experimental lines of evidence support the fractal organization of the nucleus, but the available experimental data that refer to DNA, chromatin, the nucleoplasm, or the whole nucleus and the estimates of the fractal dimension are not always consistent (Bancaud et al., 2012). In addition, the estimate of the fractal dimension cannot be assigned to a unique structure because a variety of fractal objects can be imagined to match a given fractal dimension. Notably, the link between the polymer folding properties and the fractal dimension obtained by Hi-C highlights the need for a polymer model to derive a molecular description of chromatin organization in the nucleus (Heermann et al., 2012). In the following section we therefore offer an overview of some of the main conclusions obtained by describing chromosomes with a polymer model, and we will then attempt to bridge the conclusions of the fractal and the polymer models.

5. POLYMER MODELS FOR CHROMOSOMES

5.1. Polymer models for chromosome architecture

The conformation of a polymer chain can be described by measuring the physical distance (L) between genomic loci as a function of their genomic separation (G):

$$L \propto G^{1/\varepsilon} \tag{13.3}$$

with ε the fractal dimension of the line of polymer in space. ε is *a priori* unrelated to the fractal dimension f, as shown by the thought experiment of A. Grosberg in which we consider a chromosome and freeze it instantly. We then imagine that linker DNAs are cut systematically, while the whole structure remains frozen. For the resulting disconnected system, the concepts of ε or any Hi-C type information are fully lost, because the structure is genetically disconnected, but f remains perfectly measurable. This thought experiment is particularly relevant in the case of the fractal dimension of the protein component of chromosomes. This quantity can be estimated by computing the distribution of fluorescently labeled histones in living cells, but it is not related to DNA polymer properties.

This relationship of Eq. (13.3) can be inferred by *in situ* hybridization of fluorescent oligonucleotides targeted to specific genomic sequences in fixed cells (FISH). ε was ~2 in the 150 nm to 1 μm spatial range, increased to ~3 above 1 μm (Sachs et al., 1995; Yokota et al., 1995), and a confinement (~0) was detected for distances larger than 2–3 μm (Mateos-Langerak et al., 2009).

The latter regime is consonant with the existence of chromosome territories in interphase (Cremer et al., 2006), while $\varepsilon \sim 2$ corresponds to a random walk polymer. Different models were suggested to reproduce the changes in ε with the sampling dimension, assuming that chromatin formed loops of ~1 Mb in length (Sachs et al., 1995) or of ~200 kb and bundled in groups of ~5 (Münkel et al., 1999). These models have since been challenged by a dynamic loop model, which postulates that the formation of chromosome loops is a random diffusion-driven process and that loops occur transiently (Bohn and Heermann, 2009, 2010). An appropriate set of parameters for the loop formation probability and the loop lifetime enabled reproduction of FISH and Hi-C experiments, as well as the general topography of chromosome territories (Bohn and Heermann, 2010). However, the recent identification of topological domains of ~1 Mb by Hi-C, which were shown to be an inherent property of mammalian genomes (Dixon et al., 2012; Nora et al., 2012; Sexton et al., 2012), seems to support the pioneering model of Sachs (Sachs et al., 1995), and in turn shows that the folding of chromosomes has not yet been fully clarified. Furthermore, the consistency of these models has recently been questioned because confined polymer models appeared to reproduce FISH data equally well (Emanuel et al., 2009).

Last but not least, the different loop models should be compared to the crumpled globule architecture derived from Hi-C (Lieberman-Aiden et al., 2009). The crumpling was originally imagined to explain relaxation kinetics of polymers rapidly brought into poor solvent conditions (Grosberg et al., 1988). The crumpled conformation is transient, and ultimately collapses into an equilibrium globule, which is a compact conformation favoring polymer–polymer over polymer–solvent interactions. The crumpled globule is not entangled, and is characterized by long-range intrachromosomal interactions as shown by the power-law dependence of -1 for contact probabilities in comparison to -1.5 for equilibrium globules (Lieberman-Aiden et al., 2009). While the dynamic loop model is stable thermodynamically, the crumpled globule is transient and can only be stabilized kinetically, in agreement with the analysis which showed that the organization of chromosomes in discrete territories is maintained during interphase kinetically (Rosa and Everaers, 2008). Additionally, the crumpled globule may be stabilized over long time periods by DNA sequences nonrandomly repeated along chromosomes, as suggested by the spatial correlation between CTCF binding sites and contacts between DNA fragments derived from Hi-C (Botta et al., 2010).

In another direction, FISH has been applied to characterize the folding principle of chromosomes in the yeast *Saccharomyces cerevisiae*. The physical versus genomic distance data have been analyzed with a worm-like chain (Prorod-Kratky) model which allows for derivation of the fiber persistence length l_p and its packing ratio P. The persistence length represents the length over which the fiber behaves as a rigid polymer with thermal fluctuations, and the packing ratio is the number of nucleosomes per 10 nm. Chromatin appeared to be a stiff polymer in yeast, characterized by a persistence length of 200 nm and a nucleosome density of 7–10 per 10 nm (Bystricky et al., 2004). These quantities were also assessed using 3C (Dekker, 2008), suggesting that chromatin persistence length was 66–134 nm and its packing ratio 1.1–3.3 nucleosome/10 nm. So far these results did not clarify whether chromatin is stiff or flexible and whether the density of nucleosomes is high or low. This debate is consonant with the recent questions raised about the existence of a dense and compact 30 nm fiber *in vivo*. Electron spectroscopic imaging, which distinguishes phosphorus atoms and provides a trace of the path of the DNA in a chromatin fiber (Fussner et al., 2011) and cryo-electron microscopy of frozen nuclei (Maeshima et al., 2010) have indicated that chromatin in cultured mouse embryonic fibroblasts and native mouse tissues is nearly exclusively organized in a 10 nm fiber. Altogether, despite years of research on chromatin structure, the local and global characteristics of chromosomes, which are assessed by the packing ratio, the persistence length, and the looping probability, remain the subjects of intense controversies.

5.2. Polymer models for chromosome dynamics

The physical parameters governing chromosome movements remain poorly understood, though it was established 15 years ago that chromosome loci explore a broad volume of the nucleus in yeast and in metazoans (Heun et al., 2001; Marshall et al., 1997). Nuclear constraints such as the nuclear envelope (Bystricky et al., 2004; Heun et al., 2001) or the position along the chromosome, for example, telomeres (Bystricky et al., 2005) were shown to modulate the motion of loci, but the physical parameters governing these movements were never assessed quantitatively. For instance, it has been proposed that chromatin segments undergo normal Brownian fluctuations at small time scales, and that their motion is confined in volumes of $R \sim 0.3$ μm (Marshall et al., 1997), or that the movement of the GAL1 locus on chromosome II follows an anomalous diffusive behavior characterized by an anomalous parameter of ~0.4 (Cabal et al., 2006). Detailed analysis of chromosome motion in bacteria also showed an anomalous diffusive response in the temporal regime comprised between 1 and 100 s, which has been interpreted with the Rouse model (Weber et al., 2010, 2012). The Rouse model assumes that the motion of chromosome loci is dominated by elastic interactions between nearest-neighbor chromatin segments and the viscous friction of each monomer (Doi and Edwards, 1988). Note that this model disregards hydrodynamic interactions, which are screened out in the nuclear context due to the crowding level in this compartment (Fig. 13.4A).

We have recently developed an imaging platform for 3D high-throughput fluorescence microscopy of living yeast (Hajjoul et al., 2009) that was used to track the motion of 15 loci inserted in chromosome III, IV, VI, XII, and XIV over an extended temporal range of more than four orders of magnitude (10^{-2} to 10^{3} s; Fig. 13.4B). The MSD of chromosome loci, as well as the statistical properties of the trajectories, followed an anomalous diffusive response that was characteristic of the Rouse model (Hajjoul et al., 2013). This behavior was consistently observed for seven loci along the contour of chromosome XII (Albert et al., 2013), and it appeared to be specific to the bulk of chromatin because the motion of loci in the nucleolus was reduced (Fig. 13.4C). In addition, the analysis of the extent of spatial fluctuations suggested that chromatin persistence length is highly flexible in yeast and less than 30 nm. This result is consistent with the recent indications that chromatin is folded in a 10 nm fiber *in vivo* (see above), and shows that motion analysis provides an efficient solution to assay chromosome structural properties and hence complements conventional techniques such as FISH. In the following section, we illustrate how different techniques can be combined to derive chromosome structural properties.

6. EVALUATING CHROMOSOME STRUCTURAL PROPERTIES WITH COMBINED TECHNIQUES

The haploid genome of *S. cerevisiae* is composed of 12 million base pairs arranged in 16 chromosomes. The architecture of this small genome has recently been modeled using simulations (Gehlen et al., 2012; Tjong et al., 2012; Wong et al., 2012), reproducing experimental results of Hi-C, telomere repartition, or nucleolus formation with a minimal number of physical principles, including volume exclusion and physical tethering of polymers. Therefore, the yeast nucleus constitutes an excellent model

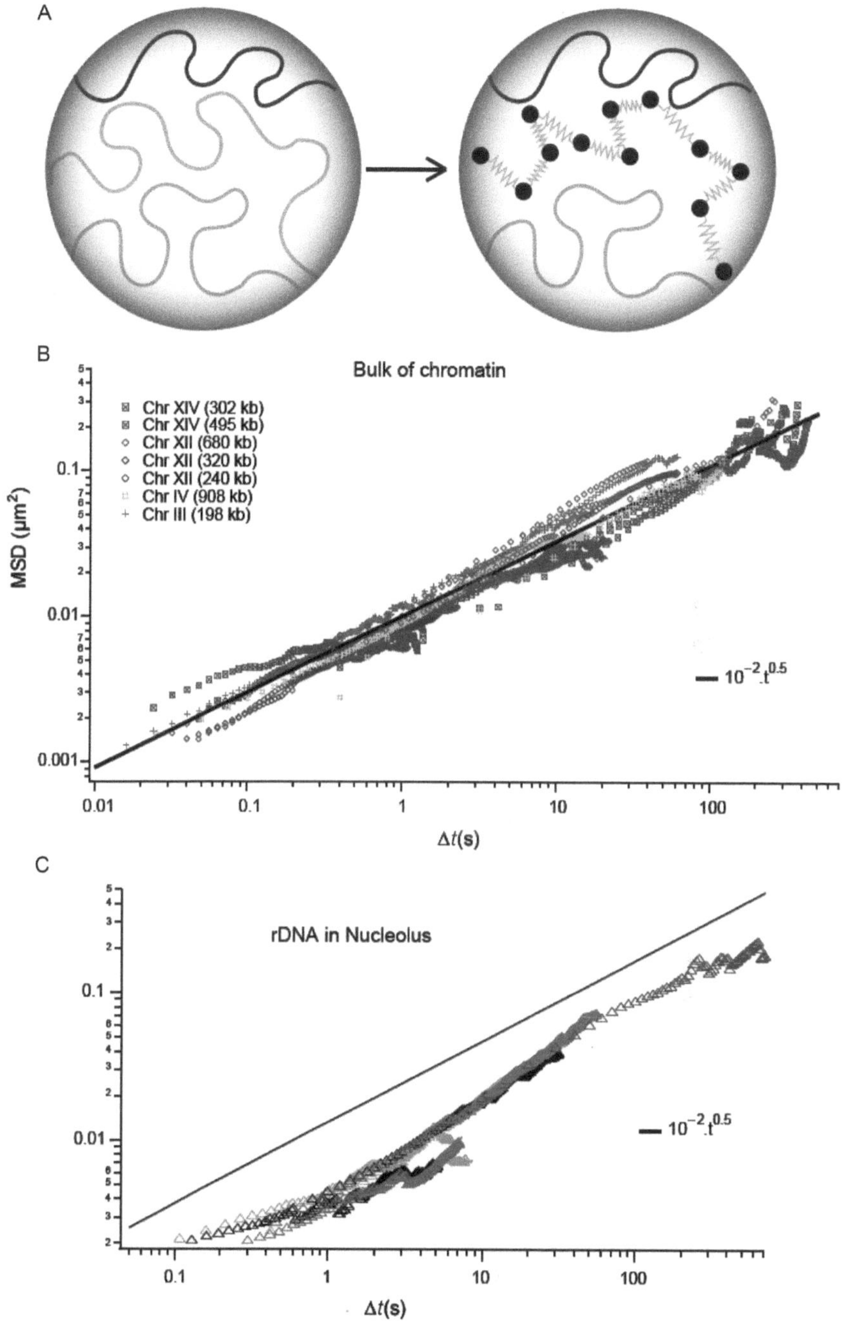

Figure 13.4 Motion analysis. (A) The nuclear volume is densely filled with chromosomes and nuclear proteins, and volume exclusion between chromosomes has been shown to

(Continued)

system to validate physical concepts, and we demonstrate here how two different assays can be recapitulated to extract chromatin persistence length and packing ratio.

Distance measurements between genomic loci on the same chromosome have been analyzed with the WLC model (Bystricky et al., 2004). The validity of this model is unclear, however, because chromosomes are assumed to be ideal chains and monomer–monomer interactions are disregarded, whereas recent simulations showed that volume exclusion was essential to model chromosome conformation (Tjong et al., 2012). These simulations enabled to reproduce the results of gene maps (Wong et al., 2012) (upper panel in Fig. 13.5A) which represent the position of one chromosome loci averaged over a cell population comprising ~2000 individuals, hinting at the relevance of polymer models to model the yeast genome. We thus propose to revisit distance measurements carried out over short distances of less than 400 nm (equivalently to 100 kb) with the blob picture (de Gennes, 1979), which describes the steady-state conformation of a confined polymer as a series of self-avoiding chains (dashed red circles in the left and middle panels of Fig. 13.5B). Note that we showed that this model was consistent with our MSD data using Langevin dynamics simulations (Hajjoul et al., 2013), which also hinted to the structural organization of chromosomes in blob-like geometries (Fig. 13.5B, right panel). According to the blob model, the power-law exponent relating the physical L versus the genomic distance G is 0.6 (Ostashevsky and Lange, 1994), and the prefactor is dependent on l_p and P (de Gennes, 1979):

$$L = \left(\frac{\alpha}{2l_p}\right)^{1/5} \times 2l_p \times \left(\frac{G}{K}\right)^{3/5} \quad (13.4)$$

Figure 13.4—Cont'd be a key mechanism of the yeast genome organization. The crowding level screens out hydrodynamic interactions between distant monomers, so that nearest-neighbor elastic interactions dictate the movements of chromosome loci (left panel). This regime of diffusion is the Rouse dynamics. (B) The motion of chromosome loci located on chromosome III, IV, XII, and XIV is characterized by anomalous diffusion, and the anomalous diffusion coefficient is 0.52, in good agreement with the Rouse model predictions. (C) The dynamics are slowed down for loci on the rDNA, which are nearly systematically in the nucleolus. In addition the anomalous diffusive motion in the short time regime departs from the Rouse regime, because the slope of the MSD response is lower than the Rouse dynamics. *(B) Reprinted from Hajjoul et al. (2013). (C) Reprinted from Albert et al. (2013).* (For color version of this figure, the reader is referred to the online version of this chapter.)

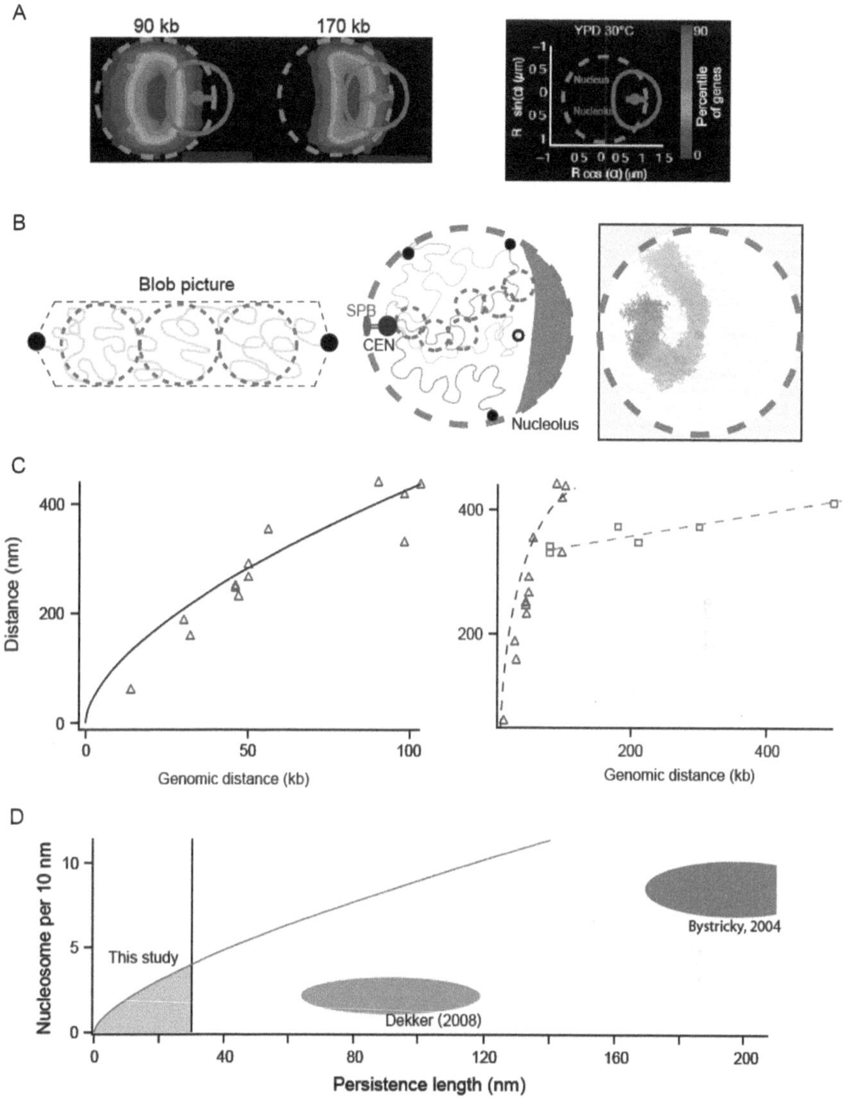

Figure 13.5 Recapitulation of experimental results with the blob picture. (A) The gene maps on the left represent the statistical distribution in 2D of chromosome loci located at 90 and 170 kb from the centromere on right arm of chromosome XII in living yeast. The plot on the right shows the scale bar, the color scale, and the reference position of the nucleolus, which are used to construct gene maps. (B) The left panel represents the blob picture, in which chromosomes are segmented in consecutive self-avoiding chains (red dashed circles). The diameter of each blob is determined by the crowding level in the nucleus, as suggested by the middle panel. The right panel is the superposition of the conformation of chromosome III in the time course of a Langevin dynamics simulation. The repulsion between chromosomes guides their folding in an elongated

(Continued)

with α, the fiber diameter in nm and K the number of bp in one Kuhn segment of length $2l_p$. Note that Eq. (13.4) is relating the distance to the number of monomers in one Kuhn segment of length $2l_p$, and we checked the relevance of this scaling prediction with Langevin dynamics simulations (not shown). K is related to the physical distance by the packing ratio:

$$K = P \times 2l_p. \tag{13.5}$$

We thus conclude that

$$L = \left(\frac{\alpha}{2l_p}\right)^{1/5} \times (2l_p)^{2/5} \times P^{-3/5} \times G^{3/5}. \tag{13.6}$$

The left panel of Fig. 13.5C shows that this power-law scaling relationship between the physical and genomic distances is consistent with FISH data (Bystricky et al., 2004). In addition, our recent data on the distance between genomic loci separated by more than ~100 kb on chromosome XII showed a linear regime (green dataset in the right panel of Fig. 13.5C) (Albert et al., 2013), indicating a different folding mechanism consistent with the formation of series of blobs in a tube-like configuration, as shown in the simulation of Fig. 13.5B. The transition between the linear and nonlinear regimes occurs for genomic and physical distances of ~100 kb and ~400 nm, respectively, indicating that the typical dimension of blobs is ~400 nm. Overall, the responses for short and long length scales of the physical versus genomic distance data tend to support the blob picture model.

Moreover, assuming that the contribution of the fiber diameter is negligible, as is conventionally done in FISH analysis (Bystricky et al., 2004;

Figure 13.5—Cont'd tube-like conformation. (C) The two graphs represent physical versus physical distance measurements obtained by FISH in fixed cells or deduced from gene map analysis on chromosome XII (black and green datasets, respectively). For short length scales, the data can be analyzed by the blob picture model (see Eq. 13.6). We observe a different response for distances larger than ~100 kb, which is characterized by a slow linear increase of the physical distance (the green dashed line is a guide to the eye), which may be attributed to the folding of chromosome XII in a directional pearl-necklace of blobs. (D) The plot representing the number of nucleosome per 10 nm as a function of the persistence length enables us to compare the results of different analysis based on FISH, Chromosome Conformation Capture, or motion analysis in combination with FISH analysis (green, orange, and gray regions, respectively). *(A) Reprinted from Albert et al. (2013). (B) Reprinted from Hajjoul et al. (2013).* (For interpretation of the references to color in this figure legend, the reader is referred to the online version of this chapter.)

Ostashevsky and Lange, 1994), we neglect the first term of Eq. (13.6), and deduce from the fit that $l_p^{2/5} \times P^{-3/5} = 0.31$ nm. Given that we obtained with another method a persistence length of less than 30 nm, we propose that the structural properties of chromatin are defined in the region of the parameter space shown in Fig. 13.5D. Obviously, more work is needed to provide the community with reliable estimates, which will involve multiple assays because one technique is generally insufficient to evaluate multiple quantities with precision.

7. UNIFORMITY OF CHROMATIN MECHANICAL PARAMETERS

The bending persistence length of chromatin has been estimated to ~30 nm based on single-molecule stretching (Bancaud et al., 2006; Bennink et al., 2001; Cui and Bustamante, 2000), 28 nm by monitoring the recombination frequency of chromatin circles *in vivo* (Ringrose et al., 1999), 50–100 nm by evaluating the circularization frequency in Hi-C (Dekker, 2008), to ~200 nm using *in situ* hybridization experiments (Bystricky et al., 2004) (Table 13.1). We first estimated the torsional persistence length on reconstituted chromatin fibers manipulated by magnetic tweezers (Bancaud et al., 2006; Celedon et al., 2009) and showed a surprisingly low value of 5 nm, which is almost 20 times lower than that of naked DNA (Strick et al., 1999). For the stretching modulus, a rough value of ~5–8 pN has been derived from the same kind of experiments (Bancaud et al., 2006; Cui and Bustamante, 2000), close to the theoretical estimate (Ben-Haïm et al., 2001) and probably reflecting the weak internucleosomal interactions disrupted during the mechanical extension of the fiber, as well as the partial unwrapping of DNA (Victor et al., 2012). Although a pure physical polymer-based model is obviously useful to get general laws of chromosome organization, the discrepancies between different evaluations of the persistence length may be associated with the intrinsic variability caused by the geometrical polymorphism of this complex nucleoprotein assembly (Ben-Haïm et al., 2001; Wedemann and Langowski, 2002). For instance, how biochemical modifications of its components affect chromatin mechanical properties is still debated (Lavelle, 2009). In this regard, recent evidence that DNA methylation affects DNA mechanical properties (Severin et al., 2011) and nucleosome structure and dynamics (Choy et al., 2010; Jimenez-Useche and Yuan, 2012) suggests that methylation may regulate gene expression not only through biochemical mechanisms (e.g., serving

Table 13.1 Methods used to measure chromatin persistence length *in vitro* and *in vivo* (transparent and gray backgrounds, respectively)

Method	Persistence length (nm)	Experimental conditions	References
Force spectroscopy	30	Purified chromatin from chicken erythrocytes	Cui and Bustamante (2000)
Force spectroscopy	28	Reconstituted chromatin on repeated 5S sequences	Bancaud et al. (2006)
Force spectroscopy	30	*X. laevis* egg extract reconstitution	Bennink et al. (2001)
Distance measurements	196–272	Chromosome 4 in G1 human fibroblasts (fixed cells)	Ostashevsky and Lange (1994) and van den Engh et al. (1992)
Distance measurements	170–220	Chromosomes 5, 6, 14 in *S. cerevisiae*	Bystricky et al. (2004)
Looping probability	28	Recombination in human cells	Ringrose et al. (1999)
Looping probability	66–134	3C on chromosome 3 in *S. cerevisiae*	Dekker (2008)
Spatial fluctuations	<30	Motion analysis on chromosomes 3, 4, 12, 14	Hajjoul et al. (2013)

as molecular "handles" recognized by various regulatory proteins), but also through physical mechanisms (changing DNA elastic and dynamical properties).

8. CONCLUDING REMARKS

We conclude from this survey of experiments and models of chromosome structure and dynamics that the field is particularly active, although there is still a lack of consensus on the analytical approaches describing chromosome properties. For a long time, FISH has remained the only technique to probe nuclear architecture *in vivo*, but there has been much progress recently with the development of high resolution optical microscopy and other nonoptical imaging techniques for the study of 3D nuclear organization (Markaki et al., 2012; Rapkin et al., 2012), as well as with the advent of high-throughput cell biology technologies. We now have the tools in hand to get a unified picture of nuclear architecture and dynamics. This objective requires the integration of experimental data coming from different approaches, for example, Hi-C and FISH as discussed in Gibcus and Dekker (2013), or FISH and motion analysis as proposed in this report. Models are mandatory to support these integrated researches and the rapid development of multi-scale simulations, which can now reproduce the behavior of the whole nucleus of *S. cerevisiae* with a minimal number of adjustable parameters (Tjong et al., 2012; Wong et al., 2012), clearly offers a great opportunity. Notably, several problems should be addressed to improve our description of the nuclear environment. For instance, it is important to link nuclear protein dynamics and chromosome structural properties, as was recently initiated by Langowski and collaborators in Fritsch and Langowski (2011). In another direction, the development of a combined approach based on polymer physics including biochemistry to take into account the heterogeneity of the polymer caused by sequence-specific DNA-binding proteins interacting at discrete loci is needed to reach a functional model of chromatin organization in the nucleus. These new efforts and models may be confronted by experiments carried out in living cells, in fixed cells, or in isolated nuclei, provided that the consequences of molecular crowding can be directly investigated with nuclei (Hancock, 2004), and that technical progress to handle and analyze these structures was recently reported (Rowat, 2009).

Several studies have shown that some chromatin "outlooping" can occur upon transcription (Volpi et al., 2000; Williams, 2003) to drive targeted gene into functional nuclear compartments such as transcription factories (Bartlett et al., 2006; Jackson et al., 1993; Osborne et al., 2004). From a mechanistic point of view, outlooping might rely on changes in chromatin conformation and topological state (Mozziconacci et al., 2006), potentially assisted by some ATP-dependent motors (Havas et al., 2000). Some evidence for an actin/myosin system in the nucleus renewed our vision of mechanisms potentially tasked with long-range chromatin movements (Chuang et al., 2006), occurring at velocities ranging from 0.1 to 1 μm/min and over distances of 1–5 μm. The role for an actin/myosin complex in the active translocation of chromatin domains has been suggested by several studies (Kumaran et al., 2008). Motors have to be anchored to generate force, so it has been suggested that nuclear myosin I (Nowak et al., 1997), which binds to RNA polymerases (Fomproix and Percipalle, 2004), interacts with certain actin polymers, although these nuclear polymers are different from cytoplasmic filaments (McDonald et al., 2006). So far, the literature on chromosome reorganization events has been mostly focused on evidence of directed movements, but the driving force of these reorganizations, the questions on how chromosomes accommodate these mechanical constraints, the distance over which mechanical constraints propagate, and the kinetics of reorganization events mostly remain elusive.

More challenging than transcription-induced chromosome reorganization events are the dynamics of the genome during DNA repair (Dion and Gasser, 2013), cell differentiation and development (Chalut et al., 2012; Fisher and Fisher, 2011; Krijger and de Laat, 2013), or cancer (Göndör, 2013). The link between genome topology and its functions (Cavalli and Misteli, 2013) is increasingly observed experimentally, but these observations have not yet been tested in quantitative models although preliminary contributions investigated the interplay between molecular crowding and DNA repair in prokaryotes (Delmas et al., 2013) or eukaryotes (Amouroux et al., 2010). The regulatory function of the genome 3D organization and its reorganization propensity cause variegated gene expression levels (Krijger and de Laat, 2013) and define cell-specific genomes, which can be understood quantitatively only with concerted efforts of physicists and biologists. This *tour-de-force* is not only appealing, but also mandatory to clarify our understanding of genome transactions.

ACKNOWLEDGMENTS

The authors are grateful to J. Mozziconacci for stimulating discussions and to O. Gadal and I. Léger-Silvestre for generating the electron micrograph in Fig. 13.1. We also thank Madeleine Jensen for critical reading of the manuscript and Jan Ellenberg for constant support on these researches. SH was supported by funds from the Agence National de la Recherche (JCJC-SVSE2-2011, ChromaTranscript project) and from the European Union (FP7-PEOPLE-2011-CIG, ChromaTranscript project).

REFERENCES

Albert, B., Mathon, J., Shukla, A., Saad, H., Normand, C., Villa, D., Kamgoue, A., Mozziconacci, J., Wong, H., Zimmer, C., Bhargava, P., Bancaud, A., Gadal, O., 2013. Systematic characterization of the conformation and dynamics of budding yeast chromosome XII. J. Cell Biol. 202 (2), 201–210.

Amouroux, R., Campalans, A., Epe, B., Radicella, J.P., 2010. Oxidative stress triggers the preferential assembly of base excision repair complexes on open chromatin regions. Nucl. Acid Res. 38, 2878–2890.

Avivi, Y., Morad, V., Ben-Meir, H., Zhao, J., Kashkush, K., Tzfira, T., Citovsky, V., Grafi, G., 2004. Reorganization of specific chromosomal domains and activation of silent genes in plant cells acquiring pluripotentiality. Dev. Dyn. 230, 12–22.

Baccarini, P., 1908. Sulle cinesi vegetative del "Cynomorium coccineum L. N. Giorn. Bot Ital. N. Ser. 15, 189–203.
Bancaud, A., Conde e Silva, N., Barbi, M., Wagner, G., Allemand, J.F., Mozziconacci, J., Lavelle, C., Croquette, V., Victor, J.M., Prunell, A., Viovy, J.L., 2006. Structural plasticity of single chromatin fibers revealed by torsional manipulation. Nat. Struct. Mol. Biol. 13, 444–450.
Bancaud, A., Huet, S., Daigle, N., Beaudouin, J., Mozziconacci, J., Ellenberg, J., 2009. Molecular crowding affects diffusion and binding of nuclear proteins in heterochromatin and reveals the fractal organization of chromatin. EMBO J. 28, 3785–3798.
Bancaud, A., Huet, S., Rabut, G., Ellenberg, J., 2011. Fluorescence perturbation techniques to study mobility and molecular dynamics of proteins in live cells: FRAP, photoactivation, photoconversion, and FLIP. Cold Spring Harb. Protoc. 2010, 67–93.
Bancaud, A., Lavelle, C., Huet, S., Ellenberg, J., 2012. A fractal model for nuclear organization: current evidence and biological implications. Nucl. Acids Res. 40, 8783–8792.
Banks, D.S., Fradin, C., 2005. Anomalous diffusion of proteins due to molecular crowding. Biophys. J. 89, 2960–2971.
Bartlett, J., Blagojevic, J., Carter, D., Eskiw, C., Fromaget, M., Job, C., Shamsher, M., Trindade, I.F., Xu, M., Cook, P.R., 2006. Specialized transcription factories. Biochem. Soc. Symp. 73, 67–75.
Beaudouin, J., Mora-Bermudez, F., Klee, T., Daigle, N., Ellenberg, J., 2006. Dissecting the contribution of diffusion and interactions to the mobility of nuclear proteins. Biophys. J. 90, 1878–1894.
Ben-Haïm, E., Lesne, A., Victor, J.M., 2001. Chromatin: a tunable spring at work inside chromosomes. Phys. Rev. E Stat. Nonlin. Soft Matter Phys. 64, 051921.
Bénichou, O., Chevalier, C., Klafter, J., Meyer, B., Voituriez, R., 2010. Geometry-controlled kinetics. Nat. Chem. 2, 472–477.
Bénichou, O., Chevalier, C., Meyer, B., Voituriez, R., 2011. Facilitated diffusion of proteins on chromatin. Phys. Rev. Lett. 106, 038102.
Bennink, M.L., Leuba, S.H., Leno, G.H., Zlatanova, J., de Grooth, B.G., Greve, J., 2001. Unfolding individual nucleosomes by stretching single chromatin fibers with optical tweezers. Nat. Struct. Biol. 8, 606–610.
Bernhard, W., 1969. A new staining procedure for electron microscopical cytology. J. Ultrastruct. Res. 27, 250–265.
Bohn, M., Heermann, D.W., 2009. Conformational properties of compact polymers. J. Chem. Phys. 130, 174901.
Bohn, M., Heermann, D.W., 2010. Diffusion-driven looping provides a consistent framework for chromatin organization. PLoS One 5, e12218.
Bolzer, A., Kreth, G., Solovei, I., Koehler, D., Saracoglu, K., Fauth, C., Mäller, S., Eils, R., Cremer, C., Speicher, M.R., Cremer, T., 2005. Three-dimensional maps of all chromosomes in human male fibroblast nuclei and prometaphase rosettes. PLoS Biol. 3, e157.
Botta, M., Haider, S., Leung, I.X.Y., Lio, P., Mozziconacci, J., 2010. Intra- and inter-chromosomal interactions correlated with CTCF binding genome wide. Mol. Syst. Biol. 6, 426.
Boveri, T., 1914. Zur Frage der Enstehung Maligner Tumoren. Gustav Fischer Verlag, Jena.
Branco, M.R., Pombo, A., 2006. Intermingling of chromosome territories in interphase suggests role in translocations and transcription-dependent associations. PLoS Biol. 4, e138.
Brasch, K., Seligy, V.L., Setterfield, G., 1971. Effects of low salt concentration on structural organization and template activity of chromatin in chicken erythrocyte nuclei. Exp. Cell Res. 65, 61–72.
Brown, R., 1831. Observations on the organs and ode of fecundation in Orchideae and Asclepiadeae. Trans. Linn. Soc. Lond. 16, 685–746.

Bystricky, K., Heun, P., Gehlen, L., Langowski, J., Gasser, S.M., 2004. Long range compaction and flexibility of interphase chromatin in budding yeast analysed by high-resolution imaging techniques. Proc. Natl. Acad. Sci. U.S.A. 101, 16495–16500.

Bystricky, K., Laroche, T., Van Houwe, G., Blaszczyk, M., Gasser, S., 2005. Chromosome looping in yeast. J. Cell Biol. 168, 375–387.

Cabal, G., Genovesio, A., Rodriguez-Navarro, S., Zimmer, C., Gadal, O., Lesne, A., Buc, H., Feuerbach-Fournier, F., Olivo-Marin, J., Hurt, E., 2006. SAGA interacting factors confine sub-diffusion of transcribed genes to the nuclear envelope. Nature 441, 770–773.

Carrivain, P., Cournac, A., Lavelle, C., Lesne, A., Mozziconacci, J., Paillusson, F., Signon, L., Victor, J.-M., Barbi, M., 2012. Electrostatics of DNA compaction in viruses, bacteria and eukaryotes: functional insights and evolutionary perspective. Soft Matter 8, 9285–9301.

Castelli, C., Rosa, G.A., 2001. Ultrastructural complexity of nuclear components during early apoptotic phases in breast cancer cells. Anal. Cell. Pathol. 23, 1–9.

Cavalli, G., Misteli, T., 2013. Functional implications of genome topology. Nat. Struct. Mol. Biol. 20, 290–299.

Celedon, A., Nodelman, I.M., Wildt, B., Dewan, R., Searson, P., Wirtz, D., Bowman, G.D., Sun, S.X., 2009. Magnetic tweezers measurement of single molecule torque. Nano Lett. 9, 1720–1725.

Chalut, K.J., Hapfler, M., Lautenschlager, F., Boyde, L., Chan, C.J., Ekpenyong, A., Martinez-Arias, A., Guck, J., 2012. Chromatin decondensation and nuclear softening accompany nanog downregulation in embryonic stem cells. Biophys. J. 103, 2060–2070.

Chambeyron, S., Bickmore, W.A., 2004. Chromatin decondensation and nuclear reorganization of the HoxB locus upon induction of transcription. Genes Dev. 18, 1119–1130.

Cho, J.C., Kim, J.S., 2012. Crowding effects on the formation and maintenance of nuclear bodies: insights from molecular-dynamics simulations of simple spherical model particles. Biophys. J. 103, 424–433.

Choy, J.S., Wei, S., Lee, J.Y., Tan, S., Chu, S., Lee, T.-H., 2010. DNA methylation increases nucleosome compaction and rigidity. J. Am. Chem. Soc. 132, 1782–1783.

Chuang, C.-H., Carpenter, A.E., Fuchsova, B., Johnson, T., de Lanerolle, P., Belmont, A.S., 2006. Long-range directional movement of an interphase chromosome site. Curr. Biol. 16, 825–831.

Condamin, S., Bénichou, O., Tejedor, V., Voituriez, R., Klafter, J., 2007. First-passage times in complex scale-invariant media. Nature 450, 77–80.

Condamin, S., Tejedor, V., Voituriez, R., Bénichou, O., Klafter, J., 2008. Probing microscopic origins of confined subdiffusion by first-passage observables. Proc. Natl. Acad. Sci. U.S.A. 105, 5675–5680.

Cremer, T., Cremer, M., 2010. Chromosome territories. Cold Spring Harb. Perspect. Biol. 2, a003889

Cremer, T., Cremer, M., Dietzel, S., Muller, S., Solovei, I., Fakan, S., 2006. Chromosome territories—a functional nuclear landscape. Curr. Opin. Cell Biol. 18, 307–316.

Cui, Y., Bustamante, C., 2000. Pulling a single chromatin fiber reveals the forces that maintain its higher-order structure. Proc. Natl. Acad. Sci. U.S.A. 97, 127–132.

Dauty, E., Verkman, A.S., 2004. Molecular crowding reduces to a similar extent the diffusion of small solutes and macromolecules: measurement by fluorescence correlation spectroscopy. J. Mol. Recognit. 17, 441–447.

de Gennes, P.G., 1979. Scaling Concepts in Polymer Physics. Cornell University Press, Ithaca.

de Vries, R., 2010. DNA condensation in bacteria: interplay between macromolecular crowding and nucleoid proteins. Biochimie 92, 1715–1721.

Dekker, J., 2006. The three 'C's of chromosome conformation capture: controls, controls, controls. Nat. Methods 3, 17–21.
Dekker, J., 2008. Mapping in vivo chromatin interactions in yeast suggests an extended chromatin fiber with regional variation in compaction. J. Biol. Chem. 283, 34532.
Dekker, J., Rippe, K., Dekker, M., Kleckner, N., 2002. Capturing chromosome conformation. Science 295, 1306–1311.
Delmas, S., Duggin, I.G., Allers, T., 2013. DNA damage induces nucleoid compaction via the Mre11–Rad50 complex in the archaeon Haloferax volcanii. Mol. Microbiol. 87, 168–179.
Delpire, E., Duchêne, C., Goessens, G., Gilles, R., 1985. Effects of osmotic shocks on the ultrastructure of different tissues and cell types. Exp. Cell Res. 160, 106–116.
Dhar, A., Girdhar, K., Singh, D., Gelman, H., Ebbinghaus, S., Gruebele, M., 2011. Protein stability and folding kinetics in the nucleus and endoplasmic reticulum of eucaryotic cells. Biophys. J. 101, 421–430.
Dion, V., Gasser, S.M., 2013. Chromatin movement in the maintenance of genome stability. Cell 152, 1355–1364.
Dixon, J.R., Selvaraj, S., Yue, F., Kim, A., Li, Y., Shen, Y., Hu, M., Liu, J.S., Ren, B., 2012. Topological domains in mammalian genomes identified by analysis of chromatin interactions. Nature 485, 376–380.
Doi, M., Edwards, S., 1988. The Theory of Polymer Dynamics. Oxford University Press, USA.
Dundr, M., 2012. Nuclear bodies: multifunctional companions of the genome. Curr. Opin. Cell Biol. 24, 415–422.
Dundr, M., Misteli, T., 2010. Biogenesis of nuclear bodies. Cold Spring Harb. Perspect. Biol. 2, a000711.
Einstein, A.J., Wu, H.-S., Sanchez, M., Gil, J., 1998. Fractal organization of chromatin appearance for diagnosis in breast cytology. J. Pathol. 185, 366–381.
Ellis, R.J., 2001. Macromolecular crowding: obvious but underappreciated. Trends Biochem. Sci. 26, 597–604.
Emanuel, M., Rajda, N.H., Henriksson, A., Schiessel, H., 2009. The physics behind the larger scale organization of DNA in eukaryotes. Phys. Biol. 6, 025008.
Fisher, C.L., Fisher, A.G., 2011. Chromatin states in pluripotent, differentiated, and reprogrammed cells. Curr. Opin. Genet. Dev. 21, 140–146.
Flemming, W., 1878. Zur Kenntniss der Zelle und ihrer Theilungs-Erscheinungen. Schr. Naturwiss. Ver. Schlesw. Holst. 3, 23–27.
Foley, P.L., Wilson, D.B., Shuler, M.L., 2010. Macromolecular crowding can account for RNase-sensitive constraint of bacterial nucleoid structure. Biochem. Biophys. Res. Commun. 395, 42–47.
Fomproix, N., Percipalle, P., 2004. An actin-myosin complex on actively transcribing genes. Exp. Cell Res. 294, 140–148.
Fritsch, C.C., Langowski, J., 2011. Chromosome dynamics, molecular crowding, and diffusion in the interphase cell nucleus: a Monte Carlo lattice simulation study. Chromosome Res. 19, 63–81.
Fussner, E., Ching, R.W., Bazett-Jones, D.P., 2011. Living without 30 nm chromatin fibers. Trends Biochem. Sci. 36, 1–6.
Gehlen, L.R., Gruenert, G.M., Jones, B., Rodley, C.D., Langowski, J., O'Sullivan, J.M., 2012. Chromosome positioning and the clustering of functionally related loci in yeast is driven by chromosomal interactions. Nucleus 3, 370–383.
Gerlich, D., Beaudouin, J., Kalbfuss, B., Daigle, N., Eils, R., Ellenberg, J., 2003. Global chromosome positions are transmitted through mitosis in mammalian cells. Cell 112, 751–764.
Gibcus, J.H., Dekker, J., 2013. The hierarchy of the 3D genome. Mol. Cell 49, 773–782.

Göndör, A., 2013. Nuclear architecture and chromatin structure on the path to cancer. Sem. Cancer Biol. 23, 63–64.
Göndör, A., Ohlsson, R., 2009. Chromosome crosstalk in three dimensions. Nature 461, 212–217.
Goodsell, D.S., 2005. Visual methods from atoms to cells. Structure 13, 347–354.
Gorisch, S.M., Richter, K., Scheuermann, M.O., Herrmann, H., Lichter, P., 2003. Diffusion-limited compartmentalization of mammalian cell nuclei assessed by microinjected macromolecules. Exp. Cell Res. 289, 282–294.
Görisch, S.M., Wachsmuth, M., Töth, K.F., Lichter, P., Rippe, K., 2005. Histone acetylation increases chromatin accessibility. J. Cell Sci. 118, 5825–5834.
Grosberg, A.Yu., Nechaev, S.K., Shakhnovich, E.I., 1988. The role of topological constraints in the kinetics of collapse of macromolecules. J. Phys. France 49, 2095–2100.
Grosberg, A.Yu., Rabin, Y., Havlin, S., Neer, A., 1993. Crumpled globule model of the three-dimensional structure of DNA. Europhys. Lett. 23, 373–378.
Grünwald, D., Martin, R.M., Buschmann, V., Bazett-Jones, D.P., Leonhardt, H., Kubitscheck, U., Cardoso, M.C., 2008. Probing intranuclear environments at the single-molecule level. Biophys. J. 94, 2847–2858.
Hajjoul, H., Kocanova, S., Lassadi, I., Bystricky, K., Bancaud, A., 2009. Lab-on-Chip for fast 3D particle tracking in living cells. Lab Chip 9, 3054–3058.
Hajjoul, H., Mathon, J., Ranchon, H., Goiffon, I., Mozziconacci, J., Albert, B., Carrivain, P., Victor, J.-M., Gadal, O., Bystricky, K., Bancaud, A., 2013. High-throughput chromatin motion tracking in living yeast reveals the flexibility of the fiber throughout the genome. Genome Res. 23, 1829–1838.
Hall, D., Minton, A.P., 2003. Macromolecular crowding: qualitative and semiquantitative successes, quantitative challenges. Biochim. Biophys. Acta 1649, 127–139.
Han, J., Herzfeld, J., 1993. Macromolecular diffusion in crowded solutions. Biophys. J. 65, 1155–1161.
Hancock, R., 2004. A role for macromolecular crowding effects in the assembly and function of compartments in the nucleus. J. Struct. Biol. 146, 281–290.
Hancock, R., 2007. Packing of the polynucleosome chain in interphase chromosomes: evidence for a contribution of crowding and entropic forces. Semin. Cell Dev. Biol. 18, 668–675.
Hancock, R., 2008. Self-association of polynucleosome chains by macromolecular crowding. Eur. Biophys. J. 37, 1059–1064.
Handwerger, K.E., Cordero, J.A., Gall, J.G., 2005. Cajal bodies, nucleoli, and speckles in the Xenopus oocyte nucleus have a low-density, sponge-like structure. Mol. Biol. Cell 16, 202–211.
Harp, J.M., Hanson, B.L., Timm, D.E., Bunick, G.J., 2000. Asymmetries in the nucleosome core particle at 2.5 A resolution. Acta Crystallogr. D Biol. Crystallogr. 56, 1513–1534.
Havas, K., Flaus, A., Phelan, M., Kingston, R., Wade, P.A., Lilley, D.M.J., Owen-Hughes, T., 2000. Generation of superhelical torsion by ATP-dependent chromatin remodeling activities. Cell 103, 1133–1142.
Heermann, D.W., Jerabek, H., Liu, L., Li, Y., 2012. A model for the 3D chromatin architecture in pro and eukaryotes. Methods 58, 307–314.
Heitz, E., 1928. Das Heterochromatin der Moose. Jahrb. Wiss. Botanik 69, 762–818.
Hell, S.W., 2003. Toward fluorescence nanoscopy. Nat. Biotechnol. 21, 1347–1355.
Heun, P., Laroche, T., Shimada, K., Furrer, P., Gasser, S., 2001. Chromosome dynamics in the yeast interphase nucleus. Science 294, 2181–2186.
Hihara, S., Pack, C.-G., Kaizu, K., Tani, T., Hanafusa, T., Nozaki, T., Takemoto, S., Yoshimi, T., Yokota, H., Imamoto, N., Sako, Y., Kinjo, M., Takahashi, K., Nagai, T., Maeshima, K., 2012. Local nucleosome dynamics facilitate chromatin accessibility in living mammalian cells. Cell Rep. 2, 1645–1656.

Hinde, E., Cardarelli, F., 2011. Measuring the flow of molecules in cells. Biophys. Rev. 3, 119–129.

Huisman, A., Ploeger, L.S., Dullens, H.F.J., Poulin, N., Grizzle, W.E., van Diest, P.J., 2005. Development of 3D chromatin texture analysis using confocal laser scanning microscopy. Cell. Oncol. 27, 335–345.

Hyman, A.A., Brangwynne, C.P., 2011. Beyond stereospecificity: liquids and mesoscale organization of cytoplasm. Dev. Cell 21, 14–16.

Iborra, F.J., 2007. Can visco-elastic phase separation, macromolecular crowding and colloidal physics explain nuclear organisation? Theor. Biol. Med. Model. 12, 4–15.

Ignatova, Z., Gierasch, L.M., 2004. Monitoring protein stability and aggregation in vivo by real-time fluorescent labeling. Proc. Natl. Acad. Sci. U.S.A. 101, 523–528.

Jackson, V., 1978. Studies on histone organization in the nucleosome using formaldehyde as a reversible cross-linking agent. Cell 15, 945–954.

Jackson, D.A., Hassan, A.B., Errington, R.J., Cook, P.R., 1993. Visualization of focal sites of transcription within human nuclei. EMBO J. 12, 1059–1065.

Jiao, M., Li, H.-T., Chen, J., Minton, A.P., Liang, Y., 2010. Attractive protein-polymer interactions markedly alter the effect of macromolecular crowding on protein association equilibria. Biophys. J. 99, 914–923.

Jimenez-Useche, I., Yuan, C., 2012. The effect of DNA CpG methylation on the dynamic conformation of a nucleosome. Biophys. J. 103, 2502–2512.

Keating, C.D., 2012. Aqueous phase separation as a possible route to compartmentalization of biological molecules. Acc. Chem. Res. 45, 2114–2124.

Konopka, M.C., Shkel, I.A., Cayley, S., Record, M.T., Weisshaar, J.C., 2006. Crowding and confinement effects on protein diffusion in vivo. J. Bacteriol. 188, 6115–6123.

Kopelman, R., 1988. Fractal reaction kinetics. Science 241, 1620–1626.

Krijger, P.H.L., de Laat, W., 2013. Identical cells with different 3D genomes: cause and consequences? Curr. Opin. Genet. Dev. 23, 1–6.

Krivega, I., Dean, A., 2012. Enhancer and promoter interactions-long distance calls. Curr. Opin. Genet. Dev. 22, 79–85.

Kumaran, R.I., Thakar, R., Spector, D.L., 2008. Chromatin dynamics and gene positioning. Cell 132, 929–934.

Lavau, C., Marchio, A., Fagioli, M., Jansen, J., Falini, B., Lebon, P., Grosveld, F., Pandolfi, P.P., Pelicci, P.G., Dejean, A., 1995. The acute promyelocytic leukaemia-associated PML gene is induced by interferon. Oncogene 11, 871–876.

Lavelle, C., 2009. Forces and torques in the nucleus: chromatin under mechanical constraints. Biochem. Cell Biol. 87, 307–322.

Lebedev, D.V., Filatov, M.V., Kuklin, A.I., Islamov, A.K., Ketzinger, E., Pantina, R., Toperverg, B.P., Isaev-Ivanov, V.V., 2005. Fractal nature of chromatin organization in interphase chicken erythrocyte nuclei: DNA structure exhibits biphasic fractal properties. FEBS Lett. 579, 1465–1468.

Lebedev, D.V., Filatov, M.V., Kuklin, A.I., Islamov, A.K., Stellbrink, J., Pantina, R., Denisov, Y.Y., Toperverg, B.P., Isaev-Ivanov, V.V., 2008. Structural hierarchy of chromatin in chicken erythrocyte nuclei based on small-angle neutron scattering: fractal nature of the large-scale chromatin organization. Crystallogr. Rep. 53, 110–115.

Levi, V., Gratton, E., 2008. Chromatin dynamics during interphase explored by single-particle tracking. Chromosome Res. 16, 439–449.

Lieberman-Aiden, E., van Berkum, N.L., Williams, L., Imakaev, M., Ragoczy, T., Telling, A., Amit, I., Lajoie, B.R., Sabo, P.J., Dorschner, M.O., Sandstrom, R., Bernstein, B., Bender, M.A., Groudine, M., Gnirke, A., Stamatoyannopoulos, J., Mirny, L.A., Lander, E.S., Dekker, J., 2009. Comprehensive mapping of long-range interactions reveals folding principles of the human chromosome. Science 326, 289–293.

Lopez-Velazquez, G., Marquez, J., Ubaldo, E., Corkidi, G., Echeverra, O., Vazquez Nin, G.H., 1996. Three-dimensional analysis of the arrangement of compact chromatin in the nucleus of G0 rat lymphocytes. Histochem. Cell Biol. 105, 153–161.
Luger, K., Mader, A.W., Richmond, R.K., Sargent, D.F., Richmond, T.J., 1997. Crystal structure of the nucleosome core particle at 2.8 A resolution. Nature 389, 251–260.
Maeshima, K., Hihara, S., Eltsov, M., 2010. Chromatin structure: does the 30-nm fibre exist in vivo? Curr. Opin. Cell Biol. 22, 291–297.
Mandelbrot, B.B., 1967. How long is the coast of britain? Statistical self-similarity and fractional dimension. Science 156, 636–638.
Mandelbrot, B.B., 1982. The Fractal Geometry of Nature. W.H. Freeman, San Francisco.
Mardis, E.R., 2007. Chip-Seq: welcome to the new frontier. Nat. Methods 4, 613–614.
Marenduzzo, D., Finan, K., Cook, P.R., 2006a. The depletion attraction: an underappreciated force driving cellular organization. J. Cell Biol. 175, 681–686.
Marenduzzo, D., Micheletti, C., Cook, P.R., 2006b. Entropy-driven genome organization. Biophys. J. 90, 3712–3721.
Markaki, Y., Smeets, D., Fiedler, S., Schmid, V.J., Schermelleh, L., Cremer, T., Cremer, M., 2012. The potential of 3D-FISH and super-resolution structured illumination microscopy for studies of 3D nuclear architecture. Bioessays 34, 412–426.
Marshall, W., Straight, A., Marko, J., Swedlow, J., Dernburg, A., Belmont, A., Murray, A., Agard, D., Sedat, J., 1997. Interphase chromosomes undergo constrained diffusional motion in living cells. Curr. Biol. 7, 930–939.
Martin, R.M., Cardoso, M.C., 2010. Chromatin condensation modulates access and binding of nuclear proteins. FASEB J. 24, 1066–1072.
Mateos-Langerak, J., Bohn, M., de Leeuw, W., Giromus, O., Manders, E.M., Verschure, P.J., Indemans, M.H., Gierman, H.J., Heermann, D.W., van Driel, R., Goetze, S., 2009. Spatially confined folding of chromatin in the interphase nucleus. Proc. Natl. Acad. Sci. U.S.A. 106, 3812–3817.
Mazza, D., Mc Nally, J.G., 2010. Fractal geometry in the nucleus. EMBO J. 29, 2–3.
McDonald, D., Carrero, G., Andrin, C., de Vries, G., Hendzel, M.J., 2006. Nucleoplasmic beta-actin exists in a dynamic equilibrium between low-mobility polymeric species and rapidly diffusing populations. J. Cell Biol. 172, 541–552.
Meldi, L., Brickner, J.H., 2011. Compartmentalization of the nucleus. Trends Cell Biol. 21, 701–708.
Meyer, B., Bénichou, O., Kafri, Y., Voituriez, R., 2012. Geometry-induced bursting dynamics in gene expression. Biophys. J. 102, 2186–2191.
Minton, A.P., 1983. The effect of volume occupancy upon the thermodynamic activity of proteins: some biochemical consequences. Mol. Cell. Biochem. 55, 119–140.
Minton, A.P., 2006. How can biochemical reactions within cells differ from those in test tubes? J. Cell Sci. 119, 2863–2869.
Mirny, L., 2011. The fractal globule as a model of chromatin architecture in the cell. Chromosome Res. 19, 37–51.
Mozziconacci, J., Lavelle, C., Barbi, M., Lesne, A., Victor, J.M., 2006. A physical model for the condensation and decondensation of eukaryotic chromosomes. FEBS Lett. 580, 368–372.
Muller, H., 1930. Types of visible variations induced by X-rays in Drosophila. J. Genet. 22, 299–334.
Münkel, C., Eils, R., Dietzel, S., Zink, D., Mehring, C., Wedeman, G., Cremer, T., Langowski, J., 1999. Compartmentalization of interphase chromosomes observed in simulation and experiment. J. Mol. Biol. 285, 1053–1065.
Muramatsu, N., Minton, A.P., 1988. Tracer diffusion of globular proteins in concentrated protein solutions. Proc. Natl. Acad. Sci. U.S.A. 85, 2984–2988.
Neumann, F.R., Nurse, P., 2007. Nuclear size control in fission yeast. J. Cell Biol. 179, 593–600.

Nora, E.P., Lajoie, B.R., Schulz, E.G., Giorgetti, L., Okamoto, I., Servant, N., Piolot, T., van Berkum, N.L., Meisig, J., Sedat, J., Gribnau, J., Barillot, E., Bluthgen, N., Dekker, J., Heard, E., 2012. Spatial partitioning of the regulatory landscape of the X-inactivation centre. Nature 485, 381–385.

Normanno, D., Dahan, M., Darzacq, X., 2012. Intra-nuclear mobility and target search mechanisms of transcription factors: a single-molecule perspective on gene expression. Biochem. Biophys. Acta 1819, 482–493.

Nowak, G., Pestic-Dragovich, L., Hozák, P., Philimonenko, A., Simerly, C., Schatten, G., de Lanerolle, P., 1997. Evidence for the presence of myosin I in the nucleus. J. Biol. Chem. 272, 17176–17181.

Orlova, D.Y., Stixová, L., Kozubek, S., Gierman, H.J., Sustácková, G., Chernyshev, A.V., Medvedev, R.N., Legartová, S., Versteeg, R., Matula, P., Stoklasa, R., Bártová, E., 2012. Arrangement of nuclear structures is not transmitted through mitosis but is identical in sister cells. J. Cell. Biochem. 113, 3313–3329.

Osborne, C.S., Chakalova, L., Brown, K.E., Carter, D., Horton, A., Debrand, E., Goyenechea, B., Mitchell, J.A., Lopes, S., Reik, W., Fraser, P., 2004. Active genes dynamically colocalize to shared sites of ongoing transcription. Nat. Genet. 36, 1065–1071.

Ostashevsky, J.Y., Lange, C.S., 1994. The 30 nm chromatin fiber as a flexible polymer. J. Biomol. Struct. Dyn. 11, 813–820.

Pack, C., Saito, K., Tamura, M., Kinjo, M., 2006. Microenvironment and effect of energy depletion in the nucleus analyzed by mobility of multiple oligomeric EGFPs. Biophys. J. 91, 3921–3936.

Porter, K.R., Claude, A., Fullam, E.F., 1945. A study of tissue culture cells by electron microscopy—methods and preliminary observations. J. Exp. Med. 81, 233–246.

Rabl, C., 1885. Über Zelltheilung. Morph. Jb. 10, 214–330.

Rajapakse, I., Groudine, M., 2011. On emerging nuclear order. J. Cell Biol. 192, 711–721.

Rapkin, L.M., Anchel, D.R.P., Li, R., Bazett-Jones, D.P., 2012. A view of the chromatin landscape. Micron 43, 150–158.

Richter, K., Nessling, M., Lichter, P., 2007. Experimental evidence for the influence of molecular crowding on nuclear architecture. J. Cell Sci. 120, 1673–1680.

Richter, K., Nessling, M., Lichter, P., 2008. Macromolecular crowding and its potential impact on nuclear function. Biochim. Biophys. Acta 1783, 2100–2107.

Ringrose, L., Chabanis, S., Angrand, P.-O., Woodroofe, C., Stewart, A.F., 1999. Quantitative comparison of DNA looping in vitro and in vivo: chromatin increases effective DNA flexibility at short distances. EMBO J. 18, 6630–6641.

Rosa, A., Everaers, R., 2008. Structure and dynamics of interphase chromosomes. PLoS Comput. Biol. 4, e1000153.

Rouquette, J., Genoud, C., Vazquez-Nin, G.H., Kraus, B., Cremer, T., Fakan, S., 2009. Revealing the high-resolution three-dimensional network of chromatin and interchromatin space: a novel electron-microscopic approach to reconstructing nuclear architecture. Chromosome Res. 17, 801–810.

Rowat, A.C., 2009. Physical properties of the nucleus studied by micropipette aspiration. Methods Mol. Biol. 464, 3–12.

Sachs, R.K., van den Engh, G., Trask, B., Yokota, H., Hearst, J.E., 1995. A random-walk/giant-loop model for interphase chromosomes. Proc. Natl. Acad. Sci. U.S.A. 92, 2710–2714.

Sanyal, A., Bau, D., Marti-Renom, M.A., Dekker, J., 2011. Chromatin globules: a common motif of higher order chromosome structure? Curr. Opin. Cell Biol. 23, 325–331.

Saxton, M.J., 1993. Lateral diffusion in an archipelago. Dependence on tracer size. Biophys. J. 64, 1053–1062.

Saxton, M.J., 1994. Anomalous diffusion due to obstacles: a Monte Carlo study. Biophys. J. 66, 394–401.
Saxton, M.J., 2012. Wanted: a positive control for anomalous subdiffusion. Biophys. J. 103, 2411–2422.
Schnell, S., Turner, T.E., 2004. Reaction kinetics in intracellular environments with macromolecular crowding: simulations and rate laws. Prog. Biophys. Mol. Biol. 85, 235–260.
Seksek, O., Biwersi, J., Verkman, A.S., 1997. Translational diffusion of macromolecule-sized solutes in cytoplasm and nucleus. J. Cell Biol. 138, 131–142.
Severin, P.M.D., Zou, X., Gaub, H.E., Schulten, K., 2011. Cytosine methylation alters DNA mechanical properties. Nucleic Acids Res. 39, 8740–8751.
Sexton, T., Yaffe, E., Kenigsberg, E., Bantignies, F., Leblanc, B., Hoichman, M., Parrinello, H., Tanay, A., Cavalli, G., 2012. Three-dimensional folding and functional organization principles of the drosophila genome. Cell 148, 458–472.
Strick, T.R., Bensimon, D., Croquette, V., 1999. Micro-mechanical measurement of the torsional modulus of DNA. Genetica 106, 57–62.
Szymanski, J., Weiss, M., 2009. Elucidating the origin of anomalous diffusion in crowded fluids. Phys. Rev. Lett. 103, 038102.
Tjong, H., Gong, K., Chen, L., Alber, F., 2012. Physical tethering and volume exclusion determine higher-order genome organization in budding yeast. Genome Res. 22, 1295–1305.
Tóth, K.F., Knoch, T.A., Wachsmuth, M., Frank-Stöhr, M., Stöhr, M., Bacher, C.P., Müller, G., Rippe, K., 2004. Trichostatin A-induced histone acetylation causes decondensation of interphase chromatin. J. Cell Sci. 15, 4277–4287.
Tripathi, V., Prasanth, K.V., 2011. Cell Nucleus. eLS, John Wiley & Sons, Ltd., Chichester.
van den Engh, G., Sachs, R., Trask, B.J., 1992. Estimating genomic distance from DNA sequence location in cell nuclei by a random walk model. Science 257, 1410–1412.
van Holde, K.E., 1989. Chromatin. Springer-Verlag, New York, NY.
van Holde, K., Zlatanova, J., 2007. Chromatin fiber structure: where is the problem now? Semin. Cell Dev. Biol. 18, 651–658.
van Steensel, B., 2011. Chromatin: constructing the big picture. EMBO J. 30, 1885–1895.
van Steensel, B., Dekker, J., 2010. Genomics tools for unraveling chromosome architecture. Nat. Biotechnol. 28, 1089–1095.
Verschure, P.J., van der Kraan, I., Manders, E.M.M., Hoogstraten, D., Houtsmuller, A.B., van Driel, R., 2003. Condensed chromatin domains in the mammalian nucleus are accessible to large macromolecules. EMBO Rep. 4, 861–866.
Victor, J., Zlatanova, J., Barbi, M., Mozziconacci, J., 2012. Pulling chromatin apart: unstacking or unwrapping? BMC Biophys. 5, 21.
Volpi, E.V., Chevret, E., Jones, T., Vatcheva, R., Williamson, J., Beck, S., Campbell, R.D., Goldsworthy, M., Powis, S.H., Ragoussis, J., Trowsdale, J., Sheer, D., 2000. Large-scale chromatin organization of the major histocompatibility complex and other regions of human chromosome 6 and its response to interferon in interphase nuclei. J. Cell Sci. 113, 1565–1576.
Wachsmuth, M., Waldeck, W., Langowski, J., 2000. Anomalous diffusion of fluorescent probes inside living cell nuclei investigated by spatially-resolved fluorescence correlation spectroscopy. J. Mol. Biol. 298, 677–689.
Walter, A., Chapuis, C., Huet, S., Ellenberg, S., 2013. Crowded chromatin is not sufficient for heterochromatin formation and not required for its maintenance. J. Struct. Biol., in press.
Walter, J., Schermelleh, L., Cremer, M., Tashiro, S., Cremer, T., 2003. Chromosome order in HeLa cells changes during mitosis and early G1, but is stably maintained during subsequent interphase stages. J. Cell Biol. 160, 685–697.

Weber, S., Spakowitz, A., Theriot, J., 2010. Bacterial chromosomal loci move subdiffusively through a viscoelastic cytoplasm. Phys. Rev. Lett. 104, 238102.

Weber, S.C., Spakowitz, A.J., Theriot, J.A., 2012. Nonthermal ATP-dependent fluctuations contribute to the in vivo motion of chromosomal loci. Proc. Natl. Acad. Sci. U.S.A. 109, 7338–7343.

Wedemann, G., Langowski, J., 2002. Computer simulation of the 30-nanometer chromatin fiber. Biophys. J. 82, 2847–2859.

Williams, R.R.E., 2003. Transcription and the territory: the ins and outs of gene positioning. Trends Genet. 19, 298–302.

Wong, H., Marie-Nelly, H., Herbert, S., Carrivain, P., Blanc, H., Koszul, R., Fabre, E., Zimmer, C., 2012. A predictive computational model of the dynamic 3D interphase nucleus. Curr. Biol. 22, 1881–1890.

Woodcock, C.L., Ghosh, R.P., 2010. Chromatin higher-order structure and dynamics. Cold Spring Harb. Perspect. Biol. 2, a000596.

Yokota, H., van den Engh, G., Hearst, J.E., Sachs, R.K., Trask, B.J., 1995. Evidence for the organization of chromatin in megabase pair-sized loops arranged along a random walk path in the human G0/G1 interphase nucleus. J. Cell Biol. 130, 1239–1249.

Zhou, H.-X., Rivas, G., Minton, A.P., 2008. Macromolecular crowding and confinement: biochemical, biophysical, and potential physiological consequences. Annu. Rev. Biophys. 37, 375–397.

Zimmerman, S.B., Minton, A.P., 1993. Macromolecular crowding: biochemical, biophysical, and physiological consequences. Annu. Rev. Biophys. Biomol. Struct. 22, 27–65.

INDEX

Note: Page numbers followed by "*f*" indicate figures and by "*t*" indicate tables.

D

E

F

FSC
www.fsc.org
MIX
Papier | Fördert
gute Waldnutzung
FSC® C083411

Kolibri 360 GmbH
Ferdinand-Jühlke-Straße 7
99095 Erfurt, Deutschland
produktsicherheit@kolibri360.de